107TH CONGRESS, 2D SESSION HOUSE DOCUMENT NO. 107–285

Commemorative Joint Meeting of the Congress of the United States

In Remembrance of the Victims and Heroes of September 11, 2001

Federal Hall, New York, NY
Friday, September 6, 2002

Compiled Under the Direction of the
Joint Committee on Printing,
Chairman Robert W. Ney

UNITED STATES GOVERNMENT PRINTING OFFICE
WASHINGTON : 2003

House Concurrent Resolution 448

Whereas on September 11, 2001, thousands of innocent people were killed and injured in combined terrorist attacks involving four hijacked airliners, the World Trade Center, and the Pentagon;

Whereas in the aftermath of the attacks, thousands more were left grieving for beloved family and friends, livelihoods were compromised, and businesses and property were damaged and lost;

Whereas the greatest loss of life, personal injury, and physical destruction occurred in and was sustained by the City of New York;

Whereas government and the American people responded decisively, through the bravery, sacrifice and toil of the fire and rescue workers, law enforcement, building trades, caregivers, Armed Forces, and millions more who through their many expressions of care and compassion brought forth comfort, hope, and the promise of recovery;

Whereas the City of New York attended to the aftermath of the destruction of the World Trade Center with profound respect for the victims and compassion to the survivors;

Whereas the City of New York has invited the Congress to meet at the site of the original Federal Hall, where the First Congress of the United States convened on March 4, 1789; Now, therefore be it

Resolved by the House of Representatives (the Senate concurring), That, in remembrance of the victims and the heroes of September 11, 2001, and in recognition of the courage and spirit of the City of New York, the Congress shall conduct a special meeting in Federal Hall in New York, New York, on September 6, 2002.

House Concurrent Resolution 487

Resolved by the House of Representatives (the Senate concurring),

SECTION 1. AUTHORIZING PRINTING OF VOLUME OF TRANSCRIPTS OF NEW YORK CITY MEETING AND STATEMENTS OF TERRORIST ATTACKS OF SEPTEMBER 11.

(a) IN GENERAL.—A volume consisting of the transcripts of the ceremonial meeting of the House of Representatives and Senate in New York City on September 6, 2002, and a collection of statements by Members of the House of Representatives and Senators on the terrorist attacks of September 11, 2001, shall be printed as a House document under the direction of the Joint Committee on Printing, with suitable binding.

(b) STATEMENTS TO BE INCLUDED IN VOLUME.—A statement by a Member of the House of Representatives or a Senator on the terrorist attacks of September 11, 2001, shall be included in the volume printed under subsection (a) if the statement—

(1) was printed in the *Congressional Record* prior to the most recent date on which the House of Representatives adjourned prior to the date of the regularly scheduled general election in November 2002; and

(2) is approved for inclusion in the volume by the Committee on House Administration of the House of Representatives (in the case of a statement by a Member of the House), or the Committee on Rules and Administration of the Senate (in the case of a statement by a Senator).

SEC. 2. NUMBER OF COPIES.

The number of copies of the document printed under section 1 shall be 15,000 casebound copies, of which—

(1) 15 shall be provided to each Member of the House of Representatives;

(2) 25 shall be provided to each Senator; and

(3) the balance shall be distributed by the Joint Committee on Printing to Members of the House of Representatives and Senators, based on requests submitted to the Joint Committee by Members and Senators.

SEC. 3. MEMBER DEFINED.

In this concurrent resolution, the term "Member of the House of Representatives" includes a Delegate or Resident Commissioner to the Congress.

Photograph by Mark Abraham.

Photograph by U.S. House of Representatives, Office of Photography.

Photograph by U.S. House of Representatives, House Recording Studio, Todd Redlin.

Contents

Commemorative Joint Meeting of the Congress of the United States

In Remembrance of the Victims and Heroes of September 11, 2001

Commemorative Joint Meeting of the Congress of the United States in Federal Hall, New York, NY

Friday, September 6, 2002

The SPEAKER. The special ceremonial meeting will be in order.

The invocation will be given by the Reverend Daniel P. Coughlin, Chaplain of the House of Representatives.

The Chaplain of the House of Representatives, the Reverend Daniel P. Coughlin, offered the following invocation:

Lord God, this is a day of history. Bless this exceptional joint meeting of the 107th Congress which commemorates the tragic events that occurred here last September 11. The gaping hole left in this city tore into the fabric of this Nation, but there was no greater suffering than in New York.

Once again, we commend to Your loving mercy, the victims, survivors, and their families. We also honor those public servants and ordinary citizens who joined professionals in healing wounds and rebuilding lives in this proud city of life and diversity.

Gathered in this historic place, You alone can renew us as You have in the past. May the vision of the Founding Fathers come alive again in this body politic to preserve the balance of power and assure the freedom of the law-abiding people of this Nation.

The Bible here, used by George Washington when sworn in as President, speaks to Your consoling word: "I am with you"

Lord God, today is Rosh Hashanah. The traditional Jewish New Year prayer is for a good and sweet year. Many things You send us, Lord, are good, but they may hurt or are hurried. So with our Jewish brothers and sisters we pray today not only for a year of good things, but a year of sweetness, a chance to relish the blessings of the world and the freedoms You give us, and to enjoy the sweet kindness and love of one another.

May this be a good year for all Americans of all faiths, backgrounds, and traditions. We pray for a good year for America and for the world.

Amen.

PLEDGE OF ALLEGIANCE

The SPEAKER. The Chair recognizes the Honorable Jerrold Nadler, Representative from New York, and the Honorable Harry Reid, Senator from Nevada, to lead us in the Pledge of Allegiance to our flag.

Mr. Nadler and Senator Reid led the Pledge of Allegiance as follows:

> I pledge allegiance to the Flag of the United States of America, and to the Republic for which it stands, one Nation under God, indivisible, with liberty and justice for all.

The SPEAKER. Our national anthem will now be sung by LaChanze.

The national anthem was sung by LaChanze.

The SPEAKER. My colleagues, we are here in Federal Hall in New York, NY, pursuant to H. Con. Res. 448 of the 107th Congress to conduct a special ceremonial meeting in remembrance of the victims and the heroes of September 11, 2001, and in recognition of the courage and the spirit of the City of New York.

When representatives of the New York delegation introduced in the House and the Senate in 2001 concurrent resolutions that suggested the Congress convene outside the seat of government to symbolize the Nation's solidarity with New Yorkers who epitomize the human spirit of courage, resilience and strength, my initial reaction of support was tempered by the realization that under article 1, section 5, clause 4 of the Constitution, "Neither House shall, without the consent of the other, adjourn to any other place than that in which the two houses shall be sitting."

There is no precedent for the convening of an actual session of Congress outside the seat of government, but on one special occasion the Congress engaged in ceremonial functions outside the seat of government. Members of both houses traveled to Philadelphia on July 16, 1987, for organized festivities surrounding the bicentennial anniversary of the Constitution pursuant to a similar concurrent resolution.

On the strength of the precedent of the uniquely historical and national significance of that occasion, it is appropriate to dedicate another ceremonial gathering to a matter of transcendent importance at another place of basic institutional relevance to the Congress.

Thus, we are gathered in Federal Hall where the First Congress met in 1789 before moving the third session of that Congress to Congress Hall in Philadelphia, PA, in 1790.

Ladies and gentlemen, we are, therefore, meeting here under that precedent.

The Chair recognizes the Honorable Richard B. Cheney, the Vice President of the United States and President of the U.S. Senate.

Vice President CHENEY. Thank you, Mr. Speaker. Each time Congress meets, we are mindful of the great charge that we have all been given as public servants. Assembled today in Federal Hall we are reminded of the ones who served before us and those who served first. It is a humbling experience to stand on the site where the First Congress met, where the first President was sworn in, where the Bill of Rights was introduced.

Every Member of the House and Senate and every citizen of this country can draw a straight line from the events in Federal Hall to the life we all know today. When Congress convened here, America was a Nation of scarcely 4 million souls. The tallest structure in the city was Trinity Church, which still stands at the corner of Broadway and Wall Street.

The roll call of that First Congress included signers of the Declaration of Independence and men who marched in George Washington's army. Two gentlemen from Virginia still in their 30s served in that Congress. Their names were Madison and Monroe. All the Members knew that great responsibilities had come to them.

As Vice President John Adams observed, "A trust of the greatest magnitude is committed to this legislature and the eyes of the world are upon you."

In their actions, the Members of the First Congress met that test. And although this city was the Nation's Capital for only a short time, from those early days, the eyes of the world have continued to be on New York. One year ago, this great center of history, enterprise, and creativity suffered the gravest of cruelties and showed itself to be a place of valor and generosity and grace. Here, where so many innocent lives were suddenly taken, the world saw acts of kindness and heroism that will be remembered forever.

When President Bush introduced Mayor Giuliani and Governor Pataki at the joint session last September, it was, said one New Yorker, as if the Members of Congress had recognized that these two men had come directly off the battlefield.

Today, Congress gathers near that battlefield to honor the character and the courage shown in New York these last 360 days, and to remember every innocent life taken in the attacks of September 11. Since the hour of those attacks, we have been a Nation at war called once again to defend our liberty and our lives and to save humanity from the worst of wars. As a Nation born in revolution, we know that our freedom came at a very high price. We have no intention now of letting it slip away.

The Members of the First Congress shaped events long into the future. The same is now asked of us. In the principles we stand for, the values we uphold, and the decisions we make we will set the course of this Nation and with it the future of human freedom and the peace of the world.

It is not given to us to know every turn of events to come. We know, however, that we are the elected servants of a good, a just, and a decent people. May we always act in that spirit, confident in our founding principles, clear in our purposes, choosing wisely and bowing only to divine providence.

The SPEAKER. The Clerk of the House of Representatives has laid upon the desk the list of representatives in attendance.

Vice President CHENEY. The Secretary of the Senate has laid upon the desk the list of Senators in attendance.

U.S. House of Representatives Member, Delegate, and Resident Commissioner attendance is as follows:

Members of the U.S. House of Representatives in Attendance

The Honorable Aníbal Acevedo-Vilá
The Honorable Gary L. Ackerman
The Honorable Robert B. Aderholt
The Honorable W. Todd Akin
The Honorable Thomas H. Allen
The Honorable Robert E. Andrews
The Honorable Richard K. Armey
The Honorable Spencer Bachus
The Honorable John Elias Baldacci
The Honorable Tammy Baldwin
The Honorable Charles F. Bass
The Honorable Ken Bentsen
The Honorable Marion Berry
The Honorable Judy Biggert
The Honorable Rod R. Blagojevich
The Honorable Roy Blunt
The Honorable Sherwood L. Boehlert
The Honorable John A. Boehner
The Honorable Henry Bonilla
The Honorable David E. Bonior
The Honorable John Boozman
The Honorable Robert A. Borski
The Honorable Leonard L. Boswell
The Honorable Kevin Brady
The Honorable Henry E. Brown, Jr.
The Honorable Richard Burr
The Honorable Dan Burton
The Honorable Ken Calvert
The Honorable Shelley Moore Capito
The Honorable Michael E. Capuano
The Honorable Brad Carson
The Honorable Michael N. Castle
The Honorable Steve Chabot
The Honorable Saxby Chambliss

The Honorable Donna M. Christensen
The Honorable Eva M. Clayton
The Honorable Mac Collins
The Honorable John Conyers, Jr.
The Honorable John Cooksey
The Honorable Jerry F. Costello
The Honorable Christopher Cox
The Honorable Philip M. Crane
The Honorable Joseph Crowley
The Honorable John Abney Culberson
The Honorable Elijah E. Cummings
The Honorable Jim Davis
The Honorable Jo Ann Davis
The Honorable Susan A. Davis
The Honorable Tom Davis
The Honorable Diana DeGette
The Honorable Rosa L. DeLauro
The Honorable Lincoln Diaz-Balart
The Honorable John T. Doolittle
The Honorable David Dreier
The Honorable John J. Duncan, Jr.
The Honorable Jennifer Dunn
The Honorable Chet Edwards
The Honorable Robert L. Ehrlich, Jr.
The Honorable Jo Ann Emerson
The Honorable Eliot L. Engel
The Honorable Phil English
The Honorable Anna G. Eshoo
The Honorable Bob Etheridge
The Honorable Lane Evans
The Honorable Sam Farr
The Honorable Mike Ferguson
The Honorable Mark Foley
The Honorable J. Randy Forbes
The Honorable Harold E. Ford, Jr.
The Honorable Vito Fossella
The Honorable Rodney P. Frelinghuysen
The Honorable Martin Frost
The Honorable Richard A. Gephardt
The Honorable Jim Gibbons
The Honorable Wayne T. Gilchrest
The Honorable Benjamin A. Gilman
The Honorable Bob Goodlatte
The Honorable Lindsey O. Graham
The Honorable Sam Graves
The Honorable Gene Green

The Honorable Mark Green
The Honorable James C. Greenwood
The Honorable Felix J. Grucci, Jr.
The Honorable Gil Gutknecht
The Honorable Tony P. Hall
The Honorable James V. Hansen
The Honorable Jane Harman
The Honorable Melissa A. Hart
The Honorable J. Dennis Hastert
The Honorable Alcee L. Hastings
The Honorable Robin Hayes
The Honorable J.D. Hayworth
The Honorable Wally Herger
The Honorable Baron P. Hill
The Honorable Earl F. Hilliard
The Honorable Maurice D. Hinchey
The Honorable David L. Hobson
The Honorable Joseph M. Hoeffel
The Honorable Rush D. Holt
The Honorable Darlene Hooley
The Honorable Stephen Horn
The Honorable Amo Houghton
The Honorable Steny H. Hoyer
The Honorable Kenny C. Hulshof
The Honorable Jay Inslee
The Honorable Johnny Isakson
The Honorable Steve Israel
The Honorable Darrell E. Issa
The Honorable Jesse L. Jackson, Jr.
The Honorable Sheila Jackson-Lee
The Honorable Christopher John
The Honorable Eddie Bernice Johnson
The Honorable Nancy L. Johnson
The Honorable Stephanie Tubbs Jones
The Honorable Paul E. Kanjorski
The Honorable Marcy Kaptur
The Honorable Ric Keller
The Honorable Sue W. Kelly
The Honorable Mark R. Kennedy
The Honorable Patrick J. Kennedy
The Honorable Peter T. King
The Honorable Jack Kingston
The Honorable Mark Steven Kirk
The Honorable Dennis J. Kucinich
The Honorable Ray LaHood
The Honorable Nick Lampson

The Honorable James R. Langevin
The Honorable Rick Larsen
The Honorable John B. Larson
The Honorable Tom Latham
The Honorable Steven C. LaTourette
The Honorable James A. Leach
The Honorable Barbara Lee
The Honorable Sander M. Levin
The Honorable Jerry Lewis
The Honorable John Lewis
The Honorable Ron Lewis
The Honorable Frank A. LoBiondo
The Honorable Nita M. Lowey
The Honorable Frank D. Lucas
The Honorable Ken Lucas
The Honorable Bill Luther
The Honorable Stephen F. Lynch
The Honorable Carolyn McCarthy
The Honorable Karen McCarthy
The Honorable James P. McGovern
The Honorable John M. McHugh
The Honorable Scott McInnis
The Honorable Howard P. "Buck" McKeon
The Honorable Michael R. McNulty
The Honorable Carolyn B. Maloney
The Honorable James H. Maloney
The Honorable Jim Matheson
The Honorable Martin T. Meehan
The Honorable Carrie P. Meek
The Honorable Gregory W. Meeks
The Honorable Robert Menendez
The Honorable John L. Mica
The Honorable Juanita Millender-McDonald
The Honorable Dan Miller
The Honorable Jeff Miller
The Honorable Dennis Moore
The Honorable James P. Moran
The Honorable Constance A. Morella
The Honorable Sue Wilkins Myrick
The Honorable Jerrold Nadler
The Honorable Grace F. Napolitano
The Honorable Richard E. Neal
The Honorable George R. Nethercutt, Jr.
The Honorable Robert W. Ney
The Honorable Eleanor Holmes Norton
The Honorable Charlie Norwood

The Honorable James L. Oberstar
The Honorable John W. Olver
The Honorable Major R. Owens
The Honorable Michael G. Oxley
The Honorable Bill Pascrell, Jr.
The Honorable Donald M. Payne
The Honorable Nancy Pelosi
The Honorable Mike Pence
The Honorable John E. Peterson
The Honorable Thomas E. Petri
The Honorable David D. Phelps
The Honorable Joseph R. Pitts
The Honorable Todd Russell Platts
The Honorable Richard W. Pombo
The Honorable Earl Pomeroy
The Honorable Rob Portman
The Honorable David E. Price
The Honorable Deborah Pryce
The Honorable Adam H. Putnam
The Honorable Jack Quinn
The Honorable Jim Ramstad
The Honorable Charles B. Rangel
The Honorable Thomas M. Reynolds
The Honorable Bob Riley
The Honorable Ileana Ros-Lehtinen
The Honorable Mike Ross
The Honorable Steven R. Rothman
The Honorable Edward R. Royce
The Honorable Paul Ryan
The Honorable Jim Ryun
The Honorable Loretta Sanchez
The Honorable Bernard Sanders
The Honorable Jim Saxton
The Honorable Bob Schaffer
The Honorable Adam B. Schiff
The Honorable Robert C. Scott
The Honorable Jose E. Serrano
The Honorable Pete Sessions
The Honorable E. Clay Shaw, Jr.
The Honorable Christopher Shays
The Honorable Don Sherwood
The Honorable John Shimkus
The Honorable Bill Shuster
The Honorable Rob Simmons
The Honorable Nick Smith
The Honorable Vic Snyder

The Honorable John M. Spratt, Jr.
The Honorable Cliff Stearns
The Honorable Ted Strickland
The Honorable John Sullivan
The Honorable John E. Sununu
The Honorable John E. Sweeney
The Honorable Thomas G. Tancredo
The Honorable Ellen O. Tauscher
The Honorable John R. Thune
The Honorable Todd Tiahrt
The Honorable Patrick J. Tiberi
The Honorable John F. Tierney
The Honorable Patrick J. Toomey
The Honorable Edolphus Towns
The Honorable Jim Turner
The Honorable Fred Upton
The Honorable Nydia M. Velázquez
The Honorable David Vitter
The Honorable Greg Walden
The Honorable James T. Walsh
The Honorable Zach Wamp
The Honorable Maxine Waters
The Honorable Wes Watkins
The Honorable Diane E. Watson
The Honorable Melvin L. Watt
The Honorable J.C. Watts, Jr.
The Honorable Anthony D. Weiner
The Honorable Curt Weldon
The Honorable Dave Weldon
The Honorable Ed Whitfield
The Honorable Roger F. Wicker
The Honorable Joe Wilson
The Honorable Lynn C. Woolsey
The Honorable David Wu
The Honorable C.W. Bill Young

U.S. Senate Member attendance is as follows:

Members of the U.S. Senate in Attendance

The Honorable George Allen
The Honorable Max Baucus
The Honorable Robert F. Bennett
The Honorable John B. Breaux
The Honorable Sam Brownback
The Honorable Maria Cantwell
The Honorable Jean Carnahan

The Honorable Lincoln D. Chafee
The Honorable Hillary Rodham Clinton
The Honorable Susan M. Collins
The Honorable Jon S. Corzine
The Honorable Tom Daschle
The Honorable Mike DeWine
The Honorable Christopher J. Dodd
The Honorable Russell D. Feingold
The Honorable Bill Frist
The Honorable Bob Graham
The Honorable Judd Gregg
The Honorable James M. Jeffords
The Honorable Edward M. Kennedy
The Honorable John F. Kerry
The Honorable Mary L. Landrieu
The Honorable Patrick J. Leahy
The Honorable Carl Levin
The Honorable Joseph I. Lieberman
The Honorable Trent Lott
The Honorable John McCain
The Honorable Barbara A. Mikulski
The Honorable Frank H. Murkowski
The Honorable Bill Nelson
The Honorable Don Nickles
The Honorable Jack Reed
The Honorable Harry Reid
The Honorable John D. Rockefeller IV
The Honorable Paul S. Sarbanes
The Honorable Charles E. Schumer
The Honorable Richard C. Shelby
The Honorable Gordon Smith
The Honorable Olympia J. Snowe
The Honorable Arlen Specter
The Honorable Debbie Stabenow
The Honorable Craig Thomas
The Honorable Fred Thompson
The Honorable George V. Voinovich
The Honorable John W. Warner
The Honorable Paul Wellstone
The Honorable Ron Wyden

The SPEAKER. The Chair recognizes the Honorable Benjamin Gilman and the Honorable Charles Rangel, Representatives from New York, and the Honorable Hillary Rodham Clinton and the Honorable Charles Schumer, Senators from New York, in a reading and presentation of H. Con. Res. 448.

Representative RANGEL. Mr. Speaker, Mr. Vice President, leaders of the House and the Senate, on behalf of Ben Gilman, Senator Schumer and Senator Clinton, and the entire New York congressional delegation, we would like to thank you for your support of this resolution that gives us in New York an opportunity to say thank you for the way you responded to the attack on our city and our State.

You give our mayor and our Governor an opportunity to be here on this historic event to say you did not treat us like New Yorkers, you treated us like Americans.

The text of the concurrent resolution was read as follows:

Representative RANGEL. "Whereas on September the 11, 2001, thousands of innocent people were killed and injured in a combined terrorist attack involving four hijacked aircraft, the World Trade Center, and the Pentagon;

"Whereas in the aftermath of the attacks, thousands more were left grieving for beloved family and friends, livelihoods were compromised, and businesses and property were damaged and lost;"

Representative GILMAN. "Whereas the greatest loss of life, personal injury, and physical destruction occurred in and was sustained by the City of New York;

"Whereas Government and the American people responded decisively through the bravery, sacrifice and toil of the fire and rescue workers, law enforcement, building trades, caregivers, Armed Forces, and millions more who through their many expressions of care and compassion brought forth comfort, hope, and the promise of recovery;"

Senator CLINTON. "Whereas the City of New York attended to the aftermath of the destruction of the World Trade Center with profound respect for the victims and compassion to the survivors; and

"Whereas the City of New York has invited the Congress to meet at the site of the original Federal Hall, where the First Congress of the United States convened on March 4, 1789: Now, therefore, be it"

Senator SCHUMER. "Resolved by the House of Representatives (the Senate concurring), That, in remembrance of the victims and the heroes of September 11, 2001, and in recognition of the courage and spirit of the City of New York, the Congress shall conduct a special meeting in Federal Hall, New York, New York, on September 6, 2002.

Passed by the House of Representatives, July 25, 2002.

Passed by the Senate, July 26, 2002."

The SPEAKER. Without objection, the Members present, on behalf of themselves and the Congress of the United States, do hereby affirm the aforesaid concurrent resolution.

Would Mayor Bloomberg and Governor Pataki please come forward and accept the concurrent resolution.

Mayor Bloomberg and Governor Pataki of New York accepted the concurrent resolution.

The SPEAKER. The Chair recognizes the Honorable Vito Fossella, Representative from New York, and the Honorable Susan Collins, Senator from Maine, in a reading and presentation of the commemorative plaque.

Reading and Presentation of Commemorative Plaque

Senator COLLINS. Mr. Speaker, Mr. Vice President, on behalf of the U.S. Congress, we present this commemorative plaque to Director Mainella for her stewardship of our Nation's treasures, especially this building, Federal Hall.

The plaque is inscribed as follows:

"Commemorative Joint Meeting of the Congress of the United States of America in Federal Hall, New York, New York, this Sixth Day of September, Two Thousand and Two."

Representative FOSSELLA. "Convened in remembrance of the victims and heroes of September 11, 2001, and in recognition of the courage and spirit of the City of New York.

"This gift to Federal Hall from the Congress of the United States of America was made from a section of Aquia Creek, VA, sandstone and used as an original building material of the United States Capitol. It was removed on the East Central Front extension in 1958."

The SPEAKER. Director Mainella, please come forward and accept the commemorative plaque.

Director Mainella accepted the commemorative plaque.

The SPEAKER. Billy Collins, Poet Laureate of the United States of America, will now read a poem written for this occasion entitled "The Names."

Reading of "The Names" by Billy Collins, Poet Laureate of the United States

Mr. COLLINS. This poem is dedicated to the victims of September 11, and to their survivors.

"The Names"

Yesterday, I lay awake in the palm of the night.
A fine rain stole in, unhelped by any breeze,
And when I saw the silver glaze on the windows,
I started with A, with Ackerman, as it happened,
Then Baxter and Calabro,
Davis and Eberling, names falling into place

As droplets fell through the dark.

Names printed on the ceiling of the night.
Names slipping around a water bend.
Twenty-six willows on the banks of a stream.
In the morning, I walked out barefoot
Among thousands of flowers
Heavy with dew like the eyes of tears,
And each had a name—
Fiori inscribed on a yellow petal
Then Gonzalez and Han, Ishikawa and Jenkins.

Names written in the air
And stitched into the cloth of the day.
A name under a photograph taped to a mailbox.
Monogram on a torn shirt.
I see you spelled out on storefront windows
And on the bright unfurled awnings of this city,
I say the syllables as I turn a corner—
Kelly and Lee,
Medina, Nardella, and O'Connor.

When I peer into the woods,
I see a thick tangle where letters are hidden
As in a puzzle concocted for children.
Parker and Quigley in the twigs of an ash,
Rizzo, Schubert, Torres, and Upton.
Secrets in the boughs of an ancient maple.

Names written in the pale sky.
Names rising in the updraft amid buildings.
Names silent in stone
Or cried out behind a door.
Names blown over the Earth and out to sea.

In the evenings—weakening light, the last swallows.
A boy on a lake lifts his oars.
A woman by a window puts a match to a candle,
And the names are outlined on the rose clouds—
Vanacore and Wallace,
(let X stand, if it can, for the ones unfound)
Then Young and Ziminsky, the final jolt of Z.

Names etched on the head of a pin.
One name spanning a bridge, another undergoing a tunnel.
A blue name needled into the skin.
Names of citizens, workers, mothers and fathers,

The bright-eyed daughter, the quick son.
Alphabet of names in green rows in a field.
Names in the small tracks of birds.
Names lifted from a hat
Or balanced on the tip of the tongue.
Names wheeled into the dim warehouse of memory.
So many names, there is barely room on the walls of the heart.

The SPEAKER. The Chair now recognizes the Honorable Richard Gephardt, Representative from Missouri and Democratic Leader of the U.S. House of Representatives.

Representative GEPHARDT. Mr. Vice President, Mr. Speaker, and my fellow colleagues of the U.S. Congress, today we speak of the unspeakable, we remember the unimaginable, and we reaffirm our utmost resolve to defend the birthright of this land and our gift outright to this world: Ideals of liberty and tolerance that will never die.

Today, we say to the families who look to this September 11 and know that they will know the pain of their piercing loss all over again, we are with you as one, as the family of America. We pray that, for you, memory will bring hope as well as tears.

We have faith that love outlasts life, and you prove it every day as you carry on the dream of a lost husband or a wife, for the child that was both of yours and, in the truest sense, always will be.

We think of those last calls on cell phones from a doomed building or plane. Those last goodbyes. Yet the life of a good person is like a wellspring that does not run dry. Nothing reminds us more powerfully of that than the rescuers of September 11, so many of them taken too, who rescued our national spirit and, amid the smoke and the darkness at noon, sent a flickering light that became a shining beacon for America.

So we have wept together, we have prayed together, given to each other, and stood side by side since September 11 in common humanity and national purpose. The sorrow has been matched by strength. America is on a mission. Not retribution or revenge, not just to defeat terrorism, but to show once again that good can triumph over evil and freedom can overcome fanaticism, as we did in different forums in a global arena twice before in the past century.

Some say that September 11, 2001, is another date that will live in infamy. Surely that is true, but it is also true that we have never known an assault like this, not just on our Armed Forces, but on our people. Not just on our buildings and our possessions, or even on the principles that we profess, but on the very foundation of this open, diverse, democratic society.

We have grown accustomed, too accustomed, to war and slaughter in our world. But most always it was "over there." One place it came before in the heartland was the homegrown terrorism that struck in Oklahoma City. Today, our caring and thoughts are there as well. And they are a half a world away with the young Americans who are on the front lines of freedom from fear.

For all our differences, how remarkably one we are all today. From Ground Zero to a sacred field in Pennsylvania, to a shattered but now rebuilt wing of the Pentagon, and all across this broad land. On the fatal flights of September 11, courage and resistance knew no bounds of party or race or status. They included a young father, a conservative columnist, and a gay man.

E Pluribus Unum.

So while we discuss and debate the next decisions, on the fundamental issue let there be no doubt. In this great and faithful struggle there are no Republicans, there are no Democrats, there are only Americans. None of us, no matter how long we live or what else marks our time will ever forget September 11. And all of us, in the name of those who were lost for a concept of liberty that must never be lost, and in the cause of civilization itself, are as determined as an earlier generation of Americans to gain the inevitable triumph, so help us God.

The SPEAKER. The Chair now recognizes the Honorable Trent Lott, the Senator from Mississippi and the Republican Leader of the U.S. Senate.

Senator LOTT. Mr. Vice President, Mr. Speaker, Members of the Congress, and distinguished guests, on behalf of the Senate and a united Congress, it is truly an honor to stand in this place in this city, New York City, today.

We are here to remember and to continue to mourn those who lost their lives, those innocent men, women, and children who were killed in that horrible event, September 11, a year ago.

We are here to show our continued appreciation for those who struggled so mightily to free and to save those who were trapped in the aftermath of the experience here in New York City and at the Pentagon.

But we are also here to express our recommitment to the people of New York and Pennsylvania and Virginia that we are with you. We will continue our efforts to help you to rebuild physically and spiritually, and to recommit ourselves to do everything in our power to make sure that America is secure against this horrible event or anything like it ever happening again.

Over the years, New York City has been called many things, from New Amsterdam to the Big Apple. Today, the Congress of the United States, Congressman Rangel, call it home. We are here, we are comfortable here. We are here to stand with the people in this city because it is symbolic of how we stand together all across America.

We came here a year ago, the week after the infamous date. We expressed our commitment and we have been working every since to keep that commitment, and we will continue to do so.

This is a special place, as has already been said, because the First Congress began the work here that we continue to this day. The work of ordered liberty, preserving, expanding the freedoms that now, as then, are the inalienable right of every person.

Two centuries ago, there were those who thought this was all nonsense. In their ignorance and arrogance, they called America a doomed folly. But history overtook them and their crowns and armies are part of the dustbin of history. There are those like them today who cannot see beyond the limits of their own hatred. It is so hard for us in America to even understand why there would be this hatred. They do not understand that in the unending struggle against tyranny, divine providence by whatever name we use is always on the side of freedom.

When the First Congress was meeting here in New York in January 1790, President Washington asked its Members for "the cool and deliberate exertion of your patriotism, firmness and wisdom." As we face today's challenge to our country, we pledge to the people of New York just what we ask of them and all Americans: The cool and deliberate exertion of your patriotism, firmness, and wisdom.

We have seen it in this city. We have seen it in America, and we are here to do our part in that effort. The duration of our present conflict and its eventual price may be in doubt, but there can be no doubt as to its outcome. From this city's day of horror, out of all the loss and sorrow, has come a strength. I have seen it all across America. A resolve. A determination which, from Manhattan, to Mississippi, now binds us together for the mighty work that lies ahead. Thank you very much.

The SPEAKER. The Chair now recognizes the Honorable Tom Daschle, the Senator from South Dakota and majority leader of the U.S. Senate.

Senator DASCHLE. Mr. Vice President, Mr. Speaker, distinguished visitors and my colleagues, the U.S. Congress has come here to commemorate a shattering experience. One that has transformed America.

The poet Yeats, after a moment of violent upheaval in his own country, wrote: "All is changed. Changed utterly. A terrible beauty is born."

As we near the first anniversary of September 11 with profound sadness, our hearts ache for those who died and for their families and loved ones. At the same time, we are filled with an abiding sense of gratitude to the people who live and work in this great city, especially the courageous workers and rescuers, for the way they inspired and stunned a wounded Nation.

In their countless acts of heroism and compassion, a terrible beauty was born. In an hour of horror and grief, they showed us how to go on.

Here in New York, at the Pentagon, and in that lonely field in Pennsylvania, the wounds the terrorist inflicted were deep. But America's resolve was even deeper.

Let history record that the terrorists failed. They sought to destroy America by attacking what they thought were our greatest strengths, but they did not understand that the true strength of America is not steel, it is not concrete, it is our belief in the ideals enshrined in our Constitution and in our Bill of Rights. It is in our shared faith in liberty and our unwavering commitment to each other.

So what happened on September 11 did not diminish our strength. It renewed it. We stand united today as proud citizens of the oldest and strongest democracy on Earth. Our faith in that democracy and in our future is absolute and unshakable.

Next Wednesday, September 11, an eternal flame will be lit in Battery Park. That flame will symbolize our determination never, ever to forget.

We will never forget the heartbreaking loss.

We will never forget the selfless heroism.

We will never forget the terrible beauty that was born here 1 year ago.

Thank you.

The SPEAKER. The Chair now recognizes the Honorable Eliot Engel, Representative from New York.

Representative ENGEL. Mr. Speaker, I rise first to thank all my colleagues for voting to bring the Congress here to New York. It's been more than 200 years since Congress last met in New York City. It is a fitting tribute that Congress has returned here at this most sorrowful time.

The past year has been a tragic and very difficult time for me and my fellow New Yorkers. We watched in horror as terrorists hijacked two commercial airliners and slammed them into the World Trade Center. We watched in horror as the Twin Towers came down and dust and debris blanketed lower Manhattan. We watched in horror as the names of the more than 3,000 people murdered that day were announced.

But, in the days and weeks that followed, we New Yorkers experienced something else. We felt the hopes and prayers of millions of Americans flow over us. We felt the pride of being an American swell and invigorate us all. We felt the determination of the greatest Nation the Earth has every known renew itself and commit to rebuilding.

The terrorists intended many things with their attack. They sought to grievously wound our Nation. And we were—we paid with the blood of our fathers and mothers, sons and daughters, and brother and sisters. They sought to disrupt our economy. And they did—billions of dollars have been lost and will be spent on recovery.

The terrorists also sought to incite fear into the hearts and souls of every American. But they failed. Instead, they inspired a Nation of freedom-loving people to stand up to those who would seek to deny them their liberties, their justice, and the American way of life. They inspired us to fight back, so that our children's children will grow up in a world where they can safely speak their views, engage in the political system, and worship in their own way.

As we meet here, in this historic location, I am reminded of one of our country's greatest Presidents, Abraham Lincoln. His words, spoken more than a century ago, are most fitting today:

> We here highly resolve that these dead shall not have died in vain; that this Nation shall have a new birth of freedom; and that this government of the people, by the people, for the people, shall not perish from the Earth.

The SPEAKER. We are gathered here today in this ceremonial session to pay tribute to the people of New York and to the people of New York City who have suffered great loss, but persevered in the face of adversity. In doing so, we pay tribute to the American spirit.

It is altogether appropriate that we meet here today in Federal Hall. After all, it was here that the First Congress met to ratify the Bill of Rights and to inaugurate our first President of the United States, George Washington.

As in 1789, when ordinary Americans did extraordinary things to create a new Nation conceived in liberty and dedication to freedom, on September 11, ordinary Americans exhibited extraordinary courage in fighting a horrific evil.

New York lost hundreds of sons and daughters in that brutal attack on our Nation's freedom. She lost firemen and custodians, stockbrokers, police officers, construction workers and executives.

We also suffered a great loss in Virginia when a plane slammed into the Pentagon, and in Somerset County, PA, when another plane that was headed for Washington, DC, was brought down by the efforts of brave passengers.

We still feel the loss of every single person who perished on that fateful day. But as we lament the loss of life, we can marvel at the bravery of those who rushed in to help.

Such bravery was on display when Battalion Chief Orio J. Palmer and Fire Marshal Ronald Bucca of the New York Fire Department climbed to the 78th floor of the World Trade Center to organize a rescue. Their efforts saved the lives of dozens of people.

Bravery was also on display when several passengers of United flight 93 decided that they would not let the terrorist complete their plans. They sacrificed themselves rather than let the terrorists win.

It wasn't just our great city that was hit, it was our great country that was attacked. And we did come together, not as Republicans and Democrats and liberals and conservatives, but we came together in support of our President and our legislative leaders to let all of our foes know that we were united in our resolve to make certain that we would do all that we could to see that this does not happen again.

And even now as we gather to praise those who fought so hard for our country, who became our heroes, we send a message to our enemies that it is our resolve to say once again: "Don't tread on the United States of America; we are prepared to do whatever is necessary to seek out and to destroy those who seek to destroy our way of life."

And we come back to where Congress met over 200 years ago, and I cannot help but be emotionally involved in believing how proud our forefathers should be of us, to come back after 200 years, and to see what we have done with their Constitution, how much we treasured it, how much we expanded it, and how much today as we meet are we prepared to protect it. How little did they know that those who picked cotton during those days, those that would come into our country to build our roads and our railroads, those who would come from foreign countries seeking religious and economic freedom, would be coming here as a part of the U.S. Congress 200 years later.

U.S. history is strange because not only are we living it, but to give New Yorkers an opportunity to say thank you to our colleagues in the House means that we're saying thank you to America. We are basically saying, as New Yorkers, "God bless this great country that gives us an opportunity to have our diversity. We will continue to believe that a part of the legacy that we are going to leave to those who follow us is that we're not going to allow terrorism to instill terror in our hearts. Our basic commitment has to be that while we would not allow an enemy to intimidate us, we're not going to allow terror to take away our basic freedoms. We're not going to strike any unknown country without knowing where the enemy actually is. The opportunities that we have been given as a people—education, Social Security, health care—we're going to make certain that, as we protect this country, we protect those civil liberties that have been passed on to us so that when the next Congress meets, no matter where they meet, they will be saying that we protected the Constitution that was given to us over 200 years ago."

My mother, your mother, everyone always said that during times of pain you'd have to seek and you can find some good in it. But the truth of the matter is that when we were struck, it was hard to believe that we could find some good.

But there was good that we found: America gave us an opportunity to say thank you to each other. America gave us an opportunity to see how blessed we were; that we could look at each other without seeing color, without seeing party label, without seeing where we came from, and recognize that we had an obligation to protect what we have.

Mr. Governor, Mr. Mayor, thank you for giving us the support of bringing us together. And now we can say that we really owe a lot to each other, because we need each other. We hope this never happens again, but thank you, Congress, for helping us when we needed you, and not withstanding our attitude, we deeply appreciate the opportunity.

Mayor BLOOMBERG. More than 2,800 people lost their lives at the World Trade Center, but the toll could have been far, far worse if it were not for the valor and professionalism of our local and regional firefighters, police officers and emergency service personnel.

Showing tremendous courage, they effected the rescue of more than 25,000 people from the World Trade Center, the largest and most successful emergency evacuation in modem history. Their heroism inspired the Nation.

Three hundred and forty-three members of the Fire Department of New York City gave their lives for freedom on 9/11. We will never forget their bravery and their sacrifice.

It is now my privilege to introduce Susan Magazine. She is the assistant commissioner in charge of the fire department's family assistance unit. She is also a woman who lost her husband Jay, who worked at the World Trade Center.

Susan.

Ms. MAGAZINE. Thank you, Mayor Bloomberg.

Mayor, Governor Pataki, distinguished Members of Congress, honored guests, I am honored to have been asked to come here this afternoon to speak with you. I came here because I think it's very important that you, our Nation's leaders, hear directly from someone who lost a loved one, a family member, last September 11.

As the mayor said, my husband Jay was one of the more than 2,800 people who perished at the World Trade Center on that day. Jay and I spent our entire adult lives together. On the 17th of next month, we would have celebrated our 20th wedding anniversary.

We have two children. Melissa is 14 and Andrew is 11. Melissa starts high school next week, and Andrew starts middle school.

Jay was the catering sales manager at Windows on the World, the spectacular restaurant at the top of the North Tower. One of our favorite shared family memories was all of us—Jay, me, Melissa and Andrew—going up to the restaurant all wearing hard hats during the construction work to reopen Windows on the World.

What a beautiful restaurant it was. When you were up there you felt like you were on top of the world. And Jay loved it. He loved working at Windows. He loved working in the Trade Center. He loved the vibrancy of downtown Manhattan.

I recently attended a dinner for a hunger relief organization that Jay was involved with to present the first annual Jay Magazine Award of Excellence. The recipient of the award was Jay's friend, Michael Lomonaco, who was the chef at Windows.

When Michael accepted the award he told a story about how he and Jay would meet almost every morning in the Windows cafeteria for coffee. And every morning as they were leaving, Jay would turn to Michael and say, "You know, we're the luckiest guys in the world to be working here."

When our kids went to visit Jay, which was often, they would look out of his office window. You felt like you could see all of New York City from there. And he would tell them that if they looked really, really hard uptown that they could see our apartment. It was so magical up there. Now, like thousands of other families, Melissa, Andrew and I are trying to figure out how to move on and how to live our lives. That doesn't mean that we will ever forget. It doesn't mean that we're trying to get back to normal. Normal does not exist anymore for any one of the families who lost someone that they loved on that day.

It means that each one of us has to find a new normal. We have no choice. And my family is doing that. We're surrounded by incredible family, wonderful friends and support of communities. And we're doing it with the assistance of our neighbors, of our communities and you, our policymakers. And we are extremely thankful for all of the support that we have received from people everywhere.

Let me illustrate with a personal story. Our family held a memorial service for Jay at the end of September. Jay had always been in the catering and restaurant businesses and had many friends, colleagues and clients all over the country.

At that service blank cards were distributed with envelopes addressed to our children. People were asked to write down their memories of Jay, to tell us stories about the Jay that they knew. The response was unbelievable. Hundreds and hundreds of cards have come back from people who knew Jay. And then, we got cards and letters from people who didn't know Jay, but had heard about him and had heard about our family, and wanted to somehow try to connect and try and give some comfort to an individual family.

Experiences like that continue to be repeated every day for the families who have been affected by September 11. As our Nation's leaders, you should know that at the family assistance unit of the fire department, we spent hours each day responding to letters and gifts from all over the country: from your States and your districts. And we respond to each one of them. We received cartons of letters from schools, camps, houses of worships, individual people from all over the country, teddy bears, quilts, pictures, books, offers of weekends away for family members, paintings, scholarships for children, songs, poems, prayers. Whatever it is that people have to give, they want to reach out to individual family members and somehow try and make a difference to each family.

And these are the people who you represent. Please tell the men and the women and the families in your home districts and your States how much it means to us that so many Americans have offered us their generosity and their kindness.

The events of September 11 were an attack on our Nation and they were attacks on individuals and individual families. Every one of the people who perished on that day was a husband, a father, a son, a wife, a mother, a daughter, a brother, a sister, a neighbor, a friend. Over 2,800 individual people were lost on that day.

And it's been remarkable to me how many Americans truly understand that each of us were real people, were real families who have experienced this enormous tragedy in very individual, very personal and very immediate ways.

Every day the people who work for the City of New York go to enormous lengths to do whatever they can for us, for the families. The city, the State and the entire Nation have given us their support. Thank you.

And when you go home, thank your constituents for their kindness, for their generosity and for never, ever letting anyone forget. Thank you.

Mayor BLOOMBERG. Susan, thank you.

And to you and to all the families, all we can really say is, "Those we lost are in our prayers and God bless."

For the terrorists, the attack on the World Trade Center, as devastating as it was, was a failure. It did not accomplish what they hoped it would. It did not weaken us. Instead, it united us. It brought us together as a Nation determined to defend our freedoms and to punish those responsible for this despicable act.

Ordinary Americans showed the goodness in their hearts. They responded to 9/11 as if their own home communities had been attacked. An unprecedented outpouring of support flooded into New York from across the Nation.

The following video you're about to see is our way of saying, "Thank you, America."

(Videotape presentation).

Mayor BLOOMBERG. I want to thank the Interpublic Sports and Entertainment Group chairman, Mark Dowley, for producing that video and donating their services.

The power, majesty and proud heritage of the United States are expressed in our national symbol, the American bald eagle. On behalf of the people of New York City, I am pleased to memorialize this historic joint session of Congress in our city by presenting a commemorative Steuben glass eagle to the House of Representatives.

Minority Leader Gephardt.

Thank you on behalf of all New Yorkers.

Representative GEPHARDT. Thank you so much. I accept this on behalf of all of our Members. And, Charlie, I don't think you have an attitude at all.

Thank you.

Mayor BLOOMBERG. The city is also proud to present a commemorative eagle to the Senate. Minority Leader Lott, would you come up to the podium, please?

On behalf of all New Yorkers, thank you, sir.

Senator LOTT. Thank you very much, Mr. Mayor. On behalf of the U.S. Senate, we express to you our appreciation for all you've done, and for this. Senator Daschle and I will find a special place for this great eagle. Thank you.

Mayor BLOOMBERG. Thank you.

The SPEAKER. And Tom Daschle.

Representative HASTERT. On behalf of the U.S. Congress, we have a unique gift: a token of that day, and a token of the strength of this Nation.

Over the Capitol of the United States flew the flag of the United States of America. And on September 11, we took that flag down. We kept it. We weren't sure exactly how we were going to use that flag. But we think it's very appropriate today to give it to the City of New York as a memento of what this Congress believes in: the ability and strength of the people of New York. The spirit of the people of New York is truly the spirit of America. Thank you.

Senator DASCHLE. On September 11, when the people of South Dakota saw what happened, they dropped everything. One ranch couple, themselves struggling right now, sold 100 head of cattle, and donated the proceeds to the victims and their families. A class of second graders collected pennies, thinking that they might be able to collect or raise a couple of hundred dollars. They raised $1,776.05.

I'm sure you could find similar stories from Speaker Hastert's constituents in Illinois, Senator Lott's in Mississippi, Congressman Gephardt's in Missouri. But in reaching out to help the people of New York, we realized it was the people of New York who were helping us. Your courage helped steady a wounded Nation.

So today, I join Speaker Hastert on behalf of all of those you inspired, to present you with this flag. We hope it'll find a home in the memorial you build to the victims of September 11, to let all New Yorkers know that they didn't just inspire a city, they inspired a nation.

Mayor BLOOMBERG. Dick, would you come up? And, Trent, and if you could come up here as well.

Thank you.

Earlier, I proudly, perhaps boastfully but accurately, referred to New York City as the Nation's cultural capital. I will now demonstrate that this was not an idle boast. It is my great pleasure to introduce a great composer, arranger, conductor, musician, and in my book most importantly an educator. The winner of the Pulitzer Prize for music, and the artistic director of jazz at Lincoln Center, Winton Marsalis.

(Musical presentation.)

Mayor BLOOMBERG. As to my boast about culture, I will rest my case.

Thank you.

Well, thank you for joining us for this historic event. The Members of Congress will now exit en masse to visit Ground Zero and to pay their personal respects to the more than 2,800 people who died for freedom. Governor Pataki and I will go with them.

But to facilitate their orderly departure, I would ask that all other guests please remain seated until the Members have left for the ballroom.

Thank you for your cooperation. And thank you for showing your support for the greatest city on Earth.

In the House of Representatives of the United States

Monday, September 9, 2002

Hon. Philip M. Crane

OF ILLINOIS

Mr. Speaker, I rise today to join my colleagues in the House and Senate in commemorating the victims and heroes of September 11, 2001, during our special session of Congress held in Federal Hall in New York, NY.

On the days following the attacks on September 11, Americans across the country came together to demonstrate the strength and resiliency of this great country. It is in that same spirit that we stand together today—both Republicans and Democrats—to reaffirm that strength and resiliency by showing a strong bipartisan expression that we are first and foremost Americans and are committed to protecting the freedoms and values that make this country great.

As we go through this week and revisit some of the darkest moments in our Nation's history, we must remember that our Nation has always been one that has triumphed over adversity. At times of great despair, America has consistently risen to its greatest hours.

In remembrance of those lives lost on September 11 and to heroes that emerged on that fateful day, I would like to close with some words from President Abraham Lincoln's Gettysburg Address:

> that from these honored dead we take increased devotion to that cause for which they here gave the last full measure of devotion—that we here highly resolve that these dead shall not have died in vain; that this Nation shall have a new birth of freedom; and that this government of the people, by the people, for the people, shall not perish from the Earth.

May God bless America.

Hon. Bob Etheridge

OF NORTH CAROLINA

Mr. Speaker, it's hard to believe that it's been almost a full year since that awful day. On September 11, 2001, all the world saw the very face of evil. And on that day and every day since, we have felt the heartbeat of America.

For me, the most enduring image of 9/11 was the sight of the Pentagon on fire after the terrorists crashed American Airlines flight 77 into our Nation's military headquarters. A thick black smoky cloud oozed from the Pentagon and hung over the banks of the Potomac River. I will never forget seeing with my own eyes that proud building engulfed in flames. Then the whole world watched television in stunned disbelief as the Twin Towers of the World Trade Center came crumbling down in a fiery wreck of twisted steel.

On that day, America was changed forever. But, the test in life is not whether or not you ever get knocked down. The true test is whether you have the courage, pride and determination to get back up again. Every day since September 11, the people of this country have gotten back up.

We Americans from all walks of life have pulled together like never before. We have stood united to tell our enemies that the spirit of America will never be broken. We will not rest until we have eliminated Osama bin Laden and his terrorist network.

In the year since 9/11, we have come to treasure the service and sacrifice of ordinary Americans as extraordinarily heroic. The selfless devotion of the firefighters, police, EMS and other public servants in New York City and the Pentagon have given us new appreciation for our hometown heroes whose everyday service does so much to strengthen our communities. The dedicated professionalism of our men and women in uniform renew our pride in our country and make us thankful for our many, many blessings. And the incredible story of the passengers of the hijacked plane who fought back and prevented the tragic events of that day from being even worse inspires us all to take charge and give back to our country.

September 11 taught us anew the immeasurable strength of the uniquely American ideal of "We, the people." As we memorialize the lives lost 1 year ago, let us also celebrate the renewed spirit of America that has been reinvigorated by the service and sacrifice of so many ordinary citizens and inspirational heroes.

Hon. Robert A. Borski

OF PENNSYLVANIA

Mr. Speaker, and my fellow colleagues of the U.S. Congress, we are here in the City of New York as representatives of a United States that is bound together as never before. It is a tragic bond, as it came at a cost of immeasurable suffering to the people of this great city, and to those who lost their loved ones in the Pentagon or on flight 93 that ended in Shanksville, PA. Today we are gathered in remembrance of the events that pierced our hearts 1 year ago.

On September 11, the terrible and violent acts perpetrated against our homeland took the lives of so many innocents. In the days after the attacks, the courage and strength of our rescue workers lifted the spirits of our Nation. In the weeks and months following, an outpouring of generosity from every corner of our Nation showed that we stand together. Thousands lined up to give blood in a gesture that Americans would share the essence of life with no regard for whom the recipient might be. The continuing work of the young men and women in our Armed Forces is a declaration that those responsible for such cowardly acts will not escape justice. Today, 1 year later, we can say that our wounds are healing. Our Nation has overcome a great deal, and it is unity that has helped us overcome our grief.

We, as public servants, have come together to realize an even greater responsibility to our Nation. These memories are a reminder that we must remain vigilant while we rebuild and that we must never allow our greatest treasure, our liberty, to be vulnerable to the will of our enemies.

We will never forget the innocent victims. We will never forget the heroes. It is with their memory in our hearts that we live each day with a greater sense of purpose and a deeper appreciation for the gifts that we in this Nation share.

Hon. Peter T. King

OF NEW YORK

Mr. Speaker, it is with great pride as an American and as a New Yorker that I commend my colleagues for taking part in this special joint meeting of Congress in historic Federal Hall.

By meeting in this venerable hall in lower Manhattan—just blocks from where the Twin Towers of the World Trade Center were destroyed less than 1 year ago—the U.S. Senate and House of Representatives have demonstrated our Government's lasting commitment to the people of New York. And by fighting back and emerging stronger than ever, New Yorkers have demonstrated their grit, their courage and their determination. On September 11, 2001, New York took our enemy's best shot and never wavered or faltered. The police officers, firefighters and all the rescue workers who raced into the inferno demonstrated unsurpassed courage and set the tone and standard for our Nation and the world. Just as significantly, the families of the brave men and women who were murdered that day just because they went to work in the World

Trade Center have demonstrated a class and dignity that defy comprehension.

None of us will ever forget where we were or what we were doing when we first heard the news of the terrorist attacks of September 11—the attack on the World Trade Center, the attack on the Pentagon and the bringing down of flight 93 in Pennsylvania by uncommonly heroic passengers. Nor will we forget how our Nation rallied behind President Bush as he commanded the war against international terrorism. That war will be waged on many battlefields and in many ways for many years to come. But we know that America will prevail. It will prevail in large part because of the fighting spirit that rose from the flames and smoke which engulfed lower Manhattan. And it is that spirit that the U.S. Congress has honored and acknowledged by holding this extraordinary session in Federal Hall. God bless America.

Hon. Tammy Baldwin

OF WISCONSIN

Mr. Speaker, today we gather here in Federal Hall to honor and recognize the courage and determination of the survivors of the horrible terrorist attacks of September 11, 2001. It is fitting that we do so.

It was here in Federal Hall that the new Congress of the United States first met in 1789 to govern our new Nation. It was here in New York that our Founding Fathers passed the laws that are the foundation for our democratic political system. From this solid foundation, our Nation has grown and prospered. Our people have excelled in science, in the humanities, in art and culture. We have grown to be a great Nation, home to a great people, with tremendous hopes and incredible dreams for the future. And it all began right here in Federal Hall.

One year ago, terrorists attacked America. Their targets were not simply the buildings they destroyed and the people they murdered. They were attacking the very ideals that define what it means to be an American. They wanted to drive us apart and make us afraid.

Mr. Speaker, I am proud to say that they failed.

Faced with tragedy and destruction, the people of the United States, and particularly the people of New York, came together as one community. We gathered the emotional resources to survive and heal, and we pulled together the financial resources to rebuild. This has not been easy. It takes incredible courage to move forward after a tragedy like 9/11. But we did find that courage within ourselves.

That is why it is so fitting that we are here in New York today. One year after September 11, we are rededicating our Nation. The terrorists have not won. They have lost. Our Nation is stronger and more united. Our freedom, our courage, our determination, our unity, our diversity, our charity and our democracy are our strengths. New York has exhibited all of these strengths in abundance in the last year. And these strengths are everywhere in America.

Hon. Jerrold Nadler

OF NEW YORK

Mr President, Mr. Speaker, today marks a historic occasion for New York and for the U.S. Congress. This is the first Congress that has convened here in New York since the First Congress convened here to watch President Washington take the oath of office and to pass the Bill of Rights.

We join here today not as Republicans or Democrats, but as Americans. The symbolic gesture of our joint meeting is both solemn and celebratory.

It is solemn because we come here today to honor a city devastated by the most terrible single attack on American soil in our history, and the thousands of innocent people lost in that attack. As the elected Representative for the area of New York most directly impacted by the attacks of September 11, 2001, I can tell you that my constituents are grateful for the act of solidarity with New York that we show here today. I can also tell you that they are even more grateful that Congress has rallied to help this city for the past year.

Our joint meeting today is also celebratory. One year ago, a group of vicious and heartless terrorists sought to cripple this city and this country by obliterating one of its great landmarks. It was their hope that not only would thousands be rendered lifeless, but that our way of life, our democracy, would be extinguished. Today we celebrate the life and vibrancy of our democracy that still lives—and do so in a city that remains the most lively, diverse, and mighty on the face of the Earth, despite the worst efforts of those terrorists.

It is only right that we seek out those who sought to destroy us. But bombs and bullets are merely the tools we use in our self-defense. Revenge against our foes will come not through bloodshed, but through acts defiant of their goals. For the last year, despite the aim of the terrorists to kill our national spirit, this Nation has proudly and defiantly displayed the flag from our homes, our cars, our community centers, and our houses of worship. Despite the murderous foes who sought to divide us, our people have joined in concerts celebrating our country and its ideals, and vigils marking our unity.

Over two centuries ago, after stumbling through a government under the Articles of Confederation, with most of the world wishing to see our demise, we gathered here, defiant of the world and its wishes, resolved to make our great democratic experiment work. It is only fitting then, that we stand here again defiant of those who wish for our demise. Let there be no doubt, today we are telling the world that New York lives on, America lives on, and her ideals live on!

Hon. Randy "Duke" Cunningham

OF CALIFORNIA

Mr. Speaker, 1 year ago, America watched with horror as the tragic events of September 11 unfolded here in New York City and in our Nation's Capital. On that morning, already aghast at the attack on the Twin Towers, I looked out the window of my Capitol Hill office and gasped in horror as I watched the black smoke billow out of the Pentagon and drift across the clear blue sky. Still stunned, I was quickly shuffled out of my office to safety.

Today, as I reflect on the profound loss that our Nation faced in the midst of these horrific attacks, I am also heartened by the selfless acts of valor, community spirit, and national unity that have followed this tragedy. Despite the stresses that our Nation has experienced, the ties that bind our diverse country together are stronger than ever.

Immediately following the terrorist attacks on America, President Bush called on all of us to volunteer to bring our country together and reestablish those local ties that oftentimes are neglected. Americans responded. The resurgence of commitment to community can be found in our homes, our houses of worship, our schools, and our workplaces. From New York City to San Diego, citizens responded with soup kitchens, mentoring programs and charitable donations of goods and time. And the proud men and women of the U.S. military responded, allowing our country to take an aggressive lead in the war on terrorism.

I have heard many stories of outstanding individuals who have gone above and beyond this call to aid those in their community through extraordinary service and exemplary acts. While we reflect on the American spirit, it is also important to highlight the measures we have taken to prevent future attacks of this magnitude.

The attacks on our Nation were motivated by intolerant and ignorant individuals seeking to forever change our way of life and destroy this great Nation. Much like Washington, Adams and Jefferson, we must seek to ensure that this country remains an example of democracy and freedom—we must be the patriots of today.

Since September 11, 2001, all levels and branches of government have cooperated to strengthen aviation and border security, stockpile more medicines to defend against bioterrorism, improve information sharing among our intelligence agencies and deploy more resources and personnel to protect our critical infrastructure.

At the same time, the changing nature of threats to our Nation requires a new and reformed government structure to protect against enemies who can strike at any time with any number of weapons. As I write this column, no single government agency has homeland security as its primary mission. In fact, responsibilities for homeland security are dispersed among more than 100 different government organizations. America needs a unified homeland security structure that will improve protection against today's threats and be flexible enough to help meet the unknown threats of the future.

President Bush has proposed the most significant transformation of the U.S. Government in over a half century by consolidating the current confusing patchwork of government activities into a single department called the Department of Homeland Security. Changing threats require a new government structure to meet these threats. The Department of Homeland Security will have in one place all the resources needed to do what it takes to protect our country. The reorganization of America's homeland security infrastructure is crucial to overcoming the enormous threat we face today.

The shocking and tragic events of September 11 reminded us of the frailty of life, but today's special session demonstrates our resilience and strength in the face of adversity. Thousands went to work on September 11 thinking about their jobs, their families, their friends—most likely not contemplating their own mortality. Yet in an instant, death and injury met them face to face. We must never forget those who perished, and we honor them again today. We owe it to their memory to ensure that we never face such a loss again. Regrouping as individuals and as a nation, we must continue rebuilding, and working to defeat the terrorists by growing even stronger as a nation.

Hon. Constance A. Morella

OF MARYLAND

Mr. Speaker, a year ago, on September 11, 2001, Americans were faced with the horrible reality of that day's heinous attacks.

As we gather here today, in the building that served as our Nation's first Capitol and witnessed the inauguration of our first President, our blessed Nation stands firm and it stands strong.

Over the past year, Americans have shown those who wished to tear our country apart that their cowardly actions only brought our Nation closer together. Here in the place where our democracy was born, we say to the world that these States of America remain united. We are united by our values, our communities, and our freedoms. Just as we will never forget what makes this Nation great, we will never forget the hardships we have endured. We will always remember September 11.

Even though America has had a year to mourn our losses, we still weep for the victims of that day. We continue to offer our prayers, our comfort, and our resolve to those who lost loved ones on that day.

Without question, the attacks of September 11 were a strike against all nations that value freedom and democracy. It was an act of war, but we were not to be intimidated. As a Congress, we remain steadfast with our Nation in the fight against terrorism. American history has always been defined by the resiliency of our people, and I stand here today to repeat our solemn pledge to defend freedom and liberty and show that we will remain resilient no matter what the threat. The freedoms and values our forefathers gathered in this hall to protect are simply too sacrosanct to ever be compromised.

Hon. Juanita Millender-McDonald

OF CALIFORNIA

Mr. Speaker, it is an honor for me as a Member of the U.S. Congress to convene here in New York today, September 6, 2002, on this historic occasion. It has been 1 year since the tragic terrorist attacks of September 11. It has been over 200 years since Congress convened here in New York at Federal Hall to elect the first President of the United States and to ratify the Bill of Rights.

We applaud the people of this great city and State for their resolve in light of the devastation thrust upon them. It is a credit to their perseverance and their total commitment to freedom that the 107th Congress stands proud today. We reflect with remorse on the past year, but also look forward with hope and optimism as we proceed in the healing of our Nation. On this historic day, the Congress has convened to mourn the loss of families and friends during the September 11 attacks, and to salute the heroism of our police officers, firefighters, and emergency responders who performed with unbridled bravery and gave so selflessly on that tragic day.

As I reflect on the tragic events of a year ago, I am reminded of the heavy hearts of so many great Americans who have grappled with this cowardly act and of those who had loved ones who perished in New York, Pennsylvania and the Pentagon. For those who lost friends and coworkers and who must now try to go on with their lives, our prayers are offered to these families and their friends and to all America and other countries who experienced loss.

Today we are expressing our gratitude to the firefighters, police officers, healthcare workers and those individuals who, on that fateful day, performed heroic deeds and helped their fellow citizens and neighbors without regard for their own welfare.

During the recent year, my colleagues and I in Congress have actively engaged in debate about how to develop an agenda that addresses the new world in which we now live.

Though our country and the world have been shaken, we continue to build bridges toward progress and strengthen the bond of patriotism and the spirit of hope. I cannot express how important it is to map out a course for our future that will sustain, inspire and protect our children. We must provide our children with a sense of optimism and hope.

Our domestic efforts and grief over the tragic events of September 11 have heightened our appreciation for the pain of others around the world who have been subjected to the brutality and inhumanness of terrorism. That is why we have supported liberation and democratization efforts in Afghanistan and seek to assist in the rehabilitation of those persecuted and those who are attempting to rebuild their lives and their country. We must complete our mission there.

As we return to Washington, our Nation and its Congress will never forget the victims of September 11. On this occasion of reflection, recommitment and rededication to freedom and democracy, we are affirming our commitment to remember and honor the men and women who paid the ultimate price—their lives. They will always be a part of our history and our hearts. Our Nation shall continue to rise to meet the challenge of terrorism and the threats posed by terrorists who seek to derail freedom and a Nation of peace.

RECOGNIZING THE HEROISM AND COURAGE DISPLAYED BY AIRLINE FLIGHT ATTENDANTS EACH DAY

Hon. Thomas E. Petri

OF WISCONSIN

Mr. Speaker, I move to suspend the rules and agree to the concurrent resolution (H. Con. Res. 401) recognizing the heroism and courage displayed by airline flight attendants each day, as amended.

The Clerk read as follows:

H. CON. RES. 401

Whereas over 100,000 men and women serve as airline flight attendants in the United States;

Whereas flight attendants dedicate themselves to serving and protecting their passengers;

Whereas flight attendants are responsible for customer service aboard an aircraft;

Whereas flight attendants react to dangerous situations as the first line of defense of airline passengers;

Whereas safety and security are a flight attendant's primary concerns;

Whereas flight attendants evacuate aircraft in emergency situations;

Whereas flight attendants defend passengers against hijackers, terrorists, and abusive passengers;

Whereas flight attendants handle in-flight medical emergencies;

Whereas flight attendants perform routine safety and service duties on board an aircraft;

Whereas 25 flight attendants lost their lives aboard 4 hijacked flights on September 11, 2001;

Whereas 5 flight attendants helped prevent United Airlines Flight 93 from reaching its intended target on September 11, 2001;

Whereas flight attendants provided assistance to passengers across the United States who had their flights diverted on September 11, 2001;

Whereas on December 22, 2001, flight attendants helped subdue attempted shoe bomber, Richard Reid, who attempted to kill all 185 passengers and 12 crew members on board American Airlines Flight 63; and

Whereas on February 7, 2002, flight attendants helped prevent Pablov Moreira, a Uruguayan citizen, from breaking into the cockpit during United Airlines Flight 855 from Miami to Buenos Aires: Now, therefore, be it

Resolved by the House of Representatives (the Senate concurring), That Congress—

(1) expresses profound gratitude to airline flight attendants for their daily service to make air travel safe;

(2) honors the courage and dedication of flight attendants;

(3) expresses support for the flight attendants who displayed heroism on September 11, 2001, and to all flight attendants who continue to display heroism each day; and

(4) directs the Clerk of the House of Representatives to send a copy of this resolution to a family member of each of the flight attendants killed on September 11, 2001.

Mr. Speaker, I rise today in strong support of H. Con. Res. 401, honoring the over 100,000 men and women who serve as flight attendants. As we near the final anniversary of the September 11 attacks, I think it is appropriate that we stop to recognize these everyday heroes.

America's flight attendants dedicated their lives to ensuring the safety and the security of their passengers. There are over 20,000 commercial airline flights each day, and on these flights, flight attendants put the well-being of each of their passengers ahead of their own. They are the first responders to all emergencies in the cabin of an aircraft. They provide in-flight medical assistance to passengers in need. They may be the only line of defense should terrorists once again attempt to take control of an airplane.

It was the flight attendants who subdued attempted shoe bomber Richard Reid aboard American flight 63 last September. Last year, 25 flight attendants lost their lives aboard the 4 hijacked flights on September 11. In recognition of their important role, the House overwhelmingly passed legislation that would significantly increase self-defense and situational training to aid flight attendants in the case of another terrorist hijacking.

I would like to express my profound gratitude to all airline flight attendants for their daily service to make air travel safe and secure, and urge the passage of this resolution.

Hon. Elijah E. Cummings

OF MARYLAND

Mr. Speaker, I, too, want to commend the subcommittee chairman, the gentleman from Florida (Mr. Mica), and the ranking member, the gentleman from Illinois (Mr. Lipinski), for their action on H. Con. Res. 401 that recognizes the largely unsung heroism that airline flight attendants display on a daily basis.

On September 11 of last year, 25 flight attendants lost their lives as a result of terrorist attacks. When they left their homes and loved ones that morning, I am sure that none of them knew what tragic events would unfold before the day's end, and what role they would play in it. Five flight attendants working on United Airlines flight 93 helped prevent hijackers from reaching their intended target of Washington, DC. I am sure that many of us have to give credit to them for perhaps saving our lives.

Every day a workforce of 100,000 flight attendants make it their mission to ensure the safety of passengers that rely on them. In the war on terrorism, they can be compared to ground soldiers. They are our front line of defense. Not only did flight attendants display acts of heroism on September 11, but they also have repeatedly reacted courageously to thwart acts of terrorism on American aircraft.

Mr. Speaker, it is an honor to acknowledge the heroism and bravery of flight attendants. After September 11, when many were afraid to return to the air, these courageous workers devotedly returned to their jobs. I admire their sense of dedication and professional attitude. I urge my colleagues to support this very appropriate resolution.

Hon. Nick J. Rahall II

OF WEST VIRGINIA

Mr. Speaker, on behalf of West Virginians and Americans as a whole, I want to express our deep appreciation to the flight attendants who provide outstanding service while ensuring public safety for thousands of Americans on a daily basis. In addition, I want to commend our Nation's flight attendants for their acts of heroism. A considerable number of them demonstrated their great courage during the September 11 attacks that occurred almost exactly 1 year ago today.

It should not be overlooked that in the days, weeks, and months following that terrible day, America's flight attendants bravely resumed their duties serving our flying public. This contribution aided the almost immediate restoration of air service, and it provides a profound demonstration of this country's refusal to let the terrorists win. Our flight attendants, and by extension, all of us would not allow a few evil doers to destroy our daily activities and our unique way of life.

As we approach the anniversary of September 11, we must remember the contributions of this group of individuals who have so ably demonstrated their importance to this country and to its citizens. They stand as an example for the brave efforts of all hard-working Americans as we cope with the events and the aftermath of that infamous day.

Hon. Rob Simmons

OF CONNECTICUT

Mr. Speaker, a great many things changed on September 11, 2001. Among them, Americans began thinking differently about air travel, and we all gained a greater respect for those who are pledged to guarantee our safety as we fly.

For its part, Congress has moved to make air travel safer, and I have no doubt we will do more. But one of the main lines of defense against events in the air rests with the pilots and flight crews.

The professionalism, courage and common sense exhibited by these individuals is clearly exemplified in the actions of Madeline Amy Todd Sweeney, who was a flight attendant aboard American Airlines flight 11 on September 11. That was the first aircraft to crash into the World Trade Center.

Showing courage under pressure, Amy was one of the first individuals to use a cell phone and notify the world of the hijackings that were under way. Her last acts of bravery were critically important in identifying and exposing those terrorists who threatened our lives, our country and our values.

Ms. Sweeney is a true American hero. She was many things to many wonderful people, a faithful wife, a loving mother, and a devoted daughter. But she will be remembered by most Americans for her extraordinary heroism and devotion to duty on a fateful day.

Consider this—those who hijacked American Airlines flight 11 had years of training and preparation for their terrible mission. They had plenty of time to consider what they were going to do. But for Amy, the decisions of a lifetime were compressed into a few terrible minutes. Yet she responded with tremendous courage, calmness and common sense. She did her duty in the face of death. And at the last moment, she called out to God for salvation.

I had the honor earlier this year to attend a ceremony in Massachusetts where she became the first individual to be awarded the "Madeline Amy Todd Sweeney Award for Civilian Bravery." It is in recognition of Amy Todd Sweeney's heroism and courageous spirit that this award was created.

Future recipients—awarded annually on the anniversary of her death—must demonstrate exceptional bravery, without regard for personal safety, in an effort to save the life or lives of another or others in actual or imminent danger. It is a fitting tribute to her conduct that this award has been established. There can be little doubt that many people are alive today because of her quick thinking and her heroism. Her actions remind us that courage is rightly esteemed as the first of human qualities because it is the quality that guarantees all others.

And we should remember that courage does not mean an absence of fear, because without fear there can be no courage. Courage is doing the thing you think you cannot do.

Mr. Speaker, I ask unanimous consent to enter into the *Record* a recent article from the *New London Day* entitled, "A Hero on Flight 11, She Put Her Job First: Madeline Amy Todd Sweeney."

May God bless and keep you, Madeline Amy Todd Sweeney, and may God bless America.

A Hero on Flight 11, She Put Her Job First: Madeline Amy Todd Sweeney
9–11: The Shadow Of A Day

(By Bethe Defresne)

Once the extended family of Madeline Amy Todd Sweeney verified that she was aboard the flight, there was no waiting for a miracle. There was no use praying that the blond, blue-eyed young mother with the dazzling smile had been caught in some pocket of hope within the ashes of the World Trade Center, and that somehow she would rise up and come home to them.

Relatives and friends of victims who worked in the towers would spend days or even weeks holding off the inevitable. But for Sweeney's loved ones, there was only the swift, burning onslaught of grief.

She was a flight attendant on American Airlines flight 11, the plane that hit the North Tower at 8:48 a.m. on Sept. 11, when the world still thought it must have been some terrible accident.

Sweeney's colleagues in air control back at Logan International Airport in Boston, where the flight had taken off that morning bound for Los Angeles, were among the few who knew better—because she had told them.

Calmly, and with painstaking attention to detail, Sweeney had explained that the plane was being hijacked. What she said would later help Federal investigators reconstruct how the plane was taken over.

Sweeney reported the seat numbers of the hijackers, including suspected ringleader Mohamed Atta, and the progress of their assault. She described the landscape below after the flight was diverted, right up until the shocking end.

Her last words were, "I see water and buildng. Oh my God! Oh my God!"

Sweeney's father, William A. Todd of Norwich, expects that some day he'll listen to a tape of conversations from the airplane. But he's not ready.

This Sept. 11, Todd will be in Boston, the point of Sweeney's departure, not New York, the site of her tragic end, to mark the anniversary of the terrorist attacks. There he will witness the presentation of the second annual Madeline Amy Todd Sweeney Award for Civilian Bravery, posthumously presented to her in February with her family in attendance.

Sweeney lived in Acton, Mass., with her husband, Michael, and their two children, Anna, 6, and Jack, 5. The award, in the form of a medallion, is to be given each year to a Massachusetts resident who exemplifies the courage that Sweeney displayed.

Her reports from the doomed airplane have been credited with helping officials make the crucial decision to ground all airplanes on Sept. 11, perhaps saving many lives.

It's good, Todd acknowledged, to have something to do and somewhere to go on this grim anniversary. And he really didn't want to be in New York.

"It's too much," he says.

Mike Sweeney, who could not be reached, is reportedly coping as best he can, and also plans to be at the presentation in Boston on Wednesday.

Reflecting today on what enabled his 35-year-old daughter to show such remarkable strength under intense pressure, Todd draws upon an apt and familiar analogy, that of a soldier in battle. As an Army war veteran who saw combat in Korea, Todd, 65, says his daughter was doing what she was trained to do in a situation like that: focus not on yourself, but on your job.

Todd treasures an American flag carried in Sweeney's honor aboard an F–16CG Falcon during a Jan. 26 combat mission over Afghanistan. The flag was sent to him along with a citation from the 332d Air Expedition Group, called "The Tip of the Spear," certifying that the flag was carried "In Memory of the Grace and Bravery of Madeline Amy Todd Sweeney, who lost her life to a terrorist attack on the WTC while serving on American Airlines flight 11 on Sept. 11, 2001."

Todd hasn't decided yet where to display the flag, which he handles reverently, like a flag that has been draped over a soldier's coffin. It was a nephew in the Air Force, Patrick Todd, who arranged through his commanding officer to have the flag carried and delivered.

Sweeney came from a large extended family, with numerous aunts, uncles and cousins. She reveled in those myriad relationships, said Todd, and will be especially missed at the Sept. 28 wedding of her brother, William Todd III, who lives in Massachusetts. He was her only sibling.

Sitting at the kitchen table in his home on Corning Road, Todd, who retired after 15 years as a welder at Electric Boat in Groton, appears to take some comfort in thinking about the lasting impress on his daughter's life has made on others. But he is not a man given to displaying an excess of emotion.

"What can you say?" he asks. "Not a day goes by that I don't think of her."

Tears well up in his tired eyes, but he won't let them go. His wife of 23 years, Doris, is more talkative and openly emotional. Sweeney was like a daughter to her, she says, although they only got together on visits, mostly during the summer. Todd and his first wife divorced when Sweeney was 10, and she continued to live with her mother in Nashua, N.H.

It's not very difficult for family members to imagine what Sweeney, whom everyone called Amy, would have been doing this past year had she not been among the 3,008 victims of Sept. 11. She would have continued to love being a wife and mother, kept in close touch with her large circle of friends and family, and, of course, kept on flying.

The 12-year veteran of American Airlines was at a point in life where, it seems, she had everything she wanted. "She loved to fly, and she loved to travel," says Todd. She especially relished trips to the Caribbean.

The flight to Los Angeles was also one of her favorites, Todd says, because she got a layover in California.

After Sweeney graduated from high school, before she married and went to flight attendant school, she took a year off to live and travel in California, says Todd. This was her one real fling with being totally carefree.

But shouldering responsibility was something Sweeney apparently did willingly, with a modest touch that endeared her to family, friends and colleagues, as well as passengers.

She was "a natural at being a flight attendant" wrote one of her peers in a tribute booklet put out by American Airlines and given to all the families of those who died on flight 11. She was a genuine people person, it was said, always the first to volunteer when help was needed.

In her heroic death, Sweeney is forever linked with one colleague in particular, fellow flight attendant Betty Ong. The two women worked as a team to alert ground officials about what was happening.

The Todds have a tape of a "Prime Time" TV segment on the two women, hosted by Diane Sawyer. Sweeney's husband, Mike, is featured along with several members of Ong's family. So, too, are home videos of Sweeney playing and singing with her children.

But most of the program is devoted to what happened on flight 11.

Doris Todd cries softly, and her husband sits stoically upright in his chair, as a Logan flight manager who got the first call recounts his conversations with Sweeney. "Amy, honey," he began, "what's going on?"

Everything after that was dark—throats slashed, orders from hijackers—but Sweeney remained purposeful and calm through it all.

The Todds find this tape difficult to watch, but say they've looked at it several times. During the program, the Ong family reports that a bone and a flesh fragment from Betty Ong were recovered from the WTC site.

"Nothing was found of Amy," says Todd, not even a bit of DNA. But this is not something to dwell on, he says. He'd rather hold the flag that was carried over a battlefield in her honor, or point out the sign in her memory attached to his truck, the one he drove in the June parade for "A Reason to Ride," an organization that raises funds for disabled and homeless veterans.

There's been talk, Todd says, that Sweeney and some other Sept. 11 heroes might be posthumously awarded the Presidential Medal of Freedom. U.S. Rep. Rob Simmons, R–2d District, and Sen. Edward Kennedy of Massachusetts are among those who have proposed Sweeney for the prestigious award.

The Todds haven't gotten involved in the debate over what kind of memorial should be erected at Ground Zero, but they do have two thoughts on the subject. "It should be tall," they say. "And it should have all the names, like the Vietnam War Memorial."

Sweeney didn't have enough information in that early hour of Sept. 11 to go on the offensive, like the crew and passengers of United flight 93, forced down in a field in Pennsylvania. But she died in service to her passengers and, it turned out, to her country.

It's almost impossible to find a source of gratitude in the horror of Sept. 11. But the Todds say they're at least glad to know that almost to the last moment Madeline Amy Todd Sweeney was not living in terror, but helping others.

PETER J. GANCI, JR. POST OFFICE BUILDING

Hon. Jo Ann Davis

OF VIRGINIA

Mr. Speaker, I move to suspend the rules and pass the bill (H.R. 5336) to designate the facility of the U.S. Postal Service located at 380 Main Street in Farmingdale, NY, as the "Peter J. Ganci, Jr. Post Office Building."

The Clerk read as follows:

H.R. 5336

Be it enacted by the Senate and House of Representatives of the United States of America in Congress assembled,

SECTION 1. PETER J. GANCI POST OFFICE BUILDING.

(a) DESIGNATION.—The facility of the United States Postal Service located at 380 Main Street in Farmingdale, New York, shall be known and designated as the "Peter J. Ganci, Jr. Post Office Building".

(b) REFERENCES.—Any reference in a law, map, regulation, document, paper, or other record of the United States to the facility referred to in subsection (a) shall be deemed to be a reference to the Peter J. Ganci, Jr. Post Office Building.

Mr. Speaker, H.R. 5336, introduced by our distinguished colleague from New York (Mr. King), designates a post office in Farmingdale, NY, as the Peter J. Ganci, Jr., Post Office Building. Members of the entire House delegation from the State of New York are cosponsors of this legislation.

Mr. Speaker, on September 11, immediately after a jet struck the first tower at the World Trade Center, Peter Ganci, chief of department for the New York City Fire Department, rushed to the scene from his command post in downtown Brooklyn and started the rescue effort.

He was in the basement of tower one when it collapsed. Miraculously, he survived, dug himself out of the rubble and went back to work.

It then became apparent that the second tower would fall. Ganci, as the highest ranking uniformed officer on the scene, directed everyone to clear out of the area, but Ganci did not himself leave while his men were inside the tower. Ganci said, "I'm not leaving my men," and advanced toward the towers to continue his life's work of saving and protecting others.

Chief Ganci was in the basement of the second tower when it collapsed.

Prior to joining the New York City Fire Department, Ganci served in the Farmingdale Fire Department as a volunteer and in the 82d Airborne Division. Ganci served in the New York Fire Department for 33 years and was decorated repeatedly for bravery.

Ganci is survived by his wife, Kathleen; his sons, Christopher and Peter; and his daughter, Danielle. His son, Peter Ganci III, now serves with Ladder Company 111 of the New York City Fire Department.

Mr. Speaker, I urge adoption of H.R. 5336.

Hon. Danny K. Davis

OF ILLINOIS

As a member of the House Committee on Government Reform, I am pleased to join my colleague today in the consideration of two postal naming bills: H.R. 5336, introduced by the gentleman from New York (Mr. King), which names a post office in Farmingdale, NY, after Peter Ganci; and H.R. 4797, introduced by the gentleman from California (Mr. Becerra), which names a post office in Los Angeles, CA, after Nat King Cole. These measures have the support of their respective State congressional delegations to name U.S. postal facilities after very deserving individuals, and I urge their swift passage.

H.R. 5336, to designate the facility of the U.S. Postal Service located at 380 Main Street in Farmingdale, NY, as the Peter J. Ganci, Jr., Post Office Building, was introduced by the gentleman from New York (Mr. King) on September 5, 2002.

An American hero, Peter J. Ganci was the New York City Fire Department's highest ranking chief who died when the World Trade Center came down. Chief Ganci had been on the radio in front of the trade center directing the rescue efforts when the building collapsed.

Who was Chief Ganci? He was 54 years old and a resident of Massapequa, NY; the chief of the department, one of the highest ranking uniformed officers in the fire department; husband to Kathleen; father to Peter III, Christopher and Danielle. He was also a 33-year-old veteran of the New York Fire Department, whose son Peter is a firefighter assigned to Ladder Company 111 in Brooklyn, NY, and Brooklyn, NY, is the place where Peter Ganci got his start in 1960.

After serving in engine and ladder companies, Mr. Ganci rose to lieutenant, captain, battalion chief, deputy chief and then acting chief. He also ran the Bureau of Fire Investigation, and was appointed the chief of operations prior to becoming chief of the department.

A hands-on man, Chief Ganci was doing his job, commanding the rescue operations at the New York World Trade Center.

Accordingly, I urge swift passage of this bill and commend my colleague for seeking to honor Chief Peter J. Ganci, Jr., in this matter.

Hon. Peter T. King

OF NEW YORK

Mr. Speaker, I am privileged to rise today in support of this legislation. Pete Ganci was a constituent of mine He was an individual who personified the leadership and the bravery which resulted last September 11 in the greatest rescue operation in the history of this country.

Twenty-five thousand people were rescued that day, and it was done primarily through the efforts of the New York City Fire Department, and Chief Pete Ganci was the chief of the department, who was also the highest ranking uniformed officer ever to die in the line of duty, and as the gentlewoman pointed out, Chief Ganci was there that day with his men. Chief Ganci was there in the line of duty.

Chief Ganci was there directing the operation against this horrific attack that was carried out against our country, and when the first tower fell, Chief Ganci barely escaped with his life, and when he saw the terrible carnage that resulted and saw the terrible danger which was still being faced by the North Tower, which had not yet fallen, Chief Ganci ordered the entire operation to be moved north.

As the entire operation, including the mayor and other officials, went north, Pete Ganci went south to be with his men, and that personified the type of leadership which Pete Ganci gave in the New York City Fire Department. As the gentlewoman pointed out, he had been a member of the department more than 33 years, rising to the highest office in that department, chief of the department. Prior to that, he had been a paratrooper in the 82d Airborne, and until his death, he was an active member of the Farmingdale Volunteer Fire Department in New York.

Mr. Speaker, at this point also, before I digress, though, I want to pay a special debt of thanks to Sal Pontillo, the Nassau County legislator who represents the district in which Mr. Ganci lived in Nassau County, and he has come to me with this request. We have worked closely on it, and he has also served as liaison with the Ganci family and helped to bring this about.

Mr. Speaker, many of us look back on the date of September 11 and we realize what a turning point it was in the history of our country. It was a turning point for many reasons. The point I would like to focus on today is that after that attack, actually as the attack was going on, as the buildings were burning, as the buildings were coming down, as people were being killed, as people were being rescued, what the eyes of the world and the eyes of the country saw was the valor of the New York City firefighters and police officers, those who were carrying out this rescue operation under the most trying circumstances, and the person who was leading that operation was Chief Ganci. Just think what would have happened if he had not done what he did, if the firefighters had not responded the way they did, instead of running into the building, Chief Ganci had run out of the building and kept going. On the contrary, he went right back in the line of fire, the battlefield commander who died with his men. It was that type of courage that was shown that day, that type of heroism that was shown that day, that type of spirit that was shown that day which I think has inspired our country to fight back, to come back and to win this war against international terrorism, and even just as important, to show that America cannot be vanquished, that we can take the best shot of the enemy and come back stronger than ever.

That is what Chief Ganci personified that day. In fact, it is the type of story that if somebody had produced a movie about it, it would not have been believed, for a person to be there when the first tower came down and somehow survive it and go back in and to be killed in the second tower, which to me is the type of courage that I cannot even begin to fathom.

This was the first battle, Mr. Speaker, and the first great war of the 21st century, and Chief Ganci died as a battlefield commander in that war, and for that, this country owes him its untiring thanks and gratitude for all that he has done.

I want to say a special debt of thanks to his wife, Kathy, who has shown tremendous courage throughout this entire 363 days; his son, Pete, who is a member of the New York City Fire Department; his son, Chris; and his daughter, Danielle.

I also want to thank the other 30 members of the New York delegation for unanimously supporting this resolution. To me, it is just another indication of the tremendous regard that all the men and women of the New York City Fire Department had for Chief Ganci and, indeed, all of the members of the New York delegation, everyone who had the opportunity to know Chief Ganci.

I can say I was at his funeral last September 15. It was one of the most moving moments I have ever been involved in, to see the tremendous outpouring of support from his community in North Massapequa and Farmingdale, all of the surrounding community, by the thousands, coming out to join in this salute to Chief Pete Ganci.

So, Mr. Speaker, with that I thank the gentlewoman for yielding me the time. I thank the House for considering this resolution today, and I extend my very best thoughts, prayers and gratitude to the Ganci family for all they have gone through.

Hon. Jo Ann Davis

OF VIRGINIA

I thank my colleague for introducing this legislation to honor the chief of the New York City Fire Department. As the wife of a professional firefighter for almost 30 years, I know the bravery that these firefighters have, and I have had people tell me why they let chiefs send men in there to their deaths, and I have told them quite clearly, if he would not have sent them in, I know the firefighters would have gone in anyway. That is just the type of bravery they do have, and again, I thank my colleagues, and I urge adoption of this resolution.

RECOGNIZING ROSELLE, MICHAEL HINGSON'S GUIDE DOG, FOR HER BRAVERY AT WORLD TRADE CENTER ON SEPTEMBER 11

Hon. Lynn C. Woolsey

OF CALIFORNIA

Mr. Speaker, I rise today to honor Roselle, a yellow Labrador dog, whose bravery on September 11 as she led her blind owner Michael Hingson from the World Trade Center is an inspiration.

Michael and Roselle have become well known as representatives of the power of the human-animal bond to build a trust that can carry them through even the greatest tragedy.

Michael was working on the 78th floor of the World Trade Center when the building was struck. After making sure everyone in the office was evacuated safely, Roselle and Michael began the long descent down the stairwell. Despite the heat and smoke, they calmly made their way from the building and started running for the subway as tower two began to collapse. In the subway, Roselle guided Michael and another woman down the stairs and led Michael to the home of a friend in mid-Manhattan.

Michael has traveled with a dog from Guide Dogs for the Blind for 37 years. Roselle's ability to lead, and Michael's to command, under such difficult circumstances embodies the success of the lifelong partnerships developed through this program.

Mr. Speaker, Roselle's story demonstrates that there were many kinds of bravery on September 11, 2001. It is appropriate to honor the bond that enabled her to save the life of Michael Hingson.

REFLECTIONS ON AMERICA

Hon. Joseph M. Hoeffel

OF PENNSYLVANIA

Mr. Speaker, recently, as the 1-year anniversary of September 11 approached, I looked back over the statement I released the day after the attacks occurred. As I read the statement that was issued even as the smoke and dust still billowed over lower Manhattan, the Pentagon and a field in rural Pennsylvania, I was struck by how easy it was to write some of the words and how difficult it was to write others.

It was tremendously difficult to put into words my feelings of disbelief and anger over the insane acts of 19 individuals and their supporters. It still is. It was heartrending to try to voice the sorrow and sympathy I felt for the victims and their families. And that, too, is still difficult.

It was not difficult, however to write the words of hope, pride and conviction I had for this country and its people on September 12. And, today, a year later, I am happy to report that the hope, pride and conviction were well-founded. On September 12, 2001 I wrote, in part:

> As I walked to work across the Capitol grounds this morning, a day after the attacks, I was struck as I often am by the incredible beauty of the U.S. Capitol Building. The dome was shining a brilliant white against a clear blue sky on a beautiful late summer day. I realized that the glorious dome, such a symbol of the strength and stability of our country, might well have been the final target of the ill-fated fourth hijacked plane.
>
> The simple truth is that even if a terrorist act had destroyed the dome, or if a thousand terrorist acts had obliterated the entire Capital, America would still be standing firm as it is today.

In the past year, we have cried and cursed. We have opened our hearts and our wallets. We have buried our dead and we continue to heal our wounded and sorrowful. We have cleared the debris and begun to rebuild. Our President has led, and Congress has indeed stood shoulder to shoulder with him in support. Our soldiers have fought bravely and we here at home have done what we do best—we went back to work.

We will never again look at the world in the same way. We are sadder. We are wiser. We are closer. And, as I said a year ago, we are still standing firm. We've been through a lot. There will be tough times ahead, but today I have even a stronger feeling of hope, pride and conviction in our country and its people.

God bless America.

H. CON. RES. 401, RECOGNIZING THE HEROISM AND COURAGE DISPLAYED BY AIRLINE FLIGHT ATTENDANTS EACH DAY

Hon. Nancy Pelosi

OF CALIFORNIA

Mr. Speaker, I rise in support of H. Con. Res. 401, a resolution to recognize the heroism and courage displayed by airline flight attendants each day.

The anniversary of the September 11 terrorist hijackings and attacks on the World Trade Center and the Pentagon is just 2 days away. All Americans remember and mourn the lives lost that tragic day.

The world has changed for all of us. Before that time, flight attendants and their families and loved ones lived and worked with the knowledge that rare instances of mechanical failure or hijackers could endanger their lives. But no one imagined the dreadful assault of September 11, when terrorists turned four airborne planes into missiles used to attack thousands of Americans.

That day, flight attendants again demonstrated their courage in the face of extreme danger. From all that we know of the final minutes on those flights, flight attendants worked to communicate with the ground, and in all likelihood helped prevent flight 93, which crashed in rural Pennsylvania, from taking many more lives.

I would like to bring my colleagues' attention to the heroism of Betty Ong, a flight attendant on American Airlines flight 11, whose family lives in my district of San Francisco. On September 11, Betty called the airline reservations center from the plane to sound the alarm and provide information about the terrorists who had taken over the plane. I am aware of the enormous pain and suffering her family has been experiencing and extend them my deepest sympathy. It is an honor to pay tribute to Betty and express my appreciation for her life and bravery in the face of enormous danger.

Now, even stepping onto an airplane is an act requiring will power and courage for many Americans. Yet flight attendants do it every day. Flight attendants deserve our respect, cooperation, courtesy, and commendation for their hard work and courage. I urge my colleagues to vote for this resolution.

PRINTING OF PROCEEDINGS OF SPECIAL CEREMONIAL MEETING OF UNITED STATES CONGRESS HELD IN FEDERAL HALL, NEW YORK, NY, ON SEPTEMBER 6, 2002

Hon. Richard K. Armey

OF TEXAS

Mr. Speaker, I ask unanimous consent that proceedings of the special ceremonial meeting of the U.S. Congress held in Federal Hall, New York, NY, on September 6, 2002, be printed in the *Record*, and that all Members have 5 legislative days to insert their remarks on the topic of the ceremonial meeting.

Tuesday, September 10, 2002

REMEMBERING SEPTEMBER 11, 2001

Hon. Mike Pence

OF INDIANA

Madam Speaker, to state the obvious, tomorrow is September 11 again. Like many Americans, I will be home taking time with family and neighbors for a solemn reflection, remembrance and prayer, and so it should be for all Americans. But my earnest hope, Madam Speaker, is that September 11 of this year not be an anxious time; that in addition to remembering the lost, we actually, Madam Speaker, have much to commemorate.

In the past 12 months, our people have responded with selfless actions of courage and generosity, our military has responded with valor, our President with moral clarity and purpose, and this Congress, Republicans and Democrats alike, have responded with resources and reform.

America is better prepared and safer this September 11 than the last. Let us also be confident in this, that He who sets this pilgrim's dream on this wilderness shore still watches over us. And I say like Americans have said throughout generations, I lift up my eyes to the hills, and where does my help come from, my help comes from the Lord.

AMERICANS URGED TO REMEMBER SEPTEMBER 11

Hon. Nick Smith

OF MICHIGAN

Madam Speaker, this week it is important to remember those who lost their lives during the attack on America last year, as well as all of our first responders, our medics, our military personnel, and the people who volunteered and tried to help.

I think our Founding Fathers would be very proud of our new diligence in our quest to preserve liberty and freedom in this country. President Bush has designated September 11 of each year to be Patriot Day, and calls on all Americans to observe it appropriately.

I urge my Michigan citizens and all Americans to spend some time thinking about what we need to do to protect our liberty and freedom, and to pray for the families of those that died in the terrorist attack in Pennsylvania, Washington and New York. This Wednesday marks the 1-year anniversary. Let us remember what our forefathers did, and what happened to us 1 year ago and our renewed vigor to make sure that we do what is important to sacrifice ourselves in the preservation of liberty and freedom.

Hon. Vito Fossella

OF NEW YORK

Every day and especially tomorrow, September 11, 2002, we honor the heroes and the victims who perished last year, our fellow American citizens, those who just want an ordinary way of life, a peaceful life, who sought to raise a family in this great Nation, who sought to do a job and do it right, and those valiant heroes from EMS and the police department and the port authority and especially the firefighters, 78 of whom were from Staten Island alone, and names like Egan and Olsen and Curatola and Esposito and Siller and Leahy and Doyle, and tragically thousands of others. They are names, yes, but they are families. They have left behind children, they have left behind wives, husbands, parents, and grandparents, and what they were seeking is all what I think we are all about, the right to live in freedom with liberty and in peace, and that was robbed from them. It was robbed from their families. And, yes, we are a stronger and better country, and we are fortunate to have brave men and women to wear the uniform to go get those people, wherever they may be across the globe, with the Commander in Chief, President Bush, leading the way.

REMEMBERING THOSE KILLED ON SEPTEMBER 11, 2001

Hon. Sheila Jackson-Lee

OF TEXAS

Mr. Speaker, in a solemn way I recognize that this week will be a week that Americans turn toward each other seeking to embrace and seeking to love, mourning those whom we lost on September 11, 2001, and celebrating the unsung heroes and the families who have survived.

I stand before this House today simply to offer my deepest sympathy and that of my constituents of the 18th Congressional District of the State of Texas to all of those who experience this great loss as we come upon September 11, 2002. It is my desire to offer these words of sympathy because I love you and appreciate the sacrifice you made for this Nation. I conclude by saying, The Lord is my shepherd, I shall not want.

PUBLIC BILLS AND RESOLUTIONS

Under clause 2 of rule XII, public bills and resolutions were introduced and severally referred, as follows:

By Mr. WILSON of South Carolina:

H. Con. Res. 463. Concurrent resolution expressing the sense of the Congress that on September 11, 2002, the people of the United States should reaffirm the principles for which the Nation was founded so that freedom may ring from every community in the Nation and be heard around the world; to the Committee on Government Reform.

RECOGNIZING LOCAL HEROES ON SEPTEMBER 11

Hon. Frank R. Wolf

OF VIRGINIA

Mr. Speaker, as our Nation approaches the 1-year anniversary of the unspeakable acts of terror against America on September 11, 2001, I join with my colleagues, Representative Tom Davis and Representative Jim Moran to recognize some of the people in the northern Virginia area who were among the first health care responders to the attack on the Pentagon.

We attended on September 3 an event recognizing those associated with the Inova Health System for their heroic efforts on September 11, and I share today the remarks of Jolene Tornabeni, executive vice president and chief operating officer for Inova Health System, as well as a copy of a resolution adopted by the Virginia General Assembly commending Inova Health System.

RECOGNIZING INOVA'S HEROES

(By Jolene Tornabeni)

As we approach the anniversary of September 11, no doubt every one of us can remember where we were and what we were doing when the planes crashed in New York, Pennsylvania and the Pentagon. Throughout Inova Health System, the memories of that day remain fresh in our memories as well. At a time when most people wanted to be home with their families and loved ones, our staff showed its true strength of character. Hundreds of employees, nurses and physicians dropped what they were doing and volunteered to help. It was an immediate response that came from the heart.

Today, I'd like to recognize just some of those people who are Inova's heroes.

From Inova Alexandria Hospital.—Shortly after the American Airlines plane hit the Pentagon, hospital staff swung into gear readying beds, operating rooms and supplies. Inova Alexandria Hospital treated more patients from the Pentagon that day than any other Inova facility, caring for 24 people who were injured at the scene.

Among the many heroes that day are Emergency Department chairman Dr. Marty Brown, vice chairman Dr. Tom Clark and the ER staff who were at the front end of caring for the patients as they arrived at the hospital. Dr. Clark cared for Virginia State Police Trooper Michael Middleton who sustained severe smoke inhalation while trying to rescue injured Pentagon workers. In addition, emergency nurse Sherry Hemby is also with us today.

I'd like to recognize pulmonologist Dr. Tom Smirniotopolous and nurse Ellen Smith. They both cared for Trooper Middleton during his long recovery at Inova Alexandria Hospital.

Also, emergency physician Dr. James Vafier. On September 11, he was working in his role as medical director for the Alexandria Fire Department at the Pentagon. On site, he was appointed the physician in charge of civilian medical response at the Pentagon.

Keeping order that day at Inova Alexandria Hospital were Steve Fuoco, the director of engineering, who served the hospital's command center, and Greg Brison, director of security. I'd also like to recognize hospital administrator Ken Kozloff for all of his efforts and a job well done by his entire staff.

All told, Inova treated 27 patients on September 11. Inova Mount Vernon hospital treated one injured civilian. Our thanks go out to hospital Emergency Department chairman Dr. Michael Shuster and hospital administrator Susan Herbert.

Many thanks also go to the emergency department staffs at Inova HealthPlex in Springfield where two patients were treated, and the staff of Inova Fairfax Hospital. As the area's level I trauma center, Inova Fairfax Hospital freed up dozens of hospital beds and readied itself to handle many, many patients that day. Sadly, their services were not needed.

Next, I'd like to recognize Dr. Dan Hanfling, the director of Emergency Management and Disaster Medicine for Inova Health System. On September 11, Dan was called to the Pentagon to assist in the search and rescue in his role as medical team manager of the Fairfax County Urban Search and Rescue team—which falls under the auspices of FEMA—the Federal Emergency Management Agency. Dan spent a week at the Pentagon helping coordinate the Federal response to the disaster. Since 9/11, Dan has helped spearhead and focus Inova's disaster preparations across our system of hospitals and emergency care centers.

Dan also serves as the medical director of Inova AirCare, our medevac helicopter program which played a critical role at the Pentagon on September 11. Minutes after the Pentagon was hit, the helicopter flight crew of nurse Margie Roche, paramedic Chuck Crocker and pilot Pete Russet flew to the Pentagon to fly out the injured patients. That initial flight is memorialized forever in a well-publicized *Reuters* photo seen around the world in magazines and now a book. The photo depicts Inova AirCare against the backdrop of a burning Pentagon.

The helicopter shuttled much needed supplies to the scene that day, and, at the request of military leaders on the scene, remained at the Pentagon throughout the night.

As that day unfolded, the community's good will became abundantly evident as hundreds of people showed up at Inova Blood Donor Services' offices ready to roll up their sleeves and donate. Thanks to the leadership of medical director Dr. Jeanne Lumadue and administrative director Terri Craddock, Inova Blood Donor Services pulled in volunteers to help handle the onslaught and keep the offices open well past normal business hours. They collected more than 700 units of blood that day, which is just amazing. In all, they handled more than 2,000 donors and returned more than 5,000 calls from interested donors in the initial days after the attack.

It was not long after September 11 that our country faced a second threat to our sense of security in the anthrax-laced letters mailed around the country. Inova Fairfax Hospital took center stage in this national drama after diagnosing two patients from the Brentwood Postal Facility in Washington, D.C., with inhalation anthrax.

Emergency physicians Cecele Murphy and Denis Pauze relied on their instinct and medical training to make a diagnosis most doctors could not imagine making in their lifetimes. Thanks to them, and to physician assistant Ashna Nayyar and the entire ER staff, both men are alive today.

All of these people mentioned today and, in fact, all of the physicians, employees and volunteers throughout Inova Health System, are our heroes for the work they did on September 11 and its aftermath.

We also have heroes outside of our organization in the men and women of our community's police, fire and EMS agencies, particularly in Fairfax County and the City of Alexandria.

Over the past year, we have been grateful for their continual support and advice on preparedness.

SENATE JOINT RESOLUTION NO. 275—ENROLLED

COMMENDING INOVA HEALTH SYSTEM

AGREED TO BY THE SENATE, MARCH 6, 2002

AGREED TO BY THE HOUSE OF DELEGATES, MARCH 8, 2002

Whereas, Inova Health System in Northern Virginia was deeply involved in the response to the attacks on September 11, 2001, and in the diagnosis and treatment of those exposed to anthrax in October of 2001; and

Whereas, on September 11, Inova Health System treated 27 patients injured in the attack on the Pentagon at its Alexandria, Mount Vernon, and Franconia-Springfield facilities; and

Whereas, within moments of the crash at the Pentagon, Inova AirCare was on the scene to transport patients, and AirCare 2, the system's back-up helicopter, transported needed supplies to the scene for use in patient triage; and

Whereas, Inova Blood Donor Services collected more than 2,000 units of blood in the first week following the attacks, and a portion of the donations were sent to New York and New Jersey to help injured patients; and

Whereas, the Inova Institute of Research and Education contacted the Food and Drug Administration to allow usage of a new drug—in its final phase of testing—in a successful attempt to save the life of Virginia State Police Trooper Michael Middleton; and

Whereas, more than a month after the September 11 attacks, Inova Health System played a leading role in the initial diagnosis and treatment of patients exposed to anthrax at the Brentwood Postal Facility; and

Whereas, Inova Fairfax Hospital emergency room physician Cecele Murphy diagnosed the first inhalation anthrax patient on October 19, 2001, before the source of the anthrax was known; and

Whereas, within two days, the hospital diagnosed the second anthrax case, and Inova physicians soon developed protocols for hospitals to follow in screening postal workers and other potential inhalation anthrax cases; and

Whereas, in collaboration with infectious disease specialists from Kaiser Permanente, Inova physicians published an anthrax case study in the *Journal of the American Medical Association*, an article that was published faster than any other case study in the journal's history; and

Whereas, Inova Fairfax Hospital held three press conferences to educate the public on key anthrax information, including the fact that it is not contagious and that patients in and visitors to hospitals are safe; and

Whereas, Inova Health System continued to take the lead in producing and distributing anthrax information to inform the public via information hotlines, websites, the press, and public meetings; and

Whereas, throughout the turbulent Fall of 2001, the medical professionals and staff of Inova Health System responded to emergency situations with great dispatch, diligence, courage, and professionalism; now, therefore, be it

Resolved by the Senate, the House of Delegates concurring, That the General Assembly hereby commend Inova Health System for its quick and effective response to the events of September 11 and the anthrax incidents in October 2001; and, be it

Resolved further, That the Clerk of the Senate prepare a copy of this resolution for presentation to Inova Health System as an expression of the General Assembly's admiration and gratitude for its dedication to the health and welfare of the citizens of Northern Virginia.

Wednesday, September 11, 2002

SESSION BEGINS—PRAYER

The House met at noon.

The Chaplain, the Reverend Daniel P. Coughlin, offered the following prayer:

Remembering the images of September 11 a year ago can still stun a nation. A reflection of that tragic day and the thousands who were taken from us can still take the breath away of some. Others feel nothing, only emptiness. Others have moved on and celebrate the distance. Today a moment of common silence can unite us all in a deeper sense of presence.

Because words have their own spin to such an overwhelming story as this past year. Silence alone is free enough to embrace all traditions and all sentiments, drawing out a language of the heart.

Only silence can interpret some of the most sacred moments of a lifetime. So let memories flow and prayers arise in the soul of America as we enter unafraid the mystery of what has happened to us in silence.

Amen.

MESSAGE FROM THE SENATE

A message from the Senate by Mr. Monahan, one of its clerks, announced that the Senate has passed without amendment a bill of the House of the following title:

H.R. 3917. An Act to authorize a national memorial to commemorate the passengers and crew of Flight 93 who, on September 11, 2001, courageously gave their lives thereby thwarting a planned attack on our Nation's Capital, and for other purposes.

The message also announced that the Senate has passed bills of the following titles in which the concurrence of the House is requested:

S. 2136. An act to establish a memorial in the State of Pennsylvania to honor the passengers and crewmembers of Flight 93 who, on September 11, 2001, gave their lives to prevent a planned attack on the Capital of the United States.

REMEMBERING SEPTEMBER 11

Hon. Rodney P. Frelinghuysen

OF NEW JERSEY

Mr. Speaker, today as a Nation we are reunited by a sense of common grief. One year ago today, America witnessed the unspeakable when our Nation was attacked and more than 3,000 innocent lives were cruelly taken from us at the World Trade Center, the Pentagon, and in Pennsylvania. From my home State of New Jersey, we lost over 700 people, and God bless their families. It was also a day when horror was met by heroism and the worst of humanity was overshadowed by the best of America. In reflections on this day, we remember the bravery of those who responded on September 11, our police, firefighters, our first aid squads, people who never gave up hope and rallied our Nation. They who responded came from all over America and across the Hudson from New Jersey.

We are grateful as well today for the service of our young men and women in uniform who, fighting in our war against terrorism, are resolved to bring justice to those who attacked us. They are fighting to right this terrible wrong in honor of the memory of those who perished and to protect our children and grandchildren, ensuring that they inherit a Nation free from further terror.

ON THE ANNIVERSARY OF SEPTEMBER 11

Hon. Michael R. McNulty

OF NEW YORK

Mr. Speaker, whenever I think of the enormity of what happened on September 11, 2001, words fail; and so I try to recall the words of our Commander in Chief on that fateful day. He basically made three suggestions. He said we should pray for all of the innocent victims and their families, and I try to remember to do that every day. He then said we should be grateful. And I will be perfectly honest, gratitude was not what was in my heart at that moment. But he went on to explain that we should be grateful to all the police officers, firefighters, and emergency personnel who, when others were running away from the buildings in terror, running for their lives, continued to run toward the buildings, into the buildings, up the stairs, many to their deaths. That was their job. And they did it well that day because they saved thousands of lives.

Finally, he suggested that we unite as a Nation in our resolve to track down the terrorist cowards who committed those acts, bring them to justice, and take away their capabilities to ever do anything like that again either in the United States or anywhere else in the world.

On this solemn anniversary may we continue to remember those three suggestions by our Commander in Chief.

TO THE CHILDREN ACROSS AMERICA REGARDING SEPTEMBER 11

Hon. Nick Lampson

OF TEXAS

Mr. Speaker, usually when I stand here to address these 1-minutes, it is on the subject of missing children. Today I want to talk about kids who are missing mothers and fathers and sisters and brothers and friends and family because on September 11 of last year more than 2,000 children lost a parent. This may not be the kind of loss that I usually address, but it is a profound loss, nonetheless. I want to send a message to these kids today that this House is thinking about them and they are in our hearts and in our prayers.

The children of this country also lost a sense of innocence and security. I want to encourage parents across the Nation to talk to their children about the tragic events of September 11 of last year to reassure them that we all want the best for them and will continue to work to keep their, our, country safe from harm.

Mr. Speaker, on behalf of the Congressional Caucus on Missing and Exploited Children, I send our thoughts and prayers out to the children who lost somebody at the Pentagon, the World Trade Center, or in Shanksville, PA, and to the kids all across America who are learning to deal with a changed and unsettled world.

God bless you and God bless America.

ON REMEMBERING THE EVENTS OF SEPTEMBER 11

Hon. Aníbal Acevedo-Vilá

OF PUERTO RICO

Mr. Speaker, on behalf of all Puerto Ricans, I want to express heartfelt sympathy for the families of the victims from the terrorist attacks against this great Nation 1 year ago. The impacts of these horrific events were immediately felt throughout Puerto Rico. Hundreds of Puerto Ricans were among the dead and Puerto Rican emergency crews were among the first to arrive to assist crews in New York and at the Pentagon. But had there been no direct tragic link to Puerto Rico through casualties or through the emergency workers, Puerto Ricans would nonetheless continue to walk in lockstep in the war against terror. Make no mistake about it, Puerto Ricans today, as throughout the last 100 years, serve dutifully in all the branches of our armed services. Our common citizenship and common devotion toward democratic principles underscore our commitment to common defense. I stand before my colleagues today to let you all know that

Puerto Rico will always be there in this effort. The cowardly acts of September 11 have caused great pain. Our suffering shall never be forgotten. But this Nation is today stronger and more committed to our principles of freedom and justice than ever before. United we stand, divided we fall. We stand together and will never, ever forget.

EXPRESSING THE SENSE OF CONGRESS ON THE ANNIVERSARY OF TERRORIST ATTACKS LAUNCHED AGAINST THE UNITED STATES ON SEPTEMBER 11, 2001

Hon. Richard K. Armey

OF TEXAS

Mr. Speaker, I ask unanimous consent for the immediate consideration of the concurrent resolution (H. Con. Res. 464) expressing the sense of the Congress on the anniversary of the terrorist attacks launched against the United States on September 11, 2001.

The Clerk read the title of the concurrent resolution.

The SPEAKER pro tempore (Mr. Simpson). Is there objection to the request of the gentleman from Texas?

There was no objection.

The Clerk read the concurrent resolution, as follows:

H. CON. RES. 464

Whereas on September 11, 2001, while Americans were attending to their daily routines, terrorists hijacked and destroyed four civilian aircraft, crashing two of them into the towers of the World Trade Center in New York City, and a third into the Pentagon outside Washington, D.C.;

Whereas the valor of the passengers and crew on the fourth aircraft prevented it from also being used as a weapon against America;

Whereas thousands of innocent Americans were killed and injured as a result of these attacks, including the passengers and crew of the four aircraft, workers in the World Trade Center and in the Pentagon, rescue workers, and bystanders, making these attacks the deadliest terrorist attacks ever launched against the United States;

Whereas when the gravest moments came, many regular Americans, relying on courage, instinct, and grace, rushed toward the flaming buildings in order to rescue or toward terrorist-controlled cockpits in order to resist;

Whereas by targeting symbols of American strength and success, these attacks clearly were intended to assail the principles, values, and freedoms of the United States and the American people, intimidate the Nation, and weaken the national resolve;

Whereas while the States of New York, Virginia, and Pennsylvania bore the brunt of the terrorist attacks, every State and Territory and all Americans were affected and mourned these tragic losses;

Whereas Americans reached out to help strangers who had lost loved ones, colleagues, and their businesses;

Whereas local, State, and Federal leaders set aside differences and worked together to provide for those who were attacked and to protect those who remained;

Whereas Americans continue to repair damage to buildings and the economy, while relishing the freedoms they enjoy as Americans;

Whereas on September 14, 2001, in Public Law 107–40, Congress authorized the use of "all necessary and appropriate force" against those responsible for the terrorist attacks;

Whereas the United States Armed Forces subsequently moved swiftly against Al Qaeda and the Taliban regime in Afghanistan, whom the President and Congress had identified as enemies of America;

Whereas, in so doing, brave servicemen and women left family and friends in order to defend the Nation;

Whereas a year later, many servicemen and women remain abroad, shielding the Nation from further terrorist attacks;

Whereas, while the passage of a year has not softened the memory of the American people, resolved their grief, or restored lost loved ones, it has shown that Americans will not bow to terrorists;

Whereas the Congress has passed, and the President has signed, numerous laws providing additional resources for the overseas effort against terrorism, as well as additional tools for Federal, State, and local law enforcement and judicial systems to protect Americans at home; and

Whereas the Government reexamined the need for domestic security and the Congress is currently considering legislation to create a Department of Homeland Security with the specific mission of preventing further attacks: Now, therefore, be it

Resolved by the House of Representatives (the Senate concurring), That Congress—

(1) recognizes September 11 as both a day to remember those taken from their families, loved ones, and fellow citizens and a day for Americans to recommit to the Nation, to their freedoms, and to each other;

(2) extends its deepest sympathies to the countless innocent victims of the September 11, 2001, terrorist attacks, their families, friends, and loved ones;

(3) honors the heroic actions of first responders, law enforcement personnel, State and local officials, volunteers, and others who aided the innocent victims and, in so doing, bravely risked their own lives and long-term health;

(4) stands in great debt with the American people to the members of the Armed Forces serving both at home and abroad;

(5) praises the people of the United States for their patriotism, compassion, prayers, and generosity in donating time and money to support the innocent victims of the September 11, 2001, terrorist attacks, their families, friends, and loved ones;

(6) expresses thanks and gratitude to the foreign leaders and citizens of all nations who have assisted and continue to stand in solidarity with the United States against terrorism in the aftermath of the September 11, 2001, terrorist attacks;

(7) discourages, in the strongest possible terms, any effort to confuse the war on terrorism with a war on any people or any faith;

(8) commends the President and the brave servicemen and women of the United States Armed Forces in the successful effort to oust the Taliban from power;

(9) remains resolved to pursue all those responsible for the terrorist attacks of September 11, 2001, and their sponsors until they are discovered and punished; and

(10) reaffirms that Congress will honor the memory of those who lost their lives as a result of the September 11, 2001, terrorist attacks and will defend bravely the citizens of the United States in the face of all future challenges.

The SPEAKER pro tempore. The gentleman from Texas (Mr. Armey) is recognized for 1 hour.

Mr. ARMEY. Mr. Speaker, I yield the customary 30 minutes to the gentleman from Missouri (Mr. Gephardt), the Democrat leader and cosponsor of the resolution, pending which I yield myself such time as I may consume.

Hon. Richard A. Gephardt

OF MISSOURI

Mr. Speaker, a year ago today the East Coast of this great land woke up to the first rays of morning sunshine. We woke up as a Nation, full of optimism, full of gratification, aware of our freedoms, but probably taking them for granted. We were concerned about the people we love, with the confidence they would all be there. We were excited about the business of the day with hopes of good outcomes, and dearly involved with our children with an understanding of their safety. We did not think about heinous deeds when we woke that morning. Our minds were busy with our plans and hopes, dreams and schemes; and we went about our day's work.

Then, all of a sudden, out of the dark depths of the evil corners of the world, hatred, meanness, despair, jealousy, greed, whatever, rose to afflict this Nation.

We were shocked at what we saw. First, most of us thought it was an accident and how tragic it was. But soon, we realized it was a vicious, premeditated attack on us as a Nation and on innocent civilians in this country.

Those streaming rays of sunshine that came through those big buildings of New York City that stood as a monument to this thing we call the practical American genius, were shattered. That peaceful field in Pennsylvania, awake with morning dew, was smashed. Our Pentagon that stands for strength and freedom was assailed in a way that it has never been assailed. And, indeed, this very building on which we stand today was saved that morning by the first response to this vicious attack.

We took the hurt and the losses, and they came early; but it did not take us long to collect our wits in this great country. Immediately upon understanding on flight 93 how vicious this was and how evil the intent, our American heroes fought back and this Nation was inspired by Todd Beamer who started the response with that phrase, "Let's roll."

Our early responders in New York City, after the tragic loss of life following American flight 11, American flight 77, and in Northern Virginia after the horrible nightmare of United flight 175, our early responders came from our communities: firemen, policemen, and emergency workers of all types. They rushed to the danger and saved lives.

We struggled through that day with doubt, uncertainty and fear. But as the day wore on, we became more a Nation of resolve and less a Nation of fear. We began to build our way back to confidence and optimism on that very same day. America had the unwelcome need to see its own heroes fight for survival and rescue on our own land, and our heroes rose to the occasion in a way that has inspired each and every one of us.

In New York City and Pennsylvania and here in Northern Virginia, they did so no more nor any less than they would have in Kansas City, San Francisco, or Houston because they displayed the character of a free people who cherish their freedom and love their neighbors.

Now we have been asked to go on with the task of ridding the world of the evil that struck that blow. We again call upon our heroes, now not so often, not so many civilians, but honorable men and women in uniform who have stood before the history of this great Nation's marvelous tradition of defending freedom, peace and respect and have said, "I will volunteer to serve this Nation in its armed services."

These new young heroes, following generations of heroes past, are now being asked daily all across this globe to incur risks and hardship to find the evil ones and remove them.

I believe the perpetrators of evil that launched this horrible attack have seen in ways they have never dared to imagine the character and strength of this great Nation. They now know the resolve of this great Nation. They all understand the courage of our heroes.

Let me say again what I said at the time. This Nation has proven it will send its heroes. Our heroes have proven they will go when asked, they will volunteer, they will do their duty. But we do not send our heroes from this great Nation out of ambition for territorial expansion or out of a sense of revenge, but out of a requirement for a just world, a world in which people who will perpetrate evil against others will be found, and they will be prosecuted.

Let me just say to those of you who are still out there plotting and scheming, do not underestimate our American heroes; they are young, they are bright, they are strong, they have courage, and they will in fact bring you down.

Now we have come, Mr. Speaker, to this day a year later. The sun rose in the east today, and this great East Coast of this great land was the first to experience this morning. We woke a little wiser, a little more aware, a little sadder, but we awoke with optimism, love, resolve and courage, and we will be that way for so long as this great Nation shall endure.

Mr. Speaker, I am honored to cosponsor this resolution with the gentleman from Texas (Mr. Armey) in the spirit of bipartisan unity on this difficult day for our wonderful and great Nation.

We gather together today to remember the victims, honor our rescue workers, and cherish the values embraced by human beings across the globe. On this first anniversary of September 11, we recall the strength, the courage and the character of innocent people who lost their lives in this horrendous attack.

To the families of September 11 victims, we say that we are with them as one. We honor the bravery they have demonstrated in the face of unspeakable suffering, and we honor the spirit of the American people, humanity at its best.

Today, we recall that amid the chaos of September 11 our rescue workers gave the last full measure of devotion so others could live another day. Hundreds of people rushed into burning buildings to save others who they had never before even met. On one of our darkest days, they sent forth a defiant ray of hope, and words alone could never do justice to their sacrifice.

Today, we also recognize this fundamental American truth: From the horror of September 11 has come incredible, unimaginable strength. Our Nation is still grieving, but make no mistake, we stand united.

As I said Friday in New York at our joint session, in this great and faithful struggle, there are no Republicans, there are no Democrats; there are only Americans, and we will remain resolved with our President to defend all those who threaten the liberty, freedom and democracy that define our Nation.

Today, as we pray for the victims' families, we also offer profound gratitude to the people who may well have saved our lives by fighting back on flight 93. This resolution is crafted in their spirit, with them as our inspiration and as our guide.

This resolution honors people like the man who went to Ground Zero after the attacks and started digging through the rubble, searching for survivors, because, he said, we are "digging for freedom."

It honors the woman whose legs were crushed by debris at the World Trade Center, who has been in the hospital every day for the past year.

It honors the firefighters in New York who, ascending the stairs, calmly told civilians, "Just keep going down, clear run. Keep going down, clear run."

It honors the doctors at the Pentagon who, rolling in water to ward off the flames, saved countless servicemen and women because of their raw courage.

It honors the ironworkers, the construction workers, the engineers and others who worked around the clock for months to clear the debris at Ground Zero.

It honors all the workers at the Pentagon who worked day and night to meet the schedule of having the Pentagon back in perfect condition before the 1-year anniversary of September 11.

It honors our law enforcement personnel who are protecting our citizens on a daily basis here at home.

It honors the members of our Armed Forces who have been fighting and are today fighting to defend our freedom and secure our Nation.

And it honors people like those we had lunch with on Friday in New York who lost their spouses on September 11.

In the face of the unthinkable, their courage, their simple courage to move to the future while they grieved about the past, was deeply moving and inspiring. They demonstrated a commitment to the values that all of us hold dear: Freedom, family, faith and friends.

Let all these deeds in the past year, and more, stand as a lasting monument to the spirit of our great Nation.

Today, we know that our most solemn obligation is to ensure that those who died on September 11 did not die in vain. In the days and weeks ahead, let us continue to work together with humility to protect our people, guard our freedoms, and report to the world that America will never be defeated.

Let us move forward as one Nation, one people, for the sake of every single person who believes in freedom and believes in civilization and believes in humanity.

Let me end with the words of an old hymn that I love so much. When we face the unexplainable, when we face evil, many of us turn to God. And, as we did 1 year ago, we ask again today for God's help.

As the hymn says, "And He will raise you up on eagles' wings, bear you on the breath of dawn, make you to shine like the sun, and hold you in the palm of His hand."

May God shed His grace on this great and wonderful country and all of our people.

Hon. Henry J. Hyde

OF ILLINOIS

Mr. Speaker, it has been a year since the world watched the impossible happen, and yet it is difficult to believe that such a year has passed so quickly. The sense and feel of time have been altered, seemingly suspended even as the calendar's relentless progress has remained unaffected.

On this first anniversary, we would only deceive ourselves if we were to believe that those events are now safely confined to the past. We will continue to live with them all our lives.

Modern communications have brought us many new and wonderful things, but they have also made possible the communal experience of tragedy. In this new age, distance will no longer spare us, nor can an absence of ties insulate us from sorrow.

All who witnessed the events of September 11 still bear the scars of seeing inconceivable images and impossible events unfold in real time. But our own experiences, however painful, cannot compare with that of the innocents who bore the horror directly, nor with that of their families and friends who were suddenly and violently severed from their former lives and from the touch of those deeply loved.

We Americans are a practical people. Instead of resigning ourselves to the difficulties of life, we instinctively seek to identify problems in order to focus our efforts and move toward solutions. In the past year we have done this.

We have come to know our enemies and direct our determination and resources to uncovering their hiding places and plans. We are now engaged in designing and implementing measures to resist their ability to harm us. The challenge is an entirely new one, but one which gains in clarity each day. I hope all of us are now aware that in addition to our successes, we must prepare ourselves for the likelihood of failures in a struggle that may have no end.

By infusing purpose, action can thus fill many voids, but the need remains to understand what happened and to comprehend the meaning of the events of that day. Here words give way to silence, for deep reflection is the predicate to understanding. Our modern, rational world once promised in time to reveal all secrets to us, but can we still cling to that belief now that we have been confronted with things we thought long past, vanquished and erased from the world by reason and light?

The modern world has seen many efforts to eliminate God from our lives, but we have not been able to eliminate evil. The last century was unparalleled in human history in its celebration of the savagery that human beings can wreak upon one another. We had hoped we might escape that fate in this century, but now we know that we will not.

We have been forcibly awakened from our dreams of an earthly heaven by the bitter knowledge that evil still roams freely in the world. We cannot allow ourselves to be paralyzed with despair or fear, and neither can we permit our natural optimism to shield us from the realities of the world. If there is any useful thing to be drawn from this terrible experience, it is that we have been given an unmistakable warning that in this new century, unknown and fearsome challenges await us, challenges that will impose the severest test of our national character.

Knowing this, we have a duty to prepare ourselves to defend not only our lives and those of our children, not only our beloved country, not only our freedoms, but civilization itself. We are Rome, beset by new barbarians who are savagely motivated by their immense hatred of us, of our happiness and our success, of the promise America represents for the world; for our enemies have no aim except destruction, nothing to offer but a forced march back to a bleak and dismal past. Theirs is a world without light; their all-encompassing hatred a repudiation of any saving grace.

Their victory would impose a new Dark Age, but this time perhaps an endless one. They are enemies of the future itself.

As we resolve ourselves to our task, as we grieve for all of those linked to us by tragedy, we may also see ourselves more truly and thereby understand that our great strengths are interwoven with many fragile things, and that being human, we have our faults and flaws to contend with as well. The threats we face have given us a greater sense of how rare and wonderful is the world we share and of our responsibility to protect it from the storms outside.

It is for these reasons that we remember those 3,000 fellow citizens who, asking nothing other than to live their lives in peace, were brutally murdered by men without conscience or mercy. We know it is right to remember our dead and commend them to the mercy of God, because should we forget them, we would only invite new acts of terror. We remember because, to quote Lincoln's haunting phrase, "the mystic chords of memory" bind us to the victims and the heroes of September 11.

And we shall not break faith with their memory.

May those who died in the attacks of September 11 rest in the mercy of God. May those of us who remain be steadfast, courageous, and live lives worthy of their great sacrifice.

Mr. Speaker, H. Con. Res. 464 expresses our gratitude to our friends and underscores the Nation's resolve to meet the enemy and defeat them. I believe passage of this resolution will commemorate those heroic actions of last September 11 and stand as an important symbolic gesture which all Members should support.

Hon. Tom Lantos

OF CALIFORNIA

Mr. Speaker, on this first anniversary of September 11, 2001, we mourn the victims, we honor the heroes, we contemplate the lessons, and we celebrate the unity of our Nation so proudly displayed since that fateful day. One year ago, we suffered a grievous wound. One year later, that wound has begun to heal and the scar it has left has toughened our skin, but it has not and will not harden our hearts or dampen our spirits. God bless this great Nation.

Hon. Porter J. Goss

OF FLORIDA

Mr. Speaker, I thank the distinguished chairman of the committee for yielding me this time. I am obviously here to join my colleagues in honoring the memory of those who died in the terrorist attacks of a year ago this day and to underscore with gratitude the amazing valor displayed by them.

I am sure every American remembers where they were and what they were doing when these ignominious attacks occurred. For me, I recall finding Speaker Hastert in his office, urgently gazing out the window down the Mall, looking at the smoke coming from the Pentagon. I urged immediate evacuation, and the Speaker said, "Stop. First, we must open the House and have a prayer." So part of my memory includes our short gathering in this Chamber and the earnest and moving prayer by the guest chaplain of that day, Reverend Gerald Creedon.

I would like to begin my remarks this day recalling that prayer. He said, "God of peace and life, send Your spirit to heal our country; bring consolation to all injured in today's tragedy in New York and Washington. Protect us and help our leaders to lead us out of this moment of crisis to a new day of peace. Amen."

What Reverend Creedon did not know, and what none of us knew here, was that more casualties were to come in a field in Pennsylvania shortly after his prayer and our very hasty adjournment that day. Actually, this was not Father Creedon's original prayer; he had prepared one on the topic of immigration. But realizing the gravity of the situation, he spontaneously gave us heartfelt, wonderful words which were suitable to the moment and which are posted on the wall of my office to this day as a daily reminder.

To the more than 3,000 people who lost their lives that day at the World Trade Center, the Pentagon, and United flight 93, we owe continuing remembrance, deep thanks, and responsive action. A year ago today, terrorists attacked Americans and citizens of scores of other countries who were on our U.S. soil. We, along with a broad coalition of nations, have taken up the challenge of combating the scourge of global terrorism. It is serious business. President Bush has left no doubt about his commitment to have our Nation lead the way.

Last year the fight came to the doorstep of each and every American citizen; and within the very first minutes of September 11, 2001, our Nation responded as one. I would like to dedicate my short time today to celebrating the men and women, ordinary Americans, who have fought back; the quiet heroes.

Let me begin with Ceecee Lyles, a flight attendant of flight 93, who was a resident of my district of Southwest Florida. A former police patrol officer detective, she had spent 6 years risking her life to protect others in that service. In December 2000, mindful of her young children and looking for a less dangerous career, she enrolled as a flight attendant school candidate; and 6 weeks later she began flying for United Airlines out of Newark.

At 9:58 on September 11, only 5 minutes after the House recessed for the day under the dire circumstances then apparent, Ceecee called her husband, Lorne, in Fort Myers, who is a police officer there, from her plane to tell him that her flight had been hijacked. Her words: "I called to tell you I love you. Tell the kids I love them." Her last words that we know of were, "I think they are going to do it. They are forcing their way into the cockpit."

And then the phone went dead.

In this Chamber, we owe a particular debt of gratitude to CeeCee Lyles and her companions on flight 93, and we all know it. That flight may very well have been heading to Washington when it crashed into Shanksville, PA. Without prompting or training, the passengers and crew fought back, and in doing so, saved many additional lives; possibly, our lives right here in the Capitol, for as we now know, many believe, and there is some evidence, that the Capitol was the intended target of flight 93.

FBI Director Mueller, speaking at the crash site in Pennsylvania, said, "We believe that those passengers on that jet were absolute heroes." Wallace Miller, the coroner for the case, called the passengers citizen soldiers. He went on to ask: "When can you think of, other than the Revolution or our Civil War or at Pearl Harbor, where American citizens died defending their home ground?"

But let us reflect a moment on our history. While many would compare 9/11 to the devastation of Pearl Harbor, there is a significant difference. Pearl Harbor was, after all, a military-against-military matter; 9/11 was a vicious attack on civilians and on freedom.

It is obvious that all of us have had to deal with new restrictions on the way we live our lives; but we have also developed a sense of pride, patience, and individual responsibility as we go about our lives to bring us closer as Americans.

The terrorists thought they would destroy our spirit; but instead, they renewed it. Destroying the will of the enemy to fight is the common measure of victory in war. Bin Laden and his depraved extremists fueled our resolve to wipe his brand of evil from a civilized world.

In addition to people all over America who have stepped up to the plate, our government has also become more alert, more focused, and more vigilant. We all must recognize the dedication and sacrifice of the thousands of individuals in government service who are out there on the front lines. They are protecting you and me, and they are making us proud. They know we are counting on them at a time when it matters.

Every new day as we wake up safe and sound in our homes here, I hope we remember to say just a little thank you to those out there for the work they are doing. Whether fighting Al Qaeda in Afghanistan and Pakistan, working the streets in Africa, Latin America, Europe, or Asia, perhaps working in a cubicle in the Washington, DC, area, or as part of the joint terrorism task force in a U.S. city, these are the intelligence officers, the military and civilian personnel who in all likelihood will not receive public recognition for the work they do; yet they do work long hours, often in places far from friends and family, and sometimes at very great personal risk. I regrettably acknowledge that sometimes they do not come home.

Johnny Michael Spann was a CIA officer, a husband, and a father of three. He went into Afghanistan in an early phase of the war to collect information crucial to defeating the Taliban and to protecting Americans at home. He was killed last November during a prisoner uprising in Mazar-e-Sharif, which was particularly brutal, as we now know.

Finally, I would like to recognize that since last September every American has been engaged in combating terrorism. We have been more vigilant, aware and alert, reporting leads to the police and FBI in record numbers. We have volunteered time and resources to our communities. We have been more patient as we have tried new security procedures at airports and public buildings, even though some of them have clearly turned out to be unworkable. We have maintained our basic freedoms and our democracy in the face of further terrorist threat. We still fly, we go to the mall, we cheer on our sports teams, we drive over bridges, we speak our minds, and we assemble where we choose.

So on this September 11, let us rededicate ourselves to honoring the memory of those who died by continuing to stand up to terror and to fear. Then let us also look to the future and the young people who are preparing to join the fray. Our youth, who some thought might be becoming a bit apathetic, or were perhaps now taking this great country for granted, are now applying in record numbers to service academies, to police and fire departments, the military, the FBI, the CIA, and other government service. They are our future, and they are ready.

September 11 will come again next year and every year thereafter. It is now part of who we are. Woe to those who would ever test us again.

Hon. Eleanor Holmes Norton

OF DISTRICT OF COLUMBIA

September 11 will always be a day to remember; but for us, it must also be a day to consider how to go on. If one lives or works here, as Members do, we remember that no sooner had September 11 come than we had October and the anthrax tragedies that occurred, beginning in the Brentwood post office, and spread even to this very Congress.

Yet, Mr. Speaker, though some may consider the capitol city a target city, I can come to the floor this afternoon and say I have never felt myself to be a target. That is probably because I am an American, and it may be in part because I was a small child in this city during World War II. Therefore, I am blessed and perhaps burdened by the notion of American invincibility. I do not believe that simply because of our military might. Somehow I believe that my country cannot and will not be defeated ever from within or without. It is simply part of the way I was raised, and it is part of the way we must raise our children.

I know how one's spirit can be broken when one goes to the funerals of three small children and their teacher who went down in the plane at the Pentagon. It can try one's spirit. But the fact is, I regard those children as representatives of all who lost their lives in September and October of last year; and somehow or other, remembering September 11 and the October anthrax tragedies through the lives of these 11-year-old children and their teachers, random targets, has instructed me how to go on.

I believe we will defeat terrorism. I tell you, it is part of my core belief. What I think we have to learn to do is to maintain an open society in the process. No society has ever faced what we have today. No society has ever had to face keeping itself wide open while understanding that terror lies within.

I am a native Washingtonian, a fourth-generation Washingtonian. We live here and feel ourselves the stewards of the Nation's Capital. As such, we cannot stand by and see the Nation's Capital ever be closed, because if it is closed or seems to close down, the rest of America will believe it must follow behind.

If this is to be an open and free society, it must begin with an open and free Washington, DC. I am proud of the Congress for keeping our Chambers open, for doing all we can to keep this city open, and for remembering that when we are open, the rest of the country will feel itself open. Finally we will, I believe, have the rest of the world believe they, too, must open their societies to us and to the rest of the world.

Mr. Speaker, I thank the gentleman for bringing this resolution to the floor.

Hon. Benjamin A. Gilman

OF NEW YORK

Mr. Speaker, I rise today in strong support of H. Con. Res. 464, commemorating the tragic events of 9/11. I commend our distinguished majority leader, the gentleman from Texas (Mr. Armey); our distinguished Committee on International Relations chairman, the gentleman from Illinois (Mr. Hyde); and our ranking minority member, the gentleman from California (Mr. Lantos) for bringing this measure to the floor at this time.

On that day, 1 year ago, our Nation was deliberately and barbarically attacked by terrorists at New York's World Trade Center, at the Pentagon, and on flight 93 over Pennsylvania. It is our solemn duty today to reflect on those terrorist events and to memorialize those who perished needlessly at the hands of those criminals.

My 20th Congressional District in New York lost more than 90 innocent lives on that terrible day. We join in extending our condolences to all of the 9/11 victims, and to their families and to their loved ones, and at the same time, in reminding them of our Nation's unrelenting determination to bring to justice all those who carried out these evil acts.

As our Nation stands together today honoring the innocent men and women who were taken from us on 9/11, we also pay tribute to our firefighters, to the police officers, to the rescue workers, and to all the citizens who bravely mounted the largest rescue operation in history under the most unthinkable conditions. Their countless heroic acts on 9/11 mark it both as a day of tragedy as well as triumph.

Let us also pause today to salute the men and women of our Armed Forces who are out there defending freedom and democracy on the front lines of our war on terrorism. Let us pray, too, for their safe return and their eventual triumph.

As we reflect today upon our extensive losses on the anniversary of tragedy and horror, let us also remember the valor, the patriotism, and the unity of our Nation in its darkest hour. That date, 9/11, was not only a turning point in the history of our great Nation, but also the world. As we seek God's blessing for our Nation and for the victims and heroes of September 11, let us all pledge to work together to make our world a safer place in which to live.

Hon. Steny H. Hoyer

OF MARYLAND

I am pleased to join the gentleman from California (Mr. Lantos) and also my very close friend, the gentleman from Illinois (Mr. Hyde), in recognizing and remembering. As well, I am pleased to follow the remarks of my friend, the gentleman from New York (Mr. Gilman). All three of these gentlemen, Mr. Speaker, have been in the forefront of focusing on the foreign policy of the United States. All three have focused on the extension of liberty and justice and freedom throughout this world. All three have focused on human rights and the recognition of the rights of individuals. How appropriate it is that these three leaders lead us in this remembrance.

Mr. Speaker, none of us will forget where we were when we learned the news. None of us will forget those 84 minutes from 8:46 a.m. on September 11, 2001, when the North Tower was struck, to 84 minutes later when the crash of the jet from Pennsylvania precluded the success of the terrorists in striking this Capitol.

None of us, Mr. Speaker, will ever forget. In an instant on that Tuesday morning, now known simply as 9/11, this generation of Americans suffered its Pearl Harbor. In an instant, or more accurately, I suppose, in 84 minutes, more than 3,000 innocent human beings, many of uncommon courage, were murdered by criminals of unbounded evil.

Words, Mr. Speaker, cannot convey the depth of pain inflicted on this Nation and its people 1 year ago. The pain endures and will remain. Today we remember all those who were taken on that horrific day. Our thoughts and our prayers are with those who survived and those who lost loved ones, as well as with the brave men and women this very hour defending freedom here at home and abroad.

In many ways, such unspeakable acts have clarified our purpose, steeled our resolve, and confirmed who we are.

We are a peaceful, tolerant and compassionate people. The evidence of that, Mr. Speaker, lies throughout our great Nation.

Since September 11 private charities have raised more than $2.4 billion to assist survivors. Former President Clinton and former Senate Majority Leader Bob Dole joined to raise $105 million to pay for college for the children and spouses of those killed or disabled. More than 3,000 people download applications for Americorps every week; more than 76,000 have requested Peace Corps applications; and more than 48,000 have signed up for Citizen Corps programs. Yes, Mr. Speaker, this is a generous, compassionate and giving Nation.

Closer to home, Donn Marshall of Marbury, MD, refused to let the savagery of 9/11 define the life and loss of his beloved wife, Shelley. Shelley was an employee of the Defense Intelligence Agency and she was one of the 20 Fifth Congressional District residents and 53 Marylanders who were taken from their family and friends by the mindless acts of savagery on 9/11. Rather than give in, however, to unblinking but justified anger, Donn has given his sorrow meaning.

In tribute to his wife, Shelley, he established the Shelley A. Marshall Foundation, an irrevocable trust that funds children's story hours at public libraries, creative writing contests at colleges, and tea parties at nursing homes that bring senior citizens and high school students together.

The aftermath of 9/11 has seen countless other acts, Mr. Speaker, of generosity, community and courage, in your district, Mr. Speaker, in mine, and in the districts of every Member of this House, which is to say in every corner of this Nation. From the local police officers and firefighters who raced toward danger at the Pentagon and Ground Zero in New York City, to the Facchina Construction Company in my district and those employees who completed their reconstruction work at the Pentagon 3½ months early, even after their La Plata headquarters was devastated and destroyed by a tornado just a few months ago, to the local artist in my district who raised $5,000 for the widows and children of firefighters from the sale of 9/11 T-shirts that he designed.

Mr. Speaker, the terrorists who sought to break our spirit only fortified that spirit. Their barbarism reminded us of our inevitable vulnerability but also reminded us that we are a part of something much greater than ourselves. We are the land of the free because, Mr. Speaker, we are the home of the brave.

More than 40 years ago President Kennedy stirred our Nation when he said, "The energy, the faith, the devotion that we bring to this endeavor will light our country and all who serve it." And he correctly concluded that, "The glow from that fire can truly light the world."

The fire of freedom, Mr. Speaker, forged the American character and it burns deep within our souls. The ashes of tragedy have rekindled that spirit; and 1 year later, the world must know freedom's light still burns brightly and its eternal truth shall never, never be extinguished. May God, Mr. Speaker, continue to bless and guide America as we continue our commitment to a just Nation and the defense and extension of freedom.

Hon. Zach Wamp

OF TENNESSEE

Mr. Speaker, this is a day of remembrance, a day of remembering over 3,000 individual persons who died at the hands of people who were so deranged that they also died in their killings of innocents in America. A day of remembering 3,000 families that will never again be the same, but the families through their sacrifice have definitely contributed to a better America because of what they have given. A day of remembering the sacrifice and courage that was demonstrated, manifesting itself as love, immediately overtook hate, incredible love that sent warm feelings throughout the world toward the United States of America. It is a day of emotion.

I was at the Pentagon this morning, as were many of my colleagues and our President and the Secretary of Defense. I was there last year right after the Pentagon was struck. The stench of war, death and destruction a year ago was replaced this morning by the smell of new construction, a breeze that blew up the dust of a construction site now complete, sparkling new windows, and a whole new day at the Pentagon. A day of sadness for great loss. A day of joy for the phoenix project and that new beginning. A day of humility.

I continue to be in awe of this job, this country, and the people of America. A day of national pride as we stand together with our President and our leadership. A day of deep and abiding love for our fellow man of all denominations from all around the world. God created each of us. But a day of righteous anger for what has happened, and a desire for justice to be served, but for our country to never extract revenge.

Mr. Speaker, today is a day of unity in this House and across this land. The Holy Scripture says that unity is a supernatural anointing of God. The words say, One can chase 1,000, in the Old Testament. Two can chase 10,000. What that means is whenever God's children get together with a common purpose, spirit of unity, God anoints that unity and supernatural things can happen. We have seen that in this country as we have come together. We need to do it more often. We need to continue the unity that this tragedy brought to this great Nation.

Mr. Speaker, today is a day of resolve. We live in a new and difficult time. We stand today, my generation, on the shoulders of the greatest generation. A generation that I grew up admiring and continue to admire. The greatest generation rightly earned their place in our country's history because of their dedication and their sacrifice. My generation has been called the "me generation." We are self-absorbed, self-consuming, lazy, maybe not even capable of having what it takes. But I believe that in the last year, thanks to our domestic warriors, our first responders, our troops on the home front, our firefighters and law enforcement personnel and EMTs, and the bravery and courage that they have shown, this "me generation" may be becoming the "we generation." More selflessness, more sacrifice, more courage than I have seen in my lifetime has been demonstrated in the last year.

I even see the joy in the eyes of the greatest generation as they look in their later years at what has happened in the last year with some amazement and incredible pride to say to this generation, you have what it takes, too. And I am glad because the days ahead are uncertain, and challenges are many. We may have many difficult times that we must go through ahead; but, Mr. Speaker, we have what it takes because we inherited a legacy of courage and honor and valor and we must answer this call to courage because what is at stake is freedom.

It is fragile. It is a powerful force much like unity and love, but it is fragile. We must not rest. We must not grow complacent over time. We must be vigilant. We must be willing to fight and to die to preserve freedom.

Hon. James R. Langevin

OF RHODE ISLAND

Mr. Speaker, today I join millions of people in solemn reflection on the tragic events of September 11, 2001.

One year has passed since that horrific day thrust the world into shock and sadness; and while some pain has subsided, our wounds have not entirely healed, particularly for those who lost loved ones in New York, Pennsylvania, and Washington. However, by responding to this tragedy with hope and determination, our Nation has grown stronger.

As the initial confusion of the attacks subsided, we became familiar with some of the names and the faces of the victims. Rhode Island was touched personally by this tragedy through the loss of several who called our State home. David and Lynn Angell, Carol Bouchard, Renee Newell, Mark Charette, Michael Gould, Amy Jarret, Kathryn Laborie, Shawn Nassaney and Lynn Goodchild. We learned about these individuals, their interests and their lives. We have prayed for their families and loved ones and responded to their grief with compassion.

As in the past, Americans offered sympathy and support to those touched by tragedy, reminding us that though the terrorists attacked our Nation, they did not and they cannot damage our spirit.

On this solemn occasion I pay tribute to those who lost their lives, to their friends and families who continue to grieve, to the American heroes who dedicated themselves to rescue and recovery efforts, to our servicemen and women who are defending our Nation against the ongoing threat of terrorism and to the spirit of America which has helped us endure these difficult times and will grow stronger from our sacrifices.

God bless America.

Hon. Michael N. Castle

OF DELAWARE

Today is a solemn day to reflect on the lives lost and the families who have been altered forever by the terrorist attacks last September 11.

In the aftermath of these devastating attacks, the American public and people around the world came to realize that thousands of innocent Americans and others from many nations and walks of life perished because evil forces wanted to strike at the heart of this great Nation.

Today and forever we will grieve for all of the victims. We have listened to the reading of the names of those lost but who, more important, are the family, friends and loved ones of those who are still on this Earth and miss them today.

September 11 will always be etched in the minds of all Americans and our families and friends throughout the world. Today we remember our heroes who gave their lives so others may live, our brave citizens, firefighters, police officers, and emergency personnel. The outpouring of emotion, generosity, and courageousness of mankind continues to touch us daily. Now is not only the time for remembrance of the past but also a time to look forward to living our lives with vigor and joy. We are a country that stands more united than ever before. Our diversity has strengthened us and our pride in America continues to grow. On this pain-filled first anniversary, we stand tall defending freedom, working for peace, and seeking justice. We must continue to support one another, and we must remain committed and united in the war against terrorism and use all of our might to bring to justice all of those involved with the attacks. Today we pledge to do everything in our power to defeat terrorism and to make our Nation stronger in every way that has made it a beacon of freedom and opportunity in the entire world.

May God bless America.

Hon. Sheila Jackson-Lee

OF TEXAS

A year ago today, many of us as Americans might have been asking the question, Who are we? And the terrorists who attacked us might have thought in that horrific act of violence that who we were would crumble. They might have expected that because we came from all walks of life, spoke different languages when we first came to this Nation and enjoyed our respective cultures, that the coalition would disintegrate, that we would no longer be America, that we would begin to join places where we might have come from or our ancestors might have started and we might have gone at each other and might have accused each other.

But we fooled them because America is a Nation of the free and the brave. It is in fact a very special place; and more than any time in our lifetime, Americans stood united. It frightened the rest of the world, I might imagine, those who wanted to perpetrate terrorism, undermine our democratic ideals, get us to attack our Constitution. We stood firm. As I sat there today at the Pentagon and I watched as our flag began to blow in the morning's wind, it reminded me of the words of Francis Scott Key, why he was so moved to write the "Star Spangled Banner," for as he looked up as those bombs were bursting and that war was going on, he felt that there was a theme and a symbol that continued, and his words were:

"Now it catches the gleam of the morning's fresh beam,
In full glory, reflected now shines on the stream;
'tis the star spangled banner. O, long may it wave
Over the land of the free and the home of the brave."

That is what we showed the world on September 11. We showed them that we could in fact survive. And today I take time to salute those first responders who helped us survive, the police, the firefighters, the paramedics, all the medical professionals, and just plain ordinary people, the volunteers, the men and women of the U.S. military who today stand at the front line of freedom and opportunity and justice, the unsung heroes, many of whom will not and did not live to tell their own story. We honor them, and yesterday belongs to the families of those who lost their lives and the families of survivors. We honor them and we thank them.

I spoke today to a family member who lost his wife in Somerset, PA; and his words were chilling to me. It reminded me of the importance of the resolve of this Nation and of this government. He said simply, "I do not understand. I'm still living through this. I do not know how I'm going to get through it." He, however, may take comfort in the way that America has come together, how we have comforted each other. He may take comfort in knowing that anytime we are attacked, we will stand unified together. Their stories may never be told, those who lost their lives, but we will stand arm in arm together.

Might I say, Mr. Speaker, as I close, that the government came together, State and local officials. Might I also say that even though we were diverse, we did not use this time to attack any religious group, any believers of any faith, any distinctive ethnic group. We came together.

I would simply say, Mr. Speaker, that this Nation is a Nation under God. I hold this Bible. I will not read it today, but I am proud as an American that if I chose, I could read this Bible because we do have the freedom which we fight for, and that is why I know in our hearts we will continue to wage this ongoing fight against terrorism; but we will do it by showing to the world our own values of democracy, freedom, justice and equality.

I believe the "Battle Hymn of the Republic" says it well:

Mine eyes have seen the glory of the coming of the Lord.
He is trampling out the vintage where the grapes of wrath are stored.
He has loosed the fateful lightning of His terrible swift sword.
His truth is marching on.

The truth of this Nation continues to march, and we can do it in a way that understands freedom.

God bless America and God bless its people, for the truth will continue to march on.

Hon. Saxby Chambliss

OF GEORGIA

Mr. Speaker, today I rise in strong support of this resolution. On the anniversary of September 11, we remember the innocent victims of last year's terrorist attack against our Nation and pay tribute to the heroes who risked their lives to save others.

This day is indelibly marked in our Nation's memory. It is a day when our Nation changed, when we were savagely attacked by malicious enemies whose hatred of freedom and democracy runs so deep that they would murder innocent men, women and children.

These attacks on our cherished principles, values and freedoms tried to weaken our Nation; but the tragic events of September 11, 2001, only strengthened the American resolve. The loss is overwhelming; but in the wake of this tragedy, we have witnessed many of the best aspects of America: the tremendous outpouring of patriotism, a common sense of grief and loss which touched every American, a bond of love and support among strangers from across the land, a renewed commitment to our respective religious beliefs, and a renewed national resolve to root out terrorism and act against the enemies of America.

Today at the Pentagon ceremony, I visited with Pat Hogan, the widow of Maj. Cole Hogan of Macon, GA. Major Hogan was an Army Green Beret who served his country bravely here at home and around the world and was killed in the attack on the Pentagon. Our hearts, prayers, and thoughts are with the friends and families like Pat Hogan who suffered such a tremendous loss a year ago.

Over the past year we have endured daunting challenges, and our lives have been forever changed by these terrorist acts and the threat that continues to confront us. We have made progress in making America a safer place and fighting the war on terrorism. More still needs to be done; but as we move forward, we have an opportunity to rededicate ourselves to do all we can to work together to preserve the memory of those who perished in making our Nation a better place, a safer place and to ensure that the spirit of freedom, democracy and our core American values continue to burn even brighter in our Nation and around the world.

Hon. Ron Kind

OF WISCONSIN

Mr. Speaker, I too rise to remember the terrible events of September 11, to honor the victims of the attack and their families and to let them know that especially today, they do not stand alone, and to pay tribute to the heroes of that day, the passengers of flight 93, the rescue efforts at the Pentagon and in New York City, the police officers, the firefighters, the first responders, emergency medical technicians and the health care providers.

That day started with great shock and honor, Mr. Speaker; but it ended with incredible unity and a sense of resolve that those responsible would be held accountable. From the ashes of that attack arose a new sense of patriotism and a solemn pledge to never forget but to learn from that terrible day.

We became unified in this Nation out of a sense of common values that we share: love of country, love of the freedom and liberties that we hold so dear, a shared sense of common vulnerability now that our splendid isolationism in the world and our sense of innocence was taken from us. That is perhaps what has made me most angry about the events of last September 11, the grief that we have to share with the families who lost loved ones, but the knowledge that my two little boys and all our children in our country will have to grow up in a 21st century with the specter of terrorism and that vulnerability hanging over their heads.

I was heartbroken to have read the story of the two little boys who lost a father at the World Trade Center when they were at home, and they got all excited and started jumping up and down when they saw their dad's car being pulled into the driveway. They were screaming, "Mommy, Mommy, Daddy's home, Daddy's home." She knew that could not be the case, and she looked out the window to only see a tow truck dropping off her husband's car in the family driveway and having to explain to her sons why their dad was never coming home again.

Or the "Nightline" story of the retired firefighters living down in Florida who came back up to Ground Zero to sift through the wreckage in order to find the remains of bodies so they could be identified. They did it out of a sense of honor to their fallen comrades but also because they had lost their own sons, the next generation of firefighters who went into those burning buildings to save lives. One of the firefighters was interviewed by Ted Koppel, and he asked them, "Why are you doing this day in and day out, from dawn to dusk every day?" He responded, "When we signed up to be firefighters and when our sons signed up to be firefighters, we all knew there was a risk in this job, but everyone deserves a decent burial."

These are the memories that will live with us for the rest of our lives and why it is so heartbreaking.

After one of our intelligence briefings, I was talking to one of the intelligence officers and commented to him how sophisticated this terrorist operation seemed, and he replied that it was not all that impressive. He said it was a low-tech operation. It is very easy to fly commercial airlines when they are already aflight. The hard part is landing them safely, and they never intended to land safely.

That is the challenge that lies before our country today. We not only need to fly the ship of state safely, but we need to land the ship of state safely; and that is why I hope that we learn from this terrible event, but do not get too intoxicated with our own military power, which is considerable.

I would hope that we realize we must maintain our good citizenship throughout the globe, that we are in this all together. This is not only about enhancing our own security interests in the United States and for our citizens abroad, but for all of the freedom-loving nations throughout the globe who have a common goal in defeating international terrorism.

We cannot do this alone. We need the help of the international community.

It is easy for our Nation, with the military power that we now possess, to accomplish so-called regime change. The hard part is nation-building that comes after. And that is why it is vitally important, I believe, that we keep our eye on the ball; that we pursue the Al Qaeda organization wherever they have scattered, to the four winds, and that we do it with the cooperation and the help of the international community.

I am confident with the deliberations in the days ahead that we will be guided with proper decisions. May God bless and may God continue to guide this great Nation.

Hon. Todd Russell Platts

OF PENNSYLVANIA

Mr. Speaker, I appreciate the gentleman yielding me this time. One year ago today, the lives of all Americans were forever changed. For the terrorists, the story of the September 11 attacks is one of immense hate, a hatred for the principles of freedom, liberty, and equality for which our great Nation stands. For Americans, however, the story of September 11 is a story of immense love, a love of country, a love for human life.

Firefighters, police officers, and everyday citizens were heroic in rescuing victims of the attacks, and later in recovering the remains so as to properly honor those lost. Countless other citizens volunteered to assist and encouraged relief workers. Across the country, flags waved, hands were clasped in historic unity, and voices joined in prayer and in patriotic song.

In response to the attacks, America has been made stronger. Americans better appreciate the sacrifices that police officers, firefighters and emergency personnel make every day to ensure our safety. And we certainly have a much deeper admiration for the courageous devotion to duty of our servicemen and women, our men and women in uniform who fight to defend our precious freedoms throughout the world.

President Bush so profoundly captured the enduring spirit of America when he stated last year, "We will not tire. We will not falter. We will not fail. United we stand."

God bless those who lost loved ones in the attacks of September 11 and those who gave their lives seeking to save the lives of others, and certainly God bless the United States of America.

Hon. Charles W. Stenholm

OF TEXAS

Mr. Speaker, I thank the gentleman for yielding me this time. It has been said that "Greater love hath no one than to lay down his life for a friend." Unless perhaps he or she lays down a life for a stranger. That is the love, the heroism, the patriotism that we pause today to remember and to honor. It is the love we witnessed on a large scale in New York, Pennsylvania and Washington 1 year ago. But it is also the love that is practiced daily in communities across this great land.

In some ways, September 11, 2001, is so indelibly imprinted on our minds and souls that it is difficult to believe an entire year has passed. As we recall each detail personally experienced on that day, we remember the horror, the anguish, the sorrow, and the fear. Certainly these emotions remain fresh to all of those who lost a loved one. But even for citizens in other parts of the country hundreds of miles away, our connection as fellow Americans meant that we all felt deep grief.

As John Donne wrote more than 300 years ago,

> No man is an island, entire of itself. Any man's death diminishes me because I am involved in mankind; and therefore never send to know for whom the bell tolls, it tolls for thee.

In the weeks following September 11, as we heard the bells tolling or the bagpipes playing "Amazing Grace" in funeral after funeral, each one of us did feel that a part of us had died. And so today, as we remember, we feel sadness, we feel grief.

For those of us who were not physically present at one of the attack sites, perhaps our greatest wound came from the fear that gripped our chests and our country. Suddenly our sense of safety, of invulnerability, of carefree security received a vicious slash from a cruel and unexpected attacker.

As a Nation, we face the question, Will we be paralyzed by this fear or will we overcome it? The answer to that question is where the story brings us today. From biblical times until today humanity has discovered three things which overcome fear: love, faith and action.

Some people are aware of that truth every day of their lives. Each community's firefighters, our police officers, our emergency and health care workers, our men and women in military uniform, every day these heroes set aside personal fear in order to do their job so that the rest of us might live safely. They risk the possibility of laying down their lives for both friend and stranger and in doing so they demonstrate some of the greatest examples of love in our society.

The year that has passed since September 11 has also helped us find meaning in and through our grief. It has given us an ability to view both world events and our personal lives with a new perspective. And what we have seen is that one of the darkest days of our history gave birth to thousands of acts of goodness, creating perhaps one of our finest hours. Ordinary men and women across the country showed extraordinary bravery, kindness and compassion as we pulled together as one united Nation.

As we look back, we learn that it was through our giving back that we are now able to move forward. So while we respect and honor those who lost their lives a year ago, and feel compassion for those who remain in grief, today is also a day of celebration. We celebrate the American spirit, the heroes who are gathered here today among us all over America, and the three antidotes to fear: love, faith and action.

The marvelous thing is that every one of us has the capacity to bring to life these fear fighters. We cannot all rush into burning buildings or stop senseless acts of violence. Few of us may actually be faced with the opportunity to save another life. But we all can be heroes by loving, by believing, and by acting to strengthen our communities. As we honor our heroes, both living and dead, we are called to find the heroic urge inside ourselves.

September 11 may have shown us the worst of humanity, but it also reminded us of the chance to become the best of humanity, by loving, believing and acting. May each of us today honor those who died by doing just that.

Hon. Michael Bilirakis

OF FLORIDA

Mr. Speaker, exactly 1 year ago today, the lives of every American citizen were changed by the cowardly acts of terrorism committed against us on our own soil. Families were broken, loved ones were lost forever, and our sense of security was shaken.

Yet even in the dark hours of September 11, 2001, hope remained. For within the hearts of Americans, we share a common bond as citizens of the greatest Nation on Earth. This bond prompts us to courage, to service, and to patriotism. This bond inspired Americans on the day of the attacks to rush to the aid of their fellow Americans without thought of possible consequences to themselves. This bond provides hope that our Nation will heal from our grief and conquer those who would threaten our liberty and our way of life.

While we choose to honor the sacrifices and tragedy of September 11 with solemnity today, we also celebrate who we are as a Nation. We celebrate our Republic, we celebrate freedom, we celebrate service, sacrifice and love for one another. We celebrate the heroic acts of ordinary citizens and commemorate the tragic events of September 11. From the soldier to the firefighter and policeman, to ordinary folks going about their daily lives, we salute you. Your sacrifices will not be in vain. A grateful Nation lives and hopes because of your love for this great country.

I close today with Professor Al G. Wright's beautiful ode to our Nation, which celebrates our country with these words:

> I am an American. That is the way most of us put it, just matter-of-factly. They are plain words, those four. You could write them on your thumbnail, or you could sweep them clear across this bright autumn sky. But remember, too, that they are more than words. They are a way of life. So whenever you speak them, speak them firmly, speak them proudly, speak them gratefully. I am an American.

Hon. Mike Ross

OF ARKANSAS

Mr. Speaker, September 11, 2001, is a day that none of us will ever forget. I was sitting in my office across from our Nation's Capitol, and from the window in my office, I literally saw smoke rise from the Pentagon. A few hours later, I would learn that a young Navy petty officer from our district named Nehamon Lyons would be among the casualties at the Pentagon. Picking up the phone and calling his mom, Mrs. Jewel Lyons, back in Pine Bluff, AR, was the most difficult call I have ever made.

This morning, I joined with other Members of Congress and with our President, united, Democrat and Republican alike, united as one America, as we remembered in a special service at the Pentagon. We remembered those who lost their lives 1 year ago today at the Pentagon, at that pasture in Pennsylvania, and, yes, at the World Trade Center in New York City. After that ceremony I returned to my office and I called Mrs. Jewel Lyons in Pine Bluff, AR, to let her know what I had just experienced in that very special and moving service at the Pentagon and to let her know that America has not forgotten that young Navy petty officer, Nehamon Lyons, and his service to this great country.

My grandfather taught me to always look for something good in everything bad. There was not anything good about September 11, but I do believe it has brought out the best in the American spirit. I see a country today that is more patriotic than perhaps ever in my lifetime. I see a country today with a greater appreciation for our veterans and our soldiers. And I see a country today that is praying a lot more. Bible sales are up, church attendance is up, and I know that means America is only getting stronger.

Hon. Doug Bereuter

OF NEBRASKA

Mr. Speaker, on the first anniversary of the tragic events that killed so many Americans at the World Trade Center and the Pentagon, it is natural for our thoughts to turn to the victims of that day. It was, of course, a world-changing event that will continue to affect all of us. Now we know, too, that an extraordinary debt of gratitude is owed to those brave passengers of United flight 93, which crashed in a Pennsylvania field. They fought the murderous hijackers in the cockpit and, thus, foiled the plot to crash that plane into the Capitol or White House.

The families of the victims will continue to grieve their losses, but the commemoration around the Nation today should focus on reinforcing America's newly heightened unity and sense of resolve that we as a Nation will dramatically increase our effectiveness in protecting our homeland and our citizens abroad from terrorist attacks. We must remain committed to meet these challenges while at the same time preserving the freedom, civil liberties, and opportunities which make America the envy of the world.

The period set aside for formally mourning our losses is long past. The victims and their families now are best served if all of us share and act upon a commitment to keep our Nation secure, strong, and a bastion of liberty. May God bless America.

Hon. James P. McGovern

OF MASSACHUSETTS

Mr. Speaker, I rise in strong support of this resolution. It is hard to believe that it has been exactly 1 year since that terrible September day, 1 year since our Nation experienced that devastating and profound loss. Words cannot begin to express the heartache that we felt that day, or the sympathy we continue to feel for those who were lost.

Like many of my colleagues, I represented several of the victims of September 11; and I appreciate this opportunity to extend my deepest condolences and prayers to their loved ones.

It has been a difficult, but necessary, week for all of us. From our session in New York City and visit to Ground Zero, to the ceremony at the Pentagon this morning, to the consideration of this resolution, Members of this House have had the opportunity to express our solidarity with the families and communities most deeply affected by September 11.

Today is also an opportunity to pay tribute to the millions of Americans who reacted with such bravery and compassion in the aftermath of the terrorist attacks, the doctors and nurses who ministered to the injured; the firefighters, police officers, ironworkers and others who refused to leave the site of the World Trade Center until everyone was accounted for; the ordinary citizens from every corner of our country, every background and religion, who donated blood, money, or who provided a kind word or a prayer.

Mr. Speaker, I also think it is important at this moment for us to express our gratitude to our men and women in our Armed Forces who, as we speak, continue to battle terrorists in Afghanistan. They serve our country with great distinction, and they are a credit to our country and, indeed, to the world.

We have accomplished a great deal in the past year by working together, but I believe we have much more to do.

Today we recall the solidarity and compassion shown our Nation and our people by other nations and other peoples around the world, and we can work with them to bring our most cherished values into reality.

Mr. Speaker, we can, I believe, make this world less violent, more peaceful, more tolerant, and more secure. We have the ability to eradicate poverty, disease, hunger and hopelessness, the things that terrorists exploit to justify the unjustifiable. What we need is the will to make it happen. We need to lead the world in pursuit of these important goals. In doing so, I believe we will demonstrate the true and magnificent character of the United States of America.

Mr. Speaker, in conclusion, it is my hope that as we remember the victims of September 11, as we offer our condolences to their families, and as we continue to bring the perpetrators to justice, that we rededicate ourselves to providing a better world for us all.

Hon. Christopher Cox

OF CALIFORNIA

Mr. Speaker, today is Memorial Day for the victims and the heroes of September 11. It is a time to honor the people who have died and the people who still live. It is important as we grieve for the victims that we keep in mind how many people are alive today because of the efforts of those who helped them, many of them still alive.

Today it was my honor, along with Senator Ben Nighthorse Campbell of Colorado, to honor and recognize 11 members of our Park Police here in Washington, DC, who were heroes and first responders on that day. It was an interesting way to recognize them. There was a cross-country motorcycle ride organized by citizens from all of our districts and States across the country. It started out in Orange County, CA, went to San

Diego, CA, 3,500 miles later it picked up hundreds of riders and came here to the Nation's Capital. Like Ben Campbell, I am a Harley rider and so I was happy to join them. That is why I found myself in a position to be able to bestow these glass plaques to our first responders.

First, the riders themselves, our constituents. They raised $1 million for the World Trade Center Miracles Foundation just from among the riders. In that they were like the people who responded heroically on September 11. They put aside their concerns and put first the needs of others. That is what people all around our country are doing in response to September 11.

Along their 3,500 mile route, they stopped in places like Wichita, KS, where they dedicated an evening to the family of Daniel Nolan. He perished in the World Trade Center. A few days later they were in Illinois at the Champaign County Fairgrounds celebrating the important work of first responders. Only yesterday, they traveled to Stoney Creek School in Shanksville, PA, where the students had built a memorial to the heroes of flight 93. At every step along the way, they honored the dead and the living, as we do today.

This coast-to-coast effort culminated in the honoring of 11 heroes who live among us in Washington, DC, from the U.S. Park Police; and I would like to mention their names on the floor.

First, two members of the motorcycle unit, Lt. George F. Wallace, commander, and Sgt. Daniel P. Beck, supervisor of the unit. They were in front of the Capitol today. They saw what had happened at the Pentagon from this side of the river, and they did what heroes do: they went right to the middle of that danger. They were two of the first uniformed police at the Pentagon on the scene. They assisted in the initial evacuation of the wounded. Their efforts helped ensure that those injured who were still alive received immediate care, and for that, obviously, those men and women, their families, and all of us are grateful.

After the arrival of other local police and fire units, these officers continued to work. They cleared the vicinity and organized the evacuation of personnel from the monumental core area.

There were nine others outside the Capitol who were similarly honored. The members of the U.S. Park Police Aviation Unit, Eagle One and Eagle Two. Eagle One includes Sgt. Ronald A. Galey, pilot; Sgt. John E. Marsh, rescue technician; and John J. Dillon, rescue team officer.

Eagle Two includes Lt. Phillip W. Cholak, aviation unit commander; Sgt. Bernard T. Stasulli, assistant aviation unit commander; Sgt. Keith E. Bohn, pilot; Sgt. Kenneth S. Burchell, copilot; and physician assistant Keith Kettell and Jason Kepp, medic, of the Uniformed Services University.

Here is what these people did on September 11. At approximately 9:40, the U.S. Park Police received an emergency call from Reagan National Airport tower notifying them that a plane had crashed in the vicinity of the Pentagon. They did not know yet what had happened. Members of Eagle One immediately lifted off, and they flew right into the center of the disaster, hovered right over the Pentagon, and they honored an FBI request and activated their microwave downlink and provided a live video feed of exactly what was going on at that moment.

Just as Eagle One began to provide this live feed, personnel at Reagan National Airport abandoned the airport tower and said they could not see anything because of the smoke. They requested that these people, the names I just gave you, take over responsibility for all of the air space, and they did. They took control over all of the tower's responsibility.

Even as Eagle One completed this assignment, Sergeant March requested helicopters from the Maryland State Police, MedStar and AirCare, and coordinated with ground units to establish a landing zone, honoring a request from the Arlington County Fire Department to transport patients to regional burn centers.

Meanwhile, Eagle Two's crew responded to its call of duty and loaded the mass casualty kit and additional equipment on board their aircraft. They landed on the west side of the Pentagon and immediately began coordinating the transport of 11 priority-one burn patients. While Eagle One directed MedStar and AirCare aircraft into the landing zone to medevac additional patients, Eagle Two transported two of these patients to the Washington Hospital Center. Those people are alive and doing much better today as a result of these heroic efforts.

As a lasting example of their heroic efforts, the U.S. Park Police Aviation Unit responded to a final request of the Arlington County Fire Department to transport a battalion chief of the Pentagon Fire Command Center for a 1-hour flight over the crash site because he was directing the firefighting efforts on the ground from their craft.

Those are just some of the heroes of September 11. They live among us here. It was, as I said, my honor to present them with citations this morning. But as we honor these first responders, I think we have to remember that, as we draw inspiration from their work, they are still there every day protecting our neighborhoods. They are still there every day protecting us from threats, whether it be fire or a terrorist attack. Because of their bravery, which we take so much for granted, we have a wonderful future to look forward to in this country.

I have no doubt that we will succeed in our efforts in the war on terror, and I have no doubt that we will owe a continuing and ever-deeper debt of gratitude to these men and women to protect and defend our country. I am grateful for the opportunity to honor these men and women.

Hon. Lois Capps

OF CALIFORNIA

Mr. Speaker, in communities across the land, we gather to commemorate the unspeakable attacks against our Nation 1 year ago today.

Even with the passage of time, there are still no words to adequately describe our pain, sorrow, and our anger. Our hearts remain heavy with grief for the 3,000 souls who were lost, and we continue to mourn with their families and their loved ones.

I feel especially heartbroken for all the children who lost their fathers and mothers on this terrible day, and for all of the babies who came into the world after their fathers were killed. These young lives have been permanently and senselessly altered by those who so erroneously believe that extremism and hate pave the road to salvation.

Now it is our duty to help these children fulfill their dreams and understand that their parents died in an attack on the freedoms and values we hold to be self-evident.

It is also important to pay tribute to the acts of courage and heroism carried out by so many people: the firefighters, police and rescue personnel, the passengers on the doomed aircraft, and by countless citizens who volunteered at the crash sites and around the Nation. Some of them included volunteers from my district on the Central Coast of California who lended their expertise and resources.

Today is a day that we should honor all of the first responders who not only risked their lives on September 11, but who are also the first on the scenes of emergencies and disasters every day in communities across this country.

I am so proud of their commitment and their determination to make our country, indeed the world, a safer place. These brave individuals deserve our highest respect on September 11 and on every day.

It is also important to remember and honor the brave American servicemen and women who are defending our freedom around the globe, joined by defenders from other countries.

They are defending the principles of democracy and security on which this country was founded, and we salute their tireless mission. This has been a year of great sorrow and mourning. But it has also been a time of great American unity, strength of spirit, and generosity.

As one widow reflected, it is as though this entire year has been stuck on September 11. Now, perhaps, we can move on to September 12.

Mr. Speaker, I hope and pray, and let us work toward the goal, that the coming year will bring peace to our families, our community, our Nation, and that the generosity of our people and the spirit of our democracy will be well known and well documented throughout the world.

Hon. Adam H. Putnam

OF FLORIDA

Mr. Speaker, this is an emotional time for this country and an emotional time for this body. All of us have participated in events in our communities, tributes to the victims and their families in New York and the Pentagon, and that is right and proper. But I want to take this opportunity on this first Patriot Day to look forward, to remember and pause and reflect on why they envy us so, why the terrorists hate us so. What is it that would cause them to bring about the death and destruction of so many innocent lives, so many hopes and dreams shattered, so many communities that have lost church members, PTA officers, rotary club presidents, chamber directors, so many children who have lost a parent?

They despise the fact that we stand for freedom; that we represent all that is great, all that is tolerant and hopeful and helpful and strong about communities.

They misunderstood us. They thought we were no stronger than the celluloid films that come out of Hollywood; that we would buckle and cave, that we would wither up and shrink from the fight.

What they did not understand was that we had an army of ordinary American citizens who would rally; who would stand in line for hours to give their blood; who would contribute their paychecks to strangers; whose children would go to schools and have penny drives and car washes and other fundraisers to send off to the victims, their families, and even the children in Afghanistan.

They hate the fact that in America little girls have the same opportunities as little boys, to dream, to hope, to be whatever they want to be, from a teacher to a firefighter to a Member of Congress, to President of the United States. They are afforded equal opportunity.

They resent the fact that different faiths have every opportunity to worship together, side by side, on the same city block, in peace.

They resent the fact that we have heroic civil servants who deliver the mail, who put out fires, who comfort victims, who run into buildings from which everyone else is running out; that we have teachers who instill values and character into the next generation of Americans; that we have health care workers, doctors and nurses, who rally to the scene and give so much of their hearts and souls to putting lives and bodies back together.

People still risk their lives to come to this country. People still see the United States as that shining city on a hill. They still risk their family's safety, they give up all of their worldly possessions to stow away in a tanker or to cobble together a rickety raft and brave the straits of Florida or the Atlantic, to become a part of this country that those terrorists tried to destroy.

If given the opportunity, most free people, most thoughtful people, would choose that way of life, would choose that equality, that tolerance, that hope, that dream that is America.

And while all of us fight on a regular basis in this Chamber over things great and small, we never question the legitimacy of the debate or the legitimacy of the leadership or of the system or of the institution. While we criticize the policies of our President or administration, we do not question his right to be there and be our leader.

That is what they hated, that is what they envied, that is what they attempted to destroy, and they have succeeded only in bringing out the best in all Americans, leaving America today stronger than she was last September 11; a little closer together, still fairly complacent, still fairly naive about the dangers this world poses, but still very much in love with all things American, very much in love with our ability to debate on this floor, our ability to hope and dream and be whatever we want to be, and to instill that in our young people.

America is stronger today than she was a year ago; a little bruised, very bloodied in some areas, but stronger.

Mr. Speaker, God bless these United States of America.

Hon. Bernard Sanders

OF VERMONT

Mr. Speaker, our Nation was changed forever on the morning of September 11. The goal of Osama bin Laden was to demoralize us, to create fear and uncertainty, to bring about instability in our country; and, obviously and thankfully, he has failed.

Last week Congress met in New York City to pay tribute to those who were killed on September 11 in that city, and this morning we assembled at the Pentagon. At these events and similar events all over this country, we were reminded about our resiliency and how strong this Nation really is.

Last September 11, we saw amazing displays of heroism and bravery. We saw what is extraordinary and best in the human spirit. None of us will ever forget the sight of firemen entering the World Trade Center, going up the stairs while other people were going down the stairs. And, as we know, most of those firemen never got out of that building alive.

We also today remember the courage of the people at the Pentagon who saved lives there, and we honor the members of our Armed Forces who are fighting terrorism in Afghanistan and elsewhere.

Mr. Speaker, we have also learned a great deal since September 11. We have learned, as we have never learned before, that we are a vulnerable Nation. Yes, we are the most powerful Nation on Earth, but what we learned on that day is that we could be attacked and that thousands of innocent men and women could be killed.

We have also learned that we must lead an international coalition against bigoted, religious fanatics who believe that they have the right to kill innocent people in order to impose their reactionary ideology on others.

Many of us have also learned that in order to maintain true American values, we must not undermine the principles and constitutional rights that make our country great and that this country was founded on.

As an American and as a Vermonter I have been extraordinarily proud of how our people responded to this crisis in terms of blood donations and financial contributions, and how we came together as a community to support the victims of September 11 and to support each other. If there is something positive out of the horrors of September 11, it is that we as a Nation, all of us, despite our backgrounds, where we come from, our religious beliefs, must continue to show that same sense of community, that love for each other, that was demonstrated in the aftermath of the September 11 attacks.

Hon. John Linder

OF GEORGIA

Mr. Speaker, I rise in strong support of the resolution. Today is the 1-year anniversary of the most horrific attack on American soil in our history. On September 11, 2001, freedom-hating terrorists took from all of us our sense of innocence. They took from all of us the idea that innocent men, women and children going about their daily lives are immune from the horrors of war. They took from all of us the sense of safety and security to which we had become accustomed.

What they did not take from us and what they could not ever take from us, no matter how hard they tried a year ago, is our freedom and liberty, our way of life, our government of the people, by the people and for the people.

New York Governor Pataki opened the memorial ceremony at Ground Zero this morning with the reading of the Gettysburg Address which President Lincoln delivered in 1863, barely 100 miles from where the passengers of United flight 93 heroically stopped terrorists from their attempt to, I believe, destroy this very building. Part of President Lincoln's address includes these words:

> that from these honored dead we take increased devotion to that cause for which they here gave the last full measure of devotion—that we here highly resolve that these dead shall not have died in vain; that this Nation shall have a new birth of freedom.

We, too, are dedicated to a new birth of freedom here in this new century.

The horrible events of last September 11 have reawakened in all Americans a new sense of patriotism; a strength, a resolve that had lain dormant for far too long. The flags we see flying in every neighborhood across America today were not put up this morning. They have been there for the last year. The support of the American people for our men and women fighting overseas has remained unwavering, and the kind words and deeds of our fellow man seem all the more common today, 1 year later.

Eugene McCarthy once observed that America can choke on a gnat, but swallow tigers whole. This is a tiger that we as a Congress and as a Nation must deal with, and we will.

There is no lack of resolution here. There is no rancor. We will continue to stand behind the President. We will continue to do what we must do to keep those who hate our values and ideals from committing evil acts against us ever again.

Today let us honor and pay tribute to those who were taken from us before their time and resolve to remember them always. But let us also resolve that our commitment to the Republic that our Founders risked their own lives to create more than 200 years ago is stronger than ever. The foundation of our Nation is solid, and so is our dedication to her.

Mr. Speaker, God bless America.

Hon. Adam B. Schiff

OF CALIFORNIA

Mr. Speaker,

> The world must be made safe for democracy. Its peace must be planted upon the tested foundations of political liberty. We have no selfish ends to serve. We desire no conquest, no dominion. We seek no indemnities for ourselves, no material compensation for the sacrifices we shall freely make. We are but one of the champions of the rights of mankind.

These were the words of President Woodrow Wilson in his war message to Congress April 2, 1917.

This week, from Los Angeles to New York, from Fairbanks to Fort Lauderdale, and in 200 million households in between, a Nation struggles to come to grips with the most vicious attack on unarmed civilians in the Nation's history.

We once again appreciate, with the force that sometimes only tragedy brings home, that we are one country. Differences of geography, language, income and ethnicity have faded away. There are no national divides, no partisan debates, no hometown rivalries, no baseball strikes, not on this day. For all too brief a time, we are simply Americans.

And we are taking stock. Much has taken place since September 11, a date that may surpass the end of the last millennium as a turning point for the country.

Some of that change has been extraordinarily positive for our own security and for the peace and prosperity of the world. The promise of mutually assured destruction that for decades we exchanged with the Soviets has been replaced with an unprecedented partnership with Russia. This is no detente, but a completely new realignment of interests, which has transformed the world landscape and dramatically reduced the possibility of nuclear catastrophe.

For all that has changed in the last year and all that has transformed since the end of the cold war, we are still at risk. Our most immediate and tangible threat comes not from interlocking engagements with a Europe that cannot overcome its historic feuding, not from nation states that are amassing colossal military forces with an eye toward territorial aggrandizement or world domination. The threat is primarily asymmetrical now, from stateless terrorist organizations and the nations that support them; from murderous psychopaths that are bent on igniting a holy war and have a blood lust for the United States. Such depravity has always existed, but with the advent of weapons of mass destruction and their terrible availability, it no longer takes a national miscalculation to inflict misery on the world. And as the lone superpower in the world, America has a target on her back.

The threat may come from new quarters, but we have one powerful bond with Americans from the beginning of the last century: This conflict is still about making the world safe for democracy.

On September 11 we were not attacked because we sought to conquer or subjugate another people. We were not attacked over a territorial dispute or a clash of national ideologies. And, notwithstanding post-attack propaganda from the terrorists, we were not attacked over our policy in the Middle East. Osama bin Laden and Al Qaeda never showed an interest in the Palestinian cause except in a post-attack effort to point their homicidal rage as a defense to the West's supposed hostility to Islam. We were attacked simply because we existed, simply because we represented and continue to represent the triumph of free institutions, a respect for the free exercise of religion, association, and expression. We were attacked because we are a democracy in a world very unsafe for democracies. And winning this war, and the long twilight struggle it has become, will require nothing less than a sustained, unswerving commitment to the propagation of freedoms around the world.

We must root out Al Qaeda and terrorist organizations wherever they exist. We must take the fight to the enemy, as the President declares, and not wait defensively at home for the next attack. At the same time, we must open a completely new front in the war on terrorism: the battle for democracy. We must attack tyranny, despotism, and the trampling of human rights around the world. We must use every instrument of our national policy to support the growth and cultivation of free institutions, a respect for the free exercise of religion, the right to associate with whom one pleases, and the right to speak one's mind. We must encourage the growth of democracies in every corner of the globe and not simply in Europe or the Americas. Democracy must come to the Arab nations, to China, and to every corner of Africa, and not simply to our adversaries. Democracy, too, must come to our allies, to the Saudis, to the Egyptians, and to the Jordanians. Democracy, not oil, will be the ultimate guarantor of our security.

This lofty ambition is not fanciful, not quaintly sympathetic, but practical. Democracies do not make needless war, democracies do not seek to terrorize or conquer, democracies do not serve as the breeding grounds for genocidal rage or terrorist madness. Democracies are better capable of eliminating the common scourges of mankind: poverty, disease, famine, and conflict. If we are to be partisans, let us be partisans of democracy.

We may never ferret out every last terrorist; the germ of madness is difficult to eradicate completely. But our peace and prosperity lie as much in changing the soil. Peace, again, must be "planted upon the tested foundations of political liberty," and a cardinal part of winning this war, as in the war to end all wars, will be our fortitude as one of the "champions of the rights of mankind."

Hon. Earl Blumenauer

OF OREGON

Mr. Speaker, at 8:46 this morning I came to this Chamber, where I was exactly 1 year ago, to reflect on the events that had transpired that terrible day and what has ensued since then. As we all shared a flood of remembrances in honor of the memory of the victims and the heroes of last September 11, Mr. Speaker, it is just as important for us to reflect on the progress of this past year. Because tragedy gave us an opportunity and a responsibility, not just to deal with additional threats to our families, but to use the vast wealth and power of the United States to be a leader and a partner around the world, while we fulfill the promise of America here at home.

I thought about how much I have been inspired by the reactions of the American public at the time of the tragedy and of what we have seen throughout the year. But, Mr. Speaker, I wonder honestly what we think we have accomplished as a Congress in this last year. Immediately, we did do, as the resolution says: local, State, and Federal leaders set aside differences and worked together to provide for those who were attacked and to protect those who remain. But since then, is America really safer because of congressional action? Do the vast intelligence and security agencies now work together seamlessly? Have we made progress, not just against the Taliban and Al Qaeda, but to promote democracy and freedom, peace and prosperity around the globe? Is this Congress working together cooperatively on issues of peace, the environment, human health, and education?

By any objective measure, we as elected officials have fallen short of that mark. We have yet as a body to provide voice, not just to the fears and frustrations, but to the hopes and aspirations of Americans that we will seize this moment.

Now, I think people on both sides of the aisle will disagree as to why this is so, but I do not think anybody can argue that we have done all that we could, or even, frankly, that we have done all that much from airline security, to reducing energy dependence in the Middle East, to giving coherence to our policies in the Middle East and around the world. We have fallen short in doing all that we could for peace and democracy.

As part of this solemn occasion, the most fitting tribute of all is not for us just to reaffirm, as the resolution suggests, an honoring of the memory of those who lost their lives and that we will bravely defend the citizens of the United States in the face of all future challenges. Part of what we need to do is to acknowledge where we have fallen short and to renew our commitment that in this next year, we in Congress will catch up to where the actions and the expectations of the American public are; that we will enter as Members of this Congress with a new spirit of cooperation and achievement, that we will take the actions that make achievements of our values real; that we will move toward making our communities and, indeed, the world more livable and our families safer, healthier, and more economically secure.

Hon. David Dreier

OF CALIFORNIA

I would like to first express my appreciation to my good friend and fellow Californian (Mr. Lantos), as well as the gentleman from Nebraska (Mr. Bereuter) for moving this very important resolution and, like everyone else, I rise in strong support of it.

We all know that today we mark this first anniversary of one of the most tragic days in America's history. As we think about the families of those victims, our thoughts and prayers go to all of them. We all have sort of mixed emotions on this day. We all, of course, recall exactly what it was like here in the Capitol a year ago today, and we think about the day first and foremost with sadness because, as I said, of all of those whose lives were lost. But we also think about today with a great deal of resolve and defiance. In some sort of strange way, we also celebrate the success that we have had in pushing back those, as the President calls them, "evil doers," those who would, in fact, bring an end to our way of life.

Just yesterday, here in the District of Columbia and in 12 other States, there was a great celebration in that we had elections where people were choosing their leaders. We are continuing with our work here in the U.S. Capitol right now. So that is why we all have mixed emotions as we deal with today.

But it is also very important, Mr. Speaker, for us to take a few minutes to look at the history of what led up to September 11 and to realize that as we, with this resolution, are remembering and honoring those who were killed on September 11, it is also important for us to realize that this is an international war on terrorism, and it is not a war that began on September 11 of last year. It is a war which has been going on for decades.

Just a few minutes ago I sat down with some of my staff members and started talking about some of the horribly tragic events that Osama bin Laden and his terrorist allies have perpetrated over the past couple of decades, and I thought it appropriate that we take a moment as we reaffirm our strong commitment, as the President has said, to win this war on terrorism, and look at what led up to that tragic day 1 year ago.

On April 1, 1983, 63 were murdered and 120 injured when the U.S. Embassy in Beirut was bombed by the Islamic jihad. On October 23, 1983, we all remember the tragic Islamic jihad bombing of the Marine barracks when we lost 242 of our Marines. On June 14, 1985, Robert Stethem, the U.S. Navy sailor, was murdered and thrown from that TWA flight 847 which was hijacked by Hezbollah terrorists who also held 145 innocent passengers hostage for 17 days. On February 26, 1993, we all remember very well the World Trade Center bombing in which 6 were murdered and 1,000 people injured. On June 25, 1996, 19 U.S. military personnel were killed and 240 injured when the Khobar Towers housing complex in Dhahran, Saudi Arabia was bombed by Osama bin Laden's allies. On August 7, 1998, 12 Americans and hundreds of Kenyans and Tanzanians were murdered in the bombing of the U.S. embassies at Nairobi and Dar Es Salaam, directed, as we all know, by Osama bin Laden, and on October 12, 2000, 17 sailors were murdered and 39 injured in the bombing of the USS *Cole*. That was a horrible, horrible day.

These events, Mr. Speaker, underscore the fact that this is a war which has been going on now for decades.

So when I think ahead to some of the challenges that we face as a Congress, we are going to be dealing with a resolution which will help us confront those who have in fact provided shelter and refuge to Al Qaeda and, of course, I am referring to Saddam Hussein and Iraq.

So this war is one that really reaches all across our globe. It is one that I am happy to see our allies support, and it is one that will continue probably beyond our lifetimes.

As I think about some of the very wonderful quotes throughout history that led to our dealing with these challenges, I am reminded of a couple. There is one that I like to recall. At the beginning of every one of Winston Churchill's volumes, he has what is called the moral of the work, which is basically four points. He says, "In war, resolution; in defeat, defiance; in victory, magnanimity; in peace, good will." I think that underscores where it is that we are headed in dealing with this challenge, which is going to continue in the future.

Shortly after September 11, President Bush delivered a speech in Cincinnati, OH. As I think back on many of the brilliant statements that have been made following September 11, one of the most poignant underscores the sacrifice that was made a year ago. President Bush said, "Terrorist attacks can shake the foundations of our biggest buildings, but they cannot touch the foundation of America." These attacks can shatter steel, but they cannot dent the steel of American resolve.

So, Mr. Speaker, as we move ahead, I again would like to thank my colleagues, the gentleman from California (Mr. Lantos) and the gentleman from Nebraska (Mr. Bereuter), for this very important resolution, and provide strong support; and, as everyone is doing, extend my thoughts and prayers to the families of those who tragically lost their lives a year ago today.

Hon. Jay Inslee

OF WASHINGTON

Mr. Speaker, we are all united in every congressional district in coming together in America today. I would like to honor two contributions from the First District of the State of Washington.

First, I would like to honor the life and heroism of Army Sergeant Larry Strickland, who lost his life while at his post on the second floor of the west wing of the Pentagon 1 year ago today.

Army Sergeant Strickland was a native of Edmonds, WA, who served his country as a personnel manager and traveled to every post around the world to which he was assigned. He was a senior adviser to the Deputy Chief of Staff, and he earned the greatest title I think anyone in the American military could have. He was a good soldier.

But he was also a good son to proud Americans Lee and Olga Strickland of Edmonds, WA; and a good husband to wife, Deborah; and a good father to Julia, Matthew, and Chris; and a good grandfather to Brendan.

To those family members, we cannot ask to beguile them from their grief; but we hope that they are left with the cherished memory of the loved and lost, and the solemn pride that should be theirs for having left such a precious contribution at the altar of freedom. That family can be assured that we will keep Sergeant Strickland's memory alive as we go forward together in a unified way to preserve the freedoms for which he died.

Second, Mr. Speaker, I would like to honor Molly Peebles, Chuck Oppermann, and Erik Lindbergh, who led the flight of 51 planes in the Flight Across America from every State in the Union that brought every State's flag and the U.S. flag to New York this morning, having begun flights all across this country on August 11 to bring a message of resolution and healing to this Nation.

They spoke with the spirit of general aviation, which is important in this country for our freedoms and our economy. I hope their contributions send a statement that we have to honor general aviation so we can continue to have both security and a viable general aviation in this country.

Mr. Speaker, America is proud of people in every district in this country, and I offer the gratitude of this Nation to these people from the First District of the State of Washington.

Hon. Jim Gibbons

OF NEVADA

Mr. Speaker, today America remembers. America just 1 year ago witnessed three heinous and senseless terrorist attacks that will be forever etched in our minds and memorialized in our history. May I say that no American will ever forget the horrific images of September 11, 2001, and no American will ever forget the thousands of innocent victims, or the hundreds of courageous heroes who will always have a place in our history and in our hearts.

Yet since that fateful day, the American people have persevered and established a new sense of normalcy. May I say to the cowards who targeted our great country last year, let me recite that terrorism against the United States, our freedom, and our people will never be tolerated. The United States stands strong, and we will fight terrorism wherever it lurks. No corner of this world will be safe for them or those who want to attack freedom, destroy liberty, and instill fear.

America's fight will not be won quickly or easily; however, it will be won. I have seen firsthand the bravery and the patriotism of the men and women in our Armed Forces, and may I say that they will not settle for anything less than victory. They are willing to make the ultimate sacrifice to protect our liberty. They deserve our strongest support and our deepest gratitude.

Americans have answered the call to help their fellow man and to defend freedom and to serve their Nation. We will not retreat. We will not be intimidated. America is strong and her strength is in her people. It is the strength of the American people that will prevail over terrorism wherever it may hide.

So on this 1-year anniversary of September 11, on this Patriot Day, may Americans again unite to remember our loss, to celebrate our freedom, and to defend liberty.

May God bless this great country and its people.

Hon. Danny K. Davis

OF ILLINOIS

Mr. Speaker, 1 year ago today, I was in Tel Aviv, Israel. So, Mr. Speaker, I rise today to pay tribute to the over 3,000 victims who died in the September 11 attacks on America; and I wish to pay tribute to all the family members of those who perished, as well as to our public officials, businesses, and private citizens who have dedicated themselves to the rebuilding of a stronger Nation over the past 12 months.

In a recent special report about life after September 11 in America's tallest building, the Sears Tower, the *Chicago Tribune* observed that what happened was more complicated than fear. It was an awakening, then a reckoning, then a change in priorities and plans in the calculation of everyday decisions. This awakening and recalculation in our everyday lives has taken place not just among those in the Sears Tower and my home in Chicago, but across this great Nation. It is a new realization of how connected and responsible we are for the protection and well-being of all our fellow citizens.

In response to the devastation caused by the terrorists, the residents of Chicago have joined with millions of others in this country and around the world to donate millions of dollars and hours to the rebuilding efforts. Their material gifts, however, reveal an even deeper resolve to let the principles of freedom for which we stand ring loud and clear.

As we contemplate the effects of September 11 and extend our deepest sympathy to those who lost their loved ones on that fateful day, let us resolve to build not just a more vigilant and stronger homeland defense, but a society that continues to protect our personal freedoms and would enable us to fulfill the American dream of liberty and justice for all.

And in pursuit of this goal, let us, Mr. Speaker, continue to be able to sing:

O beautiful for spacious skies, for amber waves of grain, for purple mountain majesties above the fruited plain. America, America, God shed His grace on thee, and crown thy good with brotherhood from sea to shining sea.

God bless America.

Hon. Gregory W. Meeks

OF NEW YORK

Mr. Speaker, I woke up this morning not knowing, really, where was the appropriate place for me to be. Being a New Yorker, I did not know whether I should be at Ground Zero this morning, whether I should be in my community with many of my constituents, or whether I should be here in Washington, DC.

I made the decision to try to do a little bit of both, so I spent this morning with my constituents in New York, but then felt that it was most important for me to come here to Washington as we reflect on what took place just a year ago today, as we memorialize the individuals who lost their lives a year ago today, as we sit with and hold hands with the family members of those who lost their lives.

It is time for reflection; and as I reflect and think about last Friday, when Members of the U.S. Congress came to New York for the first time since 1789 and went to the original Federal Hall, it struck me about this great thing that we know called democracy, and why we have to stand and fight and make sure that democracy prevails.

I think, in 1789, I as an African-American may not have been able to be part of that Congress. But democracy has prevailed, so that in 2002 I am a Member of this Congress; and we are here today where the Congressional Black Caucus is meeting.

We must preserve that democracy. What happened on September 11 of last year threatens that democracy. We must let freedom ring.

As I reflect, the words of Dr. King come back to me. In 1965, we were talking about freedom here on these shores. His words were,

So let freedom ring from the prestigious hilltops of New Hampshire. Let freedom ring from the mountains of New York. Let freedom ring from the heightening Alleghenies of Pennsylvania! Let freedom ring from the snow-capped Rockies of Colorado! Let freedom ring from the curvaceous peaks of California! But not only that; let freedom ring from Stone Mountain of Georgia! Let freedom ring from Lookout Mountain of Tennessee! Let freedom ring from every hill and every molehill of Mississippi! From every mountainside, let freedom ring.

I think we need to add now: Let freedom ring from Afghanistan. Let freedom ring from Pakistan. Let freedom ring in Israel and Palestine. Let freedom ring in China and India. Let freedom ring in Zimbabwe and Nigeria.

Let freedom ring in every country on every continent, because when freedom rings, in the words of Dr. King, when we let it ring, we will let it ring from every village and every hamlet, from every State and every city in every nation. We will be able to hasten that day when all of God's children, black men and white men, Jews and Gentiles, Protestants and Catholics, Muslims and Buddhists, will be able to join hands and sing the words of the old Negro spiritual: "Free at last, free at last, thank God Almighty, we are free at last." God bless America.

Hon. Benjamin L. Cardin

OF MARYLAND

None of us will ever forget where we were on September 11, 2001, when we first learned about the 8:46 attack on our Nation. It was a defining moment for our Nation like November 22, 1963, and December 7, 1941. Now we commemorate the first anniversary of September 11 and, quite frankly, Mr. Speaker, we are not sure what we should say or what we should do, but we want to join together as a community to show our solidarity.

Today we honor our heroes. These are not our Presidents or our sports figures. These are ordinary Americans who performed extraordinary acts. Heroes by chance and heroes by choice. Like Todd Beamer on flight 93 who gave up his life to save many others. We now know that it was likely that the plane was headed here toward the Capitol. He may in fact have saved our lives. Heroes such as John Fischer, a New York City firefighter, who went into harm's way in order to save lives and lost his life on September 11. Heroes such as Lt. Darin Pontell, a naval officer, a young man whom I had the honor to appoint to the U.S. Naval Academy. He understood the risks of serving in our armed services, but he thought he would be safe in the Pentagon. He lost his life. Over 3,000 of our fellow citizens lost their lives, casualties to our continuing effort as a Nation to maintain liberty, safety, and freedom for all of its citizens.

Each of us was personally affected by September 11. We may not have known anyone personally who died, but we still grieve for them, and we hold the members of their families close to our hearts. We shall never forget September 11. Shortly after September 11, the Congress passed a law calling on the President to designate September 11 as Patriot Day in honor of the individuals who lost their lives as a result of the terrorist attacks against the United States that occurred on September 11, 2001.

Throughout America we join together as a community in solidarity to make it clear to the world that our Nation is united and resolved to defend freedom against all enemies, any enemies. We may be Democrats or Republicans, conservatives or liberals. We may differ in religion or ethnicity, but we are united as Americans. Terrorists destroyed the New York World Trade Center buildings, but they can never destroy the character, strength and values of the American people.

Mr. Speaker, yesterday was an important day in Maryland. It was primary election day. Marylanders chose their leaders and representatives by the ballot box. Our enemies rule by fear, intimidation, and force. Because of America's leadership, freedom and democracy are winning in all corners of the world. As President Bush has said, "We are a people dedicated to the triumph of freedom and democracy over evil and tyranny."

Today we thank millions of Americans who responded to the Nation's calls, our soldiers, our firemen, our police, our postal workers, and so many more who have been on the front line for the defense of our country.

Ever since September 11, "God Bless America," our hymn, has had a special meaning. One year ago this evening, we sang it on the steps of the Capitol to make it clear to the world that we would triumph. Tonight we will sing "God Bless America" on the steps of the Capitol to make it clear to the world that we will never forget September 11.

Mr. Speaker, may God bless those who mourn, may God bless those who serve, and may God bless America and bring us peace.

Hon. Robert Wexler

OF FLORIDA

Mr. Speaker, we are here at this somber occasion to recognize September 11 as a moment in history when American heroism and patriotism prevailed over terror, adversity, and hate. This resolution pays tribute to those families personally affected by the tragic events of September 11 and honors the innocent victims of these horrific attacks.

Today America is united, drawn together by overwhelming grief, a shared commitment to freedom and unwavering resolve. Our unity as a Nation sends an unequivocal message that despite the tragic events of September 11, the American spirit remains strong, that in the face of destruction and hate, democracy, justice, and hope will prevail.

September 11 demonstrated that in a moment of unparalleled adversity and devastation, there exists an innate desire of Americans to help others that is unimaginably selfless and good.

Never was this more clear than 1 year ago today in New York and Washington where hundreds of the world's bravest and finest, including firemen, policemen, and first responders rushed to assist the victims of these tragic attacks; or above Shanksville, PA, where passengers on flight 93 actually took a vote to sacrifice their own lives to preserve those of others and prevent a potential assault on the very Chamber in which we now stand.

The vote on flight 93 to overtake the hijackers epitomizes American values, courage, and heroism at their very best. For in a moment of unprecedented darkness and despair these brave souls refused to sit idly by and bear witness to evil. They chose to take action after evoking the most basic American right and fundamental symbol of democracy that all of us hold dear.

As we commemorate the tragic events of September 11 and honor the victims lost in New York, Washington, and aboard flight 93, we must remember the extraordinary acts of heroism that took place 1 year ago today. From Washington to every community across the United States, may we derive strength and courage from the bravery demonstrated on September 11 and commit ourselves to a future free from terror, intolerance and hate to one of understanding, freedom and, above all, peace.

Hon. Tom Lantos

OF CALIFORNIA

Mr. Speaker, this has been a fine moment for the House of Representatives. We came together, not as Republicans or as Democrats, but as American citizens who are still so painfully aware of that horrible moment a year ago when lives were snuffed out, orphans were made, widows were made, family tragedies were made across this land. And the people who did it so totally and so profoundly misunderstand the nature of an open and free society. They sought to intimidate us. They sought to make us give up the good fight, but they have only succeeded in steeling our will and determination that the terrorists of this globe and the regimes that harbor them will not prevail.

This struggle will go on for a long time. Al Qaeda may have been largely defeated in Afghanistan, but they are all over the world in scores of countries planning to perpetrate additional evil deeds. We are ready for them.

Not since Pearl Harbor have the American people been as determined, as united, as committed to defending the values of our way of life, our Constitution, our pluralism, our acceptance of all religions, of all ethnic groups, of all forms of commitments to values that we so dearly cherish in this country.

The terrorists may have succeeded in bringing down two magnificent buildings but they also succeeded, contrary to their hopes and aspirations, of steeling the determination of the American people to protect this free and open and magnificent society.

As we recall the events of a year ago and as we plan the next action, it is important for us to realize that while not always will governments be with us, people who love freedom will be with us in every corner of the world.

The President will be addressing the United Nations tomorrow. He will be speaking to all freedom-loving people on the face of this planet. He will be talking on behalf of all of us because we are determined in the 21st century, as we were in earlier centuries, not only to preserve our way of life but to expand the arena of freedom for people everywhere.

It is easy, Mr. Speaker, to become discouraged. When Pearl Harbor struck, many were pessimistic. When Hitler swept across Europe, many were pessimistic. But the indomitable spirit of men like Winston Churchill knew full well that free societies will prevail. The Osama bin Ladens of this world are simply incapable of comprehending how powerful the spirit of freedom is in open and democratic societies. We may suffer setbacks, we may suffer occasional defeat, but our goal of preserving this way of life for ourselves and making it available for others if they so choose cannot be defeated either by hijacking an aircraft, spreading biological or chemical weapons, or any such means. The spirits of free men and women will prevail. And there is no doubt in the minds of any of us in this body that however long this struggle will take against terrorism and countries that support terrorists, we will prevail in the long run as free men and women always have.

The many ceremonies across this Nation, from New York to the Pentagon to Pennsylvania, to every town and hamlet in the United States, is proof that the American people have learned the lesson of a year ago. We may have lost our innocence but we have multiplied our resolve and determination. This Nation is united, strong, and conscious of the fact that our values, more than our physical capabilities, will result in our ultimate victory.

I want to commend all of my friends and colleagues who have participated in this discussion. I trust the message that we sent with many voices, many phrases, different approaches, will not be lost on those who cynically or otherwise are doubting the resolve of the American people.

This resolve is unshakeable, and this resolve will bring about ultimate victory over terrorism and totalitarian societies.

Hon. Doug Bereuter

OF NEBRASKA

Mr. Speaker, first of all, I would like to commend my colleague and friend, the gentleman from California (Mr. Lantos), for the very able way that he has handled this debate, this memorial, this commemoration, and for his cogent and insightful remarks. The gentleman suggests that, in fact, this has reignited or reawakened a powerful, diverse Nation. Indeed, that is the case. That was what the terrorists apparently did not expect.

On a personal note, may I say that my wife and I are particularly grateful that her sister, an officer worker in the Twin Towers, survived the bombing attack in 1991 and again last year and was able to come down to safety with a lot of difficulty and a lot of courage; and all of those people who had that experience, of course, will have a commitment to make this country even better and even stronger than it had been. I believe that there is not a doubt in anyone's mind in this country that what happened on September 11 has only reinforced our strengths. It has, in fact, increased our unity and our resolve.

The comments from my colleagues here today give us some very strong indication of the sentiment that their constituents feel and resolve that their constituents expressed to them in their districts. I commend all of my colleagues on both sides of the aisle for this remarkable experience and expression here today.

Hon. Juanita Millender-McDonald

OF CALIFORNIA

Mr. Speaker, I submit this statement in the *Record* in support of this resolution and applaud all Americans for their resilience as we commemorate and reflect on last year's events.

Mr. Speaker, I rise to express my support for H. Con. Res. 464 that expresses the sense of Congress on the anniversary of the horrible terrorist attacks launched against our country and our ideals of democracy.

My heart is heavy as I reflect on the tragic events of a year ago. The past year has been especially difficult for the families, friends and coworkers of individuals who perished in New York, Pennsylvania, and Washington, DC. Our prayers are offered to the victims and their families.

Today, in a small and symbolic way, we are expressing our gratitude to the firemen, policemen, health care workers and the individuals who, on that fateful day, performed heroic deeds and helped their fellow human beings without regard for their own welfare. We thank you profusely for your efforts. America owes you a debt that we can never repay. We salute you for your service and valor.

We also salute those who have served our Nation so bravely overseas and all people of good will who personify humanitarian virtues during this tenuous and volatile time in the world.

Today, we are united as a Nation to confront current world realities that have transformed the social, psychological and spiritual fabric of the world in which we coexist with our domestic and foreign neighbors.

During the recent year, my colleagues and I have actively engaged in debate about how to formulate an agenda that addresses homeland security, national spiritual salvation, and political bridge building with our international friends.

As we continue to make inroads toward progress, I cannot overemphasize how important it is to map out a course for our future that will sustain, inspire and protect our children. We must infuse them with a sense of optimism because the confidence in which we as Americans move around our country has been shaken. However, as Americans, we will not be deterred from experiencing the freedom we cherish.

My colleagues and I on the Transportation Committee recognize the importance of protecting the confidence of America's traveling public. We have worked diligently to ensure that the security needs of the flying public are paramount. We will continue to pursue the course of protecting our transportation infrastructure, and we are committed to making sure that America continues to move passengers and cargo efficiently and safely.

As ranking member of the Subcommittee on Workforce, Empowerment, and Government Programs, I recognize how small business owners around our Nation have been victimized economically and traumatized emotionally by the events of 9/11. Consequently, my colleagues and I on the Small Business Committee have worked to ensure that industries hit hard by the traumatic events of September 11, including the travel industry, are able to survive. We have held hearings and offered legislation that seeks to resuscitate small businesses throughout our Nation. Small businesses constitute the backbone of our country's economy, and with our help, they will not only survive, but flourish.

Our domestic efforts and grief over the tragic events of September 11 have heightened our appreciation for the pain of others around the world that have been subjected to the brutality and inhumanity of terrorism. And that is why we have supported liberation and democratization efforts in Afghanistan and sought to assist and rehabilitate the victims of persecution who are attempting to rebuild their lives and their country.

On this occasion of reflection, recommitment and rededication to freedom, we are affirming our resolve to salute and honor the men and women who paid the ultimate price—their lives, just because they are Americans. They will never be forgotten and our Nation shall continue to rise to meet the challenge of terrorism and the threats posed by terrorists who seek to derail freedom and the good will of humanity.

May God bless America and all people of good will.

Hon. Dan Miller

OF FLORIDA

Mr. Speaker, I rise today to remember those who lost their lives and the lives of loved ones on September 11, 2001. I also rise to submit an article which appeared in the *National Journal* on August 31, 2002, which recounts my experience on that fateful day:

President Bush had scheduled a visit to Booker Elementary School in Sarasota to give a major speech on education on the morning of September 11. I'd spent the weekend in Washington, but I flew down in order to be with the president and to fly back on Air Force One with him. The president was scheduled to arrive at Booker at 9 o'clock in the morning. I was told something had hit the World Trade Center, but I didn't think too much more about it. We all weren't focused on that.

The classroom was small, so there were a limited number of people in that room besides the students. I went into the auditorium-type room and was awaiting the president's speech. The second plane hit at 9:05 or so. We all started to realize the magnitude of what was happening. We were just in a state of shock: "This can't be true. This can't be true."

Matt Kirk, of the Legislative Liaison Office at the White House, was assigned to us, and he tried to keep us updated. Things were in a state of flux, and the Secret Service agents were moving around. There was a question about whether the president was going to make a speech to the nation from the school, or go over in front of Air Force One. The White House staff felt the quickest way was to just do that right there. We went out and got in our van in the motorcade. I had my BlackBerry with me and I would get some news, and Matt Kirk could get some news from his little pager, but it was limited. This was, say, 9:15 to 9:30. The president came out, got in his limousine, and then we just drove very rapidly over to the Sarasota-Bradenton Airport. The president got on the plane, and it took off about 10 o'clock. We were told to tighten up our seat belts very tight, because the plane has the ability to take off very steeply. When we took off, we were told there was no communication, because they didn't want anyone to know where Air Force One was, and the press was told that too.

In many ways, for most of that day, I had far less news and information than most of the people around the world. It was frustrating. Air Force One did not have, at that time, the ability to pick up television. Matt Kirk would go up and try to talk to someone else and pick up some news. Somebody would come back and share some information. The president's political adviser, Karl Rove, came back a couple of times, just giving us an update. It was very tense with the staff, because there obviously was a great security concern by the Secret Service and the crew.

Around 10 o'clock, we were heading due north. And then, you could sense a turn to the west. I would say 10:45, maybe 10:30 or so, the plane changed course. We were told we weren't going back to Washington. We didn't know where we were going. I remember looking out my window and looking down at that Gulf Coast of the Alabama-Mississippi-Florida area.

We started to pick up some TV reception. We saw that the collapse of the towers had occurred. It was so surreal. I remember Karl Rove coming back and he said, "There are 40,000 people who are working at the World Trade Center at this time." No one knew the magnitude could have been that high. And then, to hear about the Pentagon!

About 11:30, we got called up to the president's office. The president was at his desk. There's a little sofa that can seat four or five people, and a chair where Andy Card sat. Behind us was this TV screen on the wall. I didn't even know it was there until I got a photograph, and you saw the World Trade Center, a fuzzy picture of it, right over our heads.

The president was telling us that there were some other planes—six, maybe nine, planes—that were unaccounted for, and that a plane had crashed in Pennsylvania, so the decision was made not to return to Washington. He was very serious, very determined, very focused, and very collected. And I felt much more emotional at that moment than he was. You saw he was in control. I felt choked up. It was almost like you're speechless.

He said he was determined to make sure that the people who were responsible for this would be identified and punished. There was speculation on the plane, but not with him, that it was bin Laden. The belief was, the only people capable of such an evil deed were either a government—and they didn't think it was any government behind this—or the bin Laden organization.

The president was saying, "We are going off to an undisclosed location." He was able to very calmly explain where we were and what we were getting ready to do. The only one speaking was the president. And I don't remember really even asking questions.

I remember saying as we were leaving, "God bless you, Mr. President." You could see the weight on his shoulders. He had been through a lot in those couple of hours. And he obviously knew a lot more than we knew. He talked about how he had given the order—he actually said it had been while he was driving over from the school to Air Force One—to bring all the planes down from the air. He was saying how we had an AWACS and six fighters surrounding us. He was saying we were going to land at an undisclosed location, and that we would be getting off the plane there, and he was going on to another undisclosed location.

When we got to Barksdale Air Force Base, all you saw were just rows and rows of B–52 bombers. There was a van, a Humvee, there were people standing around with automatic weapons, which you don't see in the United States. We're seeing it today—but we did not see it until September 11. And you could see the president go out. We were left there on the plane. That's when we got good TV, from noon to 1:30. We could not have any contact—no cell phones or BlackBerrys. Then the president came back about 1:30. We exited the plane and stood there on the tarmac, and Air Force One took off.

I was able to call my wife, who was at home on Capitol Hill. The White House had called her and told her I was on Air Force One. I said something like, "Honey, I'm OK." I was able to tell her where I was and that there was another plane that was going to take us back to Washington. They flew us to Andrews Air Force Base, and I got home about 6 o'clock. We have to be about the only plane in the air, with the exception of the fighter planes, because everyone was grounded, I guess. I'll never forget the landing. You saw the Pentagon smoke.

I remember when I came home and walked in the house, it was very emotional. I hugged my wife. We just squeezed each other. It was hard to comprehend. I just didn't want to talk to anyone, besides my son and daughter, because it was still so emotional. I choke up sometimes just talking about it. It's just hard to comprehend that I was right there on Air Force One when this whole thing was evolving. This was obviously the most significant event during my congressional career. You realize that the U.S. is vulnerable—that we're not immune to some of the problems elsewhere around the world, and it makes you think that life is very precious. I didn't know anyone personally who perished that day, but it brought that home to me.

Hon. Elijah E. Cummings

OF MARYLAND

Mr. Speaker, I rise today as a cosponsor of this important resolution and to share in a day of grief and reflection with all Americans.

One year ago today, the course of our Nation was forever changed. As we all know, on September 11, 2001, terrorists used hijacked airplanes to inflict catastrophic damage upon the United States, taking the lives of over 2,000 Americans. And while this day 1 year ago was marked by chaos, carnage, fear, and great loss, it also served to showcase the best we as Americans have to offer. In the year since, these qualities—courage, innovation, idealism, hope—have enabled us to take great steps toward rebuilding this Nation.

In many ways, the legacy of September 11, 2001, is a patchwork of personal stories. For many of us, the world has been changed forever in very personal and very painful ways. Each child who woke up on September 12 without a parent—and each parent who suffered the unspeakable pain of awakening without their child—deserves our tireless pursuit of those responsible for these heinous crimes. All Americans deserve a country in which they feel safe.

Each and every one of us has a connection to the destruction. Stepping out of my car that morning, I could feel the ground shake below me as the third hijacked airplane struck the Pentagon. I will never forget the uneasiness that I felt—I know that I will never be able to separate my personal attachment to that day with the larger infamy of September 11. While our individual struggles to come to terms with what occurred are often frightening and lonely, they are also what bind us together. I believe I speak for millions of Americans when I say that a piece of me died with each victim of the September 11 attacks.

Mr. Speaker, I would be remiss if I did not mention a constituent in my district who also had a piece of him taken away that day. John Wesley's fiancee, Sarah, died on the plane that crashed into the Pentagon. John, like so many loved ones left behind, funnels his grief into a positive effort that is a testament to his fiancee—ensuring that her legacy will continue. I must also mention that despite his grief, John, such a positive person—is now committed to preserving Sarah's mission of exposing kids to different cultures. Mr. Speaker, there are so many John Wesley's in our Nation and I salute them for their courage.

We are now at the 1-year anniversary of a day America will never forget. It is time to take a look at where we are. Already, we have waged our "war on terrorism" to Afghanistan with great success. Brave American servicemen and women have risked their lives halfway around the world to prevent further attacks, and bring those responsible for September 11 to justice. We have radically altered the way that we look at national security by undertaking the single greatest Federal Government reorganization in 50 years. We have taken unprecedented strides in revamping our transportation security infrastructure. We have been forced to reevaluate our aging systems of immigration and naturalization. We have cleared the tons of rubble of the two largest buildings in America's largest city. We have removed and rebuilt the nerve center of this Nation's defenses. And we are far from through.

Mr. Speaker, I rise today as a patriot and a legislator. I take both roles very seriously. Since the attacks of September 11, the American people have been reminded that freedom is not free. It takes courage to both survive in a dangerous world and retain our liberties. I will continue to work to give law enforcement the authority and resources it needs to attack terrorism, while standing sentry over the fairness, justice and constitutional rights of all Americans.

I mourn the victims of September 11, and together with all Americans, I accept the challenges facing us as a truly united America. We will never forget. But we will not let the actions of a handful of zealots derail the hundreds of years of work we have put into this country—the freest, most successful, and most democratic Nation in the world.

God bless America.

Hon. Tom Davis

OF VIRGINIA

Mr. Speaker, September 11 will always remain a day of great grief and sadness, courage and hope. The sight of the destruction and evil filled us with disbelief, sadness, helplessness and anger. That day and every day since we have also been blessed to see the best of America.

Throughout the last 12 months, we have seen the greatest of our national character in countless acts of sacrifice, courage and love. Police, firefighters and port authority officers rushed into the building with more concern for finding victims than for their own safety. Two office workers, finding a disabled stranger, carried her down 68 floors to safety. Rescuers worked past exhaustion to save lives after the collapse. The men and women of our Armed Forces, Federal law enforcement and intelligence community are now prosecuting the war on terrorism abroad. The men and women on flight 93, having spoken with loved ones on their cell phones and being told of the unfolding events, stormed the hijackers and took the plane down, saving thousands of lives on the ground. These acts showed the world what we have long known—that our fellow Americans are courageous, compassionate and brave. That the true strength of our Nation is in the souls of each of us, and that is something our enemies can never take from us and never defeat.

Our Nation still grieves through our national tragedy and personal loss. To the children and parents and spouses and families and friends of the lost, we offer our deepest sympathy, our tears, our support and our love. None of us will ever forget the events of September 11, but we will continue our lives together, arm-in-arm, with a collective strength that can carry us all.

Not only is our military winning the war on the ground, we are also winning the war in the hearts and minds of individuals across the world.

The world is beginning to see exactly who America is and who our enemies are. Never has the difference been so clear. They wish to kill and destroy, we seek to assist and build. They work for division, we seek unity. They pray for and plot our failure, we hope and work for a better life for all.

In 1 year's time, we have comforted those who lost loved ones, we have completed our cleanup at Ground Zero, rebuilt the Pentagon, rallied the civilized world against terrorism, renewed our friendship with our Muslim friends and Arab partners, destroyed terrorist training camps in Afghanistan, rid the world of thousands of terrorists, put others on the run who will soon understand there are no limits to American justice, and freed a people from an oppressive regime, restoring hope and opportunity. We're working with the new Afghan Government to lay the foundation for long-term stability and to reverse the conditions that allowed terrorist regimes to take root in the first place. This Congress has taken necessary actions to strengthen our homeland, assist the families and friends who lost loved ones, and worked to stem the economic downturn that resulted. There has been much good that has come out of that tragic day that shook us to our core, but strengthened our resolve and determination to rid the world of evil. And we still have much work to do. We did not ask for this mission, but we will fulfill it to ensure that freedom endures.

We are now engaged in a struggle that we cannot and will not lose. We have come to a unique moment in history, and America must take its rightful place, leading the charge for the rights of men and the responsibility of government. Throughout our storied history, America has reaffirmed its commitment to freedom. Today, we find ourselves at the dawn of a new birth of freedom, not only for our Nation, but indeed for all of mankind. We have been given this enormous task, and we will undertake it as only America knows how—head on—and we will succeed. We will capture this opportunity for all of mankind and all time. We will continue to lead this Nation and world that we love, confident that the same God that watched over George Washington as he led our Revolutionary Army, over our Founding Fathers as they established our system of government, over Abraham Lincoln during the Civil War, and over our Armed Forces as they stormed the beaches at Normandy, is still watching over us today as we go forth to make the world safe for freedom and democracy.

America was targeted by our enemies because we are the beacon for freedom and opportunity in the world. That light still shines brightly today. Peace and freedom will prevail. Hatred and evil are ephemeral, but love and goodness have no end. The greatest people in the history of mankind have been called to defend a great Nation and the greatest of ideas, and we will continue to succeed.

As we pause to mark the first anniversary of one of the worst days in our Nation's history, we not only remember and pray for the loved and the lost, we also renew our commitment to honoring their memory by pursuing peace and justice, by upholding freedom and democracy, and by defending all that is good and just in the world.

Hon. Janice D. Schakowsky

OF ILLINOIS

Mr. Speaker, today, as we remember the victims of 9/11 and pay tribute to the fallen heroes who sacrificed their lives to save others, our Nation offers gratitude, comfort and support to the families of those who perished on that tragic day. Our Nation is also reminded of the brave men and women who are standing guard here at home and abroad—soldiers, police officers, firefighters, and first responders. We are grateful for their service. Time may heal wounds, but we will never forget.

Hon. Robert A. Underwood

OF GUAM

Mr. Speaker, it is with a most profound sense of remembrance that I rise in support of this resolution, which I am proud to cosponsor. Although Guam, the distant U.S. territory I have the privilege of representing in this body, is roughly some 10,000 miles away from the City of New York, Pennsylvania, and Northern Virginia, its people, my constituency, share in the sorrow and concern for those directly affected by the September 11, 2001, terrorist attacks on our Nation. Our geographic isolation and great distance from the mainland has not kept us from taking part in the healing process, in demonstrating our resolve, and in providing for the defense of our country. This past year has been a time of anguish and renewal for many. It has been a time of trepidation over our future in a world with division, unresolved differences and weapons of mass destruction. As we reflect on the events of 1 year ago, we are reminded of who we are as a people, of what we believe in as a Nation, and of the values that make our democracy strong.

The people of Guam are a patriotic people whose loyalty to the United States has been tested in our most darkest hours of history. As many of my colleagues recall, concurrent with the attack on Pearl Harbor on December 7, 1941, Guam was also attacked, invaded and then occupied. During World War II, our island endured a 32-month brutal occupation. Survival during this daunting and difficult period emboldened the people of Guam and taught us to cherish freedom and democracy. It was with this experience etched in our memories, that we were confronted with the events surrounding the most devastating attacks on American civilians in our Nation's history on September 11, 2001. In the days that have followed, schoolchildren on Guam, like many across the country, have penned their thoughts on paper and creatively expressed their feelings in drawings and illustrations for the families, firefighters, police officers, rescue personnel, and others that were a part of this tragedy. On Guam, like everywhere else in the country, scores of American flags waved atop cars, trucks, and buses, on backpacks of schoolchildren and in front of homes. People filled the pews in the churches and places of worship to pray for those who lost their lives and for comfort, hope and peace. The people of Guam also donated blood, time and money in support of the rebuilding efforts. Memorial concerts, performances, and vigils have been held to bring recognition of the American heroism that was so courageously displayed in the aftermath of the attacks.

Last November, I was able to travel to the New York City Office of Emergency Management where I witnessed first-hand the professional, compassionate work of the people of New York in the face of this tragedy. I was able to bring with me then hundreds of support and thank you letters and drawings from schoolchildren on Guam.

To the families of those who lost loved ones, please know that the thoughts and prayers of the people of Guam are with you. To those first responders who aided in the rescue efforts and to those individuals who have committed themselves to the rebuilding efforts, please know that the people of Guam are grateful to you for your work and commitment. As we commemorate the tragic events of September 11, 2001, and pay tribute to all the goodness that has followed within the past year since that unforgettable day, let us never forget the sacrifice. The people of Guam stand in solidarity with the rest of our country. We continue to stand ready to do our part in this national crusade. As so often has happened in the past century, Guam is ready. We made our contributions in World War II and in the Korean and Vietnam conflicts. We will do our part, and more, again and again in this new century to defend our country, to preserve our democracy, and safeguard our values of freedom and liberty.

Hon. Michael M. Honda

OF CALIFORNIA

Mr. Speaker, today marks the 1-year anniversary of one of the most tragic days in our country's history. I am proud of the way Americans have united following the horrific terrorist attacks on civilians. Neighbors who rarely spoke to each other, or people who did not even know each other shared their thoughts, compassion, and prayers for our fallen heroes. Our country is now stronger than ever.

Understandably, a lot of Americans are outraged following the attacks. Let it be clear that we will continue to hunt down those responsible for the attacks of September 11, and those who may be foolish enough to plan to harm us again. We must also remember that we are not fighting a war with people of any specific ethnic group, nationality, or religion. We must be vigilant and wary of any efforts by the U.S. Government or members of our citizenry to racially profile innocent people, as we wage our war against terrorism. I repeat the poignant words of a famous philosopher who once said, "Whoever fights monsters should see to it that in the process he doesn't become a monster."

As a child, I was a first-hand witness to civil injustice. My family and I were locked up in a U.S. internment camp during World War II only because we were of Japanese descent. There were many patriots during this war. Thousands of young Americans of Japanese ancestry fought and died for the very country that imprisoned their families. However, their loyalty to America never wavered.

Our war against terror is going to be a long-term engagement. Along the way we will need to be thoughtful and critical of many different courses of action. It will be more important than ever to ensure that we all have the freedom to ask tough questions of our government officials, cultural institutions, and citizenry. The declaration of war against terrorism, in itself, is not sufficient justification for the passage of invasive and constitutionally suspect government powers, or calls for military action against nations. A true patriot will ensure that the actions of our government are just and reasonable.

Our Constitution is rarely tested in times of tranquility, but is severely tested in times of tension, turmoil, and tragedy. We must remember to embrace the principles of our Constitution—our contract for democracy and freedom—which others seek to destroy. They shall not succeed. My heart and prayers go out to all those who lost someone dear to them. Know that I share your pain, and that I will never forget the sacrifice your families made in the name of America.

Hon. Tom Udall

OF NEW MEXICO

Mr. Speaker, a year has passed since 3,000 innocent men, women and children tragically lost their lives in a brutal and cowardly attack on our country.

While we have had a year to come to terms with the enormous tragedy of September 11, the sense of loss remains overwhelming.

On that somber day, we all watched in disbelief as terrorists hijacked planes to attack buildings that symbolized our strength and power. We will never forget that day or the people whose lives were unexpectedly taken from them.

New Mexico's families felt the pain of losing friends and loved ones. No one who knew Alamogordo's Alfred Marchand can ever forget this man who proudly served his community in the Alamogordo Department of Public Safety. After a stellar 21-year career, he followed his heart and became a flight attendant. While living his dream, he perished on United Airlines flight 175 fighting against the terrorists.

We must also remember Senior Airman Jason Cunningham who grew up in Carlsbad. Jason died on a mission in Afghanistan while trying to save another serviceman. Albuquerque's Sgt. First Class Christopher James Speer also died in an Afghanistan fire fight with suspected terrorists. He made the absolute sacrifice to protect us from terror.

Remembering the 3,000 lives taken from us reminds us that, though we continue to move forward with our own lives, we do so in a world that has been profoundly affected by the events of September 11.

As we search for ways to deal with the unspeakable horror of this tragedy, we can take guidance from the words of Robert F. Kennedy. Amid the grief and rage that followed the assassination of Dr. Martin Luther King, Jr., RFK said, "In this difficult time for the United States, it is perhaps well to ask what kind of Nation we are and what direction we want to move in."

These questions are still appropriate today. What kind of Nation are we? What direction do we want to move in? I would answer that America remains a great Nation and I would challenge us all to move in a direction that even more closely embraces the virtues and values that make us great.

September 11 has tested—and will continue to test—the resolve and the resilience of all Americans. It will test our commitment to the virtues and values on which this Nation is built: democracy, diversity, liberty and justice for all. Our response to September 11 must be to become even more passionate in our commitment to these virtues and values: to embrace and support democracy around the world; to celebrate our national mosaic of races and religions; and to shine the light of liberty even more brightly into every corner of the Earth.

America is involved. We give of ourselves—as service members, teachers, law enforcement officials, emergency workers, and volunteers for causes great and small.

Today, public service is more important than ever. Tragedy reminds us we must come together to create an America that is even more just, more democratic and more secure.

America is a great Nation. Let us look back on September 11, 2001, with sadness and respect, grieve for those we lost, and honor those who showed the courage that makes us all proud. Let us emerge from this tragedy a stronger and wiser great Nation.

Hon. Karen McCarthy

OF MISSOURI

I rise today to commemorate the anniversary of the September 11 attacks on our Nation. As I reflect on this first anniversary of the tragic attacks on our Nation I am overcome with sadness and hope. That we as a Nation have come together in a spirit of reflection, resiliency and continued renewal is a testament to the enduring greatness of the American spirit.

Though we are pausing throughout the day to remember those we lost, our resolve to protect our freedom is unyielding. Our Nation is strong and once again America and our values have persevered. Friday's special session of Congress in New York's Federal Hall invoked the history of the first Congress convened in 1789 and represented a strong message to the world that as Americans we stand together in our fight against terrorism.

Last year when I visited Ground Zero I saw evil in the devastation present. Last week at Ground Zero I saw the continuing of our healing process and the unlimited hope of the American spirit. Throughout the past year I have been working with community leaders in my congressional district to both cope with the aggression forced upon us and assess the level of preparedness in our region should another incident occur.

We need to be sure our everyday heroes: our police, fire, ambulance, and medical personnel have the adequate resources, training, supplies, materials, and equipment they need to protect our community. These brave men and women are a critical component of our homeland security for they are our front line. Because of their valor and commitment to service we are safer than we were a year ago.

As we remember the attacks on our Nation in New York, at the Pentagon, and in Pennsylvania we are reminded of the promise and the hope that has risen from the ashes of that dreadful day. From Independence to Lee's Summit to Kansas City, the people in the heart of our Nation are joining in numerous events commemorating the 9/11 anniversary. One such event, the interfaith observance, "Remembering 9/11: A Day of Hope" reflects the uniqueness of Kansas City as a place known for its river and fountains. Water gathered from the Missouri River and many of the area's fountains will be mixed with water gathered from rivers all over the world, including Tibet and Egypt. The water will be distributed to participants in the observance to be used at other events later in the day symbolically connecting all those present to a spirit of renewal and healing, regardless of religion or creed.

Another common theme throughout the Kansas City area remembrances is the number 3,000 approximating the number of victims in the September 11 attacks: 3,000 flags are to hang at the Kansas City Middle School of the Arts, 3,000 memorial candles are to be lighted at Temple

B'nai Jehudah, 3,000 names are to be read at the bell tower at Rockhurst University.

Mr. Speaker, I join my colleagues in support of H. Con. Res. 464 and echo the sentiments of Margaret Truman, daughter of President Harry S. Truman and native of Independence, MO, that "in years to come we will see September 11 as a turning point in our history as a people, a day of grief and glory that created a new dimension in the soul of America."

Hon. Dave Weldon

OF FLORIDA

Mr. Speaker, on this 1-year anniversary, I extend my deepest sympathies to the families of the countless number of innocent victims of the September 11 terrorist attacks on the Pentagon, the World Trade Center, and in a desolate field in Pennsylvania.

Nothing will ever repair the losses we suffered as a Nation 1 year ago today. The history of the United States changed forever when 19 terrorists hijacked 4 planes and killed 3,000 Americans.

While the events of that morning will forever play in our mind's eye, we endure and we are moving forward. As a united America, we have taken the needed steps to rebuild, to heal, to pursue justice, and to secure our borders.

Since that time, I have been proud to vote for legislation to make our communities safer and our military stronger as we face the challenges of the new century. I commend President Bush for his leadership and I commend the brave men and women of the U.S. Armed Forces in the successful effort to oust the Taliban from power and hunt down those terrorists who perpetrated these acts of evil. I applaud those in our Armed Forces for their continued commitment to pursue those responsible for the attacks.

The words spoken so long ago by President Lincoln seem to fit so eloquently at this moment in history.

> It is for us, the living, rather to be dedicated here to the unfinished work which they have, thus far, so nobly carried on. It is rather for us to be here dedicated to the great task remaining before us—that from these honored dead we take increased devotion to that cause for which they here gave the last full measure of devotion—that we here highly resolve that these dead shall not have died in vain; that this Nation shall have a new birth of freedom; and that this government of the people, by the people, for the people, shall not perish from the Earth.

As we humble ourselves before Him and pray for His guidance, may God continue to bless this great Nation.

Hon. Ted Strickland

OF OHIO

We all carry with us memories of September 11, 2001. It was a profoundly personal day for all of us. No matter what happens in our lives, each of us will forever carry the horror we felt as we watched the unbelievable images on television unfold that fateful day. Even if we had no family or friends who were directly involved, as Americans we were deeply touched by those unspeakable events.

One year later, we observe a day of remembrance. Not an anniversary, but a guidepost: a mark against which we measure how far our Nation has come since that horrific day 1 year ago.

That day, we witnessed the courage of hundreds of heroes who sacrificed themselves in an effort to save others. We felt rage and despair that a few evil men could purposefully steal so many lives, and at the same time we felt deep pride in the courage of our fellow Americans. We witnessed the best and worst of humanity. We dealt with our sorrow by caring for each other.

What began as one of the darkest days in our Nation's history will long be remembered as one of America's finest hours.

One year later, I see a new America. We have not experienced the enormous sea change many predicted, but we greet each day with a greater sense of responsibility to our families, our communities and our Nation. We also know that from now on we will be asked to sacrifice convenience for safety. We now know how vulnerable our freedom makes us.

But we also have a newfound appreciation for those freedoms and a renewed sense of the strength that our liberty gives us. We have been reminded that America's strength as a Nation comes not only from its military might, but from the tightly woven fabric of the American family, symbolized in the stars and stripes of our flag. It is the solidarity of our national family that I hope will once again bring comfort to the families and friends of those who lost their lives on that fateful day.

We should also remember the thousands of Americans who are nobly risking their lives in foreign lands in the effort to wipe out terrorism. This war will continue to be time consuming and costly. But we Americans are a strong and resolute people. We have lived through the dark days of wars past.

As FDR said following the attack on Pearl Harbor,

> No matter how long it may take us to overcome this premeditated invasion, the American people in their righteous might will win through to absolute victory . . . With confidence in our Armed Forces, with the unbounding determination of our people, we will gain the inevitable triumph. So help us God.

May God place his blessing on you and yours, and may God continue to bless the United States of America.

The following is a statement by Alan Wallace. Alan was a firefighter from my district who was stationed on the helipad outside the Pentagon on September 11, 2001. For his bravery that day, Alan was a corecipient of the Department of Defense Fire Fighter Heroism Award. The following is his account of the events that day.

> On the morning of September 11, 2001, I was assigned to the heliport station at the Pentagon. I was assigned there the day before as well. I should have been assigned to the Pentagon fire station all that week.
>
> Fort Myer firefighters were taking a weeklong class on air field firefighting, given at the Classrooms Building 219. Mark Skipper, Dennis Young and I had already had the training. Mike Thayer, John Pine and Ronnie Willett also had had the training. Thayer, Pine and Willett were scheduled off on 9/11. Chief George Thompson was off because his wife was ill.
>
> Mark Skipper, Dennis Young and I were the three crew members assigned to the Pentagon fire station on the morning of 9/11. We arrived there about 0730. The fire station was new and we had only been using it since January or February 2001. We also had a new crash truck assigned there, an Emergency One Titan 3000. It carries 1,500 gallons of water and 200 gallons of 3 percent foam. Our first helicopter flight was around 10 am. But we were expecting President George W. Bush to land in Marine One around 12 noon, returning from Jacksonville, FL. (He had actually left from the Pentagon the day before.) Needless to say, neither flight arrived at the Pentagon that day because of the terrorist attacks.
>
> Mark, Dennis, and I had our turnout gear either on the crash truck or in the station. About 0830 I decided to pull the crash truck outside of the fire station and place it in a position more accessible to the heliport landing site. The truck was then parked perpendicular to the Pentagon, with the rear of the truck 15–20 feet from the west wall of the Pentagon, and the truck facing west, toward the heliport pad. The right side of the truck was approximately 30 feet from the fire station's apparatus door opening. (I forget to mention the Ford van we normally use for transport between Fort Myer and the Pentagon. It is a 15 passenger vehicle which was parked west of the fire station facing north, with its rear about 10 feet north of the apparatus end of the fire station and approximately 6 feet from the side of the fire station.)
>
> The fire station is approximately 75 feet long, 35 feet wide and 16 feet tall. The flight control tower sits above the fire station. There were two other individuals at the heliport site: Sean Berger (U.S. Army Personnel) and Jackie Kidd, both active duty Army.
>
> As I said, we were expecting President Bush about noon, which would be a Code One Standby. In such situations, one of the problems I see at the heliport is that there are too many people there. Plus there are many vehicles, including Secret Service, Pentagon SWAT, U.S. Park Police, Washington, D.C., cops on motorcycles, and the two Presidential limousines. And some of these vehicles even park in front of the fire station apparatus door, blocking the fire truck from exiting the building! That is why I wanted the crash truck out of the station and parked in a good location, for easy access to the heliport in the case of an emergency.
>
> After checking out the fire truck, eating a bowl of cornflakes, and cleaning the station and apparatus area, I sat in my favorite chair in the apparatus area to read a book about opera. About 0900 Mark and Dennis were inside the fire station in the day room. Mark came out to tell me that an airplane had just crashed into the World Trade Center. I then got up and went into the day room to watch the television coverage from New York City. While we three were watching, a second aircraft struck the second tower. I think we watched the TV for about 10 minutes or so.
>
> I then went back outside. I was soon joined by Mark. We both began to work around the crash truck and were talking about the events in New York. About 0920, Chief Charlie Campbell called the Pentagon fire station to inform us of the attacks on the WTC in New York. He actually talked to all three of us: first Dennis, then me and then Skip. He wanted to be sure we were aware of the WTC disaster and that is was definitely a terrorist attack. He wanted to be sure we were aware of everything going on around the fire station. He also said Washington, D.C., could very well be a target and if that happened, our fire truck could be dispatched to an incident.
>
> Let me say this. After the NYC attack, I began to have "second thoughts" about having the fire truck parked where it was. Would it be better for the time being to return it to the fire station until around 1100 or so? But I decided not to move it.
>
> Mark and I continued to mess around the fire truck. The last minute or two before the plane hit the Pentagon, Mark and I were working in the right rear compartment where the

foam metering valves are located. Mark told me how, if you had to, you could get as much as 50 percent foam solution out of the roof turret and discharges. We laughed about cheating the government out of some foam! Mark and I then walked toward the right front corner of the truck. We were side-by-side, always within an arm's reach of each other. We had walked past the right front corner of the crash truck (foam 161) and were maybe 10–15 feet in front of the truck when I looked up toward my left side. I saw a large frame commercial airplane crossing Washington Blvd., heading toward the west side of the Pentagon! The plane had two big engines, appeared to be in level flight, and was only approximately 25 feet off the ground, and only about 200 yards from our location. I later said the plane approached the Pentagon at about a 45-degree angle but later drawings showed it was closer to 60 degrees. The airplane appeared to be a Boeing 757 or an Air Bus 320—white, with blue and orange stripes. Mark later recalled the plane was silver and even identified that it was American Airlines.

So many people think Mark and I watched the plane hit the building. We did not. We only saw it approach for an instant, I would estimate not longer than half a second. Others didn't understand why we didn't hear it sooner. We did not hear it until right after we saw it. I estimate that the plane hit the building only 1–2 seconds after we saw it.

What I am saying is, immediately after we saw it we heard the noise, the engines, I'm sure. I described that as a terrible noise—loud, scary, and horrible. At the time we saw the plane, I said "LET'S GO!" and Mark and I ran away from the area. I turned and ran to my right, going north. (I do not remember which way Mark went, since I did not see him until I crawled out from under the Ford van.)

As I recall, I had several clear thoughts and feelings as I was running: (1) the noise from the engines of the airplane; (2) awareness that now WE are being attacked; (3) planning to run until I catch on fire, then maybe dive to the ground and then figure out what to do; (4) hearing the sound of the plane crashing into the Pentagon, which I later described as a "crunch"; (5) sensation of a lot of pressure; (6) feeling very, very hot very quickly; (7) "we're certainly not going to burn up!"

Later that morning when I began to look at the distances of everything from the fire truck, I thought the plane hit the building 200 feet south of the front of the fire truck. I had only apparently run about 20 feet when the plane hit the building. I ran another 30 feet or so until I felt I was on fire. I thought I had done everything I could do for myself. I decided to get down below the fire and fireball. So I dove face first to the blacktop. At this time, it just happened that I was right beside the left rear tire of the Ford van. (I presume that the debris from the Pentagon and airplane was being propelled away from the impact site.) I immediately crawled very quickly under the van for cover and safety.

At this time, I noticed a lot of heat and decided to crawl to the end of the van. Very soon the heat was unbearable and I decided to get out from under the van and get farther away from the impact site. It was then that I saw Mark Skipper to my left—out in the field 50–75 feet away. He was standing, looking back to the impact site and seemed to be swinging his arms. I immediately ran over to him to ask if he was OK. He said he was, and then said, "I'm glad you saw that airplane!" I said, "Get your gear on—we have a lot of work to do; I'm going to the fire truck."

It was probably at this time that I first noticed the damage to the Pentagon and the crash truck. A lot of smoke was in the sky above the Pentagon. The rear of the crash truck was on fire with a large blaze. But most noticeable was that everything around the fire truck on the ground was on fire. Also the west side of the Pentagon was on fire, all the way from the first to the fifth (top) floor.

I ran about 30 yards back to the damaged crash truck, stepping carefully, not to slip on the burning debris covering the ground. I arrived at the right cab door, opened it and climbed in. I grabbed the radio and put the headset on, then jumped over the radios into the driver's seat. I immediately pushed the two engine start buttons and to my amazement the engine started. I thought if I could pull the fire truck away from the Pentagon and put it in a left turn, I could direct the roof turret nozzle into the impact site using the foam and water on board the truck. I then pushed off the emergency brake and pulled the transmission selector into the drive range and tramped on the accelerator. (I still couldn't believe the engine had started.) However, the accelerator would not make the engine run any faster and the truck would not move. (I later found out from Mark that whenever I tramped on the accelerator, the flames on the back of the truck would flare up.) The window in the left door was open and I had left the right cab door open as I entered the truck. There was a lot of smoke coming up along the left side of the truck, and blowing through this open window and filling the cab with smoke, as well as exiting the right door. There was a fire in the left side of the driver's seat back. That must have produced a lot of the smoke in the cab as well. At some point when I was in the cab, I looked to my right and saw Dennis Young walking through the apparatus area, so I knew he was OK. At another point, I called Fort Myer Fire Dispatch on the fire radio and gave the following message: "Foam 61 to Fort Myer: we have had a commercial airliner crash into the west side of the Pentagon at the heliport, Washington Blvd. side. We are OK with minor injuries. Aircraft was a Boeing 757 or Air Bus 320." It also seemed like I mumbled something else before I removed the headset, shut off the truck engine and began to egress the vehicle.

The fire station was to my right and I noticed it was trashed and there was burning material inside the apparatus area. I saw Mark outside the right cab door signaling me to shut off the engine

(Note: I feel I had the fire truck engine running in 20 seconds after the plane hit the building. This time included running, crawling, checking on Mark and running back to the burning crash truck.)

Just as I was about to get out of the wrecked truck, someone appeared at the cab door asking for a breathing apparatus. He may have been a Pentagon cop. So I handed him one of the S.C.B.A.'s and then handed another one to Mark. Before getting out of the cab. I grabbed my helmet, radio and facepiece (for my S.C B.A.). I carried these items over to the rear of the van, an area I thought would be out of the traffic and easy to find later. Dennis was attempting to use a fire extinguisher on the truck. Mark was removing some of the EMS equipment from the truck. At this time, we all probably thought the truck would be consumed by the damaging fire.

At this point, I went into the fire station through the open apparatus door area and attempted to get dressed in my turnout gear (coat, pants, boots and helmet.) I noticed my boots and pants were covered with debris, with numerous wood, rock and metal fragments filling the boots. One of my elastic suspenders was on fire, which I stamped out (or so I thought). When I was considering how best to empty the debris from my boots, I heard a voice back outside saying "we need help here." I think it was at this time that Dennis, Mark and I began to assemble at the first floor windows of the Pentagon (behind the crash truck).

I was later told by a civilian rescuer that I helped him climb into the window of the Pentagon where most of the victims exited the building. I don't remember helping him up. But I definitely remember him being there. I feel he was instrumental in organizing the rescue effort at this area of the Pentagon. At the time, I described him as a civilian 35–40 years old wearing black jeans and a black polo shirt with a red logo on the shirt.

In April 2002, I learned that the identity of this "civilian" was Blair Bozek. He turns out to be a retired lieutenant colonel from the U.S. Air Force. He was one of the SR71 spy plane pilots. Mark and I always felt 10–15 people may have exited the Pentagon at our location. All were terrified, most were burned. They had had varying amounts of clothing burned from their bodies, and some were missing shoes. We were assisted in rescuing them by several civilians as well as Armed Forces people who, having been uninjured in the attack, had come to aid their fellow employees.

I would like to describe how very hostile the working environment was following the airplane attack. We were directly up against the Pentagon building, which was on fire with smoke pouring heavily from all of the windows. The ground was burning all around us. A magnolia tree was burning, which gave a strange sensation of flaming "things" floating in the air—I later realized they were magnolia leaves. There were several times the heat was so intense that I thought my pants were on fire. It was especially difficult to breathe because of the smoke and fumes. These conditions definitely limited how long we could assist in the rescue.

I do remember helping three men carry an unconscious man all the way out to the guard rail beside Washington Blvd. While carrying him, I noticed the 4-inch firehose from our Fort Myer Rescue Engine 161. That meant our fellow firefighters were on the scene. This was a relief, because after I called them on the radio, I was certain it would be difficult for them to get to the Pentagon because of traffic. But I learned later that RE 161, RE 162 and the assistant chief did not have difficulty getting to the Pentagon.

A further comment about my radio message: I should have followed it up with a call from one of the portable radios or possibly a phone call to Fort Myer from the heliport station phone (had it been in service). I had not waited for a reply from Dispatcher Bob Connelly (more on this subject later).

Unknown to me, before my radio message, Arlington Dispatch was receiving numerous 911 calls from all around the county. Reports were varied: helicopter crash into east side of the Pentagon, tractor trailer on fire on Washington Blvd., possible airplane crash on or near the 14th Street Bridge. Many of the 911 callers could see smoke but could not determine its source. Some likely saw a low-flying aircraft or heard the impact of the crash. Arlington Dispatch advised all listening stations about some of these reports, but of course couldn't confirm the exact location, etc. In fact, it is quite possible that one of these callers, recalling the flight 90 crash into the Potomac River many years ago, was instrumental in causing National Airport to dispatch the first big crash truck. According to the firefighters from the classroom at Fort Myer, immediately after the communication from Arlington, they heard my radio message. Therefore apparently my message was successful in informing my fellow firefighters of the exact location. After victims stopped appearing at the Pentagon windows, Mark, Dennis and I began assisting the arriving Fort Myer companies on the fire ground. My next task was to get into my "fire turn gear." Returning to the rescue site behind the crash truck, again I looked at my fire boots and pants. They were still full of debris, but now the left suspender had completely burned off down to the end where it had been attached to my pants! I picked up my gear and dumped out the rocks, etc., stepped into my boots and pulled up the fire pants. With only one suspender, I must have looked like Jethro Bodine from the Beverly Hillbillies. I also got on my nylon sock-hood and fire coat. I grabbed a big lantern and two fire extinguishers (one CO_2 and the other 20# PurpleK, potassium bicarbonate).

I pulled the safety pin on the CO_2 and placed the lantern under my left arm, walked around the burning end of the crash truck, sprayed some of the CO_2 on it and under it. The extinguisher seemed only about half full, so it was quickly discharged and I threw it aside.

Pulling the pin on the PurpleK bottle, I walked behind the truck and into the Pentagon. Holding the illuminated lantern in my left hand, I immediately notice how poor the visibility was. Keep in mind I still had no gloves, no helmet and no S.C.B.A. I do not think I went into the building any further than 20 feet. I would see fire and spray the extinguisher on it. It makes a very loud noise when being discharged and I did so several times. Out of nowhere, I heard the clear voice of a woman yell "hey!" She had heard the sound of the fire extinguisher and realized she was near another person. She did not sound panicked. I yelled back "I can't see you" and she clapped her hands. I was waving my flashlight. I did not go after her, and later I questioned my courage about why I hadn't.

Several days later, I noticed an article in the *Washington Post* which mentioned me. It also described a woman, Sheila Moody, who heard the swoosh of a fire extinguisher from someone, called out, and was answered by and rescued by a firefighter. I do not remember making contact with her. I believe it was my fire extinguisher she heard, but I also believe she was intercepted by another firefighter. But had I not had the fire extinguisher but had taken the garden hose attached to the fire station, she might not have known she was very near the outside of the building and near rescuers.

I then began to assist the firefighting crews. I got a larger nozzle tip for the attack team and got 50 feet of 4-inch hose off Engine 161 so we could move the deluge gun closer to the Pentagon. Another project I undertook was to begin removing all the equipment off the crash truck: the third S.C.B.A., all the extra air bottles, power cords, floodlights, all the 1¾-inch hose (200 feet of it), tools, and fire extinguishers. At this point, the truck was still on fire and a lot of fire was right behind the truck in the Pentagon. I also noticed that the two personnel vehicles that had been parked near the impact site, belonging to the two Army flight control tower personnel. Both had been completely destroyed by flying debris and fire.

About this time hose line crews from Fort Myer were entering the building with a 2½-inch hose with a 1¼-inch solid nozzle. We added 50 feet of 4-inch hose to the deluge gun. Capt. Dennis Gilroy noticed the first collapse of a cornice above the fifth floor windows, just above the impact site. Dennis Young and I were at the deluge gun and were told to pull back and allow the deluge gun to operate unmanned.

About the time Gilroy ordered our people to get out of the building, there was a report of another hijacked airliner, allegedly heading toward Washington, DC. During this period of waiting, Capt. Gilroy was assigning firefighters to hand line teams to attack the fire, which was beginning to spread to the third and fourth floors of the Pentagon.

By now, I was feeling the effects of exhaustion from the frantic pace and severe shortness of breath from the lack of air at the impact site where we had assisted victims. I thought Mark and Dennis were in the same shape. Mark and I both told Gilroy not to count on us for the hand line crew. Our fellow Fort Myer firefighters had become aware of our injuries and Gilroy called an EMS crew to tend to us.

Our injuries were primarily second degree burns on our necks and forearms. In addition, Mark had a laceration on his hand, Dennis had a sprained ankle, and I had left shoulder pain. (Note: Mark, Dennis and I were only wearing T-shirts, work trousers and boots or heavy shoes at the time of the attack.) A medic unit arrived, Arlington, I believe. They bandaged our burns with wet dressings and wrapped them with gauze. I was given oxygen to breathe; the others weren't experiencing difficulty breathing. We were delivered to the triage area at approximately 1100.

There, we three saw Jackie Kidd and Sean Berger from the control tower. They looked to be OK. Jackie was really shaking and Sean had his forearms wrapped, much like us. When I saw them, I realized I had not thought once about them after the attack. I felt bad about this. Later I thought I would have at least told Dennis Young to "check on the people in the tower" but I guess there was just far too much to think about in the immediate response to the attack.

Sean and Jackie were both given a ride home by a nurse-bystander named Victoria Brunner, who had been working in triage. (She now works at Fort Myer Radar Clinic as a counselor.)

Mark, Dennis and I had a welcome opportunity to rest in the triage area, and were given water, bananas, apples and plums. There were probably 50 health care people there. Triage was located in the tunnel under Washington Blvd. on Columbia Pike.

By now, word of our experience had spread to the FBI who interviewed us, as well as Kidd and Berger, while we were in triage. After our interview, I wanted to return to the fire ground to see all the people from Fort Myer. We did so and spent about half an hour there. Mark, Dennis and I stayed around triage for about 2 hours. During this time, I had a chance to use a cell phone to call my mother in Ohio. She was very relieved to learn I was OK. I also called Donna Houle at the Women's Memorial in Arlington Cemetery and asked her to contact some of my friends. In the next few days, I think I called everyone in my address book.

After all the other victims had been removed to hospitals, Mark and I were taken via ambulance to Arlington Hospital by Army Chase-Bethesda Rescue Squad 1. A young medic trainee named Sandra Melnick drove the medic unit. There were 6 to 8 people in the back of the squad, with one patient placed on a cardiac monitor. I sat in the front with her to give directions to the hospital.

After being released from the hospital, I contacted one of the hospital security officers to request a ride back to Fort Myer. He provided a driver within 5 minutes. Just as we were leaving the hospital, we were questioned by one of the local TV news channels, Fox, I believe. We told them about seeing the airplane approach in time to run away from the Pentagon building.

Our driver took us as far as the Iwo Jima Memorial, just 200 yards from one of the gates into Fort Myer. Of course by now security had been increased significantly since my arrival there at 0530 earlier in the day. (The MPs had shoulder arms and a vehicle with a machine gun mounted on top was nearby.) I was wearing a hospital gown, my fire boots and carried my fire pants in a plastic bag and had no I.D. But fortunately one of the MPs recognized me and allowed us to pass. (Mark did have some I.D.)

As soon as we were allowed to pass through the gate, an Air Force Major gave us a ride back to the fire station in his Jaguar. We were home!

We immediately began to tell our story and help out at the fire station. Dennis was there when we arrived. Soon after, Howard Kelly gave Mark a ride home. Dennis drove himself to his West Virginia home. I stayed at the firehouse that night.

I enjoyed being back with my fellow firefighters and helping get the equipment back on the truck. Our people were exhausted and some were still frightened. I think all were glad they were working that day.

Remember the three firefighters who were scheduled off the day of 9/11? Willett, Pine and Thayer all came back in when they heard the news. Thayer told me later "from 25 miles from the Pentagon, I could see smoke, and I knew you three must be dead." He also said he felt bad because he was the person who had assigned us to the Pentagon heliport.

I was grateful—and am now amazed—that my injuries were minor. The burns on my forearms and neck healed quickly. My shoulder pain persisted and ultimately required surgery in November 2001. The surgery went well and the surgeon and I were pleased with my recovery from it.

I returned back to work in February 2002, glad to have a good job. I am very proud of Dennis, Mark and myself.

I am SO grateful that none of our firefighters were seriously injured or killed.

Hon. Lincoln Diaz-Balart

OF FLORIDA

Mr. Speaker, today we remember.

The pain has not subsided. The memories of those lost will not be forgotten. Today we honor their lives and their sacrifice.

The terrorists have failed. Blinded by their hatred, the true result of the attacks on September 11 were things they could never have anticipated. America responded with courage—not fear. America responded with love—demonstrated by the thousands of rescue workers. America responded with resolve—as we continue to hunt evil doers around the globe.

We must pay our highest tribute to the unsung heroes who have labored this past year—our first responders, our men and women in the armed services. I also would like to thank our President for showing true leadership in the face of this challenge thrust upon us.

Our job is not completed. While we have responded to the events of September 11 we have not taken the final step to ensure Americans are safe. We are faced with great decisions. But we do so with great resolve.

We will continue to show that the values and principles America stands for—the values and principles our brave Americans died for—will overcome those who would kill innocent civilians.

God bless and keep those who were lost 1 year ago today. God bless America—beacon of freedom.

Hon. James H. Maloney

OF CONNECTICUT

Mr. Speaker, a year ago today our Nation was brutally attacked, and thousands of Americans were murdered. Earlier today, I shared a moment of silence with the people of Connecticut's Fifth District, in New Milford, Sandy Hook and Waterbury, to honor the heroes and remember the victims of that tragic day.

On this day, the people of Connecticut's Fifth District honor the brave firefighters from the Danbury Volunteer Fire Department and the Southbury Volunteer Fire Department. Each engine crew performed search and rescue in the hostile and dangerous environment of the devastated World Trade Center. On behalf of the people of Connecticut's Fifth District, I wish to express my deepest thanks to these heroic individuals. The contributions they made to our community and country at the risk of their own peril cannot be measured.

The families and friends of those who perished have endured a year of unbearable loss. They have my deepest sympathy. Rarely have we felt hatred of terrorism perpetrated on our shores, and our response has shown the strength of character of the American people. The sadness that we all felt that day, and in the days since, has hardened into a resolve to honor the memories of those who perished, to heal our wounds so that our Nation is even stronger than before, and to bring righteous justice to those who perpetrated the attacks.

Hon. Jeff Miller

OF FLORIDA

Mr. Speaker, today marks the 1-year anniversary of the most horrendous act of terrorism ever perpetrated against any country. Our Nation will never be the same after 19 terrorists took thousands of American lives and declared war on our great Nation.

Since the infamous day last fall, an outpouring of patriotism and love for this country can be felt in cities and towns from coast to coast. The symbol of America, our flag, can still be seen flying with dignity and honor outside homes and businesses, displaying the true pride this Nation has in its freedoms and unwavering principles.

As we look back at the tragedy of that day, I know that everyone will remember where he or she was on September 11, 2001. I am sure we have all contemplated the frailty of life and that God has never promised anyone a set number of days. It is my hope that the citizens of the

United States will use this time in our history as a catalyst to advance the Nation and to return to the roots to which it was established; faith in God, democracy and patriotism.

Terrorism is still a threat to the civilized world and must be destroyed before it spreads. The primary weapons of terrorism are violence and fear. Those who have no respect for human life and seek terror through these means have no place in civilized society, and must be eliminated.

As President Reagan has said, "We will always remember. We will always be proud. We will always be prepared, so we may always be free." Let us keep the families who lost loved ones in our prayers and continue to support our deployed military personnel who courageously protect our liberties and freedoms. They are the true patriots.

I praise my colleagues and the President for the courage and resolve they have displayed during these trying and difficult times. May God grant us the wisdom to lead this country forward in a manner that would please Him and may He look favorably upon our great Nation.

Hon. Jerry F. Costello

OF ILLINOIS

Mr. Speaker, I rise today as a cosponsor of the Patriot Day resolution, which fittingly expresses the sentiments of this body on the anniversary of perhaps the worst day in U.S. history. In many ways it is hard to believe it has been a year since that awful day—the memories are so fresh, so vivid. I visited the site of the WTC 1 week after it happened and spent time talking with survivors and rescue workers. That memory is just as fresh as those of last Friday, when we returned to New York City to participate in a wreath laying ceremony at the World Trade Center site, and those of this morning, when we attended the ceremony at the Pentagon to remember the victims and families. America will never forget.

September 11, 2001, has left an indelible mark on the American landscape and on our national consciousness. We will never forget the events of that terrible morning, nor will we forget how America responded. We continue to be inspired by the heroism of firefighters, police officers and emergency first responders, our military men and women and other ordinary Americans who have answered the call for freedom.

The attacks on the World Trade Center and the Pentagon were an attack on all of us—our people, our Nation, our spirit, our way of life, our liberty and freedom. The terrorists intended to bring down and destroy the World Trade Center, the Pentagon and other targets—and the people in them. Their real goal was to instill fear, bring about disruption and to bring down and destroy our spirit. But as Rev. Billy Graham observed at a service at the National Cathedral shortly after the attacks—Their actions have done just the opposite. The terrorist attacks of September 11 could have torn our Nation apart—but they have brought us together—we have become a family.

Mr. Speaker, the United States is the greatest country in the world! We have been tested before and we will be tested again. Those we lost last September 11 will hold a special place in the history of our great country. As we gather together today in communities, churches and other places throughout our great land we remain one Nation under God, indivisible with liberty and justice for all! I urge all of my colleagues to support the resolution, and God bless the United States of America.

Hon. Carolyn B. Maloney

OF NEW YORK

Mr. Speaker, it is an honor to serve in the people's House as a representative of New York City, the greatest city in the world.

For the Nation, today is a day of solemn reflection and remembrance. We have all tried to mark this day in our own personal ways. I have just returned from a memorial service at Ground Zero, a sacred place for us all. My thoughts and prayers over this year have been for my beloved city whose residents have been fundamentally affected by 9/11.

Just a year ago, our country witnessed the evil actions of cowards that resulted in more than

3,000 people tragically dead and 2,000 children without a parent.

Mr. Speaker, there may not be another Member of Congress who lost more constituents in the September 11 attack on the World Trade Center than I did. I applaud the House for introducing this resolution celebrating America's resolve and commemorating the lives of those we lost. This resolution, in a small way, can help to continue the process of national healing and renewal.

We will never forget the hundreds of New York City firefighters, law enforcement officers, and EMS who responded to the attacks, and valiantly fought through the terrible conditions to rescue victims and to provide emergency care to the injured immediately after the attack. Tragically, the World Trade Center towers collapsed while these heroes were attempting to save innocent lives—343 firefighters and paramedics and 60 police officers made the ultimate sacrifice.

Over these past 12 months, we have witnessed countless selfless acts by public servants and private citizens, by our friends and neighbors. It is this wonderful spirit embodied by our city and our great Nation that gives us hope.

Hon. Nick Smith

OF MICHIGAN

Mr. Speaker, I remember speaking on the House floor the day after the attacks. I asked, "Will we forget? Will this sick, sinking feeling fade? Will we fail to follow through on these promises or will we demonstrate unfaltering resolve?" I am proud to say, 1 year later, that the American people remain determined to fight the war on terror, and though the road ahead will continue to be hard, we will prevail.

We should reflect for a moment on the lessons in the attacks. In my view, there are three. First, America has enemies who resent our freedom and way of life. These enemies are determined and are ignored at our peril. Second, the oceans do not provide as much protection as they did in the past. We have to be aware that threats can come from anywhere. Third, as with Pearl Harbor, first punches can be devastating. We must not allow ourselves to be taken by surprise again.

Even though we have had a year to reflect since last September, it is still hard to comprehend the magnitude of the terrorist attacks and the historic turn of events that they triggered. In our 226-year history, America has never known an assault on our homeland such as that terrorist attack.

As horrific and sickening as the attacks were, however, they brought out the best in us as Americans. They reminded us that despite all our differences, we are one—a united America. Not only a Nation of unprecedented strength, but also a Nation that exhibits great tolerance and respect for the rights of its citizens as well as those of other nations—a Nation of unshaken spirit, a Nation bound by our shared faith in the founding principles of liberty and freedom.

America was best exemplified by the actions of our first responders after the attacks. On that warm autumn day, the New York City fireman became the symbol of American freedom and American bravery to millions around the world. I believe it is really the one silver lining that shines through the cloud of horror that surrounds 9/11.

It is our job in Congress to honor those who protect us—our defense forces abroad and our first responders here in America—by fully providing the resources and guidance that they need. As President Bush said at the annual fire services dinner in Washington last spring. "There is no substitute for the raw courage of the firefighter." And we must never forget those Americans who made the ultimate sacrifice in saving the lives of others on September 11.

Hon. Jim Matheson

OF UTAH

Mr. Speaker, when we evacuated our office a year ago, and I could see the black smoke rising above the Pentagon, I knew our country was under attack. The inconceivable was taking place.

The passing days brought more heartache than many of us thought we could bear. The hijackings and plane crashes took the lives of three

Utahns—two who were on board the plane that hit the North Tower of the World Trade Center and one who was on duty at the Pentagon.

Mary Alice Wahlstrom, of Kaysville, UT, and her daughter Carolyn, died together—two talented musicians whose families still grieve for the loss of their wives and mothers. Brady Howell of Centerville, UT, died—along with 188 others—when hijackers struck the Pentagon. His family—including his wife Liz—had to endure 1 agonizing week before learning his fate. A year of sorrowful birthdays and holidays has gone by, with a much-loved husband and wife, brother and sister, son and daughter, missing from the family pictures. Their names and their lives will always remind us of the goodness that is America—goodness that the terrorists sought to destroy. But we know that the terrorists will not succeed. Their cowardly attack took the lives of 343 members of the New York City Fire Department, but not the determination of 62 search and rescue team members from Salt Lake County. They rushed to New York, working 12-hour shifts, searching the rubble at Ground Zero for more than a week.

The terrorist attacks left many children without parents—but couldn't destroy their future, as Americans rallied to support a $100 million scholarship fund to someday send these children to college. We stand for everything that the terrorists hate—courage, freedom, compassion, democracy and hope. Even as our memorials and reconstructed buildings rise from the devastation of that day, those values emerge unscathed. We have suffered a loss, but not a defeat. We still mourn, but our faith has not faltered. The wounds are still fresh but the spirit that is America shines through—a beacon of hope for better days ahead.

Hon. Tom DeLay

OF TEXAS

Mr. Speaker, we gather today within this citadel of freedom. This room is where America unites in defense of enlightened self-government. From this place our Nation draws from the strength and wisdom of our Founders.

For this reason, the terrorists targeted this temple to justice as they set out to strike a blow against self-government. And, as the terrorists attacked symbols of might and prosperity, they attacked the spirit of this building. In this way, the Capitol, the Pentagon, and the World Trade Center were all one and the same. This is where the American people exercise the fullest measure of freedom. So, we come together in this hallowed chamber to honor the brave Americans of September 11. As their Representatives, and on their behalf, we launch the debates that chart the course for this country. And, make no mistake, it was the exercise of freedom that terrorists wished to extinguish a year ago.

The terrorists failed. And, if their objective was to compel us into abandoning our principles, there may have been no more spectacular a failure in recorded history.

My friends, the flame of American freedom is burning brighter and hotter on September 11, 2002, than at any moment in our history. And that flame is sustained by the magnificent heroism in Manhattan, in Virginia, and in the air over Pennsylvania. Freedom continues unabated in many countries around the world. Americans are doing extraordinary things in dangerous places that are known and some that we can't talk about. But when we speak of the men and women defending us at this moment, we can say this for certain: The present generation of Americans stands shoulder to shoulder with our proudest generations. And we saw, in Pennsylvania, a stirring example of what it means to be an American: Out of many, one.

Strangers, thrown together by providence, facing certain death, refused to yield in the face of raw evil. Their courage, in the moment of maximum danger, is the essence of what it means to be an American. That's why anyone in the world can become an American. All it takes is a willingness to subordinate our own individual interests to the greater good of the United States. It's a proud tradition of love, tolerance, pluralism, and determination. But we would do a great disservice to the legacy of America's September 11 heroes by casually accepting the passive posture of complacency in the face of danger.

The great lesson from 9/11 is the moral imperative to address dangers before they claim the lives of additional Americans. For that reason, we must stand with President Bush as he marshals freedom-loving people to confront gathering evils. We must actively deny the aspirations of evil groups and dangerous regimes. We must bring justice to the most remote caves where terrorists plot. We must protect America by striking our enemies before they can carry out their schemes within our borders.

This is the great decision before the Congress. This is the defining measure of our future security. And on this question, all of us will be accountable to the people.

Mr. Speaker, we offer our deepest sorrow and solidarity to the families of those who lost loved ones at the hands of evil, 1 year ago. Those who now struggle with grief and loss should know, above all, that they do not stand alone. They should take heart because every American stands beside them. We offer our love and gratitude for the sacrifices and unknown acts of heroism carried out by their relatives and friends. We'll never forget them or what they did for our country.

Thank you and God bless America.

Hon. Tim Roemer

OF INDIANA

Mr. Speaker, I rise in strong support and as a proud cosponsor of H. Con. Res. 464 expressing the sense of the Congress on the anniversary of the terrorist attacks launched against the United States on September 11, 2001.

Nearly 3,000 American lives were lost exactly 1 year ago today when the United States was suddenly and deliberately attacked by Al Qaeda terrorists bent on suicide and destruction of human life. By targeting symbols of American strength and success, these attacks clearly were also intended to assail the principles, values, and freedoms of the United States and the American people, intimidate the Nation, and weaken the national resolve. Although New York, Virginia, and Pennsylvania suffered the overwhelming burden of the terrorist attacks, every State and all Americans were affected and continue to mourn that day. We are united by the events of September 11, 2001, and while passage of 1 year has not softened our memory, resolved our grief, or restored lost loved ones, it has clearly demonstrated that Americans will not succumb to terrorists.

We observe September 11 not only to recognize the tragic deaths of the innocent souls who perished or who were gravely injured in lower Manhattan, Shanksville, PA, or at the Pentagon, but we also recognize this date to honor the firefighters, police officers, rescue workers and those intrepid eyewitnesses of this tragedy who selflessly faced grave danger in order to aid the wounded and dying in the immediate aftermath of the attacks. As the gravest moments came, many Americans, relying on courage, instinct, and grace, rushed toward the flaming buildings in order to rescue people or toward terrorist-controlled cockpits in order to resist their destructive plan. Today we honor the sacrifices and continuing heroism demonstrated by our brave servicemen and women who left family and friends in order to defend our Nation. A year later, many servicemen and women remain abroad, shielding the homeland from further terrorist attacks.

As a member of the House Permanent Select Committee on Intelligence, I am proud to serve on the ongoing congressional joint inquiry. My distinguished colleagues and I have spent considerable time reviewing the material and circumstances relating to the events surrounding last year's attacks. However, many important questions about September 11, 2001, remain unanswered. That is why I support the establishment of an independent, blue-ribbon commission to conduct a thorough investigation and to make recommendations based on its findings so that we never again experience another staggering loss of life on U.S. soil. The American people deserve a more thoughtful investigation and the families of the victims of September 11 are entitled to answers about exactly what went wrong and why.

Hon. James P. Moran

OF VIRGINIA

Mr. Speaker, I rise today on the first anniversary of 9/11, a day which changed America's history.

For the past 12 months, this Nation has collectively experienced a full range of emotion, from the initial fear and uncertainty of that fateful day, to anger and outrage at the loss of American life and the violation of two of our Nation's most recognizable symbols. We have mourned and continue to mourn for the victims of this horrible attack. Their families and friends are constantly in our thoughts and prayers. Embedded firmly in my mind is the image of streams of people who came to the ridge overlooking the Pentagon to pay their respects and sanctified that hill with flowers, candles and notes of remembrance.

Yet, in the midst of all the sadness, Americans have sought an outlet for their grief by renewing their sense of community service and patriotic pride. Our country, which has a strong history of bridging many differences, has become one. In Northern Virginia alone, we witnessed friends, neighbors and colleagues coming together to help rebuild and unite. With the round-the-clock dedication of the Pentagon renovation team, the revival of the Pentagon has served as the quintessential symbol of our country's resilience and renewal. A special debt of gratitude goes to those workers and planners who orchestrated this rebuilding.

As we bear witness to the powerful images and experiences of the past year, we are proudly reminded of what it means to be an American. The heroic acts of the firefighters, police officers and emergency responders who rushed into the inferno of the Pentagon and World Trade Center towers to save lives, touches a special place in all our hearts. It is a place where love of country and for our fellow man is second nature. This unique American spirit is what wills us to go the extra mile and put our lives on the line for what we know is right.

So, Mr. Speaker, on the 1-year anniversary of September 11, let us honor the many sacrifices that have been made by our police, firefighters, emergency responders and our men and women in uniform. Their efforts to heal, protect and preserve this great Nation deserve the utmost respect and admiration.

Hon. Carolyn McCarthy

OF NEW YORK

Mr. Speaker, I rise in support of H. Con. Res. 464, a resolution to commemorate the passing of 1 year since the cowardly, brutal attacks of September 11. I would also like to express my condolences to the families who lost a loved one, along with my reverence for the heroism of New Yorkers, and the American people.

On September 11, as the horrific events unfolded, I watched brave firefighters, law enforcement and rescue personnel from New York and around the country risk their lives to save others. I watched hospitals prepare for the wounded and our Armed Forces go on high alert. I watched a stricken Nation respond by rushing to donate blood and volunteer their time to help the injured. These are acts of honor and bravery that no barbaric act of violence can penetrate. The citizens of New York, and all of America did everything within their power to respond unselfishly and effectively to the attacks. More often than not, the very last fiber of human strength was tested. New Yorkers and all Americans rose to the daunting challenge as one proud, resolute Nation.

Throughout the past year we have witnessed the rebirth of a new America. A stronger, more resilient Nation is determined to eradicate all forms of terrorism. Those who oppose our way of life may try to destroy our buildings, but they will never destroy the sense of pride and love for this country cherished by Americans.

Although the tragic events of September 11 will forever bring sorrow to the families who lost loved ones, they will also serve as a reminder of how Americans unite during difficult times. This resolution reminds us all how difficult it is to kill the American spirit. Honoring the lives lost, as well as thousands of rescue workers who worked tirelessly and bravely throughout this difficult time, is a fitting reminder of what this country stands for. We never forget our own, and we will always fight to continue our way of life.

Hon. Heather Wilson

OF NEW MEXICO

Mr. Speaker, I would like to offer my support for the concurrent resolution in honor of Patriot Day under consideration by the House today.

On this solemn day of remembrance, it is important to gather with our families, our friends, and our communities to reaffirm our love of country and our dedication to upholding the values of freedom and democracy that we hold so dear.

Today, in Albuquerque and across the Nation, we will take the time to honor those who faced danger bravely to save others: firefighters, police officers, and our soldiers overseas who now risk their lives to protect the freedom we enjoy.

In this spirit, I will be in Albuquerque on September 11 with my family and neighbors to honor the heroes, to pray for those lost and their families, to comfort the hurting and to reassure the children.

Hon. David Wu

OF OREGON

Mr. Speaker, our Nation was irrevocably changed that Tuesday morning, a year ago today. The past year has not been an easy one, but the American spirit has carried us through, and our democracy stands strong.

Whether it's Bunker Hill, Pearl Harbor, or September 11, Americans have a tradition of turning disasters into launch points for a better future. The response of Americans and Oregonians on September 11 underscores the strength of our democracy and our commitment to community and freedom. One thousand Oregonians went to NYC soon after September 11 to show that we stand shoulder to shoulder. Hundreds of Oregonians are there today.

Our Nation has endured, and will, in spite of everything, thrive. Today, as we look back and remember who and what we have lost, we feel keenly the sense of security that we no longer take for granted. But we will not trade freedom for security. We will move forward together and build a future worthy of our courageous forebears, and all that they did to bring us to where we are today.

Hon. Cliff Stearns

OF FLORIDA

Mr. Speaker, I rise today not only to commemorate the lives lost September 11, 2001, but also to celebrate the indomitable American spirit that has been displayed since, and finally to remind our enemies that we are not finished with them. It has been a year to the date since an organization of men decided to test the resolve of the American people. With blind faith, unabashed cowardice, and intentions of terror, these men, these terrorists took the lives of over 3,000 men, women, and children. Through this evil act these men hoped to strike deep at our security, to impact the very fiber of our country's spirit. But as the towers fell, American flags rose, defying those who would attempt to shake the balance of freedom and power that we enjoy in this country. On that day the most diverse country on the planet was attacked, and from the twisted and smoldering wreckage arose the bond that has connected us all.

Many claimed that we would never be the same; that an inescapable change had come over our country. On the contrary, I feel that we have changed for the better. We are a stronger country, united in the face of tragedy. The terrorists failed to realize that there is no changing the American spirit, only revealing it. After the 11th, the spirit of America revealed itself all across this country. From the thousands who donated blood, to the thousands more who donated time, resources and love to the task of not only repairing buildings but also repairing the hearts of those who lost loved ones on that tragic day. In the few days following 9/11 we wondered how we would respond to these cowardly acts. It has now been a year since that day, and I feel we have responded quickly and accurately. In the past year we have been able to witness as a country the power and resourcefulness of our armed services. Combining both new and old world tactics we have seen Special Forces mounted on horseback calling in the amazing payload of a B–1 bomber. In a foreign and alien terrain we have seen the men and women of our armed services perform and adapt in outstanding fashion. Mr. Speaker, as cochairman of the Air Force Caucus I realize the increasing importance of our Air Force in current and future campaigns. Nowhere has this importance been more felt than in the precision strikes made in Afghanistan in our war against terror. As we continue to hunt down those responsible, we also continue the rebuilding process both home and abroad. As voices rise today in freedom from the sites of these grizzly attacks, soon also shall memorials rise, as a continual reminder of that day and the way in which we, as Americans have reacted in the year since.

And, as America has reacted this year, we have struggled with the tender balance between security and freedom. While we of course must gird our Nation for safety, are we eroding freedom and curtailing civil liberties and privacy in the process? Our Federal buildings, once the most open of any nation, are becoming barricaded fortresses, with streets closed for blocks around, and loss of access. Airport travelers shed clothing, common tools in their toiletry kit, and their patience in the name of passenger screening. Future airport security measures may chillingly include smart-technology that scans a traveler's identification or body feature, and searches a database including information as personal as financial stability or neighborhood involvement, in the name of determining who is a trusted traveler. State legislatures ponder the Model State Emergency Powers Act, wherein a Governor might be granted powers to quarantine citizens, force immunizations, and seize medical records, in the name of public health. I hope that as we strengthen our Nation, we keep sight of the sublime principle of liberty on which the Nation was founded, and think about the overreaching consequences of binding the cords too tightly. The terrorists attacked our freedom; we should not attack our own freedoms.

Turning from thoughts of ourselves to those of our aggressors, I remind my colleagues that we are facing an enemy who despises our very existence. They are consumed by a hate of a country that, despite its faults, is open to all people regardless race or religion. We operate under principles of freedom, the ability to pursue life, liberty and happiness. As such, our country is fighting with hope against terror, and freedom against oppression. Our enemies will never know freedom, because they are imprisoned by hate, and for that, they have already lost. Former U.S. Secretary of Defense Casper Weinberger stated that "The will of the American people once aroused . . . is capable of accomplishing all the things that have to be done." As long as we continue to maintain a moral high ground in this campaign and take the appropriate and precise responsive measures, the will of the people of this county will know no bounds.

Much has been said and will be said today about what happened a year ago. As observers, we have an obligation to the families and the victims. We must remember our fellow sufferers; that is the salve we offer the families. We also must remember those who terrorize us, as it is judgment on our enemies. Do not forget: It was a massacre—a cold-blooded, well-organized, well-executed, carefully plotted massacre of thousands of Americans. It was perversion—of a faith that preaches peace and tolerance. It was a message—delivered by maniacal men in possession of a perverse theocratic ideology. It was a crime—that must be paid for. It was an invasion—which damaged every sense of safety in every person who tried to sleep that night. For all the many things it was, it was also the beginning of a war that is not yet over.

And as much as there is to say about this day, one thing we have learned is painfully simple. We have learned that whatever false sense of isolation we felt was an illusion. We must remain vigilant and remember that "freedom is not free."

I conclude that we have learned that our lives are but a breath. That our families are more important to us than we ever knew and that protection of our lives and our families may cost us dearly. But we are Americans, and we will prevail.

Hon. Joe Baca

OF CALIFORNIA

Mr. Speaker, I rise in support of this resolution. Today, we mourn, cry, and with clenched fists restrain the raw emotions that are pulsing through our hearts. We look to the ground in sorrow and to the sky in prayer, trying to understand this senseless tragedy. More times than I can remember, I have bowed my head and prayed, asking our Heavenly Father for spiritual and emotional comfort, for those of us that still cannot understand why. We know the names and details of the actions of those terrorists, but that still does not fill the void in our hearts. To the families of those who died last year I can only say, you have the heartfelt sympathy of an entire Nation. And, to our Heavenly Father I would ask to please guard the souls of the dead and let the comfort of His love ease the pain in the hearts of the living.

Mr. Speaker, I am proud to cosponsor this resolution because it accurately captures the emotions in our hearts. We all desperately wish it could be September 10, forever. Yet, I take solace in knowing that nothing endures but change. Things will get better. We have punished those responsible and exacted that punishment with judicious caution, and not unmeasured rage. America has proven that power in defense of freedom is greater than power on behalf of tyranny and oppression. We have demonstrated that power with spirit and resolve.

One year ago, for the second time in our modern history, our Nation was attacked. War was thrust upon us. Undoubtedly, this day will forever live in our memories. Three of my constituents lost their lives on September 11, and my community will never forget their sacrifice: Cora Holland, mother of three and grandmother; Rhonnda Sue Rasmusen, who died at the Pentagon; and Navy Yeoman Second Class Melissa Rose Barnes, who remains unaccounted for at the Pentagon. We as a Nation have pulled together to build our courage and strength, for we are united and our faith will guide us.

I salute those brave individuals, police, firefighters, emergency medical personnel and others who sacrificed of themselves for their fellow Americans.

Mr. Speaker, I ask for the prayers of the American people for those whose lives have been lost. May God grant us the wisdom to continue to steer our great Nation.

God bless America!

On behalf of the people of the Inland Empire of California, I join my colleagues in full support of this resolution.

Hon. Michael E. Capuano

OF MASSACHUSETTS

Mr. Speaker, I rise today in remembrance of those who perished as a result of the terrorist attack on September 11. It is impossible to forget the events of that dark day and difficult to comprehend the grief of the families who lost loved ones in an instant. My deepest sympathies are with them on this first anniversary of the attacks. The loss of so many innocent lives and the bravery of the rescuers will never be forgotten.

Since September 11, Americans have adapted to a new reality—a reality with additional security, higher unemployment, economic insecurity, anthrax and the ongoing war on terrorism. But with this new reality we are also witnessing renewed sense of American pride. September 11 reminded us all to treasure our freedom. American flags fill our streets. Patriotic anthems play on the radio and in stadiums from coast to coast. The principles upon which this country was founded brought us together and the strength and spirit of our Nation will endure this challenge.

Today, in every corner of this great country, vigils, prayer services and memorials will be held to honor the victims of the attacks. As we reflect on the events of a year ago, let us honor the emergency workers, firefighters, police officers, hospital employees and grief counselors who went above and beyond the call of duty that September morning and during the months that followed. We must also remember the airline employees and postal workers whose jobs were changed forever on September 11. Finally, our hearts should also go out to the thousands of children and families in New York, Virginia, and in communities across the Nation and around the world who lost mothers, fathers, brothers and sisters on that tragic day.

Mr. Speaker, I rise today in remembrance of all those who sacrificed their lives on September 11, 2001, and I honor their memory.

Hon. Ken Lucas

OF KENTUCKY

Mr. Speaker, I rise today to reflect on the events of the past year. As I do so, I am reminded that the spirit of America is unbreakable, unwavering, and unshakable. September 11 and the actions that followed have affected all of us. As a Nation, I see we are strong—if not stronger—since that great tragedy struck our homeland. The terrorist tried to break the spirit of America but they failed. As Americans, we are united and we will work together to fight the war on terrorism, to improve our lives, and the lives of our neighbors.

We will always remember the casualties of September 11, the brave firefighters, police officers, and civilians that fought to save the lives of so many Americans. They were the first casualties in the war on terrorism. The heroes of September 11 are not just located in New York, Washington, DC, and Pennsylvania, but all over this great Nation and in the military serving our country overseas. For the Americans that donated time, blood, money, and prayers, they are also the heroes of September 11. These deeds and sacrifices will not be forgotten.

I recently had the opportunity to visit our troops in Afghanistan and I am more confident than ever that we are in capable hands. There is no doubt that we have the best and most professional military in the world. And last week, I participated in a joint session of Congress in New York to honor those who lost their lives on September 11. During our visit we laid a wreath at Ground Zero. From my perspective, both events clearly demonstrated America's renewed sense of solidarity, patriotism and pride.

Although September 11 will be a difficult reminder for all Americans, this is also an opportunity for the Nation to show its strength and its unity. God bless America.

Hon. Rubén Hinojosa

OF TEXAS

Mr. Speaker, I rise today in support of this resolution honoring the patriots of September 11, 2001. A year ago, our Nation suffered a terrible blow. Thousands of our friends and neighbors were lost in an attack by terrorists who despise America and all it represents.

This morning, many of us in Washington came together at the Pentagon observance to comfort and pray for all of those who are experiencing renewed memories of the pain and anguish of that tragic day.

This afternoon, we come to reaffirm our resolve to stand strong for the ideals of liberty and unity.

September 11, 2001, was one of the worst days in our history. It was also one of our finest hours. That day, America showed the world that, through the spirit and courage of the American people, this great Nation did not and will not crumble despite those who try to tear it down.

Many people were heroes that day. Some of their stories have been told, but many acts of courage will never be known. Emergency responders braved fire and flames to climb the stairways of the World Trade Centers in New York City to help people evacuate. Ordinary office workers carried strangers down hundreds of stairs to safety.

At the Pentagon in Washington, DC, military and civilian personnel went into the inferno over and over again to rescue their coworkers who were trapped.

Heroic passengers abroad flight 93 sacrificed their lives on a field in Pennsylvania to prevent the deaths of hundreds more of their fellow Americans.

Yet even while the Pentagon burned and the World Trade Center towers fell, we were already preparing our response to this act of war. On the other side of the Pentagon, the military was making its plans. Within an incredibly short time, Congress came together in a bipartisan manner and quickly passed historic legislation to secure our homeland and our skies.

Many of our allies pledged to stand with the U.S.A. in our war against terrorists. We will always remember those nations who have fought by our side in this war.

A year later, we have accomplished much. The Pentagon has been made whole. The rubble at Ground Zero has been cleared. Al Qaeda has been defeated and stripped of its power base in Afghanistan. Our brave servicemen and women, together with Allied Forces, are far from home, but are proudly carrying out their mission of destroying what's left of Al Qaeda's terrorists.

Today, as we remember the patriots of September 11 and mourn their loss, let us never compromise the ideals of liberty for which they, like so many Americans before them, have died. Let us honor them by remaining strong in our unity and in our diversity. Let us always remember that good overcomes evil and darkness is always followed by light.

The American flag represents freedom and still proudly waves. With our strength and resolve, we shall remain united in freedom . . . "one Nation under God, indivisible, with liberty and justice for all."

Hon. Bill Pascrell, Jr.

OF NEW JERSEY

Mr. Speaker, today, all across this great land, we honor the memory of those who lost their lives as a result of the terrorist attacks of September 11, 2001.

We honor those who were taken from us by cowardly murderers, the very worst of human kind, simply because we are a free people. This 1-year anniversary is indeed a day of deep reflection and remembrance. I am not sure if the American family will ever come to terms with the visions of our brothers and sisters, our sons and daughters, our mothers and fathers, being victimized by the despicable acts perpetrated by the evil and the cowardly. But I am sure that our love of country and our love of each other will help us continue to recover and respond.

We remember and we mourn today. Those we lost will forever be in our hearts and minds. Our lives now are about making sure theirs were not lost in vain, about ensuring their values, their ideals, and their spirit always endure. We will also never forget what we saw in the immediate aftermath of the attacks. We saw—amid the carnage, amongst the destruction—the amazing heights of benevolence and decency and courage that mankind can offer.

What we saw was America.

Within moments of the first attacks, our first responders entered buildings without reservation in an attempt to save others—and they did so knowing full well that they themselves might never exit. Everyday Americans became extraordinary heroes to people they had never met before. Our eternal gratitude will also be extended to the passengers of United flight 93 who prevented it from being used as a weapon against America.

As we mourn the victims and honor the heroes of September 11, we must be resolute in our efforts to ensure that we protect and defend this Nation against all those who would do us harm. And we must never forget what it means to be an American—to cherish the principles of freedom, democracy, and human rights for all. It is what separates us from them.

Across our Nation, in synagogues, Roman Catholic churches, Presbyterian chapels, Baptist meeting houses and mosques, words of comfort, hope and grief will echo from pulpits. At dinner tables across this Nation, families will grieve, and they will love each other. It is what we should do on this day.

America is vast and diverse, but today we are united as never before in our history. The victims of September 11 came from 735 towns and cities in 40 different States, all members of one American family. My district lost wonderful people, brothers and sisters, fathers and mothers, dear friends. As our Nation pays tribute today, I think it is appropriate to enter the names of the 54 individuals from my district who died a year ago today.

You will never be forgotten.

Daniel Affilito, John Candella, Lt. Robert Cirri, Caleb Dack, Antoinette Duger, Edgar Emery, Barry Glick, Emeric Harvey, Howard Kestenbaum, David Lee, Ming Hao Liu,

Robert Murach, Eshtesham Raja, Linda Rosenbaum, John Skala, Jorge Velazquez, Leah Oliver, Paul Lasczynski.

Cesar Alviar, Kyung Cho, Robert Coll, Robert Deraney, Luke Dudek, William Erwin, Tim Graziozo, Zhutu Ibis, Lauren Kestenbaum, Craig Lilore, Joseph McDonald, Ed Murphy, Steven Roach, Daniel Rosetti, Michael Stewart, Douglas MacMillan, Dorota Kopiczko, Catherine Nardella.

Paul Aquaviva, Kirsten Christophe, Michael Collins, Georgette Deraney, John Eichler, Christopher Faughnan, John Graziozo, Donald Jones II, Franco Lalama, Ken Lira, Craig Montano, David Pruim, Leo Roberts, Norman Roosinow, Francis Trombino, Marsha Rodriguez, Robert Cordice, Linda Walker.

Hon. Edward J. Markey

OF MASSACHUSETTS

Mr. Speaker, I rise as one of the hundreds of cosponsors of this resolution today to commemorate the victims of the terrorist attacks on September 11, 2001, and to honor the families who grieve and the heroes who served on that terrible day in American history.

American Airlines flight 11 and United Airlines flight 175 took off for the West Coast early in the morning from Logan Airport in Boston. The Al Qaeda terrorists hijacked and redirected these planes into the Twin Towers of the World Trade Center, a crossroads of international trade and commerce.

In Newark, NJ, and in Washington, DC, similar teams of terrorists aimed flight 77 toward the symbol of American strength, the Pentagon, and took flight 93 toward the symbol of American democracy, the Capitol dome.

We remember and honor the brave men and women aboard flight 93, who overwhelmed the Al Qaeda operatives of flight 93 to prevent a devastating fourth blow to America—an attack on this very Capitol Building. Instead, they crashed that plane into a field in Pennsylvania and saved thousands of others from the tragedy that was visited upon the Pentagon and the Twin Towers.

The resolution we will pass today commemorates this day in American history and the more than 3,000 lives lost—some 93 from the Commonwealth of Massachusetts alone. But the resolution also honors all those who became America's new heroes—the policemen, firemen, rescue workers, medics and volunteers who toiled that day and days afterward to pull victims out of the wreckage.

President John F. Kennedy said at his inauguration, "In the long history of the world, generations have been granted the role of defending freedom in its hour of maximum danger. I do not shrink from this responsibility—I welcome it! I do not believe that any of us would change places with any other people of any other generation. The energy, the faith; the devotion which we bring to this endeavor will light our country and all who serve it and the glow from that fire can truly light the world."

Our new American mission is clear—we must never forget those who died a year ago today on September 11. We must not rest until those who committed these terrorist acts are brought to justice. And we must protect our country, all its citizens, from all that threatens democracy and freedom—for these are the fires that have lit the world.

Hon. Jack Kingston

OF GEORGIA

Mr. Speaker, on this day I believe we have the duty to remember all the ones who died on 9/11 and all the ones who have died in Afghanistan and elsewhere as a result of this war on terrorism.

As one stares at Ground Zero today, it looks like any other large building construction site in any other large city in America. There is raw dirt, a fenced perimeter, earth-moving equipment going about and hard hat workers milling to and fro, and if you didn't know better you'd keep driving by. But upon close inspection, you notice all the buildings around it have brand new facades, you notice also a new bike path, you notice the road has been redone and much of the surrounding infrastructure. Then you notice another building, a historic building, that was located right next door; it's covered with soot, its windows are cracked and it's still boarded up. This building makes a quiet but solemn statement that sets the tone because as you look at this site you know that it's not just any other construction site, there is an eerie stillness about it. Rudy Giuliani has called it a cemetery but it's more than that; it's a battleground, just like Manassas or Gettysburg. A great battle has been fought here and the feeling of reverence one gets is universal.

Like all Americans, I remember that morning's events. I was in Washington, DC. As we watched in disbelief the horror of New York City, we were soon disrupted by an explosion at the Pentagon. We evacuated our building, and went onto a chaotic street scene, where we were told that the Capitol was under attack, that the Mall area had been hit, the State Department and the Sears Tower. Later that night, Congress gathered on the steps of the Capitol and sang "God Bless America." It was a moving American moment. Later in the week, Congress attended a church service at the National Cathedral with Presidents Carter, Ford, Bush, and Clinton. President George W. Bush spoke, as did Billy Graham. Then, on September 20, President Bush addressed the Nation. The sense of Americanism had never been stronger.

But of all these moments and all these experiences, none struck me, nor it seems anyone else in America, as deeply as the photos and images of the firefighters and policemen rushing up the steps of the World Trade Center at 9:30 the morning of the attack. It was there and then at that moment that Osama Bin Laden was defeated. He had underestimated the American spirit as these brave men rushed to rescue people that they did not know, people who they did not see socially, people who probably would not even eat lunch with them, and yet they were Americans, and that was all that counted to the hundreds of firefighters, police officers and public safety workers who put their lives on the line.

Of the hundreds who died, many people don't know that 60 of them were off duty. One such fireman had a 9 o'clock tee time on the golf course. He was already on the golf course, in anticipation of a joyful day of golf, when he heard the news. Without even calling in, he threw the clubs in the trunk of his car and drove to the precinct to report. His body was found at 4 o'clock that afternoon.

At another fire station, six men were getting off duty having pulled an all-night shift. Their fresh replacements were just finishing up with breakfast when the alarm sounded. The 6 new ones and the 6 off duty all jumped on board the fire truck and, of the 12 of them, not 1 made it back. Such was the spirit of volunteerism that day. In fact, one precinct asked the mayor's office to quit sending the call for more recruits since they were already too crowded with men and women who had stepped forward to answer the call.

On this day of observance, we should remember this lesson about being on and off duty. For freedom does not wait for the on duty only. If you and I are to preserve and protect freedom for the generations to come we must do it 24 hours a day 7 days a week. That is the best way to commemorate those who died on September 11, and our soldiers who have died in Afghanistan and everyone else who has suffered and sacrificed for this great land of liberty.

Hon. Darrell E. Issa

OF CALIFORNIA

Mr. Speaker, I rise today to thank the city of Leidschendam-Voorburg in the Netherlands for their act of friendship toward their sister city, Temecula, CA.

As a way to express their sentiments of sorrow and sympathy for the events that occurred on September 11, the citizens of Voorburg have graciously donated the statue "Singing in the Rain" by Frans Kokshoorn to the city of Temecula. The residents of Voorburg donated thousands of dollars to have this statue built and shipped to Temecula for its installation on this day of remembrance.

Mr. Speaker as we reflect on the events of 1 year ago, I would like to join the city of Temecula in thanking the city and citizens of Leidschendam-Voorburg for this genuine gesture of kindness during a difficult time for every American.

Hon. Greg Walden

OF OREGON

Mr. Speaker, today we gather together as one people united in observance of the greatest tragedy in American history. We do so mournful of the staggering loss of life we suffered that terrible day 1 year ago and humbled by the heroes whose courage lifted the spirit of a grieving Nation.

The attacks of September 11 offered us a grim view of the evil capacity of mankind, just as it showed us the triumph of the human spirit and the resilience of the American people. In the heroism of the firemen and policemen of New York, who rushed into burning buildings without regard to their own lives, we saw barbarism met with humanity. In the bravery of Pentagon personnel, who pulled their wounded comrades from the fiery ruins, we saw wickedness met with honor. And in the defiance of the passengers of flight 93, who sacrificed their lives to deny victory to murderers, we saw cowardice met with valor.

While a year has passed since the Twin Towers fell and the symbol of America's military strength was breached, we remain numb to the magnitude of the suffering wrought by evil men. And while our grief subsides with time, it never leaves us completely. The emotions that swept over us that awful day—horror, sadness, fear, and anger—still come creeping back to remind us that the scars of September 11 will never fully heal.

But just as the terrorists dealt us a grievous wound, they also succeeded in uniting the American people like never before. We have renewed our faith in our system of government and reaffirmed our commitment to the spread of freedom and justice around the globe. And we have been reminded that whatever differences separate us, we remain a profoundly unified people.

Mr. Speaker, in the years ahead, the attacks of September 11 will be remembered not merely as an unspeakable tragedy, but as a date that triggered a renewal of the American spirit. As we move forward in our battle against the perpetrators of evil, we will proceed with the unshakable certainty that America's brightest days lie ahead God bless you, and God bless America.

Hon. Bud Shuster

OF PENNSYLVANIA

Mr. Speaker, I rise today to share with you and all of my colleagues a poem written by Mr. Bruce Starr of Warfordsburg, PA, in remembrance of the tragic events of September 11. Mr. Starr's poem eloquently speaks of the spirit and the sense of unity that is America.

I AM AMERICA

(By Bruce A. Starr)

I AM a most magnificent land of dreams with wondrous
opportunity of fabulous wealth.
I AM holding a vision for all of happiness and radiant health.
I AM loving and caring for children of God everywhere, and
my generous sharing is beyond compare.
I AM bringing hope and courage to many for a really fresh,
new start.
I AM the joy of freedom that beats from my heart.
I AM a powerful light of spirit which gloriously illumines
the earth.
To peace in the valley, I am graciously giving birth.
I AM patiently awaiting everyone's communion, for our
gentleness and strength abides in union.
I AM the truth and beauty that sets souls free, and
I AM guarding and protecting your God-given right to be!
For after all, "I AM America!"

Hon. Jim Davis

OF FLORIDA

Mr. Speaker, I rise in honor of the Bayshore Patriots, a group of four proud Americans from Tampa whose patriotism inspired thousands of people from across Tampa Bay to join together this morning for "Flags Along Bayshore: Tampa Remembers 9/11," an event to remember those lost in the September 11 attacks and honor those who protect and serve our Nation every day.

The Bayshore Patriots—Linda Alfonso, Julie Sargent, Julie Whitney, and Bill Hamblin—have gathered every Friday afternoon since September 11 to wave flags on Bayshore Boulevard in Tampa, a major route for servicemen and women who work on MacDill Air Force Base. Through their simple act—the waving of a flag—this group has shown their support for troops in the Tampa Bay area and sent a message that terrorism will not destroy Americans' love of country. The Bayshore Patriots' spirit and dedication has invigorated the Tampa Bay community as more and more flag wavers join the group each week and passing motorists honk their horns in support. Gen. Tommy Franks, Commander in Chief of U.S. Central Command, based at MacDill, has seen the group on this way to work and stopped to show his appreciation for the group's efforts.

When the Patriots decided to organize a September 11 tribute, with the hopes of having all 4.5 miles of Bayshore Boulevard lined with Tampa residents all waving flags, they were overwhelmed with support. Local businesses volunteered time, money and services to make the event possible, and people from every corner of our community signed up to wave flags and participate in the event, which was scheduled to include a keynote address by General Franks, patriotic songs, and remembrances. A steady downpour may have interrupted the program, but nothing could have dampened the resolve of the participants.

The Bayshore Patriots have taught us that we all can make a difference in the war on terrorism. They started as just a few voices calling out in patriotism and support for those impacted by September 11, but today, they were joined by a giant chorus of voices—men, women, and children from all walks of life singing in harmony. On behalf of the Tampa Bay community, I thank Linda Alfonso, Julie Sargent, Julie Whitney, and Bill Hamblin for their inspiration.

Hon. Rosa L. DeLauro

OF CONNECTICUT

Mr. Speaker, I rise in the strongest support of this resolution. Our Nation has endured so much pain—so much hardship and grief since we were attacked a year ago. Simply put, our world changed irrevocably. More than 3,000 lives were lost, and today, America remembers their powerful legacy of courage.

Today, there remains a profound sadness in America, a sadness that will surely endure as spouses, parents, and friends across the Nation continue to mourn their unfathomable loss. But in these last 12 months, Americans have begun the healing process—a process that continues to this day, inch by inch, hour by hour. That resilience is, perhaps, the ultimate symbol of the indomitable strength of the American spirit.

All of us were touched by the tragedy of September 11, including so many from my home state of Connecticut. It was something that once again hit home for me 2 days ago, when I attended a ceremony dedicating a garden to the memory of three brave men from Milford, CT, who perished in the World Trade Center. The ceremony was particularly moving because, in the World Trade Center bombing of 1993, one of these men, Seth Morris, had carried a pregnant woman 103 floors to safety. His was the kind of bravery we now understand is at the core of what it means to be American. It was the same heroism we saw in the firefighters and police officers who ran into the burning buildings while others ran out, and in the heroes on flight 93 who made the ultimate sacrifice to save others. These personal stories are now a part of our ongoing national story.

The anniversary of September 11 serves as a reminder to all Americans that our Nation has changed forever. We are now so much more aware of our freedoms and liberties, our strength of diversity and collective purpose. Our commitment to freedom and our strength as a Nation has never been on fuller, broader display.

As our world has changed, so too have the workings of this great body in the last year. When it comes to protecting our people, Congress has spoken with one voice—powerful, determined and compassionate. Many here will remember when this body joined on the steps of the Capitol to spontaneously sing "God Bless America" on this day a year ago. Then, we said to those who had attacked us, "You will not dampen our spirits, you will not break our will."

And now, after a year of grief, unbearable sadness and the beginnings of the healing process, we have a similar message to share with the whole world: our spirits have not been dampened, and our will will never be broken.

That is what this resolution is about—reaffirming that commitment to protecting our American way of life and our dedication to making our Nation not only safer, but stronger. For representatives of the American people, there truly is no higher calling.

Hon. Barbara Lee

OF CALIFORNIA

Mr. Speaker, as we pause to remember the horrific and tragic events of September 11, 2001, let us honor the memory of the innocent men, women, and children whose lives were lost on that fateful day. The families, victims, and survivors are in our hearts and prayers as we support efforts to rebuild and recover from such senseless, inhumane, and inconceivable attacks.

We are filled with admiration for those who willingly rushed into danger to try to save others—the firefighters, police officers, rescue workers, and ordinary Americans who proved to be most extraordinary. They raced up stairs, they ran into burning buildings, and they brought down a plane to save others.

We pray that our young men and women in our Armed Forces who are putting themselves in harm's way will return safely to their families and friends. In the last year, they have stood watch to keep us safe, and we are profoundly grateful.

To say America suffered a terrible blow is an understatement. Since that terrible day we have slowly been recovering from our profound sense of shock. The walls of the Pentagon have been reconstructed. The terrible devastation at the World Trade Center has been gradually, painstakingly cleared away.

Out of the ashes of loss, we must reshape a future, a world free from horror and hatred, one that offers security for our children and future generations. To shape the future, this better world, let us recommit ourselves to justice and peace. As we rebuild the Pentagon, memorialize the World Trade Center, and journey to a pasture in rural Pennsylvania where the men and women of flight 93 gave their lives so that others might live, let us emerge more dedicated to peace, more aware of the world around us, and more secure.

Let us maintain the spirit of unity, of neighborly concern, of friendliness toward others, and of service that was so profoundly displayed in the aftermath of 9/11, and keep it alive and well. Let us hold on to the spirit that led us to stand in line for hours in order to donate blood because we so wanted to help. Those values exemplify true patriotism and demonstrate what is best about America.

I am reminded of the words of a song which has been sung so often, by so many, which begins, "Let there be peace on Earth and let it begin with me. Let there be peace on Earth, the peace that was meant to be."

Let us remember that hymn as we remember those we lost. Let us keep them as a constant reminder to be our own best selves, to stand up for democratic ideals, to work for peace, disarmament, and security, and to continue to display the love and courage that they shared with us 1 year ago.

Hon. Lane Evans

OF ILLINOIS

Mr. Speaker, the devastating acts committed against the United States on September 11 will never be forgotten. Today we remember those who perished in the attacks and extend our continuing support to their families. We honor and thank thousands of individuals—doctors and nurses, police and firefighters, military personnel, volunteers and blood donors and others—for their incredible acts of valor and courage and service to our Nation. We salute postal workers and letter carriers who were threatened and felled by a threat they never saw. They, too, were innocent victims of these horrific acts.

Our thoughts this day are with the men and women in uniform half a world away. They are on duty to preserve and defend our Nation against the scourge of terrorism. We honor their service and thank them from the bottom of our hearts.

Since September 11, our country has stood united in its resolve to overcome these horrific acts. I and fellow Members of Congress have joined together and supported President Bush in the war against terrorism. We have taken steps to make our country safer, assist those who have been affected by these acts of terrorism, and give law enforcement and the military the resources necessary to protect us from further acts of violence.

Those who carried out these acts can try to attack our way of life and democracy, but they cannot and will not defeat it or destroy it. We will continue to work together to ensure that these acts will never be perpetrated again.

Terrorism can never undermine our national spirit and character. We are a great Nation. We are brave and courageous people. The values that guide us remain unbent and unbroken. They will endure.

Throughout our history, we have met great challenges. In every instance, we have overcome every test, every danger. And each time we have moved forward a stronger, greater Nation with a brighter future.

This solemn anniversary reminds us of a great tragedy. But it also helps reinforce our national strength and what it means to be an American. Our purpose and resolve are undeterred.

Hon. Bill Luther

OF MINNESOTA

Mr. Speaker, the tragedy of September 11 is a stark reminder of the dangerous world in which we live and the risks faced by people who are firmly committed to democracy, freedom and opportunity for all, as we Americans are.

As we recognize the 1-year anniversary of these attacks, it is important to remember and honor the victims and survivors, their families and loved ones. But we should also remember the amazing acts of bravery, kindness and self-sacrifice that took place on September 11: citizens helped each other, firefighters risked their lives to save those of others, Americans participated in food and blood drives and other efforts across the country.

The day was one of unspeakable horror, but also one of triumph. We Americans committed ourselves to gaining from this tragedy. As difficult as the time was, we resolved to work together to become stronger as a Nation.

Events of this past year since the attacks remind us that we can easily lose the spirit of September 11 as we go about our daily business. At times we may have forgotten the feelings of national unity and pride that came in the immediate aftermath of September 11. But in order to continue America's mission in the world, we must continue in that spirit and work together as Americans every day.

I, along with my family and staff, join all Americans in remembering the loss of that day and in thanking Americans for their many contributions in the face of tragedy. To truly honor them and the Nation we love, we must continue in the spirit that followed September 11 and work together to ensure peace, justice and prosperity for all.

Hon. Robert W. Ney

OF OHIO

Mr. Speaker, whereas, September 11 was a day that impacted everyone in the United States of America in a shocking and terrible way; and,

Whereas, September 11 also became a pivotal event that unified all Americans, strengthening our communities and Nation in amazing and inspiring ways; and,

Whereas, the committee of Phil Wallace, Marian Klier, Dorothy Powell, and Marian Martin are to be commended along with the community of Martin's Ferry for seeking to honor and remember those who lost their lives that day; and,

Whereas, this anniversary of September 11 calls for solemn remembrance, gratitude, patriotism, and most importantly a celebration of the indelible American spirit;

Therefore, I join with the residents of Martin's Ferry and the entire 18th Congressional District of Ohio in remembering those who died and thanking those who became heroes with perseverance and American pride.

Hon. James P. Moran

OF VIRGINIA

Mr. Speaker, I rise today on the first anniversary of 9/11, a day which changed America's history.

For the past 12 months, this Nation has collectively experienced a full range of emotion, from the initial fear and uncertainty of that fateful day, to anger and outrage at the loss of American life and the violation of two of our Nation's most recognizable symbols. We have mourned and continue to mourn for the victims of this horrible attack. Their families and friends are constantly in our thoughts and prayers. Embedded firmly in my mind is the image of streams of people who came to the ridge overlooking the Pentagon to pay their respects and sanctified that hill with flowers, candles and notes of remembrance.

Yet, in the midst of all the sadness, Americans have sought an outlet for their grief by renewing their sense of community service and patriotic pride. Our country, which has a strong history of bridging many differences, has become one. In Northern Virginia alone, we witnessed friends, neighbors and colleagues coming together to help rebuild and unite. With the round-the-clock dedication of the Pentagon renovation team, the revival of the Pentagon has served as the quintessential symbol of our country's resilience and renewal. A special debt of gratitude goes to those workers and planners who orchestrated this rebuilding.

As we bear witness to the powerful images and experiences of the past year, we are proudly reminded of what it means to be an American. The heroic acts of the firefighters, police officers and emergency responders who rushed into the inferno of the Pentagon and World Trade Center towers to save lives, touches a special place in all our hearts. It is a place where love of country and for our fellow man is second nature. This unique American spirit is what wills us to go the extra mile and put our lives on the line for what we know is right.

So, Mr. Speaker, on the 1-year anniversary of September 11, let us honor the many sacrifices that have been made by our police, firefighters, emergency responders and our men and women in uniform. Their efforts to heal, protect and preserve this great Nation deserve the utmost respect and admiration.

Hon. Eni F.H. Faleomavaega

OF AMERICAN SAMOA

Mr. Speaker, although the scope and severity of the terrorist attacks on America make it difficult to know how best to memorialize those who were lost on September 11, 2001, I rise today to pay tribute to the passengers of United flight 93 who courageously thwarted an attack on our Nation's Capital.

To the firefighters of New York City who gave their lives to rescue others, I join with my colleagues in saying that you will always be our heroes. To the World Trade Center victims, we mourn your passing. To those who died at the Pentagon, we will not forget you. To every man and woman serving in the U.S. Armed Forces, we stand by you. To our friends and neighbors across the globe, we thank you for supporting us in a time of need. For every American who has made the ultimate sacrifice and those who continue to risk their lives in order to save others, our Nation stands forever grateful.

We are one Nation, under God, united in our resolve to defend freedom in the aftermath of the September 11, 2001, terrorist attacks on America. As President Franklin Roosevelt said, "We will not only defend ourselves to the uttermost but will make very certain that this form of treachery shall never endanger us again. With confidence in our Armed Forces, with the unbound determination of our people, we will gain the inevitable triumph, so help us God."

Mr. Speaker, I commend President George W. Bush for his leadership in securing our homeland and strengthening America's resolve to triumph over terrorism. I also commend the Honorable Colin L. Powell, U.S. Secretary of State, for his untold achievement in strengthening our alliances. I commend the Honorable Donald H. Rumsfeld, Secretary of Defense, for mobilizing our troops and protecting U.S. interests overseas. I commend the Secretary of Transportation, the Honorable Norman Y. Mineta, for his decisive action to ground all planes and avoid further tragedy on September 11, 2001.

I also commend my colleagues in both the House and Senate. I commend both Republicans and Democrats. I commend all Americans united in their resolve to end the threat of terrorism for future generations.

On behalf of the people of American Samoa, I rise today to say that we will always remember the heroic actions of those who gave their lives so that we might live. We stand united in our resolve to defend freedom. Like all Americans, we join in prayer and proclaim September 11, 2002, as a day of solemn observance.

I commend the Honorable Tauese Sunia, Governor of American Samoa, for proclaiming Wednesday, September 11, 2002, to be a day of solemn observance throughout the territory. All flags will be flown at halfmast. Memorial services will start in the territory at sundown on Tuesday, September 5, 2002, and will end with the last service set for 6 p.m. on September 11, 2002.

During this time, American Samoa will participate in a worldwide choral event. Choirs in every zone around the world will perform Mozart's "Requiem" at 8:46 a.m., the exact time of the first terrorist attack on America. American Samoa will represent the last time zone on the globe and the American Samoa Community College Choir will be the last choir on Earth to sing Mozart's "Requiem" during this worldwide, commemorative service.

I applaud the volunteers from the Seattle Symphony Chorale who organized this worldwide event to pay homage to the victims, survivors, and heroes of September 11, 2001. I also commend the students, staff, and faculty members at the American Samoa Community College for representing American Samoa on this historic and solemn occasion.

I also express my deepest gratitude for those serving in the U.S. Armed Forces during this critical time in our Nation's history. I am pleased to say that the sons and daughters of American Samoa serve proudly in the U.S. military and, per capita, there are probably more soldiers in the U.S. Army from American Samoa than any other State or U.S. territory.

I thank the sons and daughters of American Samoa for answering the call to serve. I pray for them. I pray for their families. I am painfully aware of the sacrifices they are making. I am very mindful of the dangers they are facing. Some 30 years ago, I served in the Vietnam war. As a Vietnam veteran, I remember all too well what it is like to be separated from loved ones. Each day, I wondered if I would ever see my loved ones again or if I would be among the thousands to return home in a body bag.

By the grace of God, I returned home safely. I now pray that the Good Lord will watch over the brave men and women of American Samoa who are also willing to pay the ultimate sacrifice so that future generations may live in peace.

On this day of solemn observance, in this somber time of remembrance, my thoughts and prayers also go to all those whose lives have been changed by the tragic events of September 11, 2001. May we always stand together in the defense of freedom and may God bless America.

Hon. Hilda L. Solis

OF CALIFORNIA

Mr. Speaker, I rise today in remembrance of one of the most horrific events in our Nation's history.

Today we honor the thousands of innocent people who lost their lives in the World Trade Center, at the Pentagon and aboard flight 93 a year ago and salute with great pride the many rescue workers, medical personnel, and firefighters who risked their own lives to save the lives of others.

September 11 is a very emotional day for Americans of all walks of life and it is especially difficult for those who were directly impacted in one way, shape or form. My heart goes out to the survivors, their families and all who were affected—emotionally or physically—by this event. I know it has not been an easy pain to bear.

Last year's terrorist attacks shook the sense of security we have come to take for granted in our daily lives. Although our country will never be the same, we have a renewed sense of commitment and are dedicated now, more than ever, to upholding our freedoms and liberties. Our country, with all its diversity, stands united to prevent such a tragedy from happening again.

Mr. Speaker, I ask for a moment of silence in remembrance of those who were lost. May God bless America today and ever more.

Hon. Louise McIntosh Slaughter

OF NEW YORK

Mr. Speaker, I rise to join my colleagues in remembering the terrorist attacks of a year ago and paying tribute to the victims, the survivors, and the American spirit.

One year ago today, almost 3,000 Americans lost their lives in a series of despicable attacks. These acts were carried out by a group of people who hated everything our Nation stands for, and who sought to destroy the symbols of our freedom and prosperity.

Despite the destruction they were able to inflict, these terrorists actually achieved the opposite of their intended goal. Instead of dividing us, they united us as never before. The evil of a few was met by the courage of thousands, and the generosity of millions.

The day of the attacks witnessed countless instances of unflinching, selfless courage at the sites in New York, Pennsylvania, and the Pentagon. Workers in the towers and at the Pentagon helped each other through the evacuation. Firefighters, law enforcement officers, and emergency personnel rushed to the scene, heedless of the danger to themselves. Teachers shepherded children to safety; not a single child at the World Trade Center or Pentagon day care centers was harmed. And no one will ever forget the heroism of the passengers who crashed flight 93 rather than allow it to continue to its intended target. The world watched in humbled awe as ordinary Americans performed extraordinary acts.

The following days and weeks saw another quintessential American trait expressed: our generosity. Millions of Americans contributed goods, services, and funds for the rescue effort. Equipment, food, and supplies poured into the City of New York. Over a few short weeks, millions of dollars were donated to the families of the victims of the attacks. Children held penny drives and car washes; businesses had fundraisers; corporations opened their coffers. No sooner was any need made known than it was met, often to overflowing.

The trials of September 11 proved that our Nation's motto still holds true—"E pluribus unum." From many, one. From many individuals, many cultures, and many ideals we stand together as a single Nation, united in purpose and resolve. Like steel tempered in fire, the challenges of this ordeal have made us stronger.

Today, we pay tribute to all those who were lost a year ago, and to those they left behind. My district of Monroe County, NY, bears its share of the collective grief. Pittsford businessman Thomas Duffy was at an early morning meeting in the towers and perished. The Vincent family lost their 24-year-old daughter, Melissa, who had just begun a career with Alliance Consulting. Several other constituents lost children, brothers, and sisters. Many lost friends and loved ones. None of these families will ever be the same again.

The best homage we can pay is to ensure that these people did not die in vain. We have already taken significant steps by removing the Taliban regime in Afghanistan, pursuing terrorists across the globe, and improving homeland security. But we can and must do much more.

Many vital security needs still must be addressed. We must pursue a long-term strategy for rooting out terrorism and eliminating the conditions that allow it to thrive. Our public health infrastructure must be rebuilt and strengthened. Perhaps most of all, however, we must rededicate ourselves to principles of freedom and democracy. Our precious liberty can never be taken for granted. We must find the delicate balance between protecting our security and preserving our freedom.

Finally, we must pay special attention to the needs of our children in these difficult times. Too many of our children across the Nation were traumatized by the terrorist attacks, and many remain anxious and fearful. These events must not be allowed to poison an entire generation. We must ensure that our children receive the aid they need to face the future happy, healthy, and secure. Terror is indiscriminate, and the young are especially vulnerable.

I join my colleagues in paying tribute to all the heroes of September 11. Our honored dead will not be forgotten. Their families shall not be alone. We, the survivors, will carry their memories in our hearts and live their legacy through our actions.

Hon. Gerald D. Kleczka

OF WISCONSIN

Mr. Speaker, it was 1 year ago that we first shared that sense of inimitable sadness upon learning that thousands of our fellow Americans were suddenly and unexpectedly lost to us forever. They met their ends in the fields of Western Pennsylvania; at our national military headquarters in Washington, DC; in two of the world's tallest office buildings in New York City. Some were sitting down for work, or simply taking a plane trip when the unthinkable occurred. Hundreds of others, heroes, were taken from us as they selflessly struggled to bring others to safety.

We came together today to honor their memory. In churches, synagogues, mosques, schools, homes, and workplaces across the land and around the world, people will observe moments of silence in solemn remembrance. We pray for those left behind, whose lives are scarred forever with the loss of loved ones. We ask God for healing for them and for our country.

We also gather as a Nation to lift our voices in song. While uniting us across the miles, our shared music today not only pays homage to the lives lost, but reaches out to embrace the grieving. The songs give us a shared strength by allowing us to publicly reaffirm the triumphs of our humanity over terror, of community over hatred, of rebuilding over destruction.

May today's remembrances bring honor to the memory of those who died 1 year ago, healing to the wounded of body and spirit, and the blessings of courage and strength for all who remember.

Hon. Melissa A. Hart

OF PENNSYLVANIA

Mr. Speaker, on this fateful day last year, the cowardly acts of terrorists tried to divide this Nation and destroy the American way of life. Instead they united us, and Americans rose above the ashes to show the indomitable spirit that makes this Nation so great.

Thousands of lives were lost that day, but millions of us answered the call of a Nation under attack. The volunteers who helped the victims and families, the outpouring of donations—these are the shining examples of what we are capable of when our country needs us most.

President Bush asks us to honor the memory of those lost "by pursuing peace and justice in the world and security at home." While September 11 was a tragic day, we must also acknowledge the historic outpouring of charity and sacrifice by all Americans to those in need.

The war on terrorism is not concluding; it has only just begun. President Bush made it very clear that whenever there is terrorism in the world, the United States cannot rest. This is a war that we must remain united in—united and prepared for the challenge to defeat those who use terror and fear to oppress and destroy.

Those who perished on September 11 will forever remain in our memories and in our hearts. It is up to us to protect liberty and freedom for all future generations.

God bless America.

Hon. David D. Phelps

OF ILLINOIS

Mr. Speaker, on the night of 9/11/2001, when all the Members of the House and Senate gathered for that historic press conference to assure the American people that we had not run from the terrorists and were doing the business on behalf of the American people, that which we were elected to do, I was inspired to write these words and set them to music.

I believe my song expresses much of the sentiments of the American people that we will and must respond when threatened or harmed. We will defend the deepest principles of freedom and our Nation's heritage.

Especially on behalf of those brave people, our heroes: the firemen, policemen, emergency medical teams, our postal workers, the Pentagon workers, and those who were aboard the hijacked planes, Here We Come! With you in our minds, and in our hearts, and for everything which this country stands!

HERE COMES AMERICA!

The greatest of all nations, where freedom was born
through wars and sacrifices tested, tried and weary worn.
We stand for truth and justice, and our aim is strong and sure.
The red, white and blue waves on for freedom we shall endure

Here comes America on strong.
God bless America's her song.
United more than ever now.
In prayer to God we humbly bow
for freedom's cause we will not fail.
Over fear we shall prevail.
Let Old Glory wave.
HERE COMES AMERICA.

When liberty is threatened, we'll defend the people's will.
Though heroes have fallen, our resolve we shall fulfill.
A forgiving, loving people, pursuing peace and happiness
but if harmed or terrorized, comes the eagle from her nest.

Hon. Sam Graves

OF MISSOURI

Mr. Speaker, on September 11, 2001, America awoke to the worst terrorist attack in history. As we went to work and school, we left with a feeling of security that we have long since forgotten. By the time we returned to our families, our lives and our Nation had forever changed. It had been many years since America felt so insecure, so vulnerable. On that morning, the American people's resolve was put to the ultimate test. Everything appeared to be so uncertain that day. Who would do such a thing? Why would they do it? Is there more to come? How can I protect my family?

But there was much that was certain that day. America made a promise to the victims and their families, to future generations of Americans, and to the world. The American people promised that this action would not go unanswered. We promised that this action would only strengthen and unite us, not divide us. We would respond forcefully to those who were responsible while tending to our neighbors, our fellow countrymen. Together, you and people across Northwest Missouri and our Nation donated blood for the victims, and donated money for their families. Together, we prayed for those who lost so much that day. We prayed for our soldiers who stood ready—preparing to defend our freedom.

As we stop to remember that terrible day, some of the pain and fear has subsided. But our determination to defeat those who seek to terrorize us must never fall victim to the passage of time. In the coming months, the American people will face a choice: live up to our responsibility by making tough choices and sacrifices to continue our assault on terrorism, or quit now and hope that they choose to stop planning future attacks. The American people should never have to endure such a tragedy again. As we have learned over the past year, we can do something about it. We must never mislead ourselves that we have to wait to be attacked again to continue our defense from terrorism. The more than 3,000 lives lost is all the justification we need to have to defend against a certain threat of terrorism. The United States must remain vigilant and prepared so that we remain forever free.

Hon. Mike Pence

OF INDIANA

Mr. Speaker, the United States of America is founded on the fundamental principle that all citizens have the inalienable right to life, liberty and the pursuit of happiness.

The United States of America stands as a beacon of freedom and opportunity for everyone regardless of race, creed or religious belief.

The strength and vitality of the United States of America is in the diversity of its people, the diversity of its ideas, the freedom to express those ideas and the opportunity to achieve one's potential and direct one's destiny.

Mr. Speaker, these ideals and principles are absolute and will not be surrendered or weakened by the cowardly acts of terrorists who fear the sunshine of freedom and the responsibility it brings.

Let us forever remember that the date September 11 reaffirms the principles for which the United States of America was founded and that on this day each year freedom shall ring from every community in this great land and the voice of America will be heard around the world.

Hon. Xavier Becerra

OF CALIFORNIA

Mr. Speaker, a year ago I stood on this same floor of the people's House as the ruins still smoldered, the families still prayed and hoped, and all of us searched for ways to explain and prepare for the events that were to unfold. I will never forget that feeling of collective will which permeated this Congress to act to bring to justice those who committed these heinous crimes.

America will never forget September 11, 2001. In its tragedy, in its despicability, and in its lessons and impact on our way of life, it represents a singular moment of history. But what really do we remember about this event? And for today, the first anniversary of 9/11, perhaps the best question is: What should we remember?

Mr. Speaker, my answer to that is: Let us remember the mothers and sons, the brothers and sisters, who perished on September 11, by remembering, today, to touch our own mother or son, our brother, our sister.

We will never forget the firefighters, police officers, and volunteers—the heroes of September 11. Let's prove that by not forgetting to extend a hand to our heroes in uniform, who protect us today.

And let us commemorate 9/11 and honor our fallen by forever defending and living up to the ideals embodied in our Constitution. Our way of life may have been challenged, but it has not been compromised. We can prove to the world that in triumph or tragedy, we shall be a Nation of laws.

As I stood on the floor of the House a year ago I asked: How do we explain this barbarous act of terrorism to our children? I did not believe then, nor do I believe now, that it is possible to really explain—to make sense of what happened—to our little ones. But I remember my words that followed and they ring so very true to me today: "From my words and my deeds, from the way our country prosecutes this unconscionable crime, I hope {our children} will learn and they will remember how this country lives and breathes its freedom and relentless search for justice. Perhaps, then, as our children grow older and wiser, they will be better prepared to preserve life and defend America's values."

So, Mr. Speaker, as we move forward to complete the unfinished business of 9/11, let us remember our fallen, let us reflect on our tenacity and perseverance, and let us be the heroes in life that the victims of that day are in death.

I remember that feeling of hope and justice back then, and I believe it will guide us through our mission now. May God bless America.

Hon. Jane Harman

OF CALIFORNIA

Mr. Speaker, I rise in support of the commemorative resolution honoring those who died last September 11, those who came to the rescue, and those who served, and continue to serve, in the fight against terrorism.

Our minds are still seared with the images of last year's tragedy. Members knew some of the individuals who died in the attacks. In California's 36th District, where LAX—the destination of three of the four hijacked planes—is located, four of my constituents were killed.

They are: Anna Alison of Torrance, Chandler "Chad" Keller of Manhattan Beach, Stanley Hall of Rancho Palos Verdes, and John Wenckus of Torrance.

Today, we remember these individuals and join with their families in commemorating their lives and their contribution to our community and Nation.

We also remember several other individuals who worked in the 36th District but lived elsewhere. Their coworkers remember them fondly: Ruben Ornedo, who worked at Boeing; and Peter Gay, who worked for Raytheon Electronics and commuted to El Segundo weekly from the East Coast.

Mr. Speaker, our hearts are still broken and we continue to grieve. The remembrance ceremonies I attended—last week at Ground Zero and this morning at the Pentagon—were incredibly moving and heartfelt.

But as we work together to rebuild the lives shattered by the events of September 11, we also look to identify the actions we need to take at all levels of government to ensure maximum preparedness and protection against this threat—beginning with a reorganization of our government's resources.

We have learned from this horrific experience.

Sadder, but wiser and stronger, we are aided in our efforts this past year by the testimonies of those who walked through hell and by those who are prepared to do the same.

I am reminded of the words of Tennyson—

Tho' much is taken, much abides; and tho'
We are not now that strength which in old days
Moved Earth and heaven, that which we are, we are—
One equal temper of heroic hearts,
Made weak by time and fate, but strong in will
To strive, to seek, to find, and not to yield.

New Yorkers, the workers at the Pentagon and the passengers and crew aboard American flight 77 did not yield.

Nor will America.

Hon. Solomon P. Ortiz

OF TEXAS

Mr. Speaker, September 11, 2001, is a day you will always remember where you were, who you were with, and what you were doing. My personal experiences that day are vivid: going to a press conference in the Capitol at 9:30, moving fast to get out when we saw the reports of smoke at the Pentagon and seeing the monitor in the House radio-TV gallery flash, "White House being evacuated." Nothing can describe the feeling of rushing out of the building you love, hearing the attack was not concluded and seeing the smoke from the Pentagon rising behind the building in which my congressional office is located.

I went to the Pentagon 2 days later to thank the emergency workers and was struck by the smells that were still so strong there, the stench of the burning building and literally the smell of death. I'd been to that building so many times and it seemed so impenetrable. This attack served to illustrate how vulnerable we are in this country. We live in a wildly dangerous world; the security increases we see here now have been the norm in Europe and around the Middle East for decades. Our world is dangerous. We live in this world; now we appreciate that danger every day.

Our Nation was founded on the belief that God is great, that He is watching out for us. While the fear and hurt was apparent that day and the days that followed, so too, was the strength and courage of men and women who risked their own lives for their country and their American family that day. The passengers in the plane that crashed in Pennsylvania jumped into action, fighting a battle that eventually saved a Washington-area target, quite possibly the building where Congress meets.

Our legacy, our duty today is to ensure that those who died on the planes, in the buildings, and on the battlefield, did not die in vain. Our respect for their memory must be to recommit ourselves to our Nation, our freedom and each other. Today we mourn anew the lives lost in the attack 1 year ago, and the lives lost in battle since then and we offer our sympathies again to those they leave behind. We also honor the police, firefighters and rescue personnel who risked their lives to help those hurt in the attacks.

Our challenges are huge. Many people in this world carry evil in their hearts and minds. We can never change the hearts of humans, but great American men and women are doing their best every day to prevent that evil from finding its way here to our shores again. The United States is a beacon of light around the world. There are thousands of freedoms and privileges in this country that we enjoy every day but those privileges come at a cost, at a sacrifice. Our way forward will not be easy, for this is not over.

Ask God for wisdom and strength to protect us here at home and those around the world who love freedom and democracy. Always remember those servicemen and women serving in the military. They are doing difficult, dangerous work on our behalf, and on behalf of freedom and democracy. Our test is to be united at this hour of crisis in our Nation and in our world. Congress will honor the memories of all those we have lost to the war on terror by defending the United States in the face of future attacks.

Hon. George R. Nethercutt, Jr.

OF WASHINGTON

Mr. Speaker, 1 year ago on September 11, our country suffered unexpected and unjustified attacks that killed and injured innocent people from many nations of the world. While the attacks occurred in the City of New York, a peaceful field in Pennsylvania and at the Pentagon in Virginia, the effects of terrorism were felt in every corner of America and throughout the world of nations that shared our sorrow.

The resulting war on terrorism has called together the people of America to unite behind a commitment to defend our homeland and preserve our way of life against all enemies of freedom and liberty. In doing so, America has witnessed a heartening resurgence of patriotism, a deep appreciation for the ordinary heroes among us, a fervent call to prayer, and a thankful devotion to the simple blessings of family, community, faith and friendship. We are indeed a blessed people, committed to liberty for individuals and nations everywhere, but mindful that freedom too often comes with great sacrifice.

Who can doubt our American faithfulness and resolve as we grieve for those who lost life on September 11, applaud the brave government servants and military might of our great Nation, strive for economic stability and quietly pray for a peaceful world free of tyrants and violence.

The United States is no stranger to evil—our forefathers have overcome it through 225 years of proud but difficult history. The challenge for this and future generations will be to never succumb to the temptation of withdrawing from our national obligation to resist evil and fight for freedom for future generations of Americans. The war against terrorism, memorialized in those who died 1 year ago, is only the latest test of our national resolve. In lasting memory of September 11, let the legacy of this new century be one of victory for mankind as the United States of America leads the world to liberty and justice for all.

Hon. John B. Larson

OF CONNECTICUT

Mr. Speaker, I rise today, 1 year after the terrorist attacks on our Nation, as we collectively pause to reflect, remember, and memorialize those who lost their lives on September 11, 2001. We proudly remember those heroic acts of bravery and courage selflessly performed by citizens that day. It was a day when ordinary people performed extraordinary deeds.

The Nation was struck by terror, but responded with the courage and the conviction that makes us free. As we pause to recognize our precious freedoms, we will never forget September 11 and what happened in New York City, the Pentagon or in the fields of Pennsylvania.

The terrorists failed in their attempts to bring down this Nation because the United States responded with love of country and the resolve to bring them to justice, ensuring that those who gave their lives will not have perished in vain.

With great compassion and understanding, our hearts go out to the families of those who lost their loved ones, knowing life will never be normal for them. As we pause today in remembrance, we are resolved to go forward protecting and cherishing the freedoms that sustain us all.

Mr. Speaker, H. Con. Res. 464 appropriately recognizes the 1-year anniversary of September 11 and I urge my colleagues to unanimously support this resolution. God bless America.

Hon. Randy "Duke" Cunningham

OF CALIFORNIA

Mr. Speaker, a year ago today our Nation was attacked by terrorists intent on destroying the most enduring symbols of our success and our way of life. On this solemn occasion, the anniversary of those horrific attacks, I rise in strong support of the resolution recognizing September 11 as a national day of remembrance.

On this day, we mourn the loss of thousands of innocent lives, we honor the selfless acts of those who came to the victims' aid, and we pay tribute to those who willingly put their lives on the line to stop the terrorists from wreaking further destruction on our land.

September 11 will forever be remembered as one of the darkest days in our Nation's history. Today the horror of that event is still fresh, and the pain is still raw. And just as it is today, it will always be one for sadness and reflection. But the response it inspired in our Nation was uplifting. As we shudder at the recognition of devastation on this day, we can also take heart in the greatness and strength it inspired.

Time will ease our sorrow, and we will continue to prosper as a Nation. But we must never forget the lesson we were forced to relearn on September 11, 2001: there are those who are threatened by our strength, our freedoms and our way of life, and they want to destroy our Nation. We must remain united in our commitment to pursue those who threaten us, to ensure our way of life and to uphold the hope of freedom around the globe.

Hon. Rob Simmons

OF CONNECTICUT

Mr. Speaker, 1 year ago today our Nation endured an attack by a dedicated and dangerous enemy. Since that day we have seen that the principles and ideals that are the foundation of America are far stronger than any of the steel and concrete that fell that day.

We are gathered to remember the innocent lives that were taken. The district that I represent, the Second District of Connecticut, lost a number of wonderful people on September 11. The names of some of them follow:

Josh Piver, of my hometown of Stonington, worked at Cantor Fitzgerald on the 105th floor of the World Trade Center. Josh loved jazz, loved living in New York City and had an exuberant passion for life. He was a fine young man with a bright future.

Madeline Amy Sweeney, a flight attendant for American Airlines flight 11, the first jet to strike the World Trade Center. Displaying a courage while under tremendous pressure, she proved that on that day a loving wife, and a caring mother and daughter, could also be a hero. She used her cell phone to report what was happening and the information she provided helped us identify the attackers. She jump-started our investigation. Her parents live in Norwich, CT.

Ruth McCourt, a homemaker from New London, CT, and her 4-year-old daughter, Juliana, were on United Airlines flight 175 bound for Los Angeles. They were on their way to enjoy a vacation at Disneyland. Their jet crashed into the South Tower of the World Trade Center.

James Greenleaf, Jr., a 32-year-old history buff, was a football star in high school. He worked on the 92d floor of the World Trade Center. His friends said he was the type of individual who would spend a week's vacation helping an old friend build a new house.

Ed Calderon, 43, was a security guard for the Port Authority of New York and New Jersey. He loved to dance and especially loved the salsa. He had worked at the World Trade Center for 22 years and was last seen running toward the North Tower after helping dozens of workers reach safety. He was hoping to help a few more people just before the building collapsed.

Bruce Eagleson, 53, a vice president of the Westfield Group, was at a meeting on the 17th floor of the World Trade Center when a jet struck the building. His oldest son called after the first plane hit and urged him to leave. He told his son he was helping to evacuate people and promised he would get out. Mr. Eagleson had hoped to retire within 7 years. He loved fishing and golf.

Eric Thorpe, 35, was the star quarterback of his undefeated high school football team. He was one of the top salesmen at Keefe, Bruyette and Woods. But he kept success in the business world in perspective. He helped run a soup kitchen during college, worked as a Big Brother and participated in Hands Together, an anti-poverty program in Haiti. His friends knew him for his nonstop sense of humor.

Eric Evans, 31, was an easygoing individual who was determined to succeed in business. He also loved to garden and enjoyed tending his tomato and basil plants. His friends said he loved animals, except for the squirrels that would get into his tomatoes.

These fine people, and all the others who were taken from us on September 11, 2001, leave behind families and friends and lives that were full of the promise of tomorrow. Today it is appropriate that we honor their memory.

But this is more than a day of remembrance. This is a day to recommit ourselves to the values that are the foundation of America. Freedom, justice, honor and an unwavering belief in self-government—those are the values we believe in and they are the values that those we remember today believed in as well.

This is a day to keep in mind that there is a great and profound difference between the use of force to liberate and the use of force to enslave.

And this is a day to recommit ourselves to our love of our country. Let us join together and appreciate America's history and stand firm in support of our institutions and the duties of citizenship. This is a day to look to the future.

A great deal of learning in our Nation has traditionally taken place when families gather around the kitchen table at mealtime. I hope that beginning tonight all American families will take time to discuss today's events around the dinner table. Talk together about what it means to be a citizen of this great Nation. Share your thoughts with each other about what the events of today mean to you, your family and friends.

John Winthrop, one of the Pilgrims who came to this new world, described it as a "shining city upon a hill."

Today, with our prayers, we remember those who are gone. Let us also direct our efforts to ensure, for our sake and for the sake of those yet to come, that this shining city on a hill will remain a beacon of freedom and hope that will forever reach out to embrace the aspirations and dreams of all the people of the world.

Hon. Jo Ann Davis

OF VIRGINIA

Mr. Speaker, a year ago today, a radical, Islamic terrorist group seized four aircraft, turning them into deadly weapons and killing thousands of defenseless people. Today, we continue to mourn the loss of those victims, and honor those who, with great bravery and instilled sense of duty, risked their lives to protect our people.

Yet while we reflect on this day, 1 year ago, we cannot look back. Throughout the history of America, we have been a forward-looking Nation, striving for excellence and finding strength in our love of country. America is a God-fearing land, and because of this, our Nation has been blessed mightily.

These attacks have been compared to December 7, 1941, when Pearl Harbor was also struck from the sky. At the end of that day, America was devastated and struck with the great sorrow of this tragedy. Yet as history so aptly tells us, America's resurgence brought forth the liberation of the world, and global peace for decades.

Today, as we look out across our land, we will see America at its most glorious hour. We will see the spirit of America at every ceremony, in every city and on the face of all Americans. It was the spirit of America that got us through that fateful day 1 year ago, and it will be this spirit that will carry us through for years to come.

It is contagious, this American spirit. Last year, as our firefighters, police officers and rescue workers sacrificed to save lives, the American spirit shined. As our military men and women headed overseas to defend our country, the American spirit shined. As Americans across our land joined together in unity and with a renewed sense of patriotism, the American spirit shined.

This is America, and faith and freedom will always be our call. That is what makes us unique. That is what makes us a people of great pride and resiliency. And that is what makes us a target. Yet in the end, America endures.

Today the war on terrorism is progressing, but it will take time. As the President has said many times—this is a faceless enemy that we fight, and determination and perseverance will be our keys to victory. In the end our victory will once again bring about global peace.

So as we reflect on that tragic day, 1 year ago, and mourn the loss of so many of our fellow Americans, rest assured that our brave military men and women overseas are getting the job done, and making America proud.

I commend and thank our military men and women for their sacrifices and bravery as they protect the homeland. They are picking up where the heroes of September 11 left off—defending America, and fighting for the freedoms that we are willing to die for. The terrorists who attack us operate out of hatred—hatred of our freedom, hatred of our faith, and hatred of our liberty. Yet in the end, it will be our faith, freedom and love of liberty that will ultimately defeat them.

May God bless the United States.

Hon. Dana Rohrabacher

OF CALIFORNIA

Mr. Speaker, today America is reflecting on the events of 9/11, as we should; and it is a time to remember. It is a time to remember those who lost their lives and to remember those, like the firemen and the police, who gave their lives trying to save others. It is a time for rage, as we have heard, and a time for reflection, a time for pride and a time for anger. What it is not, however, is a time for mourning. That time is over.

Today, I join with all of those who solemnly commemorate this anniversary. Heartfelt commemoration, I suggest, is not enough. We must consider not just what happened a year ago, as we have been hearing for the last hour; but instead we must find and discover and talk about and we must make determinations about why 9/11 happened.

As a Nation, we are now engaged in a historic global conflict with a vile enemy who slaughters innocent people by the thousands and then makes sanctimonious references to God. Talk about blasphemy. I do not know if bin Laden is dead or alive; but I do know that when he dies he will burn in hell, and it is our job to get him there as quickly as possible.

Our President laid down a battle plan that brought the liberation of Afghanistan and will soon rid the world of threats like that of Saddam Hussein. This is a result of 9/11 one year ago, but it did not start one year ago.

The first order of business is for us to recognize that the murderous attack on us in New York and at the Pentagon was not an act of God, nor was it a natural phenomenon. It did not just happen; nor, let me add, was it just a case of bad luck.

The slaughter of our fellow citizens need not have happened. It was something that would not have happened had certain people done things differently, had certain government policies been different, had certain Federal agencies and departments been given different marching orders. In short, 9/11 need not have happened, and it is imperative that the American people look closely at the policies, the systems, and yes, the people which led to 9/11 to ensure that something like this never happens again.

What policies am I talking about? Let us start with the fundamentals or, if you will, the fundamentalists. Of the 19 hijackers on 9/11, 16 were Saudis or held Saudi passports. America's relationship with Saudi Arabia is complex but not as unfathomable as some would have us believe.

In the cold war, we worked closely with the Saudi royal family; and to be fair, they were our loyal allies. They helped us finance anti-Communist projects that were of immense importance to our national security in the days when the Soviet Union was spending billions of dollars to bury us. Saudi help was vital on a number of fronts; and, yes, there is reason today for us to be grateful.

What they did to help us in the past, however, does not excuse what they are doing today that threatens us. Times have changed, and dramatically so. If our policy toward Saudi Arabia does not change significantly, there will be a heavy price to pay in the future, if we have not already paid enough.

Relying on low oil prices and on Saudi largesse for special cold war projects left us dependent upon them, and who is them, who are we talking about? We are talking about the royal family, the royal family of Saudi Arabia that is autocratic and over the years has become fat and incompetent and in many ways cowardly. However, again, they helped us defeat an enemy intent on destroying us, communism. So we paid special attention to the Saudis.

Instead of pushing for democratic reform and human rights, we let the Saudis, and because of their influence, much of the Muslim world in general, we let them off the hook in our push for democracy and human rights.

In the short term, it makes sense. In the long term, it has had a dramatically bad impact, negative impact. Young people in that part of the world have suffered under despots and crooks; yet we Americans continually talk about stability in that part of the world, when what we should be pushing for is democratic reform and the opening of closed societies.

Entrenched regimes, royal and secular, have been brutal and corrupt. Is it any wonder that young people in a large chunk of the world turn to Islamic fundamentalism as their idealistic alternative? In their corrupt world, radical Muslims have been the only ones offering a morally based alternative, but radical Islam is not a positive force. It is tyrannical, arrogant and malevolent.

Right here we should note that most forms of religious extremism are equally reprehensible and that radical Islam should not be singled out. Although limited to a few loud voices, a drumbeat started right at September 11 to paint all Muslims as the enemy of the United States and of the West. That drumbeat started the moment those planes hit the World Trade towers; but thanks to our wise President, we did not succumb to a strategy of hate.

Bin Laden wanted us to retaliate against Muslims in general, which would have polarized hundreds of millions of people against us, many of whom would have ended up supporting bin Laden and his terrorists as their saviors. As I say, we did not fall into that trap.

By the way, just to put things in perspective, in the decade leading up to 9/11, Muslim people saw their fellow Muslims being ethnically cleansed, raped and murdered in Bosnia by thugs calling themselves Christians. They saw their fellow Muslims repressed and murdered by the tens of thousands in Kashmir by people who called themselves Hindus and were cut down in the Middle East by the Israeli Army. Hundreds of thousands of non-combatant Muslims have lost their lives due to the actions of governments controlled by people of other faiths. So from their perspective, Islamic people are no more terrorists than others.

In the West, all we see is the frightening picture of planes flying into buildings and suicide bombers blowing up Pizza Huts in Israel. So the first policy we need to change is that which has us tolerating dictatorship and corrupt governments in Muslim countries in order to maintain stability. Working with Russia, which is now our friend, and trying to build a democratic society, let us break our dependency on oil from unfriendly and democratic and undemocratic anti-Western governments. Let us seek out reformers in the Arab and Muslim world. Let us demand free elections and freedom of speech and press as well as religious tolerance in those Muslim countries.

Over the last 2 decades, the Saudi establishment has dealt with the rise of their homegrown religious extremists by ignoring them, giving them a free hand overseas and by sending them to Afghanistan.

Their extremists are called Wahabis. Those folks are on the outer limits of Islam. They are the ones who insist that women must cover themselves from head to foot. Now, that is okay if women voluntarily accept this religious mandate. Instead, however, the Wahabis act as if they have the right to control everybody, even those who do not accept their particular view, claiming to have an infallible insight about the wishes of God. They beat women with sticks if so much as their ankles are showing. They feel free to commit violence against people of other faiths and to prevent anyone with a different belief in God, even other Muslims, from worshipping and living their lives as they see fit.

These are the most radical of all Muslim sects. Instead of standing up to this religious gangsterism, the Saudi royal family allowed them to establish their base of operations in Saudi Arabia and to export Wahabi radicalism throughout the world, with the help, of course, of billions of petrol dollars.

One of the places not just influenced but under the control of the Wahabis was Afghanistan. The Taliban was not an indigenous religious sect of Afghanistan. That is the mistake so many people make. They represented a transplanted Wahabism. Transplanted from? Where else. Saudi Arabia.

These crazies did not represent the character and/or the values of the Afghan people. The Afghan people are devout in their faith but they are not fanatic. They pray and are grateful to God, but they do not feel compelled to have everyone else pray, much less feel compelled to make everyone else pray just like they pray.

I have seen this tolerance first-hand, even in the most desolate regions of that distant land. Years ago, 14 or 15 years ago, actually, I was in Afghanistan with a mujahedin unit, the mujahedin being the fighters against the Soviet occupation. During long treks across the desert, the small group of mujahedin fighters I was with would stop and pray five times a day. They would get on their knees and they would pray, and they would thank God for everything that they had. I might add that they had little. We did not even have a good clean glass of water, much less the provisions of food that could keep people healthy. Yet these people were grateful for everything.

It caused me reason to pause and think that here in the United States we have so much and how rarely people think about how grateful they should be for what we have. But here were these people, under attack by the Soviets, on their knees praying. But there were many other people in the surrounding area and with our group. About half of them were not part of the praying during those prayer sessions. They stood there.

What impressed me is that those who were praying felt perfectly comfortable. They were fulfilling their obligations to God but did not feel threatened by these others who were not praying and who were not compelled to participate. That was the essence of the Afghans—grateful to God, devoted to God, but not fanatics who were trying to suppress other people into some sort of religious dictatorship.

The attitude of the Taliban in Afghanistan, of course, was totally different from the type of attitude I am talking about. And it was not a result of the susceptibility of the people to the Taliban's form of Islam as much as it was a result, meaning the Taliban's ascension to power, of what is naturally in the Afghan people's hearts. Instead, I believe, it was the result of a deal between Saudi Arabia, Pakistan, and, unfortunately, the United States.

It, of course, goes back to the cold war, when the United States was helping the Afghan freedom fighters in their struggle against the Soviet army that occupied their country. The Saudis were helping, too. Now we helped, and we can be proud of that. The Saudis were also helping, but as I discovered, it was not quite that simple.

As I was hiking through Afghanistan with that mujahedin unit heading toward the battle of Jalalabad, which was one of the last battles the Soviets participated in in Afghanistan, we came across an encampment of white tents. These were very expensive tents. There were off-road vehicles there. The people were well fed, well clothed. And I was told by my mujahedin fellow freedom fighters to keep my mouth shut and to speak no English because this was an encampment of a crazy psychopathic killer, a Saudi named bin Laden, and bin Laden would kill all of us if he knew there was an American with the group because he hated America as much as he hated the Communists.

And much of the support that the Saudis gave to the Afghan freedom fighters was right there. It was actually bin Laden and his group there fighting against the Russians. And that was their contribution to Afghanistan in the fight against the Soviets.

Well, after the Soviets withdrew from Afghanistan, after the mujahedin victory, we can be proud we helped them fight off the Soviets by giving them the weapon systems they needed, but we did not help them at that point rebuild their country. In fact, America simply walked away and let them sleep in the rubble. We did not even help them dig up the land mines that we had given them to defeat the Soviet army. And so young kids, little kids from Afghanistan have been blowing off their legs ever since. And they cannot even treat their young people. They do not have the medicines to do so because we have not been there to help.

There was an agreement, however, as we left. It was probably not a formal agreement; probably just an understanding to let Saudi Arabia and Pakistan oversee that region. So we walked away from Afghanistan and the entire region. Instead of insisting on a government that reflected the will and values of the Afghan people, we left them in the hands of the Saudis and the Pakistanis.

For several years, there was chaos and fighting. Not as bad as before, but there was fighting that continued, and the Saudis then unleashed their ace in the hole. We had left, but the Saudis had been preparing for this eventuality. The term taliban means student and refers to those who spent most of the war against Soviet occupation not fighting the Russians. That was a whole different group of guys. That was the mujahedin. No, the Taliban were in schools, so-called religious schools, in Pakistan. Later, they emerged from these schools seemingly out of nowhere, but in fact were trained, armed and financed by Saudi Arabia and Pakistan.

Within 6 months, they had conquered over two-thirds of the country, including Kabul, the capital city. But just as it was in Orwell's *Animal Farm*, vicious dogs were surreptitiously nurtured and then suddenly unleashed to do the bidding of pigs.

Just a reminder: Many pundits fail to understand the difference between the mujahedin and the Taliban. The former fought the war against Soviet occupation troops. That was the mujahedin. The latter, the Taliban, arrived on the scene much later. And in the end, the same mujahedin who helped defeat the Soviets were our allies in this last year in driving the Taliban out of power. The mujahedin, the good people of Afghanistan, have stood with us twice. Let us pledge that we will not walk away from them again. Let us help them rebuild their country.

Let the *Record* show that I had spent a year trying to prevent the Taliban from coming to power at that time. My goal right after the end of the war with the Soviets was to try to bring the old King Zahir Shah back from his exile in Rome. Zahir Shah was one of the most beloved and pro-Western of his people. He was anxious to serve as a transition leader that would lead his country to a new political system that was based on democratic elections; on ballots instead of bullets. As I say, he was an honest, kind man, with a good heart, and respected by all the people of Afghanistan.

Instead, the King was pushed aside, or should I say he was kept on the sidelines. And I might add that our own State Department played a major role in ensuring that this positive alternative did not come to power. Instead, the Taliban assumed power with the acquiescence if not the support of the Clinton administration. Knowing there was nothing more I could do, I hoped for the best. I tried my best to prevent the Taliban from getting into power. Now they were there, and our government seemed to be going along with it, so all I could do is hope for the best.

However, within a month or so, the tyrannical ways of these religious kooks made it clear to me and to everyone that they had to go. Yes, it was clear to me, but I take that back, it was not clear to everyone, because the Clinton administration could never seem to come to that conclusion, that the Taliban had to go. In understanding who should be accountable for 9/11, we must understand that the State Department, under President Clinton, was never anti-Taliban. Our State Department, probably under the President's direction, undermined those efforts aimed at undermining the Taliban. So those of us who were anti-Taliban found ourselves the target of the State Department rather than having the State Department target the Taliban for their misdeeds.

In several personal instances I was involved with helping obtain medical and humanitarian support for people in the areas of Afghanistan that was not yet under Taliban control. I was thwarted by our own government. I was thwarted by our own State Department. NGOs with aid for Afghans who were in areas that were controlled by the Taliban, on the other hand, had no trouble with our government. They had some other troubles that, of course, the Taliban gave them themselves, but our government was perfectly happy to have NGOs operating in Taliban-controlled areas but stopping people like myself who were trying to help those people in areas that were opposed to the Taliban.

In mid-1988, however, even with this tacit support from the Clinton administration, the Taliban were incredibly vulnerable. They had overextended themselves in an invasion of the northern part of Afghanistan, and many of their best, if not most of their best, fighters were captured, along with huge amounts of war supplies.

The road to Kabul was open. And who interceded to prevent the collapse of the Taliban at this pivotal moment? Who pulled their chestnuts out of the fire? President Clinton, personally.

At this moment of maximum Taliban vulnerability, the White House dispatched Assistant Secretary of State Rick Inderfurth and Bill Richardson, then our U.N. ambassador. They flew to northern Afghanistan and convinced the anti-Taliban forces not to attack and not to retake Kabul, but, instead, to accept a cease-fire and an arms embargo.

This is the moment, and I cannot stress this more forcefully, it was a pivotal moment. The Taliban could easily have been defeated. The Northern Alliance was willing to accept a return of King Zahir Shah to lead a transition government. Instead, under the direction of the Clinton White House, these two top U.S. Government officials, Assistant Secretary of State Rick Inderfurth and U.N. Ambassador Bill Richardson, arrived on the scene to convince the anti-Taliban forces to stand back. And we thus saved this fanatical, anti-Western regime from being destroyed and being defeated.

This later led to a dramatic defeat of the anti-Taliban forces. The cease-fire lasted only long enough for the Saudis and the Pakistanis to fully rearm the Taliban. And the arms embargo that Bill Richardson and Rick Inderfurth talked about was only effective against the anti-Taliban forces, which are the people called the Northern Alliance. Think about that. We talked them into a cease-fire, which lasted only long enough for the Taliban to rearm. We talked them into an arms embargo, which was only an arms embargo against them.

Again, this was one of the major turning points that led to 9/11. Later, the Taliban, with their supplies replenished, went on the offensive and turned their country into a staging area for terrorism. So the Taliban ended up, somewhat with the Clinton administration's blessings, of taking over all but a sliver of Afghanistan. That portion, of course, that little sliver, was under the command of Commander Massoud, who stood alone in the Panjir Valley, a hero against the war on the Soviets. Now he was all that was left to resist the tyranny of the Taliban.

This is where bin Laden makes his official entrance. Behind the scenes, his foreigners, his radicals, had been there and been the Taliban shock troops for a long time. They murdered anyone and everyone who got in the way and ran roughshod over people all over Afghanistan. Bin Laden had already declared war on the United States, and had already killed military personnel and bombed U.S. embassies. The Taliban permitted them to use their country as a base of operations.

Yes, the Clinton administration repeatedly demanded that bin Laden be given up or at least kicked out of Afghanistan. They were using all of these words making demands, yet they never seemed to care enough to help Massoud or help any of the others who wanted to resist the Taliban.

So what was the Taliban leadership to think? Well, of course they thought that the U.S. Government really did not mean what it was saying. They believed it was simply posturing for domestic political consideration. This is like when the Clinton administration went to China and demanded human rights reform and then never put any type of force behind that demand.

So our government made it clear to the Taliban by our inaction to support anyone who was opposing the Taliban that our demands on them actually were just made for public consumption here, and that we were actually more concerned with our deal, whatever that deal was, with Pakistan and Saudi Arabia and that we were more concerned with that than anything going on within Afghanistan, including bin Laden. Why would the tough guys in the Taliban think that we cared about human rights abuses, about their treating women like cattle, about their harboring of terrorists like bin Laden, and about their rejection of even a consideration of free elections of any kind when we were not doing anything about it? We did not, as I say, support Massoud; and, in fact, when several of us tried to help those resisting the Taliban, it was our government, the State Department, that got in our way.

Let us be fair about it. If that is the impression the Taliban got, we should admit it. Our government at that time was not serious about democracy, human rights and such in Afghanistan. We were not serious about their form of government or even their harboring of bin Laden because our government in that administration did nothing.

What all this means is that if we stray too far from our basic principles as a country, it is going to end up hurting us. If we stray too far from the fundamental principles that make us Americans, a love of liberty and justice, a belief in the democratic procedures to guide men, and permit people to guide their own destinies and secure their own destinies through election processes, if we ignore these principles, it will come back to hurt the United States of America.

Over the years, I complained over and over again; and I will submit for the *Record* quotes of mine that warned America that we must act against the Taliban. I did this for years.

Well, obviously there was another policy. I am just a lone Congressman. I do not make policy. I try to influence policymakers. But my warnings, repeated warnings, were not heeded.

Well, who was responsible for the policies that left the Taliban free from domestic rivals, the policy that left them free from outside opposition, that left them free from the pressure to democratize and respect human rights? Who was responsible for these policies? How about Madeleine Albright? How about President Clinton? They could not get themselves to endorse any meaningful action against the Taliban even after we had been attacked in Saudi Arabia, seen in the blowing up of our military bases there, our military installations, our living quarters there, or the blowing up of U.S. embassies in Africa.

Furthermore, there is ample evidence that in the last administration they passed up promising opportunities to take out bin Laden. I, for example, several years ago during the Clinton administration contacted the CIA to let them know that I had an informant who knew exactly where bin Laden was, that he was out of Afghanistan, and that he was willing to pinpoint bin Laden for them. I gave them my contact's phone number. They never called. After a week, I called my friend back and asked, "Did the CIA get with you?" His answer was "no."

I went to the CIA again and explained that this person had impeccable credentials of knowing what was going on in Afghanistan. They said they would get to him, but they did not. A week later they still had not called. Then I complained to the chairman of the Permanent Select Committee on Intelligence, the gentleman from Florida (Mr. Goss), whom I respect; and I told him what happened.

The next day he had a meeting in this building with representatives of the NSA, the CIA, and the FBI. It was the bin Laden task force. I told them what had happened and that my friend could pinpoint bin Laden, and that he had been ignored for 2 weeks. They would get to it.

Guess what, a week later my friend still had not been contacted. By then the trail was cold. But when I went to the gentleman from Florida (Mr. Goss), it got action and my friend was called. He said it was a lackadaisical call. It looked like it was a pro forma call.

Does that sound like an administration committed to getting bin Laden? No. Let the *Record* show there were numerous opportunities to get bin Laden and not one was exploited. The Government of Sudan tried to give the United States a complete file on bin Laden and his whole gang. Madeleine Albright personally turned that down.

I know of a situation at the Defense Intelligence Agency where a young analyst felt there was a lack of information about Afghanistan and that lack of information was threatening to our national security. She wanted to get the information. She wanted to go up to Massoud's territory and find out what was going on because we did not know what was happening in Afghanistan. She was denied, and she had the gall on her own time, on her own vacation time, to go there to Massoud's stronghold to try to get that information. I think someone like that should get a medal. Instead, she was fired.

I personally asked the general who then headed up the DIA not to fire her. She got the ax anyway. By the way, there is no indication that the DIA, the Defense Intelligence Agency, warned anybody about the attack on 9/11, even though the murder of Commander Massoud 2 days prior to the attack in New York should have set off alarm bells. Of course they had fired the one person who was conscientious about Afghanistan. They had fired that person for being too conscientious, over the objection of a Member of Congress who pleaded that that was the type of responsible behavior we needed.

I say this because the death of Commander Massoud had a special significance to me. I had known Commander Massoud for many years, even before I went to Afghanistan in 1988. During my time in the White House, he sent his brother to me; and we continued a communication through third parties over the years. He was a man I deeply respected. He was a hero; not to say he did not make mistakes. Certainly he made mistakes, and he did some things wrong. But over years of fighting, everybody makes mistakes. But Massoud was a hero. He was a giant of a man.

Mr. Speaker, 2 days before they attacked us, they murdered Massoud. It took the wind right out of my sails. I had been to his stronghold 5 years before. I visited him in the mountains of Afghanistan. Our friendship was close, and I respected him. We worked out an agreement that would have King Zahir Shah return and Massoud would support that if the King would lead a transition government and have honest elections 2 years later. He was willing to support that, and then the Taliban killed him.

After I had gotten myself together after his death, I knew that it must be because they are going to attack the United States. That is why the Taliban killed him, so we could not have anyone to turn to, to rally behind in our counterattack. So the next day I called the White House. I asked to speak to Condoleezza Rice, and I wanted a meeting with her and the National Security Council because there was an attack that would soon befall the United States of America.

They got back to me, and said, Congressman, we take your opinions on Afghanistan and elsewhere very seriously, but we are very busy. Can you come tomorrow? The earliest we can fit you in is 2 tomorrow. I woke up on 9/11 expecting to have a meeting with Condoleezza Rice and the National Security Council at the White House to warn them that there was an imminent attack planned on the United States and to take seriously any possible threat that they saw. Unfortunately, at 8:45, the planes began crashing into the buildings in New York.

So here we are. One year ago our country was blindsided, attacked without warning, resulting in the slaughter of 3,000 Americans. As I have just discussed, this represents a failure of policy and a failure of the people behind that policy, primarily those in the Clinton administration, not because of politics, but because they happened to be there at the time. Who knows what would have happened if it would have been a Republican administration. It was George Bush who walked away originally and left the Pakistanis and the Saudis in charge of that region. But it was during the Clinton administration that the Taliban took over, consolidated their power in Afghanistan, and turned that country into a base of operations for anti-American terrorists. The American response is undermining those who oppose the Taliban.

This leads me to my conclusion that our policy was part of an agreement with the Saudis and the Pakistanis to keep the Taliban in power. The attack, however, reflects more than a failure of policy. It reflects more than just that policy. The attack was carried out by a terrorist organization that we had been told over and over again was the number one target of U.S. intelligence. That organization was able to launch an attack of this scope and of this magnitude requiring millions of dollars and the coordination of hundreds of people against the United States. The number one target of U.S. intelligence was able to slaughter 3,000 Americans, to blindside us. This represents a catastrophic failure of America's intelligence system; it is a failure of the DIA, the CIA, the NSA, the FBI, and the rest of the intelligence alphabet soup here in Washington, DC.

We spend tens of billions of dollars every year, and the number one target of American intelligence is able to organize and pull off an operation of this scale. The magnitude of the screw-up boggles my mind.

Now we know there were warnings. The BBC is reporting that just 2 months before 9/11, the foreign minister of the Taliban was so upset about the terrorist plot that he had heard of that he sent an emissary to an American consulate in nearby Pakistan to warn the United States of a pending attack.

But no one listened to him. Then we know of FBI field agents who were pleading that attention be paid to the terrorist ties of certain students who were being trained to fly airplanes. These FBI agents were chastised for going around channels. They had to go through channels, but they were so concerned that the people in front of them were not acting, they tried to get the attention of Washington but were chastised for not going through channels and they were ignored. The list of failures goes on and on.

I will just say on 9/11, that something like that happened to me indicates the type of mindset we are dealing with, even after the attack. On 9/11, when the planes had already crashed into the buildings, I realized, everyone realized it was an attack from Afghanistan, based on the terrorists based in Afghanistan, and I called the King of Afghanistan. I wanted to know if there was anyone there protecting him.

"Do you have any police there protecting you?"

"No."

Remember, the King of Afghanistan is in Rome, exiled in Rome. "Are there any policemen outside your door?"

"No, there aren't."

"Are there any people inside your compound with you protecting you?"

"No."

I said, "Is there anyone there with a gun to protect you?"

He said, "no."

I thought, Oh, my gosh, our number one asset, the one man who the people of Afghanistan could rally behind now that they have killed Massoud, only the King, Zahir Shah, was someone we could rally the people behind to counterattack against the Taliban, and he was hanging out there in the wind. He was totally exposed.

So I talked to a very high official in one of our intelligence agencies. He said he realized the importance of Zahir Shah and he was totally exposed, and he was vulnerable. And, guess what? Five hours later I happened to talk to that same high-level official again. I can tell you when I asked him about Zahir Shah, whether he was under guard now, his response to me was, "You don't expect us to act that fast, do you?"

Give me a break. Of course we expect our people to act that fast. You are within a phone call's distance of the Marine guards who guard our embassy in Rome. Our ambassador, or whoever was there, could have gone over and picked up the King or sent Marines over to protect him, or the agency has people in Rome, and so forth.

Instead, 5 hours later, after 3,000 of our people, at that time we thought it was 20,000 people had been slaughtered, but you do not expect us to act that fast, do you?

The people in our intelligence community are, by and large, fine and dedicated people. I will tell you that right now. I respect them, but those individuals who may have my respect as people of good hearts and are patriots, they are now part of a bureaucratic behemoth.

We are relying on what has become organizationally incompetent, a system in which individuals get fired for showing initiative, like that young analyst at the DIA, or they get reprimanded, like those FBI field agents, for begging attention on some pressing threat.

We need to reform the system and make it better. To do so we need to hold those accountable who made errors and to change the structure and mindset. Most important, we need to change the structure and the mindset of our intelligence organizations. We cannot let the cloak of secrecy be used to shield the consequences of failure and incompetence.

For that reason I voted for an investigation of 9/11, not just that it be done by our congressional oversight committees. And I have great respect for those leading those committees and members of those committees, but I believe that it should also be the responsibility of an independent commission on the level of the Warren Commission and perhaps the commission we established after Pearl Harbor to get all the facts about this historical failure of U.S. intelligence.

Let me stress again that I have tremendous respect for and trust for the gentleman from Florida (Mr. Goss) and the others in the Permanent Select Committee on Intelligence here in the House, but a redundancy like we are calling for with an independent commission looking into the problem as well cannot in any way hurt. An independent commission could do nothing but contribute to the understanding of the idea pool that is needed to reform and to fix the system.

This anniversary is with us today. We must commit ourselves to see that such surprise attacks will never again be successfully launched against the United States. We will accomplish this by making the changes in policy and the changes in personnel that are needed to keep our country secure.

We must change the way we deal with Saudi Arabia. We must evaluate how we dealt with Afghanistan and admit that it was horrendously wrong. The people behind those policies, especially those people who are still in influential positions in the State Department and elsewhere, must understand that they bear a significant share of the responsibility for the death and destruction that fell on America 1 year ago today.

The arrogant so-called experts, for example, who shoved aside exiled King Zahir Shah for years, they shoved him aside for 2 decades, claiming that he was too old to play a positive role in bringing about a better Afghanistan and peace in Afghanistan. They were so absolutely wrong. People in the State Department should find out who it was who pushed this idea that Zahir Shah could not participate, and those people should be talked to, and those people should look in the mirror and think very seriously about what they did to contribute to this loss of American life.

In essence, they kept the Taliban in power, because they prevented us from getting behind a positive alternative, whether it was Massoud or the others fighting the Taliban, or whether it was Zahir Shah. In essence, they kept the Taliban in power until 3,000 Americans were slaughtered by an attack that was launched from Taliban-controlled territory.

We were attacked a year ago today, and over these last 12 months our military has been able to launch a counterattack that has dislodged the Taliban and sent them, along with their terrorist allies, the Al Qaeda, running for cover and running to hide their heads.

Our military has done a tremendous job. They did this in a landlocked country halfway around the world. This has been a magnificent victory for our country and for its military. To the degree that we sort of have questions about the need to restructure our intelligence system, we need to praise our military and make sure that we build upon their success. They need certain changes, too, but we need to do that with the military. We can see the positive things they have done and build upon that.

This has been a magnificent victory. If bin Laden is alive today, he is in hiding and he is spending all of his hours not trying to launch some attack on us, but instead he is spending his time trying not to be captured. He could be spending his time mapping out attacks on the United States. Instead, thanks to the expertise and bravery and courage and great job our military has done, we have bin Laden and his likes in hiding, looking over their shoulders, freezing their assets, not able to launch another attack of the magnitude that we suffered 1 year ago today.

We have accomplished all of this, a tremendous accomplishment in a country on the other side of the world, landlocked. We did this with fewer than 50 American combat deaths. We dislodged the Taliban government from power, we destroyed the regime, we dislodged the terrorists, all with fewer than 50 American combat deaths.

Yes, there have been some mistakes, and in every combat situation there are. If accidentally a house or area is bombed, if we bombed some of our friends accidentally, which has happened, we just need to admit that it was a mistake and help those people rebuild. They will understand, because the Afghan people are praising us as their liberators. We have fought beside the mujahedin again, the freedom fighters of Afghanistan, to free their land from the Taliban tyranny. As I say, there have been mistakes, but compared to what has been accomplished, this mission gets an A.

Let me note that I have two complaints. They are small complaints and the Afghan people will put up with them for now, but I think that we need to pay attention.

Number one, I do not believe Karzai was the right guy to pick. He does not have a wide base of support in Afghanistan. When the loya jirga was held, we should have permitted the King to emerge, as would have naturally happened. I think there was some wheeling and dealing going on that led to Karzai's ascension, and the King could have been there. He was the natural choice.

But I believe the Afghan people have good hearts and understanding. They know we are there to help them. They know there are political considerations. But they are demanding free elections in 2 years, and that is what we should be doing, making sure that we keep that pledge and that there are free elections. And if they want to elect anybody, whether it is Karzai or a member of the royal family or whoever it is, they should have a right to do so. We should work with them and help rebuild their country, and that will be one way to really defeat the Taliban and really defeat Al Qaeda. The people of Afghanistan have looked to us as liberators.

The other concern is about drugs. We have not eliminated the drug production in Afghanistan. The poppy crop was not destroyed. We have got to do so next year. That commitment has to be there. That drug money goes into bad hands.

Finally, let us take a look at the challenge we have today and look ahead a year. The President has wisely suggested that now is the time for us to eliminate that threat that hangs over us and has hung over us for 10 years. We did not complete the job in the gulf war. We left Saddam Hussein in power. That was the gift that George Bush, Sr., gave to us. George Bush, Jr., is going to make up for that. He has committed us to eliminating the dictatorial, fascistic regime of Saddam Hussein.

We should not be weary of this. In fact, we should know that Saddam Hussein has less support in Iraq than the Taliban had support in Afghanistan. Our strategy should be to help the people of Iraq liberate themselves from this monstrous regime headed by Saddam Hussein. The people of Iraq will be waving American flags and dancing in the street because we will help them build a democratic society. We can do so with the same strategy as we did in Afghanistan, work with Special Forces teams and air support. We can support those people who want to fight for their own freedom. It worked in Afghanistan, it will work in Iraq. We should not have fear and trepidation about getting rid of this threat of Saddam Hussein. He is, as George Shultz suggested, a rattlesnake in our front yard, and we should not wait until he bites us to cut off its head.

Now we can move forward in Iraq and eliminate that threat, as we have eliminated the Taliban threat, and we can do so not by sending huge numbers of American forces, but by helping the people in Iraq, as we did in Afghanistan, to liberate themselves. That is the challenge the President is giving us. That is why we as Americans should always stand for those people who want to live in a free society and are willing with their courage and blood to fight for their freedom, but need our help logistically, need our air support, perhaps need our advice from our Special Forces teams.

So, as we remember 9/11, let us never repeat that, by being proactive in the future. Where there are dictatorships and fascist regimes like the Taliban, and if they threaten the West and the United States, we do not have to do this with all regimes that are dictatorial, but if they threaten us, let us work with the people who suffer with a boot on their face and with an iron grip around their necks. Let us work with those people to help them free themselves.

We have on the floor of the House of Representatives two pictures, one of George Washington, a great painting of George Washington, and a painting of Lafayette. Lafayette came here during the American Revolution to help us win our freedom. Let us not forget the French helped us win our freedom, and that people like Lafayette were heroes to early Americans.

We must serve that same role that Lafayette served for us. We must serve that role for those people overseas who long for liberty and justice. If we do so, we will be the light of the world. We will be the hope of all the young people in the Muslim countries who are looking for some people who believe in something, rather than people who are talking about stability and keeping the status quo.

We need to be the ones who offer moral alternatives, and the morality we offer is democratic government and a respect for human rights, treating people decently. Our flag should stand for justice and hope. If we do, rather than the type of things we were doing in the 1990s with Communist China and the Taliban and all of these regimes, where we were not doing anything to make it clear that we honestly and sincerely believed these founding principles of our society, we will be free and we will be safe.

There is a dynamic in this world between peace and freedom. Freedom tomorrow will bring peace. Just as we lived under the threat of some sort of war with the Soviet Union, the Soviet people, the Russian people were never our enemies. It was that system. As soon as we made it a fight between communism and democracy and stopped just supporting any dictatorship that was against the Communists, the Communist system itself began to crumble in Moscow, and no one was more heroic in that fight against the Soviet dictatorship than the people of Afghanistan. They fought and they bled and they gave us a more peaceful and freer world.

We did not do what was right by them. We did not help them rebuild their country at that time; we did not stick with them. We left it up to the Saudis and the Pakistanis. We have a chance now to make up for that. But we must persevere in helping them rebuild their country; and that will cement peace in that region, because people will believe in us again. We need, again, to make sure that we become the force for liberty and justice and decent treatment for people all over the world, and that is where we will find America's security. Let us have the courage to do so. Our President has charted a wise course, and we should have the tenacity and the courage to follow this through now that we have learned after 9/11 that there are consequences to pay when we do not.

IN SUPPORT OF H. CON. RES. 464, PATRIOT DAY RESOLUTION

Hon. Ken Bentsen

OF TEXAS

Mr. Speaker, I rise in support today of H. Con. Res. 464, commemorating the solemn occasion of the first anniversary of September 11, 2001, and the vicious attack on the United States and its people that day.

September 11 will long be remembered not just for the death and destruction brought upon America and too many of our people, but also for the day in which innocence was lost. The hijacking of civilian aircraft and the taking of thousands of innocent bystanders' lives will forever be among the most heinous of crimes against humanity. Yet, while the attacks, designed to shake our Nation to its deepest roots and break our spirit to be a leader of the free world, they only served to strengthen our resolve and show the world that the American sense of kindness and community could not be broken even by the most awful of acts.

In the midst of a living hell only Dante could describe, Americans rose to respond, first by the police, firefighters, and emergency medical personnel who poured into burning buildings which had become infernos, risking life and limb to save a fellow citizen and later by military force. As we now know, many made the ultimate sacrifice. Yet, rather than create panic and chaos, the attacks by the terrorists only served to underscore our resolve. Clearly, in the early hours following the attacks, the actions of otherwise ordinary Americans proved beyond a reasonable doubt that the terrorists failed in their ultimate goal. While they were able to cause pain and suffering and shatter a sense of security felt here at home, they failed in destroying the spirit which is America. And, in causing the death and destruction, they exposed the world to their own twisted ideology of anger and hate and a lack of respect for freedom and human life.

Mr. Speaker, in the intervening months, America, with our allies, has proven militarily that we can and will respond to defend ourselves and our freedom throughout the world. But even as important as it has been to respond swiftly and forcefully, it is now clear that the actions of those Americans who gave their lives on September 11, 2001, and those who responded to help them, and the faith that they inspired in the rest of us, resulted in a victory over the terrorists that very day.

Now, we must continue to battle for freedom and democracy throughout the world, not only for our own defense, but also in the memory of those who first gave their lives for the cause on September 11, 2001.

GOOD WILL OF UKRAINE HELPS AMERICA HEAL

Hon. Bob Schaffer

OF COLORADO

Mr. Speaker, this hallowed day, the anniversary of the terrible terrorist attacks upon the United States, is one that is observed throughout

America. The eloquence of our colleagues in this House has stirred our hearts, described our attitudes, and revealed the Nation's character. America's tragedy is mourned this day throughout the world, too.

It is important for Americans to understand and appreciate the outpouring of support, the gestures of solidarity, and the prayers of the faithful bestowed upon us by the people of the world. As cochairman of the Congressional Ukrainian Caucus, I rise today to speak about one country's good will toward the American people in commemoration of the September 11, 2001, terrorist attack.

Throughout my tenure in Congress, I have always been an ardent supporter of democratic development in Ukraine. The full commercial and strategic potential of a democratic Ukraine is staggering and the desire for freedom is deeply rooted in the American psyche.

Mr. Speaker, in the House the days and the weeks following the attacks, the Ukrainian people grieved with America and the rest of the world. The sentiments of the Ukrainian people were echoed in a letter from Bishop Stanislaw Shyrokoradiuk of the Roman Catholic Diocese of Kyiv-Zhytomyr. Here is what the bishop wrote:

> I would like to present to all American people, our condolences because of the terrible tragedy that has stricken the whole world.
>
> We have been deeply shaken to know about the series of acts of terrorism that happened in the United States of America yesterday.
>
> It has been an awful blow by its cruelty and scale that struck not only the U.S.A., but all humanity. I received this notice during spiritual retreats in our Higher Spiritual Seminary in Vorzel, where all priests of our Diocese came. As a sign of our unity and sympathy in your grief, we celebrated Holy Mass for the souls of the departed and prayed for all victims. May the Lord strengthen them by His grace that they may outlive this horrible disaster.
>
> There are Divine Services and prayers said for the souls of the departed victims, and for all of those who have suffered, in all of our churches.
>
> These days our hearts and prayers are with you and your people.
>
> Sincerely Yours,
>
> Bishop Stanislaw Shyrokoradiuk,
> the Vicar General of Kyiv Zhytomyr Diocese.

Bishop Shyrokoradiuk in Ukraine sent that the day after.

As a country, Ukraine immediately declared its solidarity with the United States, offering its support morally, technically, militarily and with the capabilities of its infrastructure. Ukraine's stance clearly demonstrated its friendship with the United States and the forces of freedom.

Most helpful has been Ukraine's clearance of airspace for nearly 5,000 aircraft flying in from Afghanistan and aid in transporting allied troops and materiel by air and train. Ukraine also provided planes and crews in order to provide transportation for coalition forces and humanitarian missions over Iran.

Ukrainian security forces have been cooperating with Americans offering intelligence regarding the fighting in Afghanistan and other security concerns. Ukraine's international efforts in leading the Georgia, Ukraine, Uzbekistan, Azerbaijan, and Georgia Group, called GUUAM, to secure strategic transport corridors from terrorist activity have helped secure the entire regional community. Ukraine has cooperated with us in blocking and investigating the financial transactions and accounts of suspected terrorists.

Just today, Mr. Speaker, Ukraine's Parliament, the Verkhovna (Supreme) Rada, unanimously passed three resolutions in support of America. One extends Ukraine's profound sympathy to the families and friends of victims. The other directs the Cabinet of Ministers of Ukraine to establish a memorial at the World Trade Center in New York to the Ukrainians who perished there. The third resolution outlines Ukraine's commitment to the future of freedom and reaffirms Ukraine's solidarity in the war on terrorism and its commitment to strengthening international peace and security.

This latter resolution lucidly addresses issues of global concern, including environmental problems and global health issues. Currently, Ukraine is preparing to send two IL–76 cargo planes loaded with small arms, ammunitions, and other military equipment to outfit the Afghan National Army.

This morning, the President of Ukraine, Leonid Kuchma, met with our Ambassador, Carlos Pascual. The President pledged his country's continued support for Operation Enduring Freedom and expressed his regret for the loss of lives in America 1 year ago.

This afternoon, Ukraine's Ambassador Kostyantyn Gryshchenko visited me in my office here in Washington, DC, and extended his country's condolences, support, and solidarity. He asked me to express to the House Ukraine's commitment to America's war on terrorism. He assured me that the prayers of countless Ukrainians are for the souls of the victims.

At this moment in Kyiv, Mr. Speaker, Ukrainians are paying their respects as they observe a large photo collage of Ground Zero. The display honors the emergency workers and heroes of 9/11. There is a similar exhibit at the Ukrainian National Gallery; and on Friday, they will hold a commemorative concert at the National Opera.

Ukraine's condemnation of international terrorism, its much appreciated support on the war on terrorism, its tough, newly enacted laws to combat terrorism and its commitment to fight at the side of the United States and its allies for civil society and democracy clearly demonstrates the role Ukraine and her people intend to play. Ukraine's support for America is deeply appreciated.

Mr. Speaker, I thank the people of Ukraine for their condolences and camaraderie and commend Ukraine's leaders for standing shoulder to shoulder with the United States. I will submit for the *Record* at this time the resolution that was adopted this morning by Ukraine's Parliament.

THE LESSONS OF THE TRAGEDY ARE NOT TO BE FORGOTTEN

STATEMENT BY THE VERKHOVNA RADA OF UKRAINE ON THE OCCASION OF THE ANNIVERSARY OF THE EVENTS IN THE UNITED STATES ON SEPTEMBER 11, 2001

The year that passed after the tragedy in the United States of America did not assuage anger and indignation at malevolent actions of terrorists, pain and bitterness over the loss of thousands of innocent people who died in the airplanes, skyscrapers of the World Trade Center in New York, and offices of the Pentagon. Paying homage to their memory we express once again our sympathy with families and loved ones of the victims.

Last year's tragedy has taught mankind many lessons. First of all it demonstrated that neither financial and economic power, nor possession of unprecedented arsenals of modern arms can guarantee security even for the most powerful state of the world. It has become even more obvious that only with united and coordinated actions can mankind overcome international terrorism and religious fundamentalism.

Realization of that led to creation of the anti-terrorist coalition that included dozens of countries. An important role within the coalition belongs to Ukraine, whose foreign policy major goals are strengthening of the international peace and security, maintaining peaceful, equitable, and mutually beneficial cooperation with members of the international community consonant with the generally accepted principles and norms of the international law.

The Verkhovna Rada of Ukraine considers it exceptionally important to apply the experience of the anti-terrorist coalition to strengthening cooperation of members of the international community in overcoming such global challenges as the deepening gap between a handful of the richest countries and numerous less developed nations and countries in transition; catastrophic impoverishment of hundreds of millions of people; spread of drug abuse, infectious disease, corruption and organized crime; depletion of water and other natural resources; pollution; dangerous accumulation of nuclear, chemical, and other weapons of mass destruction; and attempts to use objective processes of globalization and internationalization in the selfish interests of one state or a group of states.

United, mankind is capable of ensuring protection of the environment and biodiversity, acceleration of economic and social growth of every member of the international community. We consistently support the increasing role of the United Nations in resolving international problems, strict compliance with requirements of the U.N. Charter on conflict resolution and prevention of threats to peace and security of the nations.

The documents of the World Summit on Sustainable Development held recently in Johannesburg attracted attention in the Ukrainian society. By taking unprecedented decisions to voluntarily renounce the third largest nuclear arsenal and close up the Chernobyl nuclear power station, Ukraine, having been left alone with these very complicated problems, demonstrated to the world its devotion to the cause of peace and international security and adherence to its international commitments.

Reflecting on the lessons of last year's September tragedy, we consider it our duty to draw once again the attention of the international community to the above issues and call on all the nations to consolidate and multiply their efforts to overcome global challenges facing mankind in the 21st century.

AMERICANS STAND TALL AGAINST TERRORISM

Hon. Jim Turner

OF TEXAS

Mr. Speaker, I rise today on this September 11, the 1-year anniversary of the greatest tragedy on American soil in our history, with a heavy heart, as I know is shared by all Americans, because it was on this day 1 year ago when almost 3,000 Americans lost their lives at the hands of evil terrorists who sought to destroy our Nation, who seek to destroy our way of life.

We know from that experience that all Americans must stand tall against the threat of international terrorism. We know as Americans that we never, never, never want to experience that tragedy again. As a part of our history, we will always remember September 11 of last year.

I come to the floor today to call upon the House to pass legislation entitled the National Memorial to the Victims of Terrorism Act. This legislation would memorialize all victims of terrorism, both those who have lost their lives on September 11, as well as those who have been victims of terrorism in previous times, as well as those who have died in the cause since.

Since September 11 of last year, we have engaged in a battle in Afghanistan where 51 Americans have lost their lives. We know from experience that the war against terrorism will not be won quickly and that perhaps we may face this challenge for many years ahead.

In many ways on September 11, the American people recognized for the first time that we were in fact engaged in a war against terrorism, even though we have had ample opportunity to define those events that occurred in previous years—taking the lives of American citizens and American soldiers—as a war. We understand that this legislation we have introduced will memorialize those who lost their lives to terrorism. It was introduced by me and by the gentleman from Utah (Mr. Hansen). It is a bipartisan piece of legislation that enjoys the support of Members on both sides of the aisle.

The legislation would create a 13-member Victims of Terrorism Memorial Advisory Board appointed by the President in consultation with the Secretary of the Interior and the Secretary of Defense. Members of the advisory board would include appointees from organizations dedicated to assisting the victims of terrorism and their families. The board would begin the process of establishing a memorial not later than 1 year after the date of enactment of this legislation. The advisory board would have the responsibility of raising the necessary funds from private sources to pay for this national memorial.

Those who came and testified on behalf of this legislation in the committee last May spoke very eloquently about the importance of this national memorial. We had testimony from Lisa Beamer, the widow of Todd Beamer, who joined those brave passengers on flight 93 and fought off the terrorists and saved this Capitol and many people who were in this building.

We had testimony from Col. Ted Anderson, who saved many victims when the Pentagon was struck by the aircraft on that fateful day.

We had testimony from Joe Finley, a New York firefighter who worked hard and dedicated his efforts on that fateful day to saving the lives of those who were struck in the World Trade Center.

We also had testimony from Liz Howell, a staffer on the Committee on Resources staff who heard of this bill who lost her husband at the Pentagon on September 11.

Each of these individuals shared heartfelt testimony as to why it is important for Americans to mark this event with a national memorial, and why it is also important to reflect in this memorial the memory of all who have lost their lives to terrorism.

The war on terrorism is indeed the first war of the 21st century. It will not be one marked by any one geographic location. It is a global war. It is a global war that will be fought both at home and abroad. Though they have lost their lives in places far and near over a span of time that includes the past, the present and perhaps the future, the victims of terrorism, both civilian and military, deserve solemn tribute, for they died at the hands of the enemies of America simply because they were Americans.

I call upon the leadership of this Congress to promptly set this bill for hearing in order that we may act promptly and pass the National Memorial to the Victims of Terrorism Act.

THE MEANING OF SEPTEMBER 11

Hon. Frank Pallone, Jr.

OF NEW JERSEY

I wanted to take to the floor on the evening of September 11 to talk about the meaning of September 11 to me and particularly to my constituents in New Jersey. I wanted to particularly make reference to two events that I participated in in the last 24 hours.

Last evening, after the House adjourned during the day for votes, I went up to Middletown, NJ, which is a town in my district, in Monmouth County, that suffered more deaths in the World Trade Center building during the attack on September 11 than any other town in New Jersey or maybe in the whole metropolitan area. There was a memorial service, a dedication of a garden in memory of the 36 residents of Middletown who died on September 11. It was a very moving ceremony. We had relatives of the victims who made speeches. I would like to talk about it a little more.

The other event I went to this morning was at my own high school in Long Branch, NJ, where the entire student body from kindergarten, I believe, all the way to 12th grade, to the graduating seniors, showed up at the football stadium for a service. I think we must have had probably over 4,000 people there this morning. It also was very moving. I would like to comment on both of those ceremonies in light of what happened last year.

Mr. Speaker, last September 11 I actually was in the Capitol. Many of us know that on Mondays and Tuesdays we schedule at 9 a.m. what we call morning hour, which is very similar to the special orders that we have at the end of the day. It is an opportunity between 9 and 10 a.m. for Members of Congress to come down and give 5-minute speeches on whatever topic they desire. It is not part of the votes of the House. It is an opportunity to talk about issues or really any kind of event that you want to talk about that day.

Before morning hour on the Democratic side we have a meeting, what we call a message meeting in the morning that I chair, along with the gentlewoman from Connecticut (Ms. DeLauro), and that usually starts at 8:30 in the morning. So on that morning of September 11 last year we started at 8:30, maybe it was a little later, with a message meeting, and then we came up to the floor, some of us, including myself, to do the 5 minutes for morning hour.

I do not know exactly what time it was, probably maybe about 9:15 or 9:20, when I finished my 5-minute speech for morning hour that I walked out of this Chamber and walked over to the leader's office, the office of the gentleman from Missouri (Mr. Gephardt), and discovered that the World Trade Center had been attacked. It was on the television.

The only reason I bring this up is because over the weekend there was a report in the news media about how certain informers for Al Qaeda and the Taliban had indicated that the plane that went down in Pennsylvania on September 11 last year was actually headed for the Capitol, for the U.S. Capitol.

This morning when I was at the ceremony at Long Branch High School, the superintendent of schools, who was the MC for the ceremony, Mr. Joseph Ferraina, mentioned in his opening remarks that the people who died on September 11 basically gave their lives so that others might live.

I thought about that statement this morning, and, of course, it has a tremendous symbolic meaning, but it also had a literal meaning in a sense for me, because it is very likely that if those brave Americans who had decided to try to fight the terrorists and bring down that plane into a field in Pennsylvania had not made the decision to try to struggle and overcome the terrorists, that that plane would have headed for the Capitol and I would have been right here on the House floor and probably died as a result of that attack. So those people literally were giving their lives so that Members of Congress like myself and my colleagues could live.

It is an amazing thing to think about, that they were willing to sacrifice so that that plane would not come here and hit the United States Capitol.

But I also thought this morning that they were not only giving their lives for other Americans, possibly myself and my colleagues, but they were also sacrificing their lives, they were essentially martyrs for the cause of America in a more symbolic way. When I say the cause of America, what do I mean? I mean democracy. I mean the freedom of speech, the freedom to assemble, the freedom of religion that we find so sacred.

My wife mentioned to me this morning that the Americans who were on that plane who ended up crashing in Pennsylvania were in contact with others on the ground using cell phones.

They found out that the World Trade Center had been attacked, that the Pentagon had been attacked. They decided they would take a vote among themselves on the plane as to whether they would try to overcome the terrorists in order to veer the plane away from, in this case, the Capitol or whatever other landmark they thought it might be used by the terrorists to attack.

I thought it was terribly significant that they voted, because here we are this morning in Long Branch, yesterday in Middletown, today on the floor in Washington, talking about the meaning of democracy and how the people who lost their lives were really martyrs for the American way for democracy. Lo and behold, they were taking a vote to decide whether to overcome the terrorists, which is probably, I guess, the most basic manifestation of what democracy is about, taking a vote.

I am sure the terrorists who hijacked the plane were not taking any votes because, unlike the Americans who were willing to give their lives on that plane, they did not believe in a democratic process. They did not believe in the American way, the values that we believe in. They basically had a very different ideology, and their ideology, whatever it was, said that it was okay for them to hijack the plane, to kill innocent civilians for what would appear to be some sort of Fundamentalist religious cause.

I think that we cannot forget that so many Americans lost their lives on September 11, including the 36 in Middletown, the town that I went to last night in my district. Even though they were giving their lives for the American way, for American values, that the effect on their families, the effect on their relatives, is devastating.

It is nice to say that someone is a martyr. It is a glorious thing. But, at the same time, it is very hard to be the relative of the martyr, because your husband or your wife or your daughter or your son is no longer there. There is the huge void, if you will, that lingers. I am sure it lingers a year later or lingers 10 years later. It is never really filled.

Mr. Speaker, I will introduce into the *Record* an article that was in the *New York Times*, I think it was on September 7, just a couple of days ago, that talks about Middletown, NJ, and the grief and the difficulty that the relatives and the survivors of the victims of September 11 have been going through.

I put it in the *Record*, Mr. Speaker, not because I want to dwell on the grief. The title of the article in the *New York Times* is "Emerging From [the] Cocoon of Grief," but because I think it is important for us to understand that as much as we talk about these heroes and their families, they gave so much yet they still were people whose families now are having problems because of a void that has been left behind. I think this article sums up their courage and what they had to face, but also sums up what they face in the future.

Hon. Maxine Waters

OF CALIFORNIA

Mr. Speaker, I would like to thank my colleague for yielding to me and giving me an opportunity to voice my support for H. Con. Res. 464. I was not here when the vote was taken. I was on a plane coming from Los Angeles. I had to stay in the district to take care of some very important problems there. I tried very hard to get from the airport here to the House floor so that I might take that vote, because I think it is so important for all of us to show that we really do care and we really do honor the memory of those who lost their lives and for those families who are making sacrifices, even today, because of the devastation that they are experiencing in their families and in their homes and in their lives. So I would like it to be known that had I been present for the vote on H. Con. Res. 464, Roll Call No. 384, I certainly would have voted yes.

Again, I want to thank the gentleman from New Jersey (Mr. Pallone) for taking extra time on the floor. Having done what we all should have done, and that is voted, the gentleman has remained here, because he had something additional that he wanted to say to the families, friends, neighborhoods and communities.

Just as I walked in, the gentleman was talking about what happened on that airplane in Pennsylvania when a decision was made by a vote "to roll," to try and take the plane away from the hijackers, in an effort perhaps to prevent them from coming to this Capitol or to the White House.

The gentleman is absolutely correct, we should never forget that, and we should all know and feel that we are very blessed because there were very brave people who decided to take a courageous action in the interest of saving lives. So I thank the gentleman for the extra time that he is putting into this.

Hon. Frank Pallone, Jr.

OF NEW JERSEY

I just wanted to talk a little bit more about the Long Branch ceremony this morning and Middletown last night, if I could.

I was really happy this morning in Long Branch because there were so many young people there, about 4,000 people, as I said, maybe from kindergarten all the way to 12th grade from my home community. I think they were listening very attentively to the various speeches being made and they understood that the people who lost their lives on September 11 really were heroes to the American way.

The most important thing I think we need to do is continue to commemorate September 11. In Long Branch they said they plan on doing the ceremony every September 11 because they wanted to teach, if you will, the students and the young people about the significance of September 11. It is important that from now on, not only today, but in the future, that we continue to commemorate the day and we continue to commemorate those people who lost their lives and the heroes who tried to help the victims, the firefighters, the police. If we forget it, then we are not paying proper respect to them, and we must continue to point out that this democracy that we live in and the freedoms that we so cherish are not easy to come by, that people continue to sacrifice for them.

Obviously we must continue to do what we can here in Congress with the President and Congress working together to make sure that the terrorists do not have the opportunity to do this type of terrible act again.

As I said, there were 36 Middletown men and women who lost their lives in the World Trade Center on September 11. I mentioned the article that was in the *New York Times*, and there were a lot of news stories and national attention that was focused on Middletown over the last year because so many people died proportionately for the size of the town. But in all the descriptions in the news media about the suffering, about the people who died and their families, there was also much said about the pride of the community, the fact that the community came together in untold ways. Residents were helping each other in time of need, and the community banded together not only to help the families of people who died but also to send firefighters and police to help the efforts on September 11 and in the aftermath. I just wanted to say this evening how proud I was last night to be able to say that I represent a community like Middletown and to also have the opportunity to participate in the groundbreaking for another wonderful community effort there, the Middletown Memorial Garden.

Last night each of the elected representatives spoke briefly; they had relatives of the victims who spoke; and then they proceeded over to the new memorial garden where each of the relatives had an opportunity to to shovel some of the dirt before the garden started to be put together. Every speaker was overwhelming in terms of what they said and the significance of what they said. In particular I remember a little girl, about 9 or 10 years old, who reminded me of my own daughter. I have one daughter who is 9 years old. And she spoke about her father. She basically read a letter, I guess in a sense she was writing a letter to dad, and it was such a moving experience. And after that about 170 or so relatives that proceeded over to the garden to the groundbreaking, and there were little kids 2 years old, 3 years old, all the way up to teenagers. It was such an overwhelming experience.

I want to say in conclusion before I yield to my colleagues, the loss is always going to be there. There is no way to get away from the loss for all these relatives of what they lost on September 11, but I think if we do not forget the people who gave their lives, if we continue to commemorate their activities, if we do things like the Middletown Memorial Garden or the ceremonies like that were held at Long Branch High School this morning, then we are doing what is necessary to make sure that we never forget what happened and the significance of what happened. We need to be reminded ultimately that this battle against those who would defy America and defy America's values is never-ending and that we have to be constantly vigilant in order to protect our democracy and our freedoms. And that is why I think today was so significant to me, not only to the two ceremonies that I attended over the last 24 hours, but because I feel in general that people after a year really understand the significance of what happened on September 11 and are determined to keep in mind the lessons of that day.

Hon. Eliot L. Engel

OF NEW YORK

I am very grateful that we are able to speak on the floor about September 11 today because I have just gotten back from New York and spent my morning at Ground Zero at the World Trade Center site where we had extremely moving events. I was just out in front of the Capitol where I sang "God Bless America" with so many of our colleagues, and for me being in two places the same day, Ground Zero, the World Trade Center, and at the Capitol where we now know that the doomed flight from Pennsylvania was heading. It has been a very emotional roller coaster for me to be in both cities one day. Flying the shuttle, it was practically empty. A lot of people were obviously not flying today. But I am just so proud the way this Congress and the American people are handling the anniversary of the tragedy that happened 1 year ago today.

I saw, as I have seen in my city since September 11, just an outpouring of good will, of people just hugging each other and banding together and taking pride in being New Yorkers and taking pride in being Americans and just wanting to help each other, care about each other, be concerned with each other. We saw that again when I drove down this morning. The first thing we saw when we got near Ground Zero were people with American flags signaling victory signs and thumbs up signs and just hugging each other. When we actually got to the event, there was a platform and we started with different famous speeches that were made, and then at the exact times that the planes hit both towers at the World Trade Center, there were moments of silence, and then at the exact times that the buildings crumbled a year ago, there were bells tolling and moments of silence, and the names of all the victims of the September 11 tragedy at the World Trade Center were read from A to Z, and it was very, very moving. I was given a list of names to read to be part of the procedure, and I realized that someone had come in who was a family member of someone who had died at the World Trade Center and he had no names to read, and I gave him my list to read because it was just enough for me to be there. And I will tell my colleagues, it was a very windy day in New York City today, and it was almost as if one could just feel the spirit lifting everybody who was there. I do not think I have ever experienced anything that has been as emotional or as emotionally uplifting. It was sad, very, very sad, but at the same time it made us feel like we were all together as Americans and as New Yorkers. There were thousands upon thousands of people. There were family members making their way down to the exact spot of Ground Zero.

Last night at about 1 o'clock in the morning, there were processions, actually marches, of people from all five boroughs. New York City, of course, has five boroughs—the Bronx, where I am from; Brooklyn; Manhattan; Queens; and Staten Island. And people started from the farthest reaches of the city, from the farthest points of the five boroughs and all converged at Ground Zero. As the names were being read out, family members marched down. Yo Yo Ma was playing as he does so well, and it was just a very moving experience.

After touring the site with the President 3 days after September 11 last year, I said that I was never more proud to be an American and never more proud to be a New Yorker, and I feel that way again today. I do hope that every September 11 we can all come together. I hope we can do it 365 days of the year, but I think that September 11 is a day that we really always need to reflect and always need to understand how proud we are to be Americans, how proud we are to be New Yorkers. The terrorists think that they can destroy our way of life, but they cannot. Quite the opposite. Because what I see is a resolve among Americans, among people in this country like I have never seen before. It is almost as if a sleeping giant has awakened, and we are going to ensure that the evil of terrorism is eradicated all over the world wherever it rears its ugly head, and this country will always be in the forefront of fighting evil. We understand what it means to be an American, and we understand why it is so important that we all band together and help each other because that really gives us the meaning of what life is all about.

Those people who died on September 11 did not perish in vain because they will always be in our minds; they will always be in our hearts. The heroism that we saw from the first responders to everybody else, the untold acts of heroism that we will never know about are an inspiration to all of us.

Hon. Nancy Pelosi

OF CALIFORNIA

Mr. Speaker, I thank the gentleman for yielding and for having this very important special order this evening on this day that is etched in the hearts and minds of all Americans.

I want to commend the gentleman, my colleague from New York (Mr. Engel), and extend to him and the members of the New York delegation, especially the dean, the gentleman from New York (Mr. Rangel), for the magnificent hospitality extended to the Congress. Over 300 Members of Congress traveled to New York on Friday to the place where George Washington took the oath of office as our first President of the United States. What more suitable place could there be than for us to express our grief to all those who personally suffered on September 11 and, indeed, to our entire country which shares in that grief?

New York took a hit and all those from surrounding areas, as the gentleman said, many from New Jersey, certainly those in the plane in Pennsylvania and those at the Pentagon. In every venue the spirit of the families who suffered the loss and the spirit of the communities that were involved have lifted up our country.

This has been quite a day for our country, all over America, and in my district in San Francisco. We started at 5 o'clock this morning because it is three hours earlier, to be ready to commemorate at the exact moment the sad tragedy that our country experienced last year. But some of us were in the National Cathedral at that precise time when the great bell of the cathedral rang to observe again that sad time and to join in mourning. It is a day of mourning and memories, and it is a day to pay tribute and give thanks certainly to the New York delegation for the wonderful venue they provided for us to mourn, commiserate, and they provided us a great memory for which we are all grateful.

With the resolution that we approved today in the Congress, we expressed our utmost appreciation to those brave and courageous young men and women in uniform who are fighting the battle to root out terrorism wherever it exists. Today we remember the victims of September 11. We also remember and pay tribute to Johnny Michael Spann, the CIA officer who in November became the first American killed in combat. We honored him earlier in this Congress with his family in the gallery. His name and the names of thousands of other Americans, too many Americans, are now etched as permanently in our history as they are in the minds of their families. As the poet laureate of the United States said in New York, there are too many names for even the walls of our hearts to contain.

For some of the families of the victims, the sound of a plane flying overhead fills them with fear. Indeed, the warning of any possible terrorist act intensifies their grief, and for them and for all Americans we must do everything in our power to reduce risk to the American people. Yet as we continue to grieve, we take pride in knowing that the unspeakable events of September 11 have brought Americans closer together than ever. The gentleman from New York (Mr. Engel) referred to that, as did the gentleman from New

Jersey (Mr. Pallone). We have joined together as a community; we rush to give blood, money, and volunteer time to become more patriotic, to appreciate our freedom.

Today's resolution honored all of those affected by September 11 from whom we have learned what it means to be a member of the American family. From our first responders, our firefighters, and police officers, we have learned to be an American is to be selfless, to put honor above personal safety and the lives of strangers above your own. From cell phone calls made from crumbling buildings, we learned that to be an American is to love family with a power and a force that transcends even death. From a group of passengers in a hijacked plane over Pennsylvania, we learned that to be an American is to be brave in the face of hopelessness and to do good for others while evil is being done to oneself. And from workers at the Pentagon who went to work that day and every day, we learned that to be an American is to love freedom and to show that love everyday by serving our great country.

This morning we also went to the Pentagon and shared some sympathies with the families of those who lost their loved ones. One young man showed me the flag that was given to him in memory and honor of his father who perished that day. We also honored the workers in hardhats who rebuilt the Pentagon so that here today on that 1-year anniversary we could visit a Pentagon that was restored, a sign of confidence and pride in our country.

With these lessons in mind, we can rest assured that the assault in our heart, the heart of our Nation, will only make it beat more strongly. That strength will allow us to triumph over terror militarily, and that strength will allow us to triumph over terror in spirit. We will cherish our freedoms now more than ever and recognize, as was said at the National Cathedral this morning, that there is a high cost to freedom.

We will draw our loved ones closer and reach out further toward peace with our adversaries.

With that, I would like to once again express gratitude to all who have helped us all grieve. I hope it is a comfort to those who lost their loved ones that so many people throughout the world, and indeed, intensely in our own country, share their grief and are praying for them at this sad time.

I would also like to thank the gentleman from Illinois (Speaker Hastert) and our leader, the gentleman from Missouri (Mr. Gephardt), for making it possible for us to travel to New York, because it was a place we had to visit. We went to the heart, to where our country began, to renew ourselves and to be ready for this very, very sad day, but in a spirit of renewal and pride in our great country. God bless America.

Hon. Eliot L. Engel

OF NEW YORK

I want to add, Mr. Speaker, that the gentleman pointed out that so many people from New York City and the suburbs of New York lost their lives. Both Senators from New Jersey were there today at Ground Zero, as were the New Jersey Governor, both Senators from New York, and the New York Governor.

In my district in Rockland County and Westchester County, the suburbs of New York City, so many people lost their lives: firemen, policemen, and average citizens who went to work. So this is truly a regional feeling, and absolutely a national feeling; but of course, in the New York City metropolitan area, it is a regional feeling as well. I thank the gentleman for mentioning that.

Hon. Frank Pallone, Jr.

OF NEW JERSEY

I thank the gentleman for adding that. He is right, that we can even go beyond that. My understanding was that the plane that went down in Pennsylvania was originally headed for California, so there were probably some constituents on that plane from the district of the gentlewoman from California (Ms. Pelosi).

We know there were even foreign nationals who perished in the World Trade Center, so the tragedy was truly not only American but included people from other parts of the world.

Mr. Speaker, I include for the *Record* the article from the *New York Times* of September 9, 2002.

The article referred to follows:

[From the *New York Times*, Sept. 9, 2002]

EMERGING FROM COCOON OF GRIEF

(By Andrew Jacobs)

Middletown, N.J., Sept. 6.—Even now, a year after her husband, Louis, disappeared at the World Trade Center, Barbara Minervino struggles with the competing pulls of rejoining the living or remaining curled up in the shelter of her cream yellow ranch home with its comforting memories and distracting mounds of 9/11-related paperwork.

A photographic shrine to her husband still dominates the living room and she refuses to touch the Yankee ticket stubs and the $15 he left on a nightstand the final evening of his life. But she is also increasingly drawn into the world, both by necessity, and in recent months, the realization that she can survive as a 54-year-old widow with limited skills. "It devastates me that I was able to live without Lou for the last year," she said, sitting in Redheads, a strip-mall restaurant where hundreds of mourners gathered last year after her husband's memorial service. "I didn't change a light bulb for 29 years. I didn't buy a bedspread without consulting him."

In contrast to the unrelieved grimness of the past months, there is now a hint of levity in her voice when she talks about the road ahead. "I still don't know where I'm going, but I feel like I'm a butterfly about to come out of the cocoon," she said. "With the grace that God gives me, I look forward to October and what my place is in the world."

Since losing 36 residents on Sept. 11, this centerless hodgepodge of look-alike ranch homes and waterfront estates has become a national symbol of devastating loss and communal caretaking. Over the past year, *Vanity Fair*, "Dateline NBC" and a score of newspapers discovered that tragedy had transformed this anonymous, disjointed suburb into a model of selfless do-goodism. Local volunteers distributed more than $700,000 in cash and services to the stricken families, and many neighbors, once strangers, delivered home-cooked meals to make sure no grieving survivor would have to cook during those first terrible months. Lawn services, mechanics and plumbers donated their time, ensuring that no one would have to worry about the mundane aspects of suburban living. In a way, this community has discovered itself in its grief. But as it crosses the first anniversary, Middletown, like Mrs. Minervino, is struggling with opposing impulses: the urge to move past the trauma of last September and the need to remember.

And while both impulses have enormous force, both the individuals and the town seem intent, finally, on moving on. "Some days, I just want a normal life like other women," said Kristen Breitweiser, who lost her husband, Ronald. "I want to go food shopping. I want to bake an apple pie. I don't want to be a 9/11 widow for the rest of my life." Of course, Sept. 11 this year will be more about looking back than looking forward. By 8:46 on Wednesday morning, when the bells begin to toll at St. Mary's Roman Catholic Church, the camera crews from MSNBC, CBS and Australian television will already be broadcasting live, showcasing this township's resilience in the face of excruciating loss. Shopping malls will fly their flags at half-staff, police officers will shroud their badges in black and residents will gather for commemorative events at a fishing pier, a half-dozen churches and the Middletown train station, where township officials will break ground on a four-acre park honoring the local residents who died.

"Not an hour goes by when you don't think about it," said the police chief, John Pollinger, choking on emotion as he pulled his car into a drive-through teller. "I think all of us here have been changed, changed forever."

But neither patriotism nor civic boosterism can stop the intrusions of daily life. Mounting job losses have taken a toll on many families. The battle over a proposed megamall, dormant in the first few months after the terror attacks, has reignited with more fury than before. And although a tentative contract agreement reached Thursday means schools opened without labor strife, there is lingering bitterness from an ugly strike in December that sent 228 instructors to jail. Since then, more than 100 teachers, about one-eighth of the district's total, have left the community for other jobs or early retirement. The district's embattled superintendent moved on as well. "There are deep and painful wounds that no glossing over, no platitudes, can undo," said the union's president, Diane Swaim, a middle school teacher who has lived here most of her life.

While many families say the public outpouring of kindness helped them endure a nightmarish year, they recognize that the unlimited benevolence cannot last forever. The meals stopped coming with the onset of summer, when many families went away on vacation, and several women said they have sensed a waning tolerance for outward displays of grief. "After a very long year, people expect us to move on, to get on, to try to live life," Mrs. Minervino said.

To many family members, moving forward remains painful. Brittany Chevalier, 16, who lost her 26-year-old brother, Swede, worries that school administrators and teachers will no longer make allowances for the days she is too distraught to come to school or too upset to complete an assignment. "They were understanding on the six-month anniversary, and they'll understand if I don't come to school on Sept. 11," she said, "but they'll start to think I'm being ridiculous when the year-and-a-half anniversary arrives. I guess I'm afraid people are just going to forget and that the world will just go back to normal."

But the pull toward moving on is the dominant impulse, even for the bereaved. During the past year, Patricia Wotton was so distracted by grief she became emotionally detached from her two children, Dorothea, nearly 3, and Rod, who is named for his father, who died a week before he was born. "It was too painful interacting with them," she said, "It reminded me of what I lost. Besides, I was so focused on breathing."

Over time, Dorothea began to act out aggressively, much of it directed at her fragile brother, who was born prematurely and spent his first month in intensive care. Last month, Dorothea's therapist warned that Ms. Wotton's inattention was compounding her daughter's trauma. It was those blunt words, Ms. Wotton said, that helped her cross an invisible line.

In a burst of activity, she opened her backyard swimming pool, planted some tomato plants in the garden that was once her husband's domain and started to talk baby talk to her son. She even visited ground zero, which helped her realize that her husband was really, truly gone. "I saw where the south tower was and finally understood he couldn't have survived such hatred," she said.

In an outgrowth of her newfound strength, she has begun a campaign aimed at winning extended health coverage from her husband's former employer, Fiduciary Trust, which plans to cut off all medical benefits in December. Last month she appeared with Diane Sawyer on ABC's "Primetime Thursday," and now other networks are clamoring for an appearance. "I have this big open wound, but it's starting to form a tiny scab," she said. "I still feel the pain, but I'm doing what my husband would have expected of me."

For Elaine Chevalier, Brittany's mother, the journey back to everyday life has been powered by the earthly distractions of work and the spiritual nourishment that comes from intense faith. Those first catatonic months have given way to busy days managing commercial real estate in and around Middletown. But Ms. Chevalier says her true salvation has been her church and its support group. The crystallizing moment came one night last year in a dream, which featured Swede, the angel Raphael and her son's yellow Labrador retriever, Holly, who had also just died. "I'm trying to heal by thinking about my son in a different way, trying to think of him as a spiritual being," she said, sitting in the family's soaring great room with Brittany by her side. "Sometimes it works, but sometimes it doesn't cut it."

As she crosses the one-year mark, Ms. Chevalier believes she is entering a new phase of her life, one marked by self-reliance. (She is also seeking a divorce from her husband of 30 years). "The community has been so wonderful to us," she said, "but people can't feel sorry for us and cater to us forever."

It has been a busy year for the dozens of volunteers who came together to spoil the grieving families of Middletown. Besides raising $200,000 in cash, the group, Favor, made sure every family received overflowing gift baskets to mark Thanksgiving, Christmas, the depths of winter and the beginning of summer.

In June, the group decided it had done its job, and announced that it would disband. Several of the organizers, who set aside work and the demands of family, said it was time to return to their former lives. But Favor will not be fading away any time soon. The renewed flood of news media attention that began in recent weeks has sparked a fresh round of philanthropy, including that of a Texas millionaire who has offered scholarships to the 61 children who lost a parent last September.

At the very least, Allyson Gilbert, the group's executive director, said she and others have decided to put together one more gift basket, something small and simple, perhaps a tray of home-baked cookies crowned by a teddy bear. The baskets, she said, will probably arrive a week or two after Sept. 11, when the commemorative events and televised anniversary specials are through.

"They don't need us to deliver these huge food baskets or big checks anymore," she said. "I think they just need a reminder that we're thinking about them, that we have not forgotten, and that we're not going to go away."

MESSAGE FROM THE SENATE

A message from the Senate by Mr. Lundregan, one of its clerks, announced that the Senate has passed without amendment a concurrent resolution of the House of the following title:

H. Con. Res. 464. Concurrent resolution expressing the sense of the Congress on the anniversary of the terrorist attacks launched against the United States on September 11, 2001.

PAYING TRIBUTE TO AMANDA DAVIO AND ST. MARTHA CATHOLIC SCHOOL

Hon. Mike Rogers

OF MICHIGAN

Mr. Speaker, I rise to honor Amanda Davio and her classmates at St. Martha Catholic School in Okemos, MI, for their special efforts to thank the thousands of volunteers and emergency workers who responded to the September 11, 2001, attacks on the World Trade Center in New York City and the Pentagon here in Washington, DC.

Last year, soon after those tragic attacks, I asked children throughout Michigan's Eighth Congressional District to write letters and cards to the military men and women who were preparing for the war against terrorism. St. Martha students responded to that request along with hundreds of other students. Several of the schools, like St. Martha and Amanda Davio's kindergarten class, also sent along letters and cards for the workers at the attack sites.

These were forwarded to the Red Cross and eventually Amanda's card made it into the hands of New York City police officer Steve Tarricone. Officer Tarricone contacted the school, eventually traveled there to meet the students, especially Amanda, whose greeting has inspired him at a time when his spirits were very low.

Since then, the Davio family has visited New York and the two families have become good friends, developing a special bond born out of the shared experience of dealing with the aftermath of the tragedy of September 11, 2001.

The remarkable story of this new friendship is best told in the words of Amanda's father, Christopher Davio, who wrote:

With the approach of the anniversary of the tragic events of September 11, 2001, I'd like to relate an uplifting side to the story and how, out of such horrific happenings; blessings and new relationships can grow.

Shortly after 9/11/01, U.S. Representative Mike Rogers sent out a request to the schoolchildren of his district to write cards and letters to the rescue workers at Ground Zero, the Pentagon and in Pennsylvania. My daughters' school, St. Martha Catholic School in Okemos, like I'm sure many other schools in the area, answered the call and each class composed cards which were delivered to Mr. Rogers for forwarding to the workers.

The cards followed a circuitous route and my daughter Amanda Davio's (age 5 and in kindergarten at the time) card ended up posted in a Red Cross tent at the Staten Island landfill, along with hundreds of other cards and notes. There in early March of this year, the construction paper card caught the eye of an NYPD officer, Steven Tarricone. The card, with a 5-year-old's rendition of the American flag and the words "Thank You" on the cover had a simple but profound message inside, "You Make Me Feel Proud." Officer Torricone saw that the card, signed Amanda Davio, was stamped with the school's name and address on the back and wishing to express his gratitude for the support and comfort that the card and all the other cards and notes had given him, placed it in his pocket and took it home.

Steve showed the card to his wife and his partner. He had to put in long days and was drawing extra shifts while the city still struggled with the aftermath of the devastating attack. After a few weeks, he put together a package and sent it to the St. Martha School. Inside the package was a thank you card in which he wrote a little about himself and his job. He described the day he found Amanda's card, he was assigned to the Staten Island landfill and as he said, "My job for the day was to sift through the debris to try to find anything that would bring closure to the family members of victims of the World Trade Center." He thanked Amanda, her classmates and the staff at St. Martha School for their support and for taking the time out of their day to thank all the rescue workers. Steve included in the package five pictures he took on the days immediately following the disaster and took the time to describe on the back of each picture what was shown. He also sent along an NYPD cap, arm patch and ID cards with the request that they be given to Amanda.

Mrs. Helen Hillman, principal of St. Martha, related to me how moved she was as she opened the package and after discussing the contents with other staff members, Amanda's teacher, Kara Lampke, suggested she present the items to Amanda at the upcoming spring program that the schoolchildren were to perform the next week. When next she saw me at the school, Mrs. Hillman told me to be sure and have my camera ready at the end of the program as Amanda was going to be presented with something. Being the proud parents that we are, my wife and I videoed and photographed throughout the show and when the time came for Mrs. Hillman to present the items and relate the story, sat there stunned by the thoughtfulness of Steve's reply. For him to express his gratitude in such a way and to know that someone took the time to show all the children how important their support was to the workers and victims of September 11 was one of the most significant events in my life.

A few days later, Officer Tarricone called the school to see if they had received the package and talked to Mrs. Hillman at length about what the cards had meant to him and other workers at the sites. He said he was amazed at the outpouring of support shown by the entire county and was moved by the fact that Amanda's card had come from a kindergarten from halfway across the county.

My wife and I began to put together a few things to send back to Officer Tarricone and his family and after videoing greetings from Amanda and her sister Angela (8) and Alissa (14), included the tape of the presentation and the school program along with other cards, photos and expressions of thanks.

When Steve got that package, he called to share his excitement with us and told "in the past 24 hours, I've watched the tape at least 25 times." He had shown it to his mother and sister and they were all so happy about our reaction to his reply.

We kept in contact over the next few weeks and Steve told us that he had gone to the Policemen's Benevolent Association for permission to have Amanda named an honorary NYPD officer. After receiving permission, he had a plaque made and sent it along with more photos, commemorative pins and the arm patches from each of the units of the NYPD to Amanda.

In the meantime, Mrs. Hillman called our local papers and news outlets and a story was run on the front page of the *Community News* as well as a news segment on WLAJ which was aired as a local connection to the official closing of the cleanup effort at Ground Zero.

Since then, Mrs. Hillman has traveled to New York on a trip that she had planned long before all this developed. Steve met her at the airport with a red rose and welcomed her and her family. He arranged a visit to police headquarters and Ground Zero for all of them.

We took our family to New York at the end of August to meet Steve and his family (wife Michelle and daughter Ashley). Upon our arrival at a nearby hotel Wednesday, August 28, we called Steve and he immediately came to meet us. Greeting him for the first time was like seeing a close family member after a long absence. Amanda ran into his arms and the smiles lit up the whole lobby of the hotel. Steve took us to his house and we met Michelle and Ashley. Steve had a shirt made for Amanda in the style of his uniform, complete with her name and honorary badge number as well as NYPD arm patches and badge insignia.

We saw Steve again the next day at his house and met with a reporter and photographer from the *Long Island Catholic*, a diocese newspaper who had heard of the story from one of their staffers with family here in Okemos. After visiting with the representatives from the paper, we did a little touring locally then went to dinner with Steve and his family.

On Friday, Steve and Michelle met us at our hotel and took us into Manhattan. He had arranged a tour similar to the one given to Mrs. Hillman on her trip earlier in the month. We got to lower Manhattan about an hour before our appointment at One Police Plaza and while driving near Ground Zero, saw a fire station at the corner of Water and Wall Streets. Steve asked us to wait in the car while he went in and talked to the firemen on duty. The firemen, after hearing the story from Steve, welcomed us into the station, gave us a tour of the fire trucks and equipment and posed for pictures with the girls in fire suits and helmets. We paid our respects to the fallen members of the station at a memorial on the sidewalk in front of the firehouse, thanked the two firemen for the tour and their welcome, and went on to our appointment at police headquarters.

Pulling up to One Police Plaza was like entering a military post. Concrete barriers are placed so that only one vehicle can enter or leave the lot at a time and a large sanitation truck blocks the opening, pulling away to allow access after getting clearance from the guard post, then moving back to block the entrance. Security was tight! Upon entering the building, we passed through metal detectors and were photographed and given passes to wear on our outermost clothing.

We were all escorted to the Division of Community Affairs where we met Detective John Rowen and his daughter Ashleen. Detective Rowen took us to a conference room where we also met Detective Eugene Canapi. Gene had heard the story and came in on his day off to show us a presentation that the department had put together as a historical documentary on the events of September 11, 2001. Both men expressed their welcomes to us and told us how much the cards and letters from across the Nation had meant to them. John said that of all the cards he had seen, it never occurred to him to answer back and he was glad that Steve had shown such thoughtfulness to reply to Amanda.

After the presentation, reporters from the *New York Post* and *Newsday* interviewed us and took photos of Steve and Amanda in their "uniforms!" We met Deputy Commissioner Patrick (Division of Community Affairs) and were greeted and treated like VIPs by everyone.

Detective Rowen and his daughter took us all down to the waterfront near Battery Park where we boarded a Police Harbor Patrol boat and were given a tour of the area from the water. We rode under the Brooklyn Bridge, out to Ellis Island and the Statue of Liberty. I'm not usually an emotionally demonstrative person but I had tears in my eyes when I saw the statue. It was my first visit to New York City and I had never seen it before. I remember thinking back to 9/11/01 and hearing the threats made to her after the horrible attacks. I was sure at that time that I'd never see it in person and was truly overcome at seeing her from the boat. I imagine that is how my ancestors felt coming over from Italy at the turn of the last century.

While on the harbor patrol boat, I talked with one of the officers that made up the crew of three. He had no idea who we were and when I told him the story he told me how glad he was that Steve had replied in the way that he had. He described the events of that day and how they ferried survivors and rescue workers to the site and told me how much they all appreciated the support shown by the rest of the country after the attack. He said that the Red Cross had given him a box of cookies sent by some schoolchildren from New Jersey and that he still kept the note that they had enclosed in his wallet, nearly a year afterward.

After the harbor tour, we returned to One Police Plaza and were told that Police Commissioner Ray Kelly would like to meet Amanda. We were absolutely floored! Steve was really nervous as we waited to be escorted into the Commissioner's office, as he had never met him before either. Detective Rowen remarked that he doubted that many officers with Steve's experience (he's been on the force for 7 years) had been invited to meet the Commissioner in his office. Commissioner Kelly greeted us warmly and posed for pictures with Amanda and Steve and a group photo with all of us. He gave us a brief tour of his office and explained that his desk was Teddy Roosevelt's from his tour as Commissioner in the late 1800s.

After leaving the Commissioner's office, Detective Rowen took us to see Ground Zero. After seeing it so often in news coverage, it was an uncanny feeling to actually be at the site. I said a quiet prayer for the victims and listened while Detective Rowen related his memories from the day of the attack. He was at the command center when the first plane hit and was helping victims at the foot of the buildings when the first tower fell. There is a brief shot of him running up the street in one of the CNN videos as the cloud of debris follows behind. He said it was like a wall of water, he ducked into a side street and the wave of dirt followed him around the corner. A nearby visitor asked what is was like to be surrounded by the smoke and he said it wasn't smoke, it was more like dirt and fine particles of concrete dust that followed him and eventually covered him like so many of the photos we all saw from the news that day.

The site now looks more like a construction site and an individual happening along on it today would probably wonder what was going to be built there. The sides of the hole go straight down for probably five or six stories and you can see each level of the substructure of the underground areas across the way. We could see where the subway tunnel was going north from the site. For someone who had never seen the World Trade Center, it was hard to imagine just how tall it was. Standing at Ground Zero now, you are surrounded by skyscrapers, the tallest of which is 54 stories. I tried to explain to my daughters that if they took that building and placed another one just like it on top that would have been about the same height as the 110 stories of the Twin Towers. Many of the buildings still show damage from the attack and collapse of WTC. Still the cleanup has been a heroic effort in and of itself.

My family thanked Detective Rowen as we left and my middle daughter, Angela (8) exchanged e-mail addresses with Ashleen planning to stay in contact. As we drove back to Long Island we all were just amazed at the events of the day. Reflecting on the tragedy of last year and remembering the expressions of welcome and gratitude from each and every person we met.

When we had first discussed going to New York City to meet Steve and his family, he had told us that when we got there he was going to throw a big party. As the plans for the trip grew, we found out that his daughter Ashley's second birthday (September 1) would be celebrated on Saturday, August 31 and that he was having his whole family over. We picked that weekend to go so that we could meet all of them and make sure that they knew just how special we thought that Steve and Michelle were. Saturday dawned with a story in the *New York Post* about Amanda and Steve and how a small thank you card and its magnificent reply reached halfway across the country to bring them together. As we arrived at Steve and Michelle's, we were welcomed by all of their family and friends as a new part of the family. Steve's mother Linda had gifts for each of our girls, as did his Aunt Val & Uncle Len. We got to meet his partner and other friends form the force. About halfway through the party, we all heard the sound of bagpipes tuning up. Steve had told me that there was another surprise coming and as he called us all to the patio, he announced his lieutenant and other members of the Emerald Society, NYPD's pipe and drum corps. They gave a short concert for all of us in honor of Ashley's birthday and Amanda's visit. Once again the emotions overflowed to hear the patriotic medleys and other songs from the bagpipes and drum. It was a great way to cap off another extraordinary day.

Our last full day in New York was spent touring the city like normal visitors. We left Steven and his family to give them a chance to spend Ashley's birthday alone while we went to the observation deck of the Empire State Building, drove down Broadway and explored Manhattan by ourselves. For visitors coming to New York for the first time, you're struck by the size of all the buildings, the number of people and you quickly come to see that there is no place like it in the country. I've had the privilege of visiting 46 of the 48 contiguous United States as well as Hawaii and Puerto Rico. Truly New York City has no equal.

After having dinner with Steve, Michelle, Ashley, and Steve's sister Lisa's family, we headed back to our hotel and the next morning left for home. Our trip to New York City was way more memorable than we could ever have imagined. It gave my children an experience that they will never forget. Steve and his family are in the process of planning a trip here for a visit to St. Martha School in early November. We are all looking forward to seeing them again. The police officers and firefighters of New York showed the rest of the Nation that heroes arise from adversity. My family has been fortunate to meet one of those heroes and to get to know him as a good friend. People like Steve Tarricone are around us every day, and sometimes the small gestures, like a thank you card sent by a kindergartener can bring them into our lives. The next time your child comes home from school and says that they sent cards to someone, be it at a local nursing home, or to the President of the United States, I hope that you will recognize the importance of those messages to all who see them.

Mr. Speaker, Christopher Davio is right. Small gifts of kindness can have immeasurable benefits. Today, we wish to extend our appreciation to Amanda Davio and her family, Officer Steve Tarricone and his family and colleagues, and all the students from the Eighth District who helped our Nation begin its recovery. I now ask that our colleagues in the U.S. House of Representatives join us in recognizing this remarkable American story.

TRIBUTE TO THE HEROES OF THE 14TH CONGRESSIONAL DISTRICT

Hon. Anna G. Eshoo

OF CALIFORNIA

Mr. Speaker, I rise today to honor the extraordinary women and men of the distinguished 14th Congressional District for their heroic responses to the tragic events of a year ago.

On the first anniversary of the attacks on our Nation, we reaffirm our commitment to the ideals that have made the United States of America the greatest Nation on Earth. We have grieved for our lost loved ones, and now we continue the work of a freedom-loving Nation. We take pride in and are in awe of what the American people have done in response to the attack on our Nation. They have been brave and generous and the entire world has witnessed the strength and the decency of our people. Americans respond with open, brave and generous hands and hearts to those who are in need.

Mr. Speaker, the 14th Congressional District lost two extraordinary people, Naomi Solomon and Andrew Garcia, who enriched the lives of everyone they knew and loved.

I ask my colleagues to join me in once again offering our deepest sympathy and that of our entire Nation to the Solomon and Garcia families.

This Nation had many heroes on September 11, 2001. We all know of the supreme sacrifice made by so many firefighters, police officers and others in their response to the attacks on the World Trade Center and the Pentagon. We know of the heroism of those on the hijacked planes who prevented even more calamitous attacks on our Nation's Capital. We've learned of the heroism of people all over our country who pitched in to give service to others.

Mr. Speaker, it is with a great sense of honor and pride that I ask my colleagues to join me in paying tribute to all who have emerged from the tragedies of that fateful day to embody what it means to be true American heroes by giving so much of their time, talents and resources to heal the wounds of September 11, 2001. There are too many individuals, organizations and companies to name each separately, but I'd like to honor in our Nation's *Record* a few examples of some of the many people of the 14th Congressional District who rose to the occasion in extraordinarily generous ways:

The Town of Woodside Firefighters—held a "Fill the Boot" fundraiser on September 30 for the New York Fire 9/11 Relief Fund.

YWCA and the Mountain View City Human Relations Commission—held a "Building Community, Understanding and Respect" forum and dialog in response to September 11.

The Menlo Park Community Chorus and the Foothill Orchestra—organized a program of patriotic and inspirational music on December 15 which benefited the "victims and heroes" of September 11.

Banks, financial institutions and credit unions including San Mateo Credit Union and Stanford Credit Union—maintained and kept open their financial networks for their customers despite the resulting chaos of the attacks; and organized fundraising campaigns among their employees and customers to benefit the families affected by September 11.

Gallery Europa in Palo Alto owners Louise Erricson and David Himmelberger organized a special exhibit in which the sale proceeds were donated to families of victims of September 11.

Hyland Hogan and Lane Lees of the Half Moon Bay Fire District—following September 11, they boarded a plane and were adopted by NYFD Ladder Company 3 where they helped the company after it lost 12 of its members and assisted the families of lost firefighters. In May, the district presented the New York company with a memorial handmade case holding an ax recovered by 1 of the firefighters and pictures of the 12 who perished.

All the teachers and school administrators (like Jill Ballard and Sherry Fulton who teach American Literature and Studies at Half Moon Bay High School)—who changed their curricula and schedules to help students understand and cope with the events of September 11.

Law enforcement agencies and organizations like the Santa Clara County Sheriff's Office and the San Mateo County Deputy Sheriff's Association—set up funds to help the families of police officers killed in the September 11 attacks.

California Task Force 3 Urban Search and Rescue—deployed local firefighters and other task force members to New York to assist at Ground Zero.

Harold Schapelhouman, Menlo Park Fire District
Randall Shurson, Menlo Park Fire District
Paul Cole, Half Moon Bay Fire District
Gerald Kohlmann, San Jose Fire Department
Phil White, South San Francisco Fire Department
Troy Holt, Menlo Park Fire District
Brian Beadnell, Menlo Park Fire District
Ben Marra, Menlo Park Fire District
Carl Kustin, San Mateo Fire Department
Rexford Ianson, Menlo Park Fire District
John Preston, NASA Ames DART
Bill Trolan, Physician
Jared Strote, Physician
George Berry, Civilian—Communications Specialist
David Larton, Civilian—Communications Specialist
Martin Mijangos, Civilian—Technical Information Specialist
Mike Bavister, Civilian—Technical Information Specialist
Gordon Coe, Menlo Park Fire District
Hollice Stone, Civilian—Logistics Specialist
Mark Meyers, Civilian—Logistics Officer
Paul Brown, NASA Ames DART
Joe Zsutty, Structural Engineer
Raymond Lui, Structural Engineer
Harry Jackson, San Jose Fire Department
Kelly Kasser, NASA Ames DART
Crane Rigger, San Mateo County CDF
Robert Simmons, Civilian
James Stevens, Menlo Park Fire District—Medic
Kenneth Oliver, Menlo Park Fire District—Medic
Eric Haslam, South San Francisco Fire Department—Medic
Kevin Banks, Santa Clara Fire Department—Medic
John Wurdinger, Menlo Park Fire District—Technical Search Specialist
Roger Miller, NASA Ames DART—Technical Search Specialist K–9
Shirley Hammond, California OES, K–9
Jeff Place, California OES, K–9
Patricia Grant, California OES, K–9
Carol Herse, California OES, K–9
Tom Marinkovich, Menlo Park Fire District
Philip Snyder, NASA Ames DART
Don Chesney, Burlingame Fire Department
Michael Shaffer, Menlo Park Fire District
Rodney Brovelli, Menlo Park Fire District
Keith Slade, Menlo Park Fire District
Charles Sturtevant, Menlo Park Fire District
Jeff Schreiber, Menlo Park Fire District
Bill McFarland, Menlo Park Fire District
Mark Tagney, NASA Ames DART
Jeffrey Maxwell, Milpitas Fire Department
Chris De La Osa, Mountain View Fire Department
Daniel Horton, Redwood City Fire Department
Gerald Pera, Redwood City Fire Department
Steve Ehlers, Burlingame Fire Department
Bruce Barron, Burlingame Fire Department
Patrick Brown, Santa Clara Fire Department

Rod Villa, San Jose Fire Department
David Lerma, San Jose Fire Department
Greg Campbell, San Mateo Fire Department
Dave Rovetti, San Mateo Fire Department
Jesus Magallanes, South San Francisco Fire Department
Chris Campagna, South San Francisco Fire Department
Thomas Calvert, Menlo Park Fire District
Alex Leman, Civilian—Incident Support Team
Frank Fraone, Menlo Park Fire District—Incident Support Team
BK Cooper, Civilian—Incident Support Team
David Hammond, Civilian—Incident Support Team
John Osteraas, Civilian—Incident Support Team

The children of the Payvand Cultural School of Cupertino, an Iranian community-based school—filmed a special video after 9/11 to spread the message of tolerance and peace. The video is named "Hand in Hand" and it was sent to President Bush.

Local media—reporters like Mark Simon and Tom Abate with the *San Francisco Chronicle*, Loretta Green, Leigh Weimers and Jim Puzzanghera with the *San Jose Mercury News*, Don Kazak, *Palo Alto Weekly*, Dave Price with the *Palo Alto Daily*. Reporters from *The Almanac, Half Moon Bay Review/Pescadero Pebble, San Mateo County Times, Redwood City Independent, Los Altos Town Crier, Mountain View Voice, Silicon Valley Business Journal, Gentry, San Jose Magazine, Sunnyvale Sun, Cupertino Courier* all provided critical information, told our collective stories, shared our thoughts and helped to underscore a message of hope and tolerance.

Silicon Valley companies and businesses—Silicon Valley companies came together with their employees immediately after 9/11 to raise millions of dollars for charitable organizations. Many of the contributions made by companies were matched by employees, which brought aid to the affected families including those of firefighters and police officers. The senior executives at Sun Microsystems raised $1 million and the company matched dollar per dollar all employee contributions. That effort raised an additional $500,000. Sun Microsystems, like many Silicon Valley companies also participated in Ebay's Auction for America, donating over 1.3 million dollar's worth of products. Hewlett-Packard employees gave $1 million to support relief efforts. HP itself contributed $2 million, and matched its employees' gifts with another $1 million. In addition, HP like many other Silicon Valley companies, donated equipment to assist in the September 11 relief efforts. Companies like National Semiconductor not only made monetary contributions but also organized employee blood drives. Paypal through their members helped raise $2.35 million for the National Disaster Relief Fund of the American Red Cross. Cadence, under the leadership of CEO Ray Bingham, raised over $1.6 million in contributions to the American Red Cross and to the New York Firefighters' 9/11 Disaster Relief Fund. Cadence and its employees also held a special flag raising ceremony commemorating the tragic events of 9/11. Apple too went above and beyond to assist the victims. In addition to financial contributions to the Red Cross, Apple donated iBooks to the children of the rescue workers who lost their lives in the line of duty. These are but a few examples of the many contributions made by the employers and employees of the 14th Congressional District.

CHUMS—Children United Morally and Spiritually—designed an interfaith holiday card which they sold and donated the proceeds to victims of 9/11.

VA Palo Alto National Center for Post Traumatic Stress Disorder (NCPTSD) staff including Director Fred Gusman, Gregory Leskin, Robyn Walser, Sherry Riney, and Ken Drescher—who traveled to the Pentagon to provide the Department of Defense guidance and assistance for the psychological response efforts following 9/11.

The men and women of the California Highway Patrol—whose continuing vigilance helps ensure the safety of our bridges, airports, and other infrastructure.

The members of the Reserves and California National Guard who have been mobilized and their families—many of these dedicated individuals have taken deep pay cuts and will endure long separations from their families to prosecute the war on terror.

Palo Alto Medical Foundation and Stanford University (particularly Eric Weiss, MD and Lou Saksen)—formed a bioterrorism preparedness group to respond immediately and appropriately to any suspected cases of bioterrorism.

USPS—dealt with the aftermath of the anthrax attacks and continued to provide excellent service by delivering mail and keeping their offices open to the public.

American Red Cross Palo Alto Area Chapter—Deployed September 11 volunteers: The chapter was second in the State of California for percentage of response based on chapter population and serves 250,000 people in Palo Alto, Mountain View, Los Altos, Los Altos Hills, at Stanford University and Moffett Federal Airfield. The following Red Cross-trained volunteers and staff were deployed for assignments which lasted up to 3 weeks at a time following the September 11 attacks:

Ginny Anderson, WTC New Jersey, Disaster Mental Health
Vinnie Biberdorf, WTC New Jersey, Local Chapter Liaison
Ruth Anderson, WTC New York, Disaster Mental Health
Judy Boore, WTC New York, Disaster Mental Health
Rita Castro-Hawkins, WTC New York, Voluntary Agency Liaison
Don DeJongh, WTC New York, Family Services
Miriam DeJongh, WTC New York, Family Services
Ted Easley, WTC New York, Staffing for Disaster Services
Paige Filomeo, WTC New York, Disaster Mental Health
Adriana Flores, WTC New York, Disaster Volunteers
Lynne McCreight, WTC New York, Records and Reports
Edwin Ou, WTC New York, Logistics
Laura Quilici, WTC New York, Disaster Mental Health
Peggy Rogers, WTC New York, Disaster Mental Health
Richard Wing, WTC New York, Disaster Mental Health
Ann Ziman, WTC FMA Center, Family Services
Geoff Ziman, WTC FMA Center, Family Services
Karen Duncan, WTC NHQ Support, Public Affairs

American Red Cross Palo Alto Chapter (locally): Palo Alto Area Red Cross Chapter led by Executive Director Patricia J. Bubenik, staff members and volunteers assisted four local families with issues related to the September 11 disaster, including counseling the family of a victim of the Pennsylvania air crash. Mental health disaster volunteers went to schools, PTAs, and church groups requesting help in the aftermath.

Volunteers delivered materials to schools to assist with the conversations with children, teachers and parents.

The chapter staff also processed a total of $1,168,737 in donations designated for the National Red Cross (between September 11, 2001, and June 30, 2002).

At the same time, the chapter continued to respond to an increased interest in first aid and CPR classes and trained an increased number of disaster service volunteers who came in response to the September 11 tragedies. They also registered and placed an unprecedented number of volunteers who wanted to be of service within the community.

CENTRAL NEW JERSEY SHARES A POEM ON FREEDOM BY WORLD TRADE CENTER VICTIM DAVID SCOTT SUAREZ

Hon. Rush D. Holt

OF NEW JERSEY

Mr. Speaker, I rise today to share with you excerpts from a story that World Trade Center victim David Scott Suarez wrote about two hiking trips he had taken several years apart, and a poem he wrote about climbing as a metaphor for life and for freedom. David writes about freedom, both in terms of the struggle to attain it and the unparalleled joy of having it. In a sense, David's story reminds us that freedom is not free. It requires hard work and undaunting perseverance. Freedom can only be attained when people work together with others, exerting all of the collective strength of the unified group, to ascend its peak. One could interpret David's story to say that freedom is not even a choice, but rather a requirement for the realization of human potential, and that freedom should be our example to the world that we shout from the mountaintops.

David's parents, Ted and Carol Suarez, have so far had his poem translated into over 90 languages, including 3 of the major languages spoken in Afghanistan. They offer their son's story and poem in hopes that they will show all of the people of the world how much they have in common, so that we will always choose to communicate with each other rather than fight, and so that their son's death and the death of so many others on September 11 will not have been in vain. The following are excerpts from "Return to Freedom," by David Suarez.

My legs burned. My heart pounded. A bead of sweat ran down my forehead to the tip of my nose. I wiped it off with the back of my dirty arm just before it dripped to the ground. The air was cool and the wind grew fiercer the higher into the atmosphere we climbed. It froze sweat to my skin and blew my hair every which way, occasionally stinging my eyes. I looked up past Bob, who was directly in front of me, but I could not see our destination. The peak was covered in clouds .... Hail pelted my raincoat. The trail we'd been hiking quickly turned to a swiftly flowing stream. The sky lit up. Thunder cracked simultaneously. I began to hear the slow cracking of, not thunder but wood. We all turned abruptly. A tree fell across the path 50 yards behind us, its top shattered and smoking.

Only nine miles to go, but the weather showed no signs of letting up. At night we were going to make camp on top of Mount Philip at 11,711 feet .... It was thirteen miles away and a strenuous climb from where we broke camp in the morning. Unfortunately, that day was worse than any other had been. Like myself, the other guys in the expedition were pretty melancholy .... All I could think about was the 40 pound pack and the ice covered ground that kept me from moving forward with any sort of speed .... It continued to storm.

I trudged on. Stepping one foot in front of the other .... If I'd had a choice I'd have stopped, but there was no choice. Stopping meant hypothermia, which was worse than walking. Hours later, we reached the top. My hands were red. The tips of my fingers were almost white; they were completely numb. The clouds were so thick I couldn't see more than a few feet ahead. Everyone else was in the same condition, some worse .... There were fourteen of us, only six were able to pitch tents .... We pitched one after another. I thought each one along the way would be my last. Finally we finished and everyone was safe. Then, miraculously, the moment our tent was ready for sleep the clouds blew away and the warm sun came out ....

That was it, I was the last man standing. I was so excited I started to run to the peak .... I reached it minutes later .... I leaned back against the flagpole that stood higher than everything else. A smile of contentment crossed my face. I shut my eyes and fell asleep to the sound of the American Flag snapping in the wind. I was free.

Three thousand miles and five years later I was feeling the same thing. Freedom, what a strong word it is. Millions of people had died in its name. Do people fully understand and appreciate this single word? Do I? A month earlier I sat out on the lawn under the shade of a tall oak attending my Asian philosophy class. After class I walked past a preacher yelling that all of my peers (and myself) were doomed to hell. I walked farther and saw a stand with pictures of marijuana leaves all around, apparently fighting for its legality. I sat down and watched a couple walk past hand in hand and smiling. It was July 3. The impact of what was occurring before me hit me like a blow. I was living the dream that so many had died for. I belonged to a select group of people that could enjoy life as it should be enjoyed. In day to day life I often didn't realize that .... My mind and my talents marked the limits of where I could go. No one else dictated them.

Those thoughts reentered my mind as I climbed to the top of Sugarbush Mountain in central Maine. Climbing became a metaphor for life. We were almost at the top and the wind was blowing fiercely. We had entered the clouds and couldn't see a thing. At one point I opened my jacket and leaned into the wind. It supported my weight for awhile. Together we reached the top. We raised our hands and screamed loudly for the world to hear. We'd conquered this mountain. Although the steep slopes tried to keep us down, they couldn't. Although our lives threatened to trap us in dull routine, we escaped. We were in charge of our destinies, only us. For a moment the clouds cleared. It seemed as if we could see the entire world at once. In silence we watched. We were free.

YOU ARE FREE

(By David S. Suarez)

The air is cool, the sky is dark, your muscles relax, while
nature's breath fills your lungs
You have accomplished your tasks, felt the pain, and endured
the pressure, a pressure so immense that you lived
to escape
You have climbed to the very peak of the mountain and now
relax on a rock, high above the trees while others
sleep
You are enveloped by nature's beauty for just a moment you
abandon your incarcerated body wholly relinquishing
your ties to human nature and for only an instant,
you become part of God you are free

THE CRANBURY LIONS CLUB MARKS SEPTEMBER 11

Hon. Rush D. Holt

OF NEW JERSEY

Mr. Speaker, Wednesday, September 11, 2002, marks the 1st anniversary of the heinous attack on the United States of America by terrorists. On Saturday, September 14, 2002, the Cranbury Lions Club will remember the heroic actions on September 11 of a Cranbury, NJ, resident, Mr. Todd Beamer, with the dedication of a memorial in the township's Heritage Park. Mr. Beamer was aboard flight 93 on September 11, 2001, when it was hijacked by terrorists and crashed in Western Pennsylvania.

The memorial honors the uncommon service of Todd Beamer and his fellow flight 93 passengers whose selfless acts of courage saved countless lives and helped reunite our country. It also provides a permanent symbol to underscore the invaluable role of all citizens in protecting our unalienable rights of life, liberty, and the pursuit of happiness.

The Todd Beamer Memorial contains two symbols of strength, a boulder and an oak tree. On the boulder is a plaque that reads:

"LET'S ROLL"

These are the memorable words spoken by Todd Beamer, a Cranbury resident, who was aboard United Flight 93, when it was hijacked by terrorists on September 11, 2001, as he joined with his fellow passengers in a final act of resistance, sacrificing their lives to save countless others.

A man described as ordinary to the world, extraordinary to his family, he shall forever be remembered for his uncommon act of bravery. This memorial celebrates the faith and heroism of Todd Beamer—husband, father, son, brother, friend, civilian—an American.

Americans have read or heard about the heroic actions of Todd Beamer, and will always remember his simple, inspiring words: "Let's roll." As we memorialize his actions and words, it is equally important that we reflect on the life of Todd Beamer.

A native of Illinois, Mr. Beamer was born in Glen Ellyn, the middle child of David and Peggy Beamer, and spent his young adulthood in this suburb of Chicago. He was raised in a caring environment where value was placed on family, hard work, strength of character, and faith in God. In high school, he starred in soccer, basketball and baseball, serving as a team captain. Mr. Beamer continued to excel in athletics at Wheaton College where he earned a degree in business in 1991. He was later awarded an MBA from DePaul University.

In 1993, Mr. Beamer married Lisa Brosious, and they moved to Central New Jersey, soon settling in Cranbury to start their family. His prior success in athletics and academics was mirrored in his professional pursuits on behalf of Oracle Corporation.

Mr. Beamer's faith and commitment to his church was always evident. He was a member of the Princeton Alliance Church in Plainsboro. He served as a Sunday school teacher, participated on the church softball team, and mentored young adults.

Admired and loved by family, friends, and colleagues, the legacy of Todd Beamer will be his unwavering commitment to serving God and his fellow man. On behalf of all Americans, we extend our deepest gratitude to Todd Beamer's parents, David and Peggy; his wife, Lisa; his three children, David, Drew, and Morgan; and his two sisters, Melissa and Michelle.

Todd Beamer was a special man who made the supreme sacrifice for his country, and left a lasting mark on the people whom he touched. The Todd M. Beamer Foundation will ensure that his selfless act of giving to others in need continues in the future. He will be remembered by all.

CENTRAL NEW JERSEY SHARES THE ACCOUNT OF TRADE CENTER VICTIM FAMILY MEMBER SARAH VAN AUKEN

Hon. Rush D. Holt

OF NEW JERSEY

Mr. Speaker, I rise today to share with you an article written by Sarah Van Auken, 13-year-old daughter of World Trade Center victim Kenneth Van Auken. It was published last week in her local newspaper, and it presents a straightforward account of how the nightmare of September 11 unfolded before the eyes of a young person who found herself thrust suddenly onto the front lines of a war she didn't even know was taking place. It shows us not just how deeply painful and terrifying it is for a child to lose a parent, but also how this young woman's own feelings of fear, confusion and uncertainty as the day unfolded were magnified by the fact that she saw just the same feelings among the adults around her. Sarah Van Auken's life since that day became a swirling tapestry of endless tears, helpless longing for her father, and newfound celebrity born of the worst set of circumstances she could possibly have imagined. Out of her pain, she wrote a song in honor and memory of her father. The song paints a picture that perhaps we all might see ourselves within. A picture of a person, standing, quietly, waiting, listening for the faintest sound on the wind of the guiding hand that will come back and show us show how to get through this, the guiding hand that we can grasp so that we'll find ourselves together again, safely, home. This has been a year of deep searching and painful discovery for us all, and I would like to share Sarah Van Auken's account of it with you.

This past year has been very hard for me. You see, my father, Kenneth Van Auken, was in the World Trade Center on September 11, 2001. No, he did not escape—but he did leave a message saying, "I love you. I'm in the World Trade Center. The building was hit by something. I don't know if I'm going to get out but I love you very much. I—I hope I'll see you later. Bye." That was the single most horrible thing I had ever heard in my life. He was trying to stay calm for us—trying to let his last words be "I love you." Somehow, I wish I could go back in time and erase all that happened. Maybe even stop him from going to work. I wish I could have one last goodbye. But I guess it's too much to ask.

You're most likely wondering how I found out. Well, I was having a regular day at school. You know, boring—yet I was with my friends. Anyway, I was in study hall minding my own business when someone yelled out, "Is it true that a plane crashed into the World Trade Center?" Knowing my dad worked there, I wrote a note to my friend next to me saying, "If that's true, my dad would be dead!" I didn't believe what he said because the teacher acted like nothing happened. Also, I wouldn't trust that kid. So as the day went on, I felt weird. You know like when you know that something is wrong, but you really don't think about it? At eighth period, around 1:30 p.m., an announcement came on saying there is a "little accident" in New York—and if we get home and one of our parents is not there, we should not worry. If you get scared, we should call 911 or talk to the police. That's when I got scared. When I was walking down the hallway, I almost started crying, but held back my tears. When I got in the car to go home, my neighbor who drives me tried to get one of my classmates to stop talking about the announcement. She was obviously trying to stay away from the subject. Then, when we got to that boy's house, his dad started talking about it. He didn't say what happened, but gave me a weird look. I got home and saw my grandparents' car. I knew they weren't supposed to be there. I saw my mom with a tear-stained face, and I ran up to her and she didn't have to tell me. I just cried.

From that day on, nothing has been the same. Nobody has treated me the same. Nobody wanted to talk about it—yet they couldn't help asking me questions about what had happened, and how I was doing. When I knew for sure, after 3 days, that my father was dead, I cried harder than I have ever cried in my life. My father, my superman, was dead. We had a memorial, and went on "Oprah." I wouldn't eat. I couldn't sleep in my own bed. I would cry about the smallest things. I was wearing one of his shirts to feel close to him. I was looking at family pictures. Of course I was still crying. I couldn't figure out what would make me stop being so depressed and irritable. I had to get it out. I wanted to scream, run, jump—but I couldn't. I just didn't have the strength. I cried too much.

So, I did what I usually did to get out my feelings: I wrote a song. I sang it to my mom and she called my godmother, who called her brother-in-law, who told me to record myself singing and send it to him. Exactly a month after September 11, I recorded it in a studio. The song titled "Daddy's Little Girl" was on a local radio station twice, once in California and on "Larry King Weekend." I always wanted publicity because I wanted to be famous—but not this way. Today I am still crying when nobody's around. I think about what happened constantly, but can't really talk about it. And though I may sound selfish, somehow I think nobody knows how I really feel. My life is turned upside down. The things I used to do I either can't do anymore, or I've lost interest, or they seem so much harder. I'm trying to "move on," but I don't want to. My mind has accepted that he's dead, but my heart hasn't. And somehow, I don't think my heart will. Because I'll never stop crying, not in a million years.

Sometimes, it will hit me that he's gone forever—that he's never coming home. I recently had a bat mitzvah. It was very hard, just like the 11th of every month is hard, and Father's day, my mom's birthday, my brother's birthday, my birthday, my dad's birthday, and most of all next week's September 11 anniversary. I know most of the teens that are reading this might often think about what it would be like if you lost a parent. I used to wonder too. Except now I don't wonder. I know.

DADDY'S LITTLE GIRL

(By Sarah Van Auken)

Standing-daddy's little girl (just); Standing (yeah)-daddy's little girl . . . .

I wonder, wonder through the trees, blow the wind, blow the wind to me. Control, controlling my fears, somewhere, behind these tears. And may, maybe you'll appear, somehow whisper in my ear (my ear, my ear!)

CHORUS

If you were just standing here, I could erase these tears of mine! And all these words would disappear, oh! Standing-daddy's little girl (just); Standing (yeah)-daddy's little girl . . . .

Can it, can it be, that the wind is guiding me! Daddy are you there? 'cause I've, I've looked everywhere I need, I need you! What should, what should I do! And may, maybe you'll appear, somehow whisper in my ear (my ear, my ear!)

CHORUS

If you were just standing here, I could erase these tears of mine! And all these words would disappear! I just want to find you, but there's nothing I can do. Where do you roam? I just want you home!!!!

Standing-daddy's little girl (just); Standing (yeah)-daddy's little girl . . . .

EXPRESSING THE SENSE OF CONGRESS ON THE ANNIVERSARY OF TERRORIST ATTACKS LAUNCHED AGAINST THE UNITED STATES ON SEPTEMBER 11, 2001

Hon. Deborah Pryce

OF OHIO

Mr. Speaker, I rise today on the anniversary of the day our great Nation was forever changed when terrorists attacked and killed thousands of Americans simply because they were American.

Today, first and foremost, our Nation pays respect to the victims and their families. We stand united and remind them that 1 year has passed and still, we will never forget September 11.

What happened to the United States on that infamous day brought out the best of the American spirit.

The enemies who struck us grossly miscalculated the strength and resolve of the American people.

They didn't know that our bonds of liberty, our bonds of freedom, and our bonds of democracy are stronger and run deeper than any individual, than any building, than any monument.

As President Bush said,

> This country will define our times, not be defined by them. As long as the United States of America is determined and strong, this will not be an age of terror; this will be an age of liberty, here and across the world.

During this unprecedented time of great challenge, there will be no corner of the Earth where the demons of September 11 will be safe from justice.

America will continue to fight for the security of our great Nation, and for peace in the world.

We will never forget every firefighter, flight attendant, father and friend that died that infamous day. May God watch over their families and continue to bless America.

Hon. William O. Lipinski

OF ILLINOIS

Mr. Speaker, December 7, 1941, is the worst day in the history of our Republic in the 20th century, and September 11, 2001, is the worst day in the history of our land of liberty in the 21st century. Both days cost this Nation thousands of lives; mothers, fathers, daughters, sons, grandmothers, grandfathers, aunts and uncles perished on these days because they were Americans. Their families and friends left behind have never been the same—nor will they ever be the same—and the same can be said for our Nation.

On both occasions these victims were victims because of what America stands for: liberty, freedom, justice, human rights, opportunity, and a faith in a caring and loving God. But out of this criminal act perpetrated upon the citizens of this Nation and on this fortress of freedom that we call the United States of America, a fierce determination arose to destroy those forces of evil that without cause or warning attacked the United States. We brought those who attacked us on December 7, 1941, to justice, and we are well on our way to bringing those who attacked us a year ago to the same fate. But today, September 11, 2002, we stop to remember in a formal way the victims and their families who perished on these very, very dark days in our Nation's history. Today we stop to honor them, remember them, pray for them, and rededicate ourselves to seeing to it that this never happens again in America or anyplace else in the world.

Hon. Stephen F. Lynch

OF MASSACHUSETTS

Mr. Speaker, I rise today to recognize and pay tribute to the victims of the tragic events of 1 year ago. Last year on September 11, Americans awoke to a brutal attack on our country on its own soil. Throughout the course of this one tragic day, something that at once seemed inconceivable became a horrific reality. No one feels the pain of this day more acutely than the families and friends of the more than 3,000 people who lost their lives at the Pentagon, the World Trade Center and in Shanksville, PA. All of these people and their families are in our thoughts and prayers on this somber occasion.

Although there is little that we in Congress can do to ease those families' suffering, by adopting this resolution, we are reaffirming our commitment to honor the memory of the people who were lost that day, while also paying tribute to those individuals who unselfishly risked their own lives to protect others.

Mr. Speaker, September 11 was one of the most difficult days in American history. But in the darkness of that day, an incredible spirit of bravery and hope emerged. Hundreds of emergency rescue personnel descended upon the scene at both the World Trade Center and the Pentagon with the sole purpose of assisting others. At the same time, ordinary people demonstrated amazing courage by trying to help others escape while putting themselves in peril and in fighting back against the terrorists on United flight 93. It is truly remarkable how many people gave their lives trying to protect others. The bravery and generosity of these people is a lasting mark of September 11.

In responding to these extraordinarily trying times, the true fabric of American society was illustrated to the world. Americans around the world came together and generously gave of themselves in a myriad of ways. Rescue workers spent countless hours at Ground Zero searching for survivors and then shuffling through the debris. Construction workers, ironworkers and other personnel tirelessly worked their way through the wreckage in an effort to clean up the site. Their important task would not be interrupted by exhaustion, injury or inclement weather. However, far away from Ground Zero, and across the globe, people generously gave their time, energy, money and caring to help support the loved ones of the lost victims. Today, we honor these selfless contributions.

As we gather now, 1 year later, it is my hope that we never forget the spirit that pervaded this country in the weeks and months following the attacks. As we continue to rebuild and to heal, we will need to draw upon that strength. The American people demonstrated amazing resolve and resilience in the last year, and it is a resolve that we must continue to maintain, day by day, week by week, this year and for many years, that we will preserve our freedoms, protect our families, and work to cleanse the world of the scourge of terrorism.

Mr. Speaker, in the wake of September 11, Congress rallied in a strong bipartisan manner to quickly pass legislative measures to protect our country. It was this remarkable unity of purpose that most struck me when I was sworn into this body in October of last year. Over time, this unity has dissipated some, but our goal should remain clear. We in Congress owe it to the American people to do all that we can to make sure that the necessary resources are available to protect our country. This is a serious responsibility and not one that should be burdened by partisan debate. In the coming months, we must act responsibly and decisively to ensure that the people of the United States once again feel safe in their own cities and towns. I commend the leadership of both the Republican and Democratic parties for recognizing this important anniversary and for introducing this thoughtful resolution.

Hon. Edolphus Towns

OF NEW YORK

Mr. Speaker, we are commemorating the terrible attack on America on September 11 last year. This was an event in which about 3,000 people lost their lives. A year later, they are in our prayers.

Also in our prayers are the other victims—those who were subjected to violent, unfair attacks in the aftermath of September 11. One of these was Balbir Singh Sodhi, a gasoline station owner from Arizona. He was murdered at his gas station by someone who apparently mistook him for a follower of Osama bin Laden. His brother, Sukhpal Singh Sodhi, a cab driver in the San Francisco Bay area, was recently killed in his taxicab. I am sure that we would all like to extend our sympathies to the Sodhi family.

No one should be killed because of his religion. Even if Mr. Sodhi had been a Muslim and a follower of bin Laden, that would not justify murdering him. But what makes this crime even more disturbing is that this perception was a mistake. Mr. Sodhi was a Sikh, not a Muslim.

Sikhism is an independent, monotheistic, revealed religion that believes in the equality of all people, including gender equality. It is not part of either Hinduism or Islam, yet because of the turbans they wear, which are required by their religion, Sikhs are sometimes mistaken for Muslim followers of bin Laden.

The violence has mostly ended, but unfortunately, there are still some unrelated violent incidents. I call for an end to all these attacks and for full and prompt prosecution of all the people responsible.

Mr. Speaker, I would like to place the Council of Khalistan's recent press release on the anniversary of September 11 into the *Record* at this time.

IN MEMORY OF THOSE KILLED IN LAST YEAR'S ATTACK ON UNITED STATES

SIKHS SUFFERED THE MOST AFTER THE ATTACKS; COUNCIL OF KHALISTAN CONDEMNS ATTACKS, CALLS FOR END TO VIOLENCE AGAINST MINORITIES

WASHINGTON, D.C., September 11, 2002.—Dr. Gurmit Singh Aulakh, President of the Council of Khalistan, today remembered the attacks on America a year ago that killed almost 3,000 Americans. He also condemned the violence against Sikh Americans and other minorities that broke out in the wake of the September 11 attacks.

"On behalf of the 21-million strong Sikh Nation and especially on behalf of more than 500,000 Sikh Americans, we remember with sadness and outrage the attacks on America a year ago and offer our prayers and sympathies on this sad anniversary to the people of the United States for the terrible attack on the United States and for the loss of life it entails," Dr. Aulakh said. "We especially pray for the families of those who have departed."

"America must do what it can to eradicate terrorism from the world," Dr. Aulakh said. "We support all the efforts to do so and we must do our part as American citizens," he said. "This sad anniversary reminds us that we stand together as a nation. We must show unity on this occasion."

"We also concemn the violence against Sikhs and other minorities that took place last year after the September 11 attacks," Dr. Aulakh said. "Sikhs suffered the most in the post-September 11 violence," he said. "The very first victim of this violence was Balbir Singh Sodhi, a Sikh gasoline station owner from the Phoenix area," he noted. "Recently, his brother was killed in his taxicab. All this violence must stop," Dr. Aulakh said.

"Nobody should be killed for his or her religion, whether Sikh, Muslim, Christian, Jewish, Hindu, or whatever religion one may follow," Dr. Aulakh said. "But it is important to note that Sikhs are not Muslims nor followers of bin Laden. "We condemn bin Laden," he said. "Unfortunately, because of the turbans we are required to wear, many people mistake

Sikhs for bin Laden followers," he said. "The Sikh religion is an independent, monotheistic, sovereign religion that believes in the equality of the whole human race, including gender equality," he said. "Daily we pray for the well-being of the whole human race."

In the wake of the September 11 attacks, a couple of young Sikhs were attacked in Brooklyn. Sikh businesses have been stoned and cars have been burned. A Sikh boy was even shot in New York. Many Muslims and other minorities were also subjected to violent attacks.

"We hope that there will not be any more of these incidents in connection with the anniversary of the attacks. "Violence against innocent people of any religion or ethnicity is unacceptable," said Dr. Aulakh. "It must be condemned and the violence must be ended."

Hon. Bart Stupak

OF MICHIGAN

Mr. Speaker, I wish to add my voice to the multitude of Members honoring our Nation and its heroes on September 11, 2002.

Mr. Speaker, although I was back in my district taking part in events commemorating the impact September 11 has had on all of us, I would have voted "yes" on passage of H. Con. Res. 464. Due to a technicality, my name was not added as a cosponsor of this worthy bill, and I wish to state my intention here that I fully support this resolution and its sentiments.

As a former law enforcement officer, I know too well the toll such tragedy takes on individuals—their lives, their families, their future—and I know too well how difficult, yet how necessary it is to ensure like-minded individuals are prevented from carrying out further attacks.

This resolution makes it clear that while the passage of a year has not softened our memories, it has shown that we will not bow down to terrorism.

We must find those responsible for the deaths of so many—including my constituent Army Major Kip Taylor who perished in the Pentagon on that day a year ago—and ensure they face the consequences of their actions.

September 11 brought out the worst in our enemies. Yet it also brought out the best in our citizens. That is what we are honoring today.

Hon. Philip M. Crane

OF ILLINOIS

Mr. Speaker, Last week it was my high honor and privilege to join my colleagues in the House and Senate for a commemorative joint session of Congress in New York City to honor the victims and heroes of September 11.

While we Members of Congress are often engaged in abrasive confrontation, today I look around and see total unity, total recognition that whether Republican or Democrat, we are first and foremost Americans, and the common values we share far outweigh those we do not.

This is the same expression of unity demonstrated by Americans across the country on the days following the terrorist attacks on September 11. I find comfort in the knowledge that it represents a promise that we will not back down from preserving our freedoms and protecting our homeland from those who wish to destroy our way of life.

And as we revisit some of the darkest moments in our Nation's history, we must remember that our Nation has always been one that has triumphed over adversity. Indeed, I think it is fair to say that at times of great despair, America has consistently risen to its greatest hours.

So in remembrance of those lives lost on September 11, I would like to conclude with some words from President Lincoln's Gettysburg Address:

> that from these honored dead we take increased devotion to that cause for which they here gave the last full measure of devotion—that we here highly resolve that these dead shall not have died in vain; that this Nation shall have a new birth of freedom; and that this government of the people, by the people, for the people, shall not perish from the Earth.

Hon. Betty McCollum

OF MINNESOTA

Mr. Speaker, I rise today in memory of the events of September 11.

One year ago America suffered a horrible act of terrorism in New York, at the Pentagon and in Pennsylvania. Four planes, filled with innocent Americans, were turned into weapons at the hands of men filled with anger and hate, intent on bringing death and destruction to our great country. It is a day none of us will ever forget.

As the United States moves forward, we must remember those who died on September 11, as well as the acts of heroism, valor and courage displayed on that day and the weeks and months that followed. I continue to find inspiration in the efforts of all Americans who risked their lives to save and heal their neighbors, coworkers, and strangers in need.

Let us also not forget the men and women in our Armed Forces who today are engaged in a campaign against terrorism, fighting to protect our freedom and seeking justice against those who attacked us. Their valor is a testament to the will and resolve of our great Nation.

We will continue to pray for the victims and their families as we rebuild the communities affected by those terrible acts of violence. Today, 1 year after this horrific act of terrorism, we, as Americans, reaffirm our highest beliefs in freedom, democracy and justice.

Hon. Dave Camp

OF MICHIGAN

Mr. Speaker, I rise in support of this joint resolution and in doing so, I stand in solemn remembrance of the tragic events of a year ago and a pride in America's response.

While the loss of life was immense, and the impact of the terrorist attacks was felt far from New York, Washington, DC, and Pennsylvania, the terrorists failed. The group of men, and the larger organization they represented, wanted to break the will of a proud and strong country. The world stands in witness to their failure.

Instead of falling apart, our country united. Our brave first responders worked tirelessly to help survivors; we saw ordinary citizens involved in heroic efforts; and all across the country Americans joined together to offer assistance.

The outpouring of support and unity could be seen in every flag that was flown with pride across this country. The United States of America rose to the challenge presented to it with a resolve that was felt around the world.

Now, on our first Patriot Day, when we see our flag at half-mast, let us not only remember the tragic events of a year ago, but also the strength exhibited by all Americans. I urge my colleagues to join me in supporting this resolution by expressing solidarity on this day of remembrance.

PUBLIC BILLS AND RESOLUTIONS

Under clause 2 of rule XII, public bills and resolutions were introduced and severally referred, as follows:

By Mr. ARMEY (for himself, Mr. Gephardt, Mr. Hastert, Mr. Barton of Texas, Mr. Bereuter, Mr. Bilirakis, Mr. Bonilla, Mr. Boozman, Mr. Brown of South Carolina, Mr. Buyer, Mr. Cannon, Mr. Cox, Mr. Crenshaw, Mrs. Cubin, Mr. Cunningham, Mr. Tom Davis of Virginia, Ms. Dunn, Mr. Fletcher, Mr. Forbes, Mr. Gibbons, Mr. Gilman, Mr. Goss, Mr. Graham, Mr. Green of Wisconsin, Mr. Hayworth, Mr. Hilleary, Mr. Hoekstra, Mr. Horn, Mr. Jenkins, Mr. Sam Johnson of Texas, Mrs. Kelly, Mr. Kerns, Mr. Kingston, Mr. Kirk, Mr. LaHood, Mr. McHugh, Mr. McInnis, Mrs. Morella, Mr. Ose, Mr. Pickering, Mr. Petri, Mr. Portman, Ms. Pryce of Ohio, Mr. Putnam, Mr. Riley, Mr. Schrock, Mr. Sherwood, Mr. Simmons, Mr. Simpson, Mr. Smith of Michigan, Mr. Taylor of North Carolina, Mr. Toomey, Mr. Walsh, Mr. Wamp, Mr. Watkins, Mr. Watts of Oklahoma, Mr. Weller, Mr. Wicker, Mr. Wilson of South Carolina, Mr. Wolf, Mr. Young of Florida, Mr. Goodlatte, Mrs. Emerson, Mr. LaTourette, Mr. Ramstad, Mr. Rohrabacher, Mr. Shays, Mr. Saxton, Mr. Rogers of Kentucky, Mr. Fossella, Mr. Ballenger, Mr. Royce, Mr. Reynolds, Mr. Walden of Oregon, Mr. Linder, Mr. Mica, Mr. Castle, Mr. Stearns, Mr. Calvert, Mr. Thomas, Mr. Dan Miller of Florida, Mr. Gutknecht, Mr. Blunt, Mr. Rehberg, Mr. Nethercutt, Mr. Ehlers, Mr. Brady of Texas, Mr. Hobson, Mr. LoBiondo, Ms. Pelosi, Mr. Frost, Mr. Menendez, Ms. DeLauro, Mr. Holden, Mr. Cramer, Mr. Turner, Mr. Hoyer, Mr. Hastings of Florida, Mr. Thompson of Mississippi, Ms. Harman, Mr. Holt, Mr. Wexler, Mr. Markey, Ms. Velázquez, Mr. Green of Texas, Mr. Barcia, Ms. Norton, Mr. Wu, Mr. McDermott, Mr. Phelps, Mr. Acevedo-Vilá, Mr. Blumenauer, Mr. Tanner, Mr. Hinojosa, Mr. Pallone, Mr. Underwood, Mr. Meeks of New York, Mr. Crowley, Ms. Kilpatrick, Mr. Ackerman, Ms. Hooley of Oregon, Mr. Davis of Illinois, Mrs. McCarthy of New York, Mr. Gutierrez, Mr. Borski, Mr. Bishop, Mr. Jackson of Illinois, Mr. McGovern, Mr. Berry, Ms. McCollum, Mr. Lynch, Mr. Ross, Mrs. Napolitano, Mr. Weiner, Mr. Blagojevich, Mr. Lantos, Mr. Pascrell, Mr. Filner, Mr. Waxman, Mr. Peterson of Minnesota, Mr. LaFalce, Ms. Carson of Indiana, Mr. Ford, Mr. Moran of Virginia, Mr. Matsui, Mr. Berman, Ms. Millender-McDonald Mr. Olver, Mr. McNulty, Mr. Oberstar, Mrs. Maloney of New York, Ms. Watson, Mr. Lucas of Kentucky, Ms. Rivers, Mr. Costello, Mrs. Capps, Mr. Sabo, Mr. Meehan, Mr. Capuano, Ms. Kaptur, Mr. Boyd, Mr. Larson of Connecticut, Mr. Baca, Ms. Schakowsky, Mr. Kanjorski, Mr. Hall of Texas, Ms. McCarthy of Missouri, Mrs.

Meek of Florida, Mr. Cummings, Mr. Pomeroy, Ms. DeGette, Mr. Barrett, Mr. Doyle, Mr. George Miller of California, Mr. Lipinski, Ms. Roybal-Allard, Mr. Sanders, Mr. Rangel, Mr. Shows, Mr. Wynn, Mr. Baird, Mr. Schiff, Mr. Boucher, Mr. Murtha, Mr. Spratt, Mr. Deutsch, Mr. Kildee, Mr. Farr of California, Mr. Luther, Mr. Tierney, Mr. Clay, Mr. Engel, Mr. Gonzalez, Mr. Honda, Mr. Becerra, Ms. Jackson-Lee of Texas, Mr. Inslee, Ms. Slaughter, Mr. Ortiz, Mr. Serrano, Mr. Davis of Florida, Mr. Israel, Mr. Smith of Washington, Mr. Rothman, Mr. Osborne, Mr. Gekas, Mr. Thornberry, Mr. Hill, Mr. Stenholm, Mr. Baldacci, Mr. Rahall, Mr. Watt of North Carolina, Mr. McIntyre, Ms. Lofgren, Mr. Hoeffel, Mr. Maloney of Connecticut, Mrs. Jones of Ohio, Mr. Sawyer, Ms. Solis, Mr. Dooley of California, Ms. Baldwin, Mr. Nadler, Mr. Kind, Mr. Kleczka, Mr. Sandlin, Mr. Boswell, Mr. Coyne, Mr. Udall of Colorado, Ms. Berkley, Mr. Kennedy of Rhode Island, Mr. DeFazio, Mr. Moore, Ms. Eshoo, Mrs. Wilson of New Mexico, Mr. Cooksey, Mr. Tancredo, Mr. Barr of Georgia, Mr. Kolbe, Mr. Moran of Kansas, Mr. Johnson of Illinois, Mr. Ferguson, Mr. Culberson, Mr. Baker, Mr. Ryun of Kansas, Mr. Ehrlich, Mr. Peterson of Pennsylvania, Mr. Radanovich, Mr. Foley, Mr. Duncan, Mrs. Biggert, Mr. Pastor, Mr. Dingell, Mr. Thompson of California, Mr. Gordon, Mr. Cardin, Mr. Roemer, and Ms. Eddie Bernice Johnson of Texas):

H. Con. Res. 464. Concurrent resolution expressing the sense of the Congress on the anniversary of the terrorist attacks launched against the United States on September 11, 2001; considered and agreed to.

Thursday, September 12, 2002

HONORING CONGRESSIONAL STAFF AND GOVERNMENT EMPLOYEES

Hon. Brian Baird

OF WASHINGTON

Mr. Speaker, yesterday our Nation and this House paid a fitting tribute to those who lost their lives on September 11, to the heroic rescuers, and to their families.

Today, I would like to take just a moment to honor another group of people who serve this Nation in a less dramatic way, but who are heroes in their own right and in their own quiet ways.

Last year, just 1 day after our Nation came under attack, and this very building was among the targets, the men and women who work here in this building, in our offices and in countless other government offices throughout this land, came right back in to work to serve this great Nation. When they came in to work on that September 12 morning, they knew then and they have known each and every day since then that they work in a potential target.

Scarcely a month later, they then faced a new challenge when anthrax entered our buildings, and for some of our staff, entered their bodies. The Capitol Police, the janitors and maintenance workers, the grounds crews, the people who serve food, the secretaries, the Parliamentarians, the clerks, the young pages, our legislative and our committee staff, our field and caseworkers, and all the other dedicated and courageous people who make this place and our government run all deserve our thanks and our praise.

With tears in their eyes, with sadness and with fear in their hearts, but with indomitable courage they came right back to work to serve this country we all love.

A year has passed now, and the immediate danger may have been diminished; but it remains in our awareness. Still, our staffs and the rest of the employees come to work, and in doing so, they serve our country.

In these times, this takes courage. So, and for that courage, I am grateful and this country is deeply fortunate.

LESSONS LEARNED FROM SEPTEMBER 11

Hon. Mark Foley

OF FLORIDA

Mr. Speaker, as we continue our reflection on September 11, I wanted to take a moment to enter into the *Record* a piece written by the majority leader, Speaker pro tempore of the Florida

House of Representatives, Sandra Murman from Tampa, FL:

When I hear the date September 11, images immediately flood my mind. I see the plane hit the second tower. I see the Pentagon on fire and I can hear the sickening crunch as the towers fall while throngs of people run to escape the thick gray cloud. I also remember the utter horror I felt when I realized this was not simply one plane off course but rather a planned attack. This was our generation's Pearl Harbor. But unlike Pearl Harbor, terrorists hijacked planes full of innocent civilians and crashed those planes into buildings filled with more innocent civilians. On that day we saw the face and felt the hand of evil, but we also saw extraordinary goodness through the lives of heroic Americans in Washington, New York, and a Pennsylvania field.

As we gather here to mark the 1-year anniversary of the attack, I would like to share my thoughts on what I have learned since last September.

Lesson one: I have been reminded that life is short and precious. That argument with a spouse, the concern over which car to purchase on September 10, suddenly seemed so petty after the attacks. As I evaluated my own life, I realized what mattered most was my relationship with God, my family, loved ones and community. Everything I do now needs to have meaning, purpose, and positively impact those around me.

Lesson two: Before September 11 we knew we had enemies and lived in a dangerous world, but September 11 we discovered that organized groups of terrorists had both the desire and the ability to create devastation within our country. We can no longer take this security for granted. There is our new reality.

Lesson three: On September 11 America showed that we are still a Nation of heroes. Incredible courage was shown by the New York City firefighters who slapped on their gear and charged into the burning buildings to help victims escape. New York lost 343 of its finest that day. Hundreds of workers in the World Trade Center helped one another escape. I remembered hearing the story of one man who, instead of escaping Tower Two, chose to remain behind with a disabled colleague who could not make it down the stairs. They both perished that day. And, of course, we all heard the story of flight 93, those extraordinary men and women who said their goodbyes to their loved ones, prayed the Lord's Prayer, and with the words of "Let's roll," charged the cockpit to save countless lives in Washington, D.C.

In an instant these ordinary Americans became legends. All the sacrifices on September 11 have left us speechless with gratitude.

Lesson four: We have the responsibility to ensure that the lives lost on September 11 were not lost in vain. We were attacked because of who we are. The principles on which our country was founded, freedom, equality and the dignity of the individual, are a threat to Islamic extremists. They view open, democratic societies as the enemy and want to create a society where there is no religious freedom and no civil liberty. As defenders of liberty we stand in their way.

At this very moment our servicemen and women are defending the cause of freedom throughout the world. Here on the home front, we, too, have a responsibility. Our defense involves upholding the values of America. We have a civic duty to participate in our democratic institutions. We have a responsibility to instill in our children a love of liberty, a love of country, the difference between right and wrong and the willingness to make sacrifices in this ongoing struggle between freedom and tyranny.

Let me close by reading President Bush's September 20 speech to the Nation:

"Great harm has been done to us. We have suffered great loss. And in our grief and anger we have found our mission and our moment. Freedom and fear are at war. The advance of human freedom, the great achievement of our time, and the great hope of every time, now depends on us. Our Nation, this generation, will lift the dark threat of violence from our people and our future. We will rally the world to this cause by our efforts, by our courage. We will not tire. We will not falter. We will not fail."

Thank you. May God bless you all.

HONORING MR. ERIC MORELAND JONES

Hon. Barbara Lee

OF CALIFORNIA

Mr. Speaker, I rise today to honor and commend bravery and dedication demonstrated by my constituent, Mr. Jones, who was a first responder at the attack on the Pentagon on September 11, 2001.

The memories of the horrific terrorist attacks will remain with us forever. Yet, through the pain and adversity of these tragedies, heroes were also born.

We witnessed the attacks on the World Trade Center in New York City, we learned of the terrible airline crash in the Pennsylvania countryside, and we witnessed what was once an unfathomed attack on our Nation's center of defense, the Pentagon.

I have known Mr. Jones' family for many years. In the footsteps of his parents, he carries on a legacy of commitment to humanity though public service. On September 11, Eric was driving by the Pentagon when it was hit by American Airline flight 77. He immediately went to the Pentagon site and quickly began to aid in evacuating injured and dying personnel from the building; he carried and helped people to safety and medical triage. Eric remained at his volunteer post for more than 72 hours.

On July 15, 2002, Mr. Jones was one of two people to receive the Office of the Secretary of Defense Medal of Valor for his actions. As we commemorate the anniversary of the 9/11 attacks, we also pay tribute to thousands of first responders and volunteers like Mr. Jones who risked their own lives to ensure that others were saved.

I am deeply moved by Eric's heroism and want to extend my sincere appreciation to him. As we take time to reflect on the events of 9/11 on this anniversary day, we must also resolve and recommit ourselves to peace and security.

H. RES. 5367—TO NAME THE DEPARTMENT OF VETERANS AFFAIRS OUTPATIENT CLINIC IN HORSHAM, PA, THE "VICTOR J. SARACINI DEPARTMENT OF VETERANS AFFAIRS OUTPATIENT CLINIC"

Hon. Joseph M. Hoeffel

OF PENNSYLVANIA

Mr. Speaker, I rise today in strong support of the resolution that Representative Greenwood and I introduced yesterday, which will name the new veterans clinic in Horsham, PA, after Victor J. Saracini, a distinguished veteran and victim of the attacks on September 11.

Victor J. Saracini served his country with great pride as an exemplary technical coordinator aboard S–3A fighter jets on the USS *Saratoga*. He served in the Naval Reserve at the Naval Air Station Joint Reserve Base, Willow Grove, PA, until his honorable discharge as lieutenant in 1985. Victor Saracini was the recipient of the National Defense Service Medal, the Navy E Ribbon, and the Expert Marksmanship Ribbon.

On the morning of September 11, 2001, terrorists hijacked the Los Angeles-bound airplane that Captain Saracini was piloting, and reset course for the South Tower of the World Trade Center, killing everyone on board and murdering hundreds of other innocent civilians inside the building. These innocent victims, Mr. Saracini included, represent our Nation's first casualties in this war on terror.

To honor the life of Victor Saracini, devoted aviator, distinguished veteran, and proud defender of America's freedom, is to honor all victims of September 11 and their families. I urge my colleagues to support this resolution, and I call on the House Veterans' Affairs Committee to pass this resolution and bring it to the floor of the House of Representatives as soon as possible.

Hon. Vito Fossella

OF NEW YORK

Mr. Speaker, I am not recorded on roll call No. 384, Expressing the Sense of the Congress on the Anniversary of the Terrorist Attacks Launched Against the United States on September 11, 2001. I was with my constituents of Staten Island and Brooklyn on this sad anniversary. Had I been present, I would have voted "aye."

For the past year, our Nation has grieved over the loss of nearly 3,000 brave men and women who were cruelly and unfairly taken from God's Earth much too soon. These past 365 days have been a time of immense sadness for our Nation. We have buried too many innocent souls—too many mothers, too many fathers, too many sons and too many daughters.

Today is officially known as Patriot Day as a result of legislation that I introduced in Congress. I chose this name because I thought it best described the victims of September 11—men and women who loved their country and who died in its name. While they were not soldiers, they certainly were patriots.

Indeed, no one among us will ever forget the indelible images of brave firefighters, police officers and other emergency services personnel entering the burning towers bound by honor, duty and courage. Or the pictures of ordinary Americans leading their friends, coworkers and even strangers out of the rubble because they were taught to help those in need. In an age when the word heroism is bandied about much too often, we watched true heroes in action.

And so today, we remember these patriots—recall their smile, their laugh, their kindness. Their loss is an injustice to humanity. And while they can never be replaced, they must be remembered and honored for making the greatest of all sacrifices.

The American story is far from finished. Indeed, the best chapters are yet to come. We must believe that, for I know in my heart that it is our destiny.

We also must believe that there is a just God directing our people in a just cause of liberty. That cause, like others before, which crushed fascism and communism, is now to forbid the tyranny of terrorism. The terrorists sought to destroy America by crushing brick and twisting steel. They didn't understand that the source of America's strength is its people, and that its people embody a spirit of optimism and hope that can never be destroyed. Our hearts may still be heavy, but our soul is stronger and more vibrant than ever. The values of America will forever stand firm and resolute.

My prayers go to every family that lost a loved one on September 11. My words cannot ease your suffering, so I simply tell you that you remain in my thoughts. God bless you and God bless America.

I ask unanimous consent that this statement be printed in the appropriate part of the *Congressional Record*.

DEDICATION OF THE SEPTEMBER 11 MEMORIAL IN ORADELL, NJ, AND PRESENTATION TO MRS. TRACY WOODALL

Hon. Marge Roukema

OF NEW JERSEY

Mr. Speaker, I rise today to call to the attention of my colleagues the dedication of a memorial in Oradell, NJ, to honor and commemorate those who lost their lives in the tragic events of September 11, 2001.

Let me say, Mr. Speaker, that I have been deeply moved by the outpouring of support and dedication that we all have seen throughout our Nation over the past year. In the days and weeks after the tragic events of September 11, we heard and read the stories of countless family members, neighbors, and friends who went to work on that day and never came home.

In my own district, our Bergen County community was particularly hard hit. We all know someone who was lost. Their stories are heart wrenching . . . and still remain nearly unbearable in their sadness.

Over the past year, I have spoken to many families in my district in an attempt to bring them some consolation. Even though there are no words to relieve their anguish, I have told each family that they should take some comfort in the knowledge that the hearts and prayers of the entire Nation were with them.

As we dedicate this memorial in Oradell this evening to all of those who lost their lives on that tragic day, we pause to remember each of the men and women whose lives were so tragically cut short by a brutal and senseless act of terrorism.

In particular, we commemorate the tragic loss of one of Oradell's own residents, Brent Woodall.

In their deaths, the victims of the World Trade Center attack have come to symbolize all that we love in America. The terrorists attacked the towers because they represented America's democracy, freedom, diversity, and economic prosperity.

Brent Woodall embodied these ideals in his work and in his life. Whether in his work in the stock market—the nerve center of America's economic freedom—or as a talented athlete, or simply as a man deeply devoted to friends, family, and those whom he loved, Brent's life exemplified the American values which have made our country great.

The loss of every life that day was tragic. The loss of Brent touches each of us, as he and Tracy were just beginning so much of their life together. They had just bought a home, and were beginning a family together.

I did not know Brent personally, so I will not presume to elaborate upon his life and times beyond that. But as I have come to know the nearly 100 residents of my congressional district who never came home on September 11, so I have come to know Brent Woodall.

In every way, Brent's life was a life that is easy to celebrate.

This evening we will commemorate our losses and send a message of heartfelt sympathy and support to Brent's family and friends, particularly his wife, Tracy, and their son, Pierce Ashley, who came into this world on April 22, 2002, only a few short months after his father had perished. How proud Brent would have been of his son ... and how proud Pierce will someday be of his father, whose good nature, humor, and zest for life live on in him.

At tonight's memorial, I will be honored to present to Tracy Woodall an American flag that was flown over the U.S. Capitol in Brent's honor.

Our flag has long stood as the symbol of our core values of freedom and liberty. It now stands also as a symbol of our national resolve to bring those responsible for this atrocity to justice, and, tonight, as a tribute to Brent, and all of those who lost their lives in one of America's darkest hours. Let it serve also to let Tracy, Pierce, and all of their family know that the support of extended family, friends, community, and the Nation, are with them now and always.

Mr. Speaker, I ask my colleagues to join me in asking that God bless Tracy and Pierce Woodall, the rest of their family, and all those who lost friends, family, or loved ones in this national tragedy. And, as Brent Woodall would have wanted, we ask that God bless the United States of America.

CONGRATULATING H. BYRON MASTERSON ELEMENTARY SCHOOL OF KENNETT, MO, ON WINNING THE KIDS ARE AUTHORS CONTEST FOR "SEPTEMBER 12TH ... WE KNEW EVERYTHING WOULD BE ALL RIGHT"

Hon. Jo Ann Emerson

OF MISSOURI

Mr. Speaker, I come to the House floor today with the memories of September 11 forever etched on my mind. I remember thinking that the world would never be the same again after that fateful day. I was right. Now, 1 year later, I am touched and moved by the courage, compassion and character that people across our Nation have shown in the days and months since the attack on America.

I am nearly moved beyond words by the ways our children have responded to the new challenges facing our Nation. Immediately following September 11 I visited classrooms all over the Eighth Congressional District. I listened and spoke with students, teachers and parents and felt—for the first time I can remember—a bond and sense of purpose that was somehow missing in the days before.

I have never been as proud to be an American as I was when I visited with the children at those schools in my district. To be honest, I wasn't sure exactly how to talk about the tragic events of September 11, because I wasn't sure how much they understood about why this tragedy happened to us. Instead of comforting them, they comforted me. Instead of me telling them what happened, through their patriotic songs, intelligent questions, cheers of pride, patriotic bulletin boards, and their hugs and tears, they shared what they had learned and seen.

One of those schools was H. Byron Masterson Elementary School in Kennett, MO. The students shared their feelings, but they did more than that. They took action. And this week, a year later, the results of their actions were heard and seen in New York City. The message from the children is one of comfort. Their story is summarized in a story reported by the Associated Press and I would like to share it with all of you.

Darlene Robertson says that on some days, the rut is the best place to be. On Sept. 12th, it was the daily rut of life in Robertson's southeast Missouri town that provided the stability her first-grade students needed in that insecure time after the terrorist attacks. "September 11 upset the routine of America, and these little children felt it," Robertson said. "That's why the rut was so important for us that day." Those students, now second-graders at H. Byron Masterson Elementary School, wrote about their experience in a book entitled, September 12th ... We knew Everything Would be All Right. The book, which the children also illustrated, won the Kids Are Authors contest sponsored by Scholastic Books. Now Scholastic is publishing the book and distributing it nationwide.

When Robertson first heard of the contest, she began talking with her husband about topics for a book her students could write. They knew that a lot of children would be writing about the terrorist attacks of September 11, so her husband suggested September 12 as a topic. So the idea for the book was born, and in March, Robertson and her students began to recall what September 11 and 12 were like—how they had discussed what they saw on television, and how she assured them they were safe inside their classroom. "September 11 shook us all up. As a faculty, we had to be careful not to show our true feelings about the day. We had to do things just like we had been doing the day before. It gave the kids security." The book takes readers through the day after the attacks and how the students' daily routine was a comfort to them: "The sun rose again, and the students traveled to school as usual. They still had homework. And two plus two still added up to four," they wrote in the book. "On September 12, our parents still tucked us in our warm, safe beds," they wrote. "We knew we would be all right because our parents said they loved us."

My favorite quote from the book is one that I used recently in my weekly column about September 11. The children wrote, "We knew everything would be all right because the stars and moon came out and America went to sleep. And the next morning the sun came up again."

The students, together with their parents, were recognized for their achievement. They along with teacher, Darlene Robertson and her husband, Dennis, and Masterson Principal Elsie Heller, left for New York City early Monday morning, September 9.

The group of approximately 40 spent three days in the Big Apple including the 1-year anniversary of September 11. The trip, sponsored by NASDAQ, ended with the group taking part in the ceremonial opening and ringing of the bell at the NASDAQ market on September 11. During their stay in NYC, the group visited various sights including every child's dream, Toys R Us, New York. They also toured the Empire State Building and the New York Public Library. And they took a trip to the company, Scholastic, whose contest made all of this possible.

Scholastic will be at H. Byron Masterson Elementary School on September 12th for a banquet, where they will present the students with medals. The school also will receive 100 copies of the book and an autographed copy of the book will be sent to President Bush.

As their teacher Mrs. Robertson said "We're just a little small town of 11,000 in the Bootheel of Missouri but here we are . . . It is an honor to be chosen."

It is an honor for me to represent these children and their families in Congress. Congratulations on this remarkable and special milestone in your lives. You children have inspired me. You have shown your compassion for others. You have displayed the true character of America. You have shown me and other parents and adults your maturity and depth of understanding about our great Nation. You have given us resolve. You have given us courage. And you will help us show the world that no act of terror will ever bring us to our knees. We will be stronger than ever in the face of adversity. We will be one. We will be tougher. We will prevail.

CONGRATULATIONS TO LEAH A. CUNNINGHAM

Hon. Rob Simmons

OF CONNECTICUT

Mr. Speaker, I rise today to congratulate Leah A. Cunningham of Niantic, CT. Leah Cunningham was named a national winner in the 2002 Voice of Democracy Program and received the $1,500 Department of Colorado and Auxiliary Award. Leah was sponsored by VFW Post 5849 and its Ladies Auxiliary in East Lyme, CT.

I applaud the achievements of Leah Cunningham and ask that her award-winning essay be submitted into the *Congressional Record.*

[From the 2001–2002 VFW Voice of Democracy Scholarship Contest]

Reaching Out to America's Future

(By Leah Cunningham)

Yiyia, what is your advice for me and my role in helping America to have a better future?

"I came to this country at only 12 years old, alone and frightened of the unraveling journey ahead. I emigrated from Greece, but I soon became a loving citizen of this great land called America. I have learned that for America to have a better future, we must trust and learn from the issues of the past."

And then, my Yiyia (which is Greek for grandma) would smile in her strong oak rocking chair, gazing out the window of her apartment. Yiyia would have faith in the youth of America and their love for a country. She had seen the beginning stages of World War I as torpedoes were launched at her boat; she had watched America slowly enter World War II, and thankfully, she died before her eyes would witness the devastation of the worst terrorist act to ever assault American soil: The destruction of the New York trade center towers by two hijacked airplanes.

These horrific events of September 11 have sparked a new found interest in our past and pride. Have we perhaps become more aware of our duty to create a peaceful life for our youth? The idea is to reach out to America's future, enabling our children to create a better world, providing them with knowledge, insight. Someday as a grandmother, I hope to share with my grandchildren the knowledge with which a nation has touched my existence. I will reach out to America's youth—empowered in good faith to help America's future.

Our Nation has indeed suffered tragedy but at the same time, we have been blessed by devoted American citizens striving toward a common goal: to make America a peaceful nation. Firemen, Red Cross volunteers, policemen, and average American citizens are diligently working in New York City to defy evil and restore the site of utter human devastation. We have refocused our priorities, acknowledging kindness, not only kindness for our friends and relatives, but a rejuvenated sense of benevolence toward strangers and fellow Americans. I see a confident nation, converging together, providing that we will not fall, we will not falter, we will not fail in a time of unforeseen cruelty toward our freedom land. We are reaching out to America's future in quiet and bold ways. I even see a rebirth of historical values and national pride.

President George Washington wrote in his 1796 farewell address: "The unity of government which constitutes you one people is also now dear to you. It is justly so, for it is a main pillar in the edifice of your real independence, the support of your tranquility at home, your peace abroad; of your safety; of your prosperity; of that very liberty which you so highly prize . . ." Washington's words ring across more than 200 years to reach America today and in the future. The future of America depends on our ability to secure unity and influence the well-being of active American citizens. Simply requiring a civics or history course for high school graduation is not enough. We should urge our youth to become involved in the social fabric of the community. The little things truly aid in reaching out to America's future: encouraging youth to register for voting, involving teenagers in mock political systems, having children understand the American flag, and ensuring appreciation toward war veterans and their roles in providing long-standing freedom in America.

I have come to think of it as my responsibility, my mission, to in some way reach out to America's future. My grandparents remember where they were when Pearl Harbor was attacked by the Japanese. My parents remember where they were when John F. Kennedy was assassinated. And, I will forever remember exactly where I was on September 11, 2001, when terrorists attacked our Nation, killing thousands. My greatest achievement will be if a defining moment of my grandchildren's life is not a catastrophic preempt to war, or a brutal disheartening assassination of a loved President, or an act of horrific human destruction. But rather, their moment of true American unity and love for a nation will be when their grandmother reaches out to their curious eyes and big hearts, and tells them of her experiences as an American and what they must do to hopefully follow in her patriotic footsteps.

As Thomas Jefferson suggested in his first Inaugural Speech, our principles for peace in the future depend on the ability to historically look back in order to look forward. Jefferson states, ". . . Let us hasten to retrace our steps and to regain the road which alone leads to peace, liberty, and safety." The youth of America will bloom with bright hearts and clear visions if they are mindful of America's pursuits and "retrace their footsteps" of answers.

America is living and breathing, and within this country there is embedded a recipe for survival and for peace. Our youth need the support and encouragement of patriotic citizens. We must trust in the goodness of people, and work toward a humane world, with the youth of America as leaders toward peace and justice. We must start with the seeds of tomorrow, the children of America's future, to not only establish a long-term remedy for terrorism, but to maintain strength, pursue unity, and forever sustain national loyalty.

Tuesday, September 17, 2002

IN REMEMBRANCE OF THE VICTIMS OF THE KATYN FOREST MASSACRE AND THE WORLD TRADE CENTER ATTACKS

Hon. Robert Menendez

OF NEW JERSEY

Mr. Speaker, I rise today to honor the victims of senseless and unspeakable atrocities. The New Jersey Division of the Polish American Congress sponsored a memorial service to remember those who lost their lives during the tragic Katyn Forest Massacre 62 years ago, and those killed during the attack on America, September 11, 2001. The service was held at the Katyn Monument site in Jersey City, NJ, on September 15, 2002.

After Nazi Germany and the Soviet Union maliciously invaded Poland in 1939, the Polish citizenry fought bravely against both adversaries on two fronts. Unfortunately, in the process of valiantly defending their homeland, over 15,000 Polish soldiers, officers, intellectual leaders, prisoners of war, and other Polish citizens were brutally murdered. Perhaps one of the most unforgettable acts committed by the Soviet Union against Poland was later uncovered with the discovery of 4,500 bodies found in a single mass grave at the Katyn Forest, near Smolensk in the Soviet Union. This horrendous discovery became known as the Katyn Forest Massacre.

And in an egregious attack against humanity, over 3,000 Americans and citizens representing more than 80 nationalities were lost at the World

Trade Center, the Pentagon, and the fields of Somerset County, PA, on September 11, 2001. The heinous attacks on American soil reaffirmed our commitment to democracy in defense of a free and open society, threatened by evil, injustice, hatred, and tyranny.

Today, I ask my colleagues to join me in honoring the lives lost in these tragedies. We shall never forget these acts of barbarism. And we shall never forget the innocent lives lost as we strive, as a people, to create a peaceful world.

SEPTEMBER 6, 2002: A TIME TO MOURN

Hon. Mike Pence

OF INDIANA

Mr. Speaker, the Good Book tells us that there is a time for every purpose under heaven. There is a time to weep and a time to mourn. On September 6 I joined some 250 of my colleagues in this body as we traveled to Federal Hall in New York City to do just that.

We gathered at a place in which this Congress met and even adopted the Bill of Rights in 1789. We mourned with those who mourn and we wept with those who weep regarding the September 11 tragedy. The last time I was in New York City was September 21, 2001. I stood in the ashes and on the periphery of the devastation at Ground Zero.

As we join to pray, Mr. Speaker, let us ever remember that we are also told that there is a time for peace and there is a time for war. As we pray for the bereaved let us also pray for wisdom for our President and our leadership in this institution as we choose the times and the days ahead for war.

IN MEMORY OF ADEL A. ZAKHARY

Hon. Robert Menendez

OF NEW JERSEY

Mr. Speaker, I rise today to remember Adel A. Zakhary, a friend and neighbor to many, who lost his life in the World Trade Center tragedy on September 11, 2001. A memorial service took place in his honor on Saturday, September 14, 2002, at Saint George Catholic Orthodox Church in Jersey City.

An immigrant from Egypt, Adel lived the American dream, making America his home with his wife, Nagat, son, George, and daughter, Mariam. He was dedicated and tireless in his work, and in providing for his family. On September 11, he went to work on the 92d floor of tower one of the World Trade Center, as he had for 18 years.

In one of the most unforgivable acts against humanity, over 3,000 people were lost at the World Trade Center, the Pentagon, and the fields of Somerset County, PA, on September 11, 2001. Adel was among those individuals who were tragically lost. The horrific attacks have strengthened us in our resolve to fight evil and intolerance in pursuit of freedom, justice, and democracy.

Today I ask my colleagues to join me in remembering Adel A. Zakhary, a loving husband and father, who will never be forgotten. Let us join together not only to grieve this tremendous loss, but also to celebrate the remarkable accomplishments in his life. I extend my deepest sympathies to the family and friends of Adel.

SEPTEMBER 11, 2002: TRIBUTE TO FREEDOM

Hon. Mike Pence

OF INDIANA

Mr. Speaker, the United States of America is founded on the fundamental principle that all citizens have the inalienable right to life, liberty and the pursuit of happiness.

The United States of America stands as a beacon of freedom and opportunity for everyone regardless of race, creed or religious belief.

The strength and vitality of the United States of America is in the diversity of its people, the diversity of its ideas, the freedom to express those ideas and the opportunity to achieve one's potential and direct one's destiny.

Mr. Speaker, these ideals and principles are absolute and will not be surrendered or weakened by the cowardly acts of terrorists who fear the sunshine of freedom and the responsibility it brings.

Let us forever remember that the date September 11 reaffirms the principles for which the United States of America was founded and that on this day each year freedom shall ring from every community in this great land and the voice of America will be heard around the world.

IN MEMORY OF SEPTEMBER 11 AND ITS FORGOTTEN VICTIMS

Hon. Dan Burton

OF INDIANA

Mr. Speaker, we are commemorating the terrible attack on America on September 11 last year. This was an event in which about 3,000 people lost their lives. A year later, they are in our prayers.

Also in our prayers are the other victims—those who were subjected to violent, unfair attacks in the aftermath of September 11. One of these was Balbir Singh Sodhi, a gasoline station owner from Arizona. He was murdered at his gas station by someone who apparently mistook him for a follower of Osama bin Laden. His brother, Sukhpal Singh Sodhi, a cab driver in the San Francisco Bay area, was recently killed in his taxicab. I am sure that we would all like to extend our sympathies to the Sodhi family.

No one should be killed because of his religion. Even if Mr. Sodhi had been a Muslim and a follower of bin Laden, that would not justify murdering him. But what makes this crime even more disturbing is that this perception was a mistake. Mr. Sodhi was a Sikh, not a Muslim.

Sikhism is an independent, monotheistic, revealed religion that believes in the equality of all people, including gender equality. It is not part of either Hinduism or Islam, yet because of the turbans they wear, which are required by their religion, Sikhs are sometimes mistaken for Muslim followers of bin Laden.

The violence has mostly ended, but unfortunately, there are still some unrelated violent incidents. I call for an end to all these attacks and for full and prompt prosecution of all the people responsible.

Mr. Speaker, I would like to place the Council of Khalistan's recent press release on the anniversary of September 11 into the *Record* at this time.

In Memory of Those Killed in Last Year's Attack on United States

Sikhs Suffered the Most After the Attacks; Council of Khalistan Condemns Attacks, Calls for End to Violence Against Minorities

Washington, D.C., September 11, 2002.—Dr. Gurmit Singh Aulakh, President of the Council of Khalistan, today remembered the attacks on America a year ago that killed almost 3,000 Americans. He also condemned the violence against Sikh Americans and other minorities that broke out in the wake of the September 11 attacks.

"On behalf of the 21-million strong Sikh Nation and especially on behalf of more than 500,000 Sikh Americans, we remember with sadness and outrage the attacks on America a year ago and offer our prayers and sympathies on this sad anniversary to the people of the United States for the terrible attack on the United States and for the loss of life it entails," Dr. Aulakh said. "We especially pray for the families of those who have departed."

"America must do what it can to eradicate terrorism from the world," Dr. Aulakh said. "We support all the efforts to do so and we must do our part as American citizens," he said. "This sad anniversary reminds us that we stand together as a nation. We must show unity on this occasion."

"We also condemn the violence against Sikhs and other minorities that took place last year after the September 11 attacks," Dr. Aulakh said. "Sikhs suffered the most in the post-September 11 violence," he said. "The very first victim of this violence was Balbir Singh Sodhi, a Sikh gasoline station owner from the Phoenix area," he noted. "Recently, his brother was killed in his taxicab. All this violence must stop," Dr. Aulakh said.

"Nobody should be killed for his or her religion, whether Sikh, Muslim, Christian, Jewish, Hindu, or whatever religion one may follow," Dr. Aulakh said. "But it is important to note that Sikhs are not Muslims nor followers of bin Laden. "We condemn bin Laden," he said. "Unfortunately, because of the turbans we are required to wear, many people mistake Sikhs for bin Laden followers," he said. "The Sikh religion is an independent, monotheistic, sovereign religion that believes in the equality of the whole human race, including gender equality," he said. "Daily we pray for the well-being of the whole human race."

In the wake of the September 11 attacks, a couple of young Sikhs were attacked in Brooklyn. Sikh businesses have been stoned and cars have been burned. A Sikh boy was even shot in New York. Many Muslims and other minorities were also subjected to violent attacks.

"We hope that there will not be any more of these incidents in connection with the anniversary of the attacks. "Violence against innocent people of any religion or ethnicity is unacceptable," said Dr. Aulakh. "It must be condemned and the violence must be ended."

FOSELLA-WATT MOTION

Hon. Carolyn B. Maloney

OF NEW YORK

Mr. Speaker, as we continue our war on terrorism, we must remain vigilant in our efforts to decimate the terrorists who are threatening our very existence and have singled out Americans who represent freedom and democracy, so cherished by our citizens.

I strongly support the Fossella-Watt motion, which will finally allow American victims of international terrorism to receive compensation from blocked assets—judgments they were already awarded.

Last week, we commemorated the anniversary of 9/11, a day that marked the most devastating acts ever committed on U.S. soil.

There may not be another Member of Congress who lost more constituents in the September 11 attacks on the World Trade Center than I did. I knew countless numbers of victims and their families. One year later, the pain and hardship go on. No amount of money can bring back our loved ones, but this motion can work to prevent future tragic acts of terrorism.

The Fossella-Watt motion paralyzes the financial resources of those terrorist organizations and increases our ability to go after the sources of funding for these organizations and cells. It sends a message to terrorists that we will not stand for the murder of innocent Americans. And those who target Americans will be punished.

The United States must use every tool in its arsenal—military, diplomatic, and legal—to protect Americans and other innocent parties against these random acts of terror.

The Fossella-Watt motion is a tool to weaken the terrorist grip.

I urge my colleagues to retain this provision in the final version of the terrorism insurance bill.

Thursday, September 19, 2002

IN MEMORY OF META FULLER WALLER

Hon. James P. Moran

OF VIRGINIA

Mr. Speaker, I rise today to honor the memory of Meta Fuller Waller, a dear friend to many, a dedicated public servant, and athletic team captain who tragically lost her life in the Pentagon on September 11, 2001.

Born into a family steeped in the civil rights movement, Meta Waller learned at a very young age an appreciation for the arts and the value of a good education. Her two famous grandparents, Meta Warrick Fuller, an African-American sculptor and Solomon Carter Fuller, the first African-American psychiatrist in the United States, inspired Meta to pursue her dreams regardless of what stood in her path. These instilled values guided Meta throughout life, especially during the sorrowful loss of some of her closest family members.

With a bachelor's degree from the University of Michigan and a master's degree from the prestigious Harvard Kennedy School of Government in 1982, Meta worked hard to meet the many challenges she faced as the Special Programs Manager for the Administrative Assistant to the Secretary of the Army. In her 12 years at the Pentagon, Meta was heavily involved in the Combined Federal Campaign (CFC), the annual fund-raising drive conducted by Federal employees on behalf of numerous non-profit charities. She diligently served as the Army CFC administrator for 14 years and helped raise in excess of $30 million to benefit the least fortunate in our society.

An avid writer and poet, Meta charmed those fortunate enough to witness her literary talent. Meta's active imagination made her a gifted storyteller whose vividly refreshing tales could keep an audience spellbound for hours. Always in search of new challenges, Meta picked up the game of tennis much later in life than most. Despite a lack of past exposure to the sport, she rose to become captain of her women's tennis team, holding the position for 3 years.

Ever conscientious and adventure seeking, Meta's passions led her to travel the world often. Her most recent trip took her to Durban, South Africa, for the World Conference on Racism. Traveling with a group of schoolchildren, Meta gained a first-hand knowledge of the continuing struggle to end racism across the globe. Upon returning home, Meta told family members that the experience had changed her life.

Mr. Speaker, Meta's life serves as a testament to us all that with love and determination we can overcome any odds and lead inspired lives. Everyone misses her dearly but the memory of her indomitable spirit will never be forgotten.

TERRORIST ATTACKS OF SEPTEMBER 11, 2001

Hon. Timothy V. Johnson

OF ILLINOIS

Mr. Speaker, last Wednesday our Nation commemorated the terrorist attacks of September 11, 2001. While these attacks were committed on the World Trade Center and the Pentagon, they were in fact directed at our Nation as a whole. Our freedom, our way of life, the very foundations of our great democracy, were ruthlessly targeted by an unprecedented force of evil. Now, 1 year later, our Nation is stronger and more unified than ever to rid the world of terrorism in all of its forms, as well as its root causes including poverty, injustice, and despair. It is my sincere hope that America never forgets the terrible atrocities committed within our borders. These acts were a direct attack upon freedom-loving people everywhere and we have a duty to ensure that freedom and democracy prevail in this struggle against tyranny and oppression.

RACING REMEMBERS

Hon. Ernie Fletcher

OF KENTUCKY

Mr. Speaker, it is an honor for me to recognize the American racing industry for its response to the terrible tragedies our Nation suffered a year ago. I am deeply gratified to note that the Nation's horseracing industry, which is of such great importance to the Commonwealth of Kentucky, shared in our Nation's ceremonies of remembrance on September 11. Yesterday, all across the country, our racetracks, owners, trainers and jockeys all stood together to remember what happened a year ago and to honor those who were lost and those who showed such great courage in the aftermath of the terrorist attacks.

The National Thoroughbred Racing Association requested that all racetracks operating on September 11 cease normal business operations to share in a 10-minute, nationally simulcast observance at 4:10 p.m. eastern time. All across the country, there was no racing or related activity at any NTRA-member facilities during the brief, dignified and patriotic service which included a flag ceremony, a moment of silence, the singing of the national anthem and a video tribute.

The nationwide ceremony allowed racing and its fans to remember September 11 together, even though they were at many different locations, because the observance was broadcast via simulcast to many different facilities from Del Mar Thoroughbred Track in California. It was hosted by Emmy Award-winning broadcaster Dick Enberg.

This observance was the culmination of a year-long effort by the racing industry to raise funds for individuals and families devastated by the attacks. Over the past year, members of the international thoroughbred horseracing community, including tracks, horse owners, trainers, grooms, jockeys and veterinarians, have contributed more than $12 million to assist the families of those lost on September 11.

I am proud that the American racing and breeding industry has responded so patriotically to our Nation's ordeal and assisted so many Americans hurt by those tragic attacks.

IN HONOR OF FRANKIE M. MENO

Hon. Robert A. Underwood

OF GUAM

Mr. Speaker, today I share with you an open letter written to the American public by my constituent, Frankie Michael Meno. This letter was composed to recognize the losses of September 11, 2001, on the 1-year anniversary of the terror attacks against America. Mr. Meno's letter was accompanied by a CD containing a song, "America," which he wrote, and performed in the company of his stepchildren, nieces and nephews: Jessica, Sarah, and Mason Inder, and Shay, Daverin, and Davin Diaz.

Mr. Meno, a resident of Inarajan, GU, began writing songs in 2000. He finds the process simple as the melody and words come to him almost automatically. His song "America" was inspired by the pride he felt watching the closing ceremonies of the Winter Olympic Games in Salt Lake City, UT, where people of all nationalities, languages, and colors came together as one. Mr. Meno hopes "America's message of peace and freedom can be extended to all corners of the world."

In speaking of Mr. Meno, I wish to convey to you his pride in America and his 16 years of service with the U.S. Marines, his love of his family and children Christelle, Joseph, Antonia, and Jessica, and his grandchildren Isaiah and Jaythan, and his desire to use his songmaking abilities to help the victims of the terrorist attacks and to assist rebuilding Guam's educational system. Mr. Meno's song is one patriotic American's expression of our Nation's feelings of loss, recognition of our citizens' heroism, and the ultimate hope that America's freedom can be shared with the world. These sentiments are held by all of us, and I am glad to be able to share this letter with you today.

September 11, 2002

DEAR FELLOW AMERICANS, on this day, we join you in remembering your loved ones who left us on September 11, 2001. We would like to join with you in recognizing and remembering the brave men and women of the New York Fire Department, the New York Police Department, and the other heroes who sacrificed their lives to save another's. It is these extraordinary deeds from ordinary people that make us all proud to call ourselves Americans; your voices and deeds will never be forgotten.

My family and the people of Guam salute and embrace the American people and the noble ideas they stand for. I dedicate this song to the mothers and fathers, the sons and daughters, and the men and women who made the ultimate sacrifice to bring freedom and democracy to the island of Guam during World War II. I would also like to dedicate it to the American servicemen and to the people all over the world who long for freedom and democracy. I dream of the day when all the children of the world will be able to enjoy liberty's blessings. God bless Guam, God bless America, and God bless the world.

Semper Fidelis.

FRANKIE MICHAEL MENO.

Tuesday, September 24, 2002

HONORING HEROISM AND COURAGE DISPLAYED BY AIRLINE FLIGHT ATTENDANTS

Hon. John L. Mica

OF FLORIDA

Mr. Speaker, I move to suspend the rules and concur in the Senate concurrent resolution (S. Con. Res. 110) honoring the heroism and courage displayed by airline flight attendants on a daily basis.

The Clerk read as follows:

S. CON. RES. 110

Whereas over 100,000 men and women in the United States serve as flight attendants;

Whereas flight attendants dedicate themselves to serving and protecting their passengers;

Whereas flight attendants react to dangerous situations as the first line of defense of airline passengers;

Whereas safety and security are the primary concerns of flight attendants;

Whereas flight attendants evacuate passengers from an airplane in emergency situations;

Whereas flight attendants defend passengers against hijackers, terrorists, and abusive passengers;

Whereas flight attendants handle in-flight medical emergencies;

Whereas flight attendants perform routine safety and service duties on board the aircraft;

Whereas 25 flight attendants lost their lives aboard 4 hijacked flights on September 11, 2001;

Whereas 5 flight attendants helped to prevent United Flight 93 from reaching its intended target on September 11, 2001;

Whereas flight attendants provided assistance to passengers across the United States who had their flights diverted on September 11, 2001;

Whereas flight attendants on American Airlines Flight 63 helped to subdue Richard Reid on December 22, 2001, thereby preventing him from detonating an explosive device in his shoe intended to bring down the airplane and kill all 185 passengers and 12 crew members on board; and

Whereas flight attendants helped to prevent Pablo Moreira, a Uruguayan citizen, from breaking into the cockpit on February 7, 2002, during United Flight 855 from Miami to Buenos Aires: Now therefore be it

Resolved by the Senate (the House of Representatives concurring), That Congress—

(1) expresses its profound gratitude for the faithful service provided by flight attendants to make air travel safe;

(2) honors the courage and dedication of flight attendants;

(3) supports all the flight attendants who continue to display heroism on a daily basis, as they had been doing before, during, and after September 11, 2001; and

(4) shall send a copy of this resolution to a family member of each of the flight attendants killed on September 11, 2001.

Mr. Speaker, this afternoon we take up as a body S. Con. Res. 110. That resolution addresses a long overdue requirement of this House, and that is to honor the heroism and also the courage displayed by our flight attendants not only on September 11 but on a daily basis.

Mr. Speaker, on September 11, 2001, more than 3,000 innocent people lost their lives. This devastating number includes some 25 flight attendants who were on board the four hijacked aircraft on that day. As a result of the tragic events of that fateful day, the vital role that these men and women play as a very first line of defense became painfully evident. Flight attendants react and they also provide essential guidance to passengers during emergency situations. Flight attendants are in fact responsible and primarily concerned with passenger safety. Once the aircraft door is closed, they not only provide safety for the traveling public but also become our very first line of defense in aircraft security. They also guarantee that there are in fact additional eyes and ears on guard for suspicious and threatening behavior.

Examples of their acts of heroism and service include the actions of the 25 flight attendants who lost their lives on September 11 in the 4 aircraft that were hijacked that day. Mr. Speaker, we have honored firefighters, police, pilots, and ordinary citizens. Today it is fitting that we take some time to recognize those flight attendants who served both the aviation industry, the public, and America so well. It is also important to note that we also have the remarkable assistance that these flight attendants provide every day and particularly on the day that those flights around the Nation and around the world were diverted.

The flight attendants on American Airlines flight 63 last December recognized the terrible threat that that aircraft faced. It was not security guards. It was not air marshals. It was not a large force. It was flight attendants who helped subdue the attempted and now somewhat infamous shoe bomber, Richard Reid. They acted. They saved the lives of countless passengers on that aircraft. We are indebted to those flight attendants.

Also flight attendants helped prevent another tragedy on a flight. United flight 855 in February of this year, a deranged individual attempted to break into a cockpit. I believe that was on a flight from South America to Miami. They also acted with heroism.

In recognition of their vital role as a first line of defense, the House voted in July to strengthen the flight attendant training program, and those reforms are long overdue because sometimes these flight attendants are left at bay to fight these terrorist acts and other disruptions on aircraft. H.R. 4635, which is primarily devoted to arming pilots and allowing pilots to defend themselves, also requires that flight attendants receive much needed hands-on training in self-defense so they too can defend themselves, the passengers, the aircraft, and again serve as a first line of defense.

As H.R. 4635 demonstrates, the House supports these brave men and women and wants to ensure their safety and their security along with that of the flying public.

So today we are considering S. Con. Res. 110. The House passed a similar version of this concurrent resolution earlier this month. The concurrent resolution recognizes the over 100,000 airline flight attendants who have dedicated themselves to serving and protecting our passengers, the flying public, on a daily basis. It also recognizes the courage and heroism of those who lost their lives on September 11. It expresses Congress' profound gratitude to airline flight attendants and it rightfully honors their courage and dedication. For all these reasons and many more, I encourage my colleagues in the House to pass the Senate concurrent resolution so rightfully honoring our flight attendants.

Hon. Jim Matheson

OF UTAH

I want to thank the gentleman from Florida (Mr. Mica), our subcommittee chairman. I have enjoyed serving on the Subcommittee on Aviation under his leadership, and I appreciate his leadership on this bill today.

Mr. Speaker, I rise in support of S. Con. Res. 110. It is such an appropriate thing for us to be doing in terms of recognizing the service and honor and courage of over 100,000 flight attendants that fly in the skies above this country. It is important that we acknowledge their service. And something that has helped us focus on this service is the actions that took place on September 11 and actions that took place subsequent to September 11.

As was mentioned, 25 flight attendants lost their lives that day. That same day let us not forget that all the other planes that were up in the sky were ordered down on the ground by the FAA and a number of flight attendants on all those flights on that day faced some real challenges. They faced the fear that we all felt that day, but they also faced the job of having to work with a number of passengers on all those airplanes that were being diverted and asked to land on emergency notice, and the flight attendants in this country served us well that day in terms of dealing with that difficult situation, and that applies to the days following September 11. We had a public that was nervous, and the flight attendants represented the face of courage. They were the first line of defense, quite frankly, in maintaining security on those airplanes.

We know the stories about how they caught the shoe bomber, Richard Reid. We know that the flight attendants are keeping their eyes open.

So it is appropriate that we honor them in this context, but we ought to honor them also for all the work they have done. Let us face it. When they get on that airplane, they are the face of the airline for which they work. The traveling public at times experiences some frustrations maybe through delays, maybe through the weather or what not. Sometimes those frustrations are unfairly directed toward flight attendants because they are the ones who are there interacting with the public, and I think that as a profession they deal with that situation so well and they certainly deserve our gratitude and our respect.

It is important that we do not forget the folks who lost their lives September 11, those 25 flight attendants, and we owe them a lot. We owe them this resolution today to honor what they have done but we owe them more. We owe them the commitment that we are going to continue to move forward and try to encourage as good a security situation as we can get in our aviation industry. That is the commitment we need to make to the flight attendants as well. They are on the front lines, and we are going to do whatever we can do to back them up to make sure this is a secure situation in our aviation industry.

So it is with great pleasure that I advocate support of S. Con. Res. 110.

Hon. Carolyn McCarthy

OF NEW YORK

Mr. Speaker, I rise in support of S. Con. Res. 110. The women and men who make up America's flight attendant workforce deserve recognition for their role as safety professionals. As the eyes and ears of the aircraft cabin, flight attendants have historically provided detailed information on countless safety issues.

Flight attendants are a highly trained, highly skilled workforce, charged with the safety and security of passengers, other crewmembers, and the aircraft itself.

On September 11, 2001, 25 flight attendants on board the 4 hijacked flights provided the government with vital information, and with little more than their own ingenuity and bravery, fought the armed hijackers and performed their duties as safety professionals to the end.

Since that day, safety in the air is of paramount concern to millions. We now understand the vulnerability that flight crews have felt for years. But today, more than 12 months after the attacks on our Nation, flight attendants are no more prepared to defend the flying public today than they were on the morning of September 11, 2001.

Air Tran flight attendant Susan Cosby began developing her airline's voluntary defense training program within days of September 11. In a visit to my office, Cosby posed this question, "Flight attendants have always been expected to save lives in emergencies. Why should defending the flying public from the threat of terrorism be any different?"

Now more than ever, it is crucial for us to recognize the importance of flight attendants. It is my hope of America's 100,000 flight attendants, that the Congress will quickly pass meaningful security training legislation.

VICTIMS OF TERRORIST ATTACKS MEMORIAL ACT

Hon. James V. Hansen

OF UTAH

Mr. Speaker, I move to suspend the rules and pass the bill (H.R. 2982) to authorize the establishment of a memorial within the area in the District of Columbia referred to in the Commemorative Works Act as "Area I" or "Area II" to the victims of terrorist attacks on the United States, to provide for the design and construction of such a memorial, and for other purposes, as amended.

The Clerk read as follows:

H.R. 2982

Be it enacted by the Senate and House of Representatives of the United States of America in Congress assembled,

SECTION 1. AUTHORIZATION OF MEMORIAL.

(a) In General.—The Advisory Board established in section 2(a) is authorized to establish a memorial (referred to hereafter in this Act as the "Memorial") in accordance with this Act on Federal lands administered by the National Park Service in the District of Columbia and its environs (as defined in section 2(e) of the Commemorative Works Act (40 U.S.C. 1002(e)) to victims who died as a result of terrorist acts against the United States or its people, at home or abroad, except those individuals identified by the Attorney General of the United States as participating or conspiring in terrorist-related activities.

(b) Detail of Employees.—The Secretary of the Interior (referred to hereafter in this Act as the "Secretary") shall detail to the Advisory Board such support staff as are necessary to assist the members of the Advisory Board in carrying out its responsibilities.

(c) Relationship to the Commemorative Works Act.—The Commemorative Works Act (40 U.S.C. 1001 et seq.) shall apply to the Memorial, with the exception of section 3(c) of that Act which shall not apply to the Memorial.

SEC. 2. ADVISORY BOARD.

(a) Establishment.—There is established an advisory board to be known as the "Victims of Terrorism Memorial Advisory Board" (referred to hereafter in this Act as the "Advisory Board").

(b) Members.—The Advisory Board shall consist of 13 members who shall be appointed, not later than 3 months after the date of the enactment of this Act, by the President (in consultation with the Secretary of the Interior and the Secretary of Defense) from interested persons, including representatives of organizations dedicated to assisting victims of terrorism and their families.

(c) Chairperson.—The Chairperson of the Advisory Board shall be one of its Members elected by a majority of the Members at the first meeting of the Advisory Board.

(d) Terms; Vacancies.—Members of the Advisory Board shall serve for the life of the Advisory Board. The President shall make appointments to fill any vacancies that occur.

(e) Duties.—The Advisory Board shall—

(1) raise necessary funds to establish, design, construct, and maintain the Memorial; and

(2) begin consultation under section 7 of the Commemorative Works Act not later than 1 year after the date of the enactment of this Act.

(f) Donations.—The Advisory Board may accept donations on behalf of the United States for the establishment, design, construction, and maintenance of the Memorial.

(g) Termination.—The Advisory Board shall terminate not later than 120 days after completion of the Memorial.

(h) FACA.—The Federal Advisory Committee Act (5 U.S.C. App.) shall not apply to the Advisory Board.

SEC. 3. DEPOSIT OF EXCESS FUNDS.

If, upon payment of all expenses of the establishment of the Memorial (including the maintenance and preservation amount provided for in section 8(b) of the Commemorative Works Act), or upon expiration of the authority for the Memorial under section 10(b) of that Act, there remains a balance in the funds received under section 3(f) for maintenance of the Memorial, the Chairperson of the Advisory Board shall transfer the amount of the balance to the Secretary of the Treasury for deposit in the account provided for in section 8(b)(1) of that Act.

H.R. 2982, introduced by the gentleman from Texas (Mr. Turner), who has worked so tirelessly on this legislation, and sponsored by myself and over 121 Members of the House of Representatives, would establish a memorial to the victims who died as a result of terrorist acts against the United States or its people.

The gentleman from Texas (Mr. Turner) went out of his way to do an exceptionally fine job on this legislation. One of the most interesting hearings we have had in the Committee on Resources was put on by him. It included Lisa Beamer whose husband Todd was part of flight 93 that crashed in Pennsylvania and coined that phrase, "Let's roll," with the President standing there in front of the Chair where he sat.

Mr. Brady Howell, whose wife, Liz, works for us in the Committee on Resources. Brady was the all-American boy. He was an Eagle Scout. He was a 4.0 student. He was the quarterback. He was the student body president. He was a missionary for his church. He did everything one can imagine and had a great sense of humor.

Mr. Joe Finley, a New York firefighter, who most of his squadron was killed on that tragic day.

Lt. Col. Terry Andersen, who went into the Pentagon and saved many people and worked diligently.

So many illustrations of honor.

This bill would authorize a memorial to victims who died as a result of terrorism against the United States or its people at home or abroad, except those individuals identified as participating or conspiring on terrorist-related activities.

Mr. Speaker, the great memorials that dot the landscape of our Nation's Capital reflect the course of American history and are a constant reminder of our commitment to freedom, justice and democracy. We see these shared values in our monuments to great leaders, and we see them in our memorials to the soldiers who died in great wars fought in Europe, in the Pacific, in Korea and in Vietnam.

The new war of the 21st century, this war on terrorism, will not be marked by one geographic location. It is a global war that has been, is, and will be fought at home and abroad. Already thousands of people have lost their lives to terrorist attacks on the United States. These victims of terrorist attacks deserve solemn tribute, for they died at the hands of enemies of America simply because they were Americans.

Mr. Speaker, H.R. 2982 would authorize the establishment of an advisory board to raise funds for the design, construction, and maintenance of a living memorial, and to work with the National Capital Memorial Commission and the Secretary of the Interior on the placement of the memorial.

H.R. 2982 will honor those Americans whose lives have been lost to terrorism and will symbolize the great struggle in which we are now engaged. Someday this memorial will mark the time and the course of history when freedom and respect for the dignity of man overcame tyranny and hate and evil. Indeed, it will stand for the age when America faced its greatest challenge.

I urge my colleagues to support H.R. 2982, as amended.

Hon. Jim Turner

OF TEXAS

Two weeks ago we observed the 1-year anniversary of the September 11 attacks on our Nation. It was on that date in 2001 that our Nation was made acutely aware of the threats posed by those who seek to destroy our way of life.

Since then we have come together as a Nation and remembered those who lost their lives and we have pledged jointly an unwavering resolve to win the war on terrorism no matter what the cost and no matter how long it takes. This legislation, which was introduced by the gentleman from Utah (Mr. Hansen) and me, would authorize the establishment of a national memorial to all the victims of terrorism in attacks against the United States or its people including those who died on September 11.

It has been said that the war on terrorism may be known as the first war of the 21st century.

It will not be marked by any specific geographic location because it is global. It has been and will be fought at home and abroad. This national memorial created by this bill will be dedicated to the memory of those victims who lose their lives and have lost their lives at the hands of terrorists. The memorial will reflect, in my view, the history of the struggle in which we are engaged, and will remind future generations of the challenges that we faced and the challenges that we have overcome through courage and bravery of the American people.

As the gentleman from Utah (Mr. Hansen) mentioned, this legislation creates a 13-member advisory board appointed by the President and includes representatives from organizations dedicated to assisting the victims of terrorism. The advisory board is charged with the duty of raising the funds from private sources to establish, design, construct and maintain this memorial. In accordance with the Commemorative Works Act, the advisory board will consult with the appropriate commissions already provided by existing law regarding the site selection and design of this memorial.

When the House Committee on Resources held its hearing on this bill, we were honored, as the gentleman from Utah (Mr. Hansen) mentioned, to hear from several outstanding Americans whose lives, like so many Americans, were personally and forever changed on September 11. Their testimony spoke eloquently of the significance of a national memorial to the victims of terrorism.

Liz Howell, who is on the staff of the Committee on Resources, whose husband, Brady, died at the Pentagon, said so eloquently "I believe a national monument to terrorism would become a hallowed place for the people of this generation to remember and grieve. Perhaps even more important, it will teach future generations about the heroism, the sacrifice, and the patriotism that surrounded the deaths of people who simply died because they were Americans."

That quote from Liz Howell represents, I think, so very eloquently the purposes embodied in this legislation.

We heard from Lisa Beamer, whose husband Todd was among the heroes of flight 93. We heard from Joe Finley, a New York firefighter, who testified on behalf of this legislation, that he lost many in his firehouse who sacrificed their lives at the World Trade Center. Lt. Col. Ted Anderson, who rushed into the fiery flames of the Pentagon, saving many of his friends and colleagues there, spoke of a need for a national memorial. They all shared their hopes that this memorial would not only be a tribute to those who lost their lives but a constant reminder to the American people of the importance of courage, bravery, and patriotism.

It is my hope that at some point in the future this memorial will mark a time in the course of our history when freedom and respect for the dignity of man overcame hate and evil. It will stand for the period in our history when our country stood tall, persevered and protected peace and civility for all mankind.

Mr. Speaker, we hope that the House will join us in unanimously supporting this legislation to create this memorial to the victims of terrorism.

Hon. Nick J. Rahall II

OF WEST VIRGINIA

Mr. Speaker, while the text of this legislation refers to a memorial intended to honor anyone who has ever been the victim of a terrorist attack on the United States, our hearings on this measure focused on the victims of the September 11 hijackings and attacks on the World Trade Center and the Pentagon. Those events are the catalysts for this legislation.

While it has been said often, it cannot be said often enough: our thoughts and prayers continue to go to those affected by the events of that awful day. While we as a Nation have attempted to go on with our lives, the tragedy and loss of that day are never far from our hearts or minds.

The scope and severity of that terrible tragedy make it difficult to know how best to memorialize those who were lost.

Mr. Turner's bill is one approach and we will support it, but there may be others.

It is our hope that, over time, we may all gain the wisdom and perspective to devise a memorial, or series of memorials, that will tell the story of these attacks, the people who were lost, their families and our resulting efforts to end the threat of terrorism, in such a way that future generations will never forget these events.

Better yet, if we do it right, perhaps such a memorial could, in some small way, reduce the chance that a future generation will have to endure such a tragedy.

Hon. Eleanor Holmes Norton

OF DISTRICT OF COLUMBIA

Memorializing the victims of the war on terrorism is the least we can do.

I am going to support this bill with some reservation because it continues to violate the 25-year rule. I want to give Members some context for that. A bill was brought forward early in this Congress to establish a Ronald Reagan memorial on the mall. The Bush administration did not support it because it did not want to set that precedent and it is sufficient to say did not support violating the 25-year rule. Nobody who remembers the tenure of Ronald Reagan both in this city and out of the country can doubt that his legacy will withstand the 25-year rule. The World War II memorial is going up now, more than 25 years after World War II. No one has pressed forward an Oklahoma City memorial though that was one of America's great tragedies of the 20th century. The Martin Luther King memorial could not be built until 25 years after his death.

I say all this because the House needs to understand the context and why it is this way. We lost a number of residents in 9/11. A number of those working in the Pentagon came right from the District of Columbia. The most visible ones were three children and their three teachers. I certainly want to see them memorialized on the mall. The context, though, we have to understand is what is happening to the mall. We literally are in danger in one generation of using up virtually all the prime space of the mall. Yet the mall was meant for eternity. That is why there is a 25-year rule. We must not get to where London is now. London is having to tear down memorials in order to allow memorials to go up. That is not the only reason that the Commemorative Works Act has a 25-year rule that says to Congress, step back, and the reason that Congress has stepped back in each and every occasion, the pain of a memorial may be fresh, the reputation of a personage needs time to gel. History needs to inform us and see in context what it is we want to do.

We know that memorials are best when there is a nationwide competition, when people who have looked at memorials over time can bring their expertise to bear, and that is going to happen here. We know that we cannot sit here and say where a memorial should be built, that we have got to trust those with whom we have delegated this responsibility. That is gone.

I will submit for the *Record* the letter of Mr. Cogbill so that the *Record* can be fully informed with respect to the problems that the initial bill has.

Mr. Speaker, I want to once again express my appreciation to the gentleman from Utah (Mr. Hansen) and certainly to the gentleman from Texas (Mr. Turner) for the great sensitivity they both have shown in designing this bill. I support the bill with the reservations I have noted.

The letter previously referred to follows:

NATIONAL CAPITAL PLANNING COMMISSION
Washington, DC, March 19, 2002.

HON. JAMES V. HANSEN, CHAIRMAN,
Committee on Resources,
House of Representatives, Washington, DC.

Re Terrorism Memorial

DEAR CHAIRMAN HANSEN: I am writing to express the concerns of the National Capital Planning Commission with regard to H.R. 2982, a bill that would authorize a memorial in Washington, DC, to victims of terrorist attacks on the United States.

The Commission mourns the tragic loss of life resulting from terrorist attacks on the United States, including the atrocities committed on September 11, when terrorists orchestrated the most devastating attack on American soil in our Nation's history. We share the desire to find appropriate ways to remember and honor the victims, but we urge that, consistent with the Commemorative Works Act (CWA), sufficient time be allowed to pass so that these tragic events are put in proper historical perspective before commencing the process of locating and designing such an important national memorial.

As you are aware, the process for establishing memorials in the Nation's Capital is governed by the CWA. By setting forth criteria for the subject matter, location, and design of memorials, the CWA is intended to preserve the integrity of the L'Enfant and McMillan Plans for the Nation's Capital, while protecting and maintaining the limited amount of open space available on and around the Mall.

The Commission is concerned that H.R. 2982 circumvents one of the key provisions of the CWA—the 25-year waiting period for the authorization of new memorials in the Nation's Capital. The purpose of this provision, which states that a memorial "shall not be authorized" by Congress until at least 25 years after the death of the individual or event, is to ensure that enough time passes following an event for policymakers and historians to gain an appropriate historical perspective before establishing a permanent memorial in the Nation's Capital.

Just as with other tragedies in American history—from Gettysburg to Pearl Harbor to Oklahoma City—a more meaningful and appropriate place to honor victims at this time might be at the sites of the tragedies themselves. Congress has already authorized legislation for a memorial at the Pentagon, and permanent memorials at the World Trade Center and at the Pennsylvania crash site are currently being considered.

Other provisions of H.R. 2982 are also inconsistent with the CWA. In order to help preserve the limited number of sites available in area I (sites on or near the Mall), the CWA requires passage of a separate act of Congress, following a recommendation by the National Capital Memorial Commission, before locating a memorial in this prominent area. Yet this bill directly authorizes the memorial to locate in area I, overriding the requirement for a second round of consideration by Congress. In addition, the bill suggests that any specific location for the memorial be "approved by the Congress," again contrary to the CWA, which delegates to the Federal land-holding and review agencies decisions as to the specific location and design of new memorials.

The Monumental Core of our Nation's Capital has evolved over the centuries into a powerful expression of our Nation's values, achievements, losses, and challenges. By respecting the process established by the CWA, we can ensure that the victims of terrorist acts against our country are properly memorialized and, at the same time, the historical integrity of our grand Monumental Core is preserved.

Sincerely,

John V. Cogbill, III,
Chairman.

Hon. Benjamin A. Gilman

OF NEW YORK

Mr. Speaker, I rise in strong support of H.R. 2982, the Victims of Terrorist Attacks Memorial Act, which will establish a memorial in Washington, DC, to honor those Americans whose lives were tragically taken as a result of terrorism.

At the Murrah Building in Oklahoma, the World Trade Center, the Pentagon, and in the fields of Pennsylvania, our Nation has witnessed the best and the worst of humanity. These despicable and cowardly terrorist acts were valiantly countered with the incredible heroism and courage of not only our firefighters, law enforcement officers, and emergency personnel but also our fellow citizens.

Accordingly, it is incumbent upon our Nation to honor those departed heroes. Establishing a memorial in honor of those deserving men and women will be a fitting tribute to their memory and their contribution to our Nation's freedom. Moreover, it will act as a permanent reminder to our Nation and the world that our Nation is engaged in an ongoing battle in the name of those who were taken from us to rid the world of these most heinous of crimes. Accordingly, I urge my colleagues to support this important measure.

Hon. Steve Israel

OF NEW YORK

Mr. Speaker, I rise today in strong support of H.R. 2982. This legislation will authorize a memorial in Washington, DC, to commemorate American victims of terrorist attacks. The desire to memorialize the victims of the September 11 terrorist attacks was the driving force behind this bill, and as an original cosponsor, I applaud my colleagues bringing this bill to the floor today.

I have met with families in my district who lost loved ones in the attack on the World Trade Center and the Pentagon, and I have tried to help them with legislation that will ease their financial burden. I am hopeful that this bill will help them in a different way. I want them to know that the American people support them, and this memorial will show them that their husbands, wives, sons, daughters, brothers and sisters will never be forgotten.

This memorial, however, is not only for the victims of terrorist attacks and their families. It is a memorial for every person in our Nation. It will give the people of the United States a site to pay their respects to the victims. It will serve as an area for mourning. It will also be a place to remember. I am hopeful that the memorial will also be a space where people can see the American spirit, which cannot be defeated, to take comfort in America's resolve and the inevitable triumph of freedom.

The September 11, 2001, attacks changed every American life. It was the saddest and most enraging day in many of our lives. It is appropriate that we build this memorial to commemorate not the attacks, not response, not the war, but the victims of the attacks.

WAGING THE PEACE

Hon. Major R. Owens

OF NEW YORK

One letter I submit for the *Record* is not so simple, but it is written by one of my constituents, and obviously she has given a great deal of thought to this letter, and I appreciate the thinking here. I want my colleagues to hear the connection here with September 11 and how she weaves all of this together and understands very clearly the mood of America. The mood of America is anger; the mood of America is hurt; the mood of America is fear. But we should not let the mood of anger, hurt, and fear drive us into reckless actions that will make matters worse.

BROOKLYN, NY,
August 9, 2002.

REP. MAJOR OWENS,
House of Representatives,
Brooklyn, NY.

DEAR REP. OWENS: I am writing to you, because I feel so helpless to stop what seems to be inevitable—War with Iraq.

Like you and every New Yorker, I tasted war on September 11. It wasn't pleasant and I'm not eager to experience it again. For hours I couldn't find my husband who worked across the street from 1 World Trade Center. Fortunately he returned home safely after witnessing unspeakable carnage, but many of our friends and neighbors weren't so lucky. That evening, I walked down 7th Avenue in Park Slope, Brooklyn, to get a handle on the losses. The stench from burning buildings, computers, and bodies was pervasive and the smoke cast an eerie haze over our little community. Everywhere I went I learned of more losses—12 firemen from Squad 1 on my block, loved ones of students and a teacher at the Park Slope Dance Studio, parents with kids at 321, Berkeley Carroll, and St. Ann's, members from church, a former colleague, and many of our neighbors were all among the missing. At 7 p.m. that day, we foolishly held out hope that some would be found in area hospitals, but unfortunately they weren't.

Weeks later I attended the memorial service for my friend, Jeff Hardy, who was killed because he happened to be working on the 101st floor of Tower 1. Hours after I attended Jeff's service, a woman at 7th Avenue and Carroll approached me and asked me to sign a petition opposing the war in Afghanistan. I refused. I supported the war in Afghanistan and have been grateful that our allies have worked with us to round up terrorists worldwide.

However, I have seen absolutely no evidence that Iraq had anything to do with this attack. The rumor that Mohamed Atta met with an Iraqi intelligence agent has been denied by the Czech government. I am not aware of one Iraqi who fought with the Taliban, although I know the citizens of many of our allies fought with the Taliban, are members of Al Qaeda, were on those planes September 11, and continue to threaten Americans and other foreigners every day, particularly in Pakistan.

My hope is to destroy Al Qaeda and stop the spread of Islamic religious fundamentalism and hatred for the United States, Christians, and Jews. To fight the Islamists, we need the cooperation of all of our allies and all countries in the Middle East. I am afraid that this fragile alliance will dissolve if we attack Iraq without provocation and we may not get the help we need. Invading Iraq will only inflame anti-American rhetoric and could even jeopardize our allies in the Middle East. I'm deeply worried about the welfare of President Musharraf and concerned that if anything happens to him, religious fanatics could take control of Pakistan, which we know has both nuclear weapons and Al Qaeda members. Musharraf is already under attack in his country because of his support of the U.S. and the *New Yorker* reported this week that a recent car bomb that killed 12 people was intended for him. I truly think declaring war on Iraq will put more U.S. citizens in harm's way than containment.

To me this administration's warmongering is further evidence of the "Kremlinization" of Washington under Bush. This administration thrives on secrecy. In the beginning of the term we saw cronyism and secret agreements among the elites in government and business. Now there is lavish Federal spending in Florida where the President's brother happens to be running for reelection. According to a recent *New Republic* article, even questionable SBA loans are being made in Florida at a time when several businesses with which I have worked that were located at or near Ground Zero have been denied SBA assistance.

After September 11, we had secret arrests and detentions of more than 1,000 individuals. Even Reagan-appointed, Federal judges have been appalled by this. We have seen civil rights being applied arbitrarily with some American citizens who happen to be poor and of color like Jose Padilla being denied the right to legal counsel and the American justice system, while prosperous Americans like John Walker Lindh, who actually fought American soldiers, received them. No investigation has been allowed into the intelligence failures before September 11. *Time* magazine this week has a scathing article about how this administration ignored terrorist threats prior to the attacks, but we can't examine this. Free speech has been chilled because any elected official who dares criticize or stand in the way of the administration has been called unpatriotic and obstructionist and in some cases compared to Saddam Hussein in newspaper ads. The government is asking ordinary citizens to spy on one another, reminiscent of something out of a Solzhenitsyn novel. The attorney general has ignored the Supreme Court's 1939 opinion on the Second Amendment and has decided to apply his own, wildly different interpretation and also won't allow gun checks on suspected terrorists. I won't even get into what started all of this, the election of 2000 and how the voter registration lists were "scrubbed" and the failure of the Supreme Court to honor a Presidential candidate's request to count votes as allowed under Florida law. Now this administration is invading countries without adequate discussion or support.

Following the tragedies of September 11, we were a city in mourning. We spent months going to funerals and neighborhoods completely shut down when funerals for firefighters were held. The physical and emotional damage contributed to economic downturn here. I run a small, but successful public relations firm and I booked 93 percent of my revenues in 2001 on projects completed before September 11 and only 7 percent after September 11. My situation was not unusual. Small businesses—graphic designers, contractors, beauticians, photographers, etc.—everywhere in the metropolitan area suffered the same fate. Large companies like my husband's were evacuated from lower Manhattan never to return. His company had to rebuild complete systems within days to be able to compete when the markets opened the following Monday and use AOL or other carriers to communicate by e-mail because the company's servers were destroyed. We all limped along. Our woeful city tax revenues are enduring evidence of the economic damage we experienced.

This country and especially this city have not yet digested the economic and emotional fallout from September 11. New York City is still struggling to get back on its feet and continues to get hammered by low tax revenues, the recession, stock market volatility, and corporate scandals. The economy can't take another shock like a war with Iraq and its unknown consequences.

We have so much unfinished international business that to go forward with a war with Iraq right now would be irresponsible. I share the same concerns that King Abdullah of Jordan has that invading Iraq could lead to a further destabilization of the Middle East, including possibly a civil war, at a time when we need to be rebuilding Afghanistan and seeking a solution to the Israeli/Palestinian War. Even the Kurds are begging us not to invade. We still haven't found Osama bin Laden and Mullah Omar yet, how are we going to round up Saddam Hussein and his secret weapons, particularly without the support of our allies? It's suicidal. I'm reminded of our many unsuccessful attempts to oust Fidel Castro. Besides the economic and diplomatic problems of a war with Iraq, I have a serious moral problem with killing innocent people in the country. I know what it feels like when innocent lives are lost. Even Rep. Dick Armey was quoted today in the *Times* as saying that an unprovoked attack would violate international law. However, this administration will not listen to its allies and is only fueling anti-Americanism worldwide.

I am a conservative Democrat and was highly supportive of President Clinton and particularly his economic policies because he gave everyone a seat at the table of opportunity, cut budget deficits, and supported free trade. (Unlike Bush who has caved to special political interests on steel, the farm bill, tax cuts, energy, the environment, etc.) I don't trust these people in the White House now. Unfortunately, they seem to be unstoppable. Please help stop them.

Sincerely,

GAIL DONOVAN.

REMEMBERING SEPTEMBER 11, 2002

Hon. Ralph M. Hall

OF TEXAS

Mr. Speaker, as we reflect on the tragic attacks on America on September 11, 2001, and the remarkable heroism of those who responded selflessly to those attacks, the weapon we need now more than any time in the history of the Republic is prayer.

We need prayer for our brave men and women in the military who are fighting the war on terrorism and for those who will go, prayer for our President and our leaders as they chart our Nation's course, prayer for the families and friends of the more than 3,000 citizens who lost their lives in New York City, on the place that crashed in Pennsylvania and at the Pentagon, and prayer for the safety and security of our Nation.

We need prayer as well for the selfless and hard-working members of our police, firefighter and emergency response teams. The harsh reality of the danger of their jobs came crashing down on America on September 11 when 343 firefighters, 23 police officers and 37 port authority officers in New York City lost their lives at the World Trade Center.

With a year to ponder the terrible losses suffered that day and the threat of terrorism and evil still looming over our Nation, we realize now more than ever the debt of gratitude that we owe to those who place themselves in harm's way to protect our Nation and our citizens.

September 11 unified America and reawakened a spirit of determination and patriotism not seen since World War II. On that day we were reminded that freedom is not free—that there is a price that must be extracted to defeat tyranny and terrorism and to preserve the precious freedoms for which previous generations of heroes have fought so valiantly.

The war on terrorism will be a long war—but it is a war that we can and will win. So as we remember those who paid the ultimate sacrifice for their country and those still on the front lines in the battle for freedom, we are filled with renewed resolve that those who perpetrated these terrible acts will be brought to justice and that such a tragedy will never happen again in America.

May God bless the families of the victims of September 11, and may God continue to bless America.

H.R. 2982

Hon. Jerry F. Costello

OF ILLINOIS

Mr. Speaker, H.R. 2982 authorizes the establishment of a memorial within the District of Columbia to the victims of terrorist attacks on the United States.

Words are generally inadequate to give voice to the loss we suffered on September 11, 2001. Today, we will try to leave a more worthy token.

The enormity of what happened last year is still difficult to grasp, especially to those of us in the communities most directly affected. The scope of the casualty list is particularly overwhelming when considering each individual that was taken from us. Each had a name, a face, a family, a personality, a legacy that they have left behind.

Susan Conlon said goodbye to her 6-year-old daughter, Kimberly, before going to work that day in the World Trade Center, in an office she had occupied for less than 3 months. Robert Curatolo was a newlywed who charged into danger as one of all too many firemen who never returned that day. Vassill Haramis was a hero of the 1993 WTC bombing, an engineer who loved working there as he had since the 1970s, not long after coming to this country.

These stories, times a thousand and more, can only begin to trace the outline of the victims of the 9/11 attacks. I offer them as examples from my own district of heartbreaking losses.

I believe what we are proposing today will be an eloquent testament to the memory of the victims. By acting today, while the memory of that terrible day has not yet faded, we can be sure future generations will have a better understanding of the victims and heroes of September 11, and their legacies will never be forgotten.

Wednesday, September 25, 2002

WE ARE "GREAT BECAUSE WE ARE GOOD"

Hon. Zach Wamp

OF TENNESSEE

Mr. Speaker, as we observe the remembrance of September 11, it is my hope that the citizens of the United States will honor the legacy of those who lost their lives and pay tribute to their survivors in time-honored American ways . . . like helping others in need, saying a kind word to a stranger, volunteering at a homeless shelter or sending relief to people around the world who we may never even meet. After all, our country is not great because of our military strength, our free enterprise system or even our right to vote (as awesome as these qualities are!). America is great as we give more than we take and as we are willing to serve and sacrifice for others.

We now know countless stories of heroism and remarkable bravery—passengers on flight 93 that had the courage to stand up to terrorists giving their lives to protect hundreds of others, a lieutenant colonel who died trying to get his coworkers to safety or a firefighter who ran up the stairs of a building that was coming down on top of him. Although they didn't sign up to fight in the trenches of the war on terrorism, firefighters, EMTs, law enforcement officers, medical professionals and even airline passengers were willing to lay down their lives for people they had never even met.

The sacrifice and courage of our first responders on September 11 caused a swell of pride in all Americans of every generation. What we witnessed when America came under attack was comparable to the noble actions of the "greatest generation"; veterans on D-day when they stormed the beaches of Normandy or in the lonely courage of American heroes in the jungles of Vietnam.

The United States of America is at her best not when the Dow Jones average is above 10,000 points, or when we land on the Moon, but when our citizens are willing to sacrifice themselves so that others might be secure.

I participated in a historic joint session of Congress at Federal Hall in New York City, laid a wreath at Ground Zero and spoke at a memorial service in a Brooklyn church. On Wednesday, September 11, I attended the National Memorial Service at the Pentagon with President George W. Bush and Defense Secretary Donald Rumsfeld. That evening I listened with the rest of the world to President Bush speak about this past year and America's security in an unstable world.

As we bow our heads in respect, let us all be committed to honoring our country and those who have gone before us by giving of ourselves to help others. After all, every day of life is a gift from God and none of us know which might be our last. Let us stay united and make the most of every day!

REMARKS DELIVERED ON THE FIRST ANNIVERSARY OF SEPTEMBER 11, 2002

Hon. Steven R. Rothman

OF NEW JERSEY

Mr. Speaker, we come here today, as one community, to reflect on the events of September 11, 2001. On that terrible day, a group of evil men murdered more than 3,000 innocent American men, women, and children—here on American soil—as their coconspirators attempted to kill thousands more.

Today, we still mourn the loss of our fellow citizens: those trying to reach safety and those who deliberately placed themselves in harm's way, who saved literally tens of thousands of their fellow Americans in the process. We will also never forget those who were injured and who are still suffering from the wounds, physical and emotional, that were inflicted upon them. We will never forget the heroism and sacrifice of those—many of whom are with us today—who responded immediately and selflessly, who prevented a terrible ordeal from being even worse.

While we will always continue to remember what happened, we must also continue our Nation's effort to bring to justice and punish those who perpetrated these terrible acts and those who are planning new ones. Government's first priority is, after all, to protect the people, and as your Representative in Congress, I assure you that Congress is working to see that our government meets our country's needs for our homeland security and for our national defense: from strengthening our borders, to improving law enforcement and intelligence capabilities, to ensuring that our military is fighting with superior forces and weapons. We never forget that we Americans depend on our government to protect us.

We are forever grateful to the men and women in law enforcement and in our Armed Forces, here and around the world, who put their lives at risk so that we may keep our country and her people safe and free.

Is America a perfect nation? Are we as individuals perfect people? No, America is not perfect, and none of us has ever met a perfect person. But what we have in America is the greatest nation the world has ever known—a country committed to freedom, democracy, and equal justice under the law. We are an imperfect country, but one whose principles of freedom of speech and expression allow us and even demand us to continually seek to make our Nation more perfect in its realization of our founding principles. We are still the shining beacon of hope and liberty for every nation in the world and every man, woman, and child on this planet.

Finally, we must always remember that in our 226-year history, America has prevailed over many more powerful enemies than the ones we face today. It took the lives and sacrifices of countless numbers of Americans. It took money. It took time. It took patience. And it took perseverance. But we prevailed. Make no mistake about it, my fellow Americans, America will prevail again today.

God bless you, my friends, and God bless the United States of America.

COMMEMORATION OF SEPTEMBER 11, 2001

Hon. Nick J. Rahall II

OF WEST VIRGINIA

Mr. Speaker, "We must consider that we shall be as a city upon a hill," the Puritan preacher John Winthrop proclaimed, as he and his followers sailed for America and freedom. "The eyes of all people are upon us." And so they have remained for nearly four centuries. Many have looked to us in awe, inspired by a Nation rooted in liberty. Others have hated the ideal we embody, and wished us ill. But none can remove us from their gaze.

Today America's economic prosperity, military power, and technological advancement are without peer. Our daily comforts and conveniences exceed those available to most of the six billion people who inhabit the Earth. But the ease of our lives does not render us soft, or reluctant to retaliate when attacked. A year ago, all the world watched in horror as a small gang of wicked men took 3,000 innocent lives in New York, Washington, and Pennsylvania.

Since the moment the first airplane struck the first tower, Americans have shown, both on the battlefield and at home, the strength of our spirit, the mettle of our souls, and the force of our arms. From the firefighters climbing to their deaths, to the airline passengers who battled back, to the precious West Virginia sons and daughters who gave their lives in Afghanistan, the world has witnessed acts of American selflessness and bravery that rival the most revered in the annals of human history.

Just as Winthrop defined America's place in the world, he described how we must live to maintain it. "We must delight in each other," he instructed. "Make others' conditions our own; rejoice together; mourn together; labor and suffer together." Our whole Nation suffered the same grievous wound on September 11. Those who delivered the blow hoped it would inaugurate our destruction. Instead, they inspired America's return to the community values and mutual commitment upon which our country was built.

The attacks, the ongoing war, and the continuing threats spur us to embrace again our founding ideas: that all men and women are created equal; that America's destiny is the world's destiny—to secure life, liberty, and the pursuit of happiness; that we cannot allow the centuries-old, worldwide fight for freedom to falter. This recollection of our original rights and responsibilities is a fitting tribute, is an apt memorial, to the lives that were lost and devastated on that sad September day.

Hon. Henry E. Brown, Jr.

OF SOUTH CAROLINA

Mr. Speaker, I will never forget the tragic events of September 11, 2001. Although this unprovoked attack on our Nation by faceless cowards sought to damage American will, there can be no doubt that we are more determined than ever to fight for our freedom and preserve our way of life We have sent our sons and daughters into battle in Central Asia and throughout this world to bring the perpetrators to justice and to eradicate the scourge of terrorism from the face of the Earth. I know that we will succeed.

During the past year, we have pulled together as Americans with a renewed sense of patriotism and pride in all of our institutions. Each of us has made a tremendous difference in so many ways like donating blood or food to relief efforts and flying the American flag outside our homes as a sign of solidarity. In the Congress, members of both parties worked together in a bipartisan fashion like never before to demonstrate our resolve to the world community and to care for the victims and their families. When we sang "God Bless America" on the Capitol steps that same night, it was an incredibly emotional moment that truly touched my soul.

It was a true honor to be in New York City at the special joint session of Congress. A couple of weeks after the attacks, I went to Ground Zero with other Members to witness first-hand the devastation that had been wrought. The heroic determination of the firefighters, police officers and rescue workers will be etched into my mind for the rest of my life. When I returned to New York City, I was amazed at the progress that the people of this great city have made in the area where the Twin Towers once stood. It is truly a testament to the strength and heart of the citizens of New York and all Americans. It makes me proud to serve in the Congress.

Like so many other Members of Congress, constituents from the First District of South Carolina and their families were among the victims on that tragic day. They will be sorely missed, but we will never forget them. As we commemorate the unity of this great Nation on the first anniversary of these terrorist attacks, I pray for these families and for all Americans. The foundation of this great land is strong, and we will never waiver from our cause. God bless America.

SEPTEMBER 11

Hon. John F. Tierney

OF MASSACHUSETTS

Mr. Speaker, On this somber anniversary of the terrible attacks on our country last September, we pause in remembrance of all those who died, and we stand in solidarity with the many families here in our communities and elsewhere who continue to live every day with the grief and pain of their unspeakable loss. Their lives and ours will never be the same, but we come together today in communities large and small across our Nation not only to comfort one another and remember but to proclaim anew our values as Americans—values that we as a Nation have rediscovered in ourselves and each other since last September 11; values that challenge us to live better, nurture our relationships, and serve our community; values that command us to respond to tragedy as all of these brave families have—with courage and resolve, undaunted by acts of cowardice and hatred.

This gathering today is yet another step that we as a community, indeed we as a nation, are taking together to win this battle against the assault on innocent civilians living in a free society. While we continue to experience competing emotions of sorrow, anger and frustration, we refuse to allow these acts to rob us of our values and our spirit.

My colleagues and I will continue to work together with the President to bring about the end of terrorism. We have the ability and the wherewithal to confront this challenge as we have met so many others in the past so that when future generations pause in remembrance of this day in our history, they will do so in the shelter of a just and free and united country.

Hon. John E. Peterson

OF PENNSYLVANIA

Mr. Speaker, 1 year ago, our Nation was attacked by terrorists who believed that by taking innocent life, they could destroy our spirit and tear down the principles, values, and freedoms that we hold dear. Despite our initial shock and horror on that fateful September morning, Americans from all walks of life proved the terrorists wrong by immediately joining hands to search for survivors, comfort those who lost loved ones, and bring about healing and renewal. There has never been a time when the world witnessed greater heroism, compassion and unity.

Under the leadership of President Bush, our Nation has made great strides to bring justice to those who perpetrated this evil and improve our Nation's defenses against future terrorist attacks. Our men and women in uniform responded valiantly, toppling the Taliban regime and bringing freedom to a Nation that had served for many years as a haven for terrorism and oppression. The effort to protect our Nation from terrorism is ongoing, and patience will be necessary as we work to establish a permanent

Department of Homeland Security and thwart the continued efforts of those who seek to kill innocent Americans in order to advance their political agenda.

Looking back over the past year, it is clear that the events of September 11 have strengthened our Nation and given us a greater appreciation for freedom. Americans have demonstrated that we are committed to working together to preserve our freedom so that we will continue to be a beacon of hope to freedom-loving people around the world.

Thursday, September 26, 2002

IN MEMORY OF THE TRAGEDY OF SEPTEMBER 11, 2001

Hon. Deborah Pryce

OF OHIO

Mr. Speaker, I rise today, just over 1 year after the tragic events that touched the life of every American, to give solemn remembrance to that darkest of days.

As do all Americans, my heart continues to ache when I think about the countless victims and families struck by the sad and shocking attacks of September 11, 2001. But there is also pride in my heart for our great Nation and the men and women who have responded so resolutely and valiantly to this challenge to our very way of life.

The stories of tragedy, and the compelling stories of heroism that emerged from the smoke and shattered buildings will forever be a part of our memory that day.

When I rose to the House floor 1 year ago filled with so many deep and powerful emotions, I pledged that we would not let the days that followed be remembered just for our sadness and anger, but for our national resolve. As a nation, we have pulled together in so many ways to overcome the vicious attempt to break our national spirit.

We are living in historic times, and I have been so proud of the American people. New York City has been the personification of American strength and resiliency.

I have also been proud of our work here in Congress to put aside politics to provide our Armed Forces and law enforcement officers with the resources and tools they need to fight the war on terrorism, and to make this Nation safer than it was on September 10, 2001. This war is not an issue of politics, it is an issue of patriotism.

Those responsible for last year's horrific events seriously miscalculated the strength and resolve of Americans. Our sense of security may have been temporarily unsteadied, but our unity is unwavering. Our bonds of liberty, our bonds of freedom, our bonds of democracy are stronger and run deeper than any individual, than any building, than any monument. No act of violence, no sharpened razor, can sever them.

America has been committed through this last year to the difficult realities of living in the shadows of war. We have gone on living our lives because to do otherwise would be giving in to the evil behind September 11. But there should be no doubt that we will remain committed until those responsible learn the steep cost of taking innocent lives—innocent American lives—on American soil. We will never stop working to make America safe and secure.

The flame of liberty remains bright and will continue to shine upon the world, casting deep into the dark shadows of violence, intolerance, and extremism. This is a time of remembrance. But it is also a time to renew our dedication to fighting until America is free from the threat of terrorism.

Hon. John Elias Baldacci

OF MAINE

Mr. Speaker, like every American, I will always remember September 11. Today we gather to mourn our Nation's losses and to demonstrate our resolve that America will not be slowed or diminished by terrorist attacks.

September 11 was an unsettling day for each of us, wherever we were. Nearly 3,000 Americans lost their lives that day. All of us lost the sense of security that we as Americans had long taken for granted.

The victims came from all across the country and every walk of life. They had one thing in common—they were simply going about their business. It was a crisp, clear, sunny morning. I will never forget the contrast between the beautiful weather and the acrid smoke, dust and rubble at each of the impact sites.

Each of us shares the grief of families who lost loved ones. Each of us mourns the loss of innocence that resulted from the attacks.

Even in those darkest hours, however, America's light shined through. Millions of us joined together to donate blood and money to help the victims and their families. I visited the Pentagon to encourage rescue workers, and worked with the FAA to ensure that medical supplies would continue to arrive at Maine hospitals during the shutdown of airline service. Maine businesses and individuals donated food and supplies for workers and for displaced families.

The great irony of September 11 is that the terrorists sought to drive America apart, but instead brought us together as a nation. Our people have once again shown an incredible resilience and an ability to come together in times of need. America is, indeed, one Nation, under God, indivisible.

The acts of terrorism perpetrated against our country have reminded us of the precarious nature of life and of the lives of those around us. We will always carry in our hearts the memories of those who were lost on September 11. May we also always remember the patriotism and unity that we have experienced in its aftermath.

We stand together today as Americans, united in mourning and also in our resolve to triumph over factions that would tear us apart. Together, we will ensure that hope, freedom and justice will prevail.

THE WILL OF THE AMERICAN PEOPLE

Hon. Cliff Stearns

OF FLORIDA

Mr. Speaker, recently I came across an e-mail that had been circulated last year shortly after the September 11 attacks. It was sent by a former military professor in which he stated that the will of the American people is the fulcrum of this Nation's war on terrorism. He stated that the terrorists who attacked on 9/11 counted on a "soft and spoiled" America, who would eventually quit if retaliation did not result in immediate success.

What the author of the e-mail discovered was an outpouring of "what can I do" to help in response. And I believe this amply illustrates what we continue to see across the United States. What the terrorists actually demonstrated with their attacks on 9/11 was their profound ignorance of the American people and of history.

Military theorist Clausewitz terms war a "remarkable trinity composed of the primordial passions of the people, the rational policies of the state, and the combination of incidents in battle."

The "passions of the people" were awakened after Pearl Harbor and again were awakened after 9/11. One year later, the passions are still high. Americans are aware that what we are facing is that which America has never seen up close. We were attacked on our own soil by an organization of individuals bent—not on removing our presence from certain parts of the world—but on our utter destruction.

We are facing an enemy who despises our very existence. They are consumed by hatred for the United States, that despite its faults, is open to all people regardless of race or religion. We operate under principles of freedom, the ability to pursue life, liberty and happiness. As such, our country is fighting with hope against terror and freedom against oppression. Our enemies will never know freedom, because they are imprisoned by hate; and for that they have already lost.

Former U.S. Defense Secretary, Caspar Weinberger, stated "The will of the American people once aroused . . . is capable of accomplishing all the things that have to be done." As long as we continue to maintain a moral high ground in this campaign against terrorism and its supporters and take the appropriate and precise responsive measures, the will of the people of this county will know no bounds.

TRIBUTE TO THE VICTIMS AND HEROES OF SEPTEMBER 11, 2001

Hon. Ken Calvert

OF CALIFORNIA

Mr. Speaker, I rise today to honor and pay tribute to the victims and heroes of September 11, 2001. Over the past year, this country has tried to come to terms with the tragedy we experienced on September 11. Although I have written or spoken about that day many times, I have found that words often fail to describe the magnitude of that day. There are the haunting stories of loss and grief as well as stories of heroism and triumph. We learned about ourselves and our country that day and while we grieved for those we lost; we also cultivated a new sense of unity and patriotism. As a Nation we renewed our belief in the American spirit and in the bravery of fellow Americans who would willingly risk their lives for a stranger. The stories of what ordinary men and women did under extraordinary circumstances continue to amaze me.

As we remember September 11, 2001, let us remember all the brave men and women who not only saved lives but saved our sense of brotherhood. Let us extend our prayers to all our brave men and women in the Armed Forces who right now are protecting our way of life and let us extend our deepest gratitude to our fire and police forces who have redefined the word "sacrifice."

Hon. Dennis R. Rehberg

OF MONTANA

Mr. Speaker, Americans should be proud of how they've handled the past 12 months. Yes, September 11 changed America forever, but not in the way the terrorists anticipated.

Those who committed this horrible act of war were intent on destroying America and our way of life—but they failed miserably. They tried to make us question our dedication to democracy but they only increased our resolve in preserving the greatest form of government the world has ever known. They tried to destroy our economy—but the whole world knows America is still open for business, for farming, for travel. And they tried to extinguish the flame of liberty and hope in our country, but they only made it burn brighter. Yes, America has changed—we're stronger than ever.

Hon. Sam Graves

OF MISSOURI

Mr. Speaker, on September 11, 2001, America awoke to the worst terrorist attack in history. As we went to work and school, we left with a feeling of security that we have long since forgotten. By the time we returned to our families, our lives and our Nation had forever changed. It had been many years since America felt so insecure, so vulnerable. On that morning, the American people's resolve was put to the ultimate test. Everything appeared to be so uncertain that day. Who would do such a thing? Why would they do it? Is there more to come? How can I protect my family?

But there was much that was certain that day. America made a promise to the victims and their families, to future generations of Americans, and to the world. The American people promised that this action would not go unanswered. We promised that this action would only strengthen and unite us, not divide us. We would respond forcefully to those who were responsible while tending to our neighbors, our fellow countrymen. Together, you and people across Northwest Missouri and our Nation donated blood for the victims, and donated money for their families. Together, we prayed for those who lost so much that day. We prayed for our soldiers who stood ready—preparing to defend our freedom.

As we stop to remember that terrible day, some of the pain and fear has subsided. But our determination to defeat those who seek to terrorize us must never fall victim to the passage of time. In the coming months, the American people will face a choice: Live up to our responsibility by making tough choices and sacrifices to continue our assault on terrorism, or quit now and hope that they choose to stop planning future attacks. The American people should never have to endure such a tragedy again. As we have learned over the past year, we can do something about it. We must never mislead ourselves that we have to wait to be attacked again to continue our defense from terrorism. The more than 3,000 lives lost is all the justification we need to have to defend against a certain threat of terrorism. The United States must remain vigilant and prepared, so that we remain forever free.

Hon. F. James Sensenbrenner, Jr.

OF WISCONSIN

Mr. Speaker, since September 11 last year, life has taken on new meaning. For some, that day's devastation has caused them to become more cynical, changed by the events of a few hours. For others, each day since then has taken on more significance as they realize what it means when people say that you can't take life for granted. But for all of us the memories of that day will live on as we not only think about the people trapped in the World Trade Center, the Pentagon, and the hijacked planes, but also remember their loved ones who were helpless to prevent the tragedies. As the President said in his proclamation earlier this month, "Those whom we lost last September 11 will forever hold a cherished place in our hearts and in the history of our Nation. As we mark the first anniversary of that tragic day, we remember their sacrifice; and we commit ourselves to honoring their memory by pursuing peace and justice in the world and security at home."

Life after September 11 took on new meaning for Members of Congress too. We reacted by changing our priorities, and began work on legislation to respond to that day's horrific events. One of the first things we did was pass legislation authorizing the use of U.S. Armed Forces against those responsible for the attacks. Since then, Congress passed numerous bills dealing with the issues that are widely believed to have allowed the events of September 11 to occur. They include: strengthening airline security and our Nation's borders, restructuring the Immigration and Naturalization Service to make it a more efficient agency, and passing the Patriot Act to improve information sharing between law enforcement and intelligence communities. More recently, the House passed legislation to create a new Department of Homeland Security in response to the President's request for a flexible, effective department, with the singular mission of protecting our Nation. Financially, we have also passed legislation to provide the Department of Defense with the resources it will need to address the new challenges that now face the Nation.

Many individuals have changed their priorities too. Spurred by our war against terrorism and the words of the administration and various law enforcement agencies, Americans have begun to pay more attention to their surroundings and take better note of anything that appears out of the ordinary, particularly in airports. Gone are the days when one can easily dismiss peoples' errant behavior as harmless without making sure that that is indeed the case. As we learned, the price to pay for not checking can be awfully steep.

After the events of that Tuesday, life in Washington, DC, returned to some semblance of normalcy. Yet, a year later, although life appears the same as always, there is a difference. Certainly, Congress is in the middle of its appropriations debates as it almost always is this time of year and Republicans and Democrats are embroiled in many of the same arguments that typically take place around now. However, there is now an underlying sense of wariness in our Nation that didn't exist before—but this is good, as it shows that we have learned from last year's events. It demonstrates that as a nation, we have grown. September 11, 2001, wasn't that long ago, but America has lived a lifetime in the year since that fateful day.

SPECIAL JOINT MEETING OF CONGRESS IN NEW YORK ON SEPTEMBER 6, 2002

Hon. James H. Maloney

OF CONNECTICUT

Mr. Speaker, it was a historic experience to join with my colleagues in the special joint meeting of Congress held in New York City. I have visited New York many times since the terrorist attacks on September 11, 2001. It is a city that has recovered from, but not forgotten, the events of that tragic day.

The families and friends of those who perished have endured a year of unbearable loss. They have my deepest sympathy. Rarely have we felt vile acts of terrorism perpetrated on our shores, and never on the scale of September 11, 2001. Our response has shown the strength of character of the American people. The sadness that we all felt that day, and in the days since, has hardened into a resolve to honor the memories of those who perished, to heal our wounds so that our Nation is even stronger than before, and to bring righteous justice to those who perpetrated the attacks.

The congressional session in New York was a fitting salute to that great metropolis, and a dramatic affirmation that all Americans stand united with the people of New York as we move forward to root out terrorism and build a free and secure world community.

Hon. Nita M. Lowey

OF NEW YORK

Mr. Speaker, today I rise in recognition of the 1-year anniversary of the tragic events of September 11, 2001.

Last September, the lives of all Americans were forever changed. Loved ones have been lost, and the grief we feel is as sharp now as it was 1 year ago. There is an emptiness in families, in offices, in communities, where sons and daughters, husbands and wives, parents, siblings and friends once were.

Our country has changed. We have focused intently on the task of ensuring our homeland security—a term seldom heard before this past year. We have invested billions of dollars in securing our ports and borders, water and food, and airways. We have enhanced the strength of our military and intelligence capacities, undertaking an unprecedented campaign to end the threat of international terrorism. We have a new appreciation for the hard work of our men and women in uniform—whether they are soldiers, police officers, emergency medical technicians, or firefighters.

We joined together with the President, the Governor, and former Mayor Giuliani in passing an emergency spending bill which provided $21 billion to fund the rescue and recovery efforts at the World Trade Center site and cleanup in lower Manhattan. This funding has also helped alleviate some of the economic ramifications of the terrorist attacks, provide counseling to New York schoolchildren affected by 9/11, and it is now being used to modernize the transportation systems that were devastated by the towers' collapse. The wounds remain, but our community has shown extraordinary courage in dealing with the challenges before us. We will never forget, but we will recover.

Shattering as this ordeal has been, the fundamental character of America has remained the same. And for this we should all be proud. We are still a strong and diverse nation, focused on the pleasures and challenges of everyday life, caught up in the struggle to provide good homes for our children, achieve meaning in our lives, and leave this world a better place after we've gone. We still believe in the importance of our democratic ideals—the foundation on which our country was built, and continues to thrive.

As a Nation, we have joined together to provide support for our neighbors, friends, coworkers and fellow Americans. As I have traveled around New York, I have seen remarkable displays of the resilience of the American spirit as we have worked in the recovery effort, giving from both our hearts and our wallets. The tragedy of September 11 was once unimaginable as were the courage and empathy that were displayed that terrible day. Now, it is this strength and this concern for our fellow citizens that redefine us as Americans.

This is what gives us hope. And this is what gives us confidence that, despite the dangers of the world and the challenges our country faces, we will prevail in perpetuating the values we hold dear. I am humbled and honored to stand before you today in remembrance of the tragedy of September 11 and the heroism and patriotism of New Yorkers and all Americans over the past year.

Hon. Bernard Sanders

OF VERMONT

Mr. Speaker, our Nation was forever changed on the morning of September 11. The goal of Osama bin Laden was to demoralize us, create fear, uncertainty and instability in our country—he failed. Last week Congress met in New York to pay tribute to those who were killed, and I was reminded how strong and resilient our country truly is. Last September 11 we saw amazing displays of heroism and bravery. I will never forget the sight of firemen entering the World Trade Center risking their lives to save others.

We have also learned a great deal since September 11. We've learned that we are a vulnerable nation, and that we must lead an international coalition against bigoted, religious fanatics who believe they have the right to kill in order to impose their reactionary ideology on others. Many of us also have learned that in order to be true to American values, we must not undermine the fundamental principles and constitutional rights that our country was founded on.

As an American and a Vermonter, I was enormously proud of how our people responded to this crisis in terms of blood donations, financial contributions and coming together as a community to support the victims and each other. It is my hope that we will continue to show that same sense of community that we demonstrated in the aftermath of the September 11 attacks.

Hon. Michael R. McNulty

OF NEW YORK

Mr. Speaker, on Friday, September 6, of this year, a special joint session of the U.S. Congress convened in New York City to reflect on the terrorist attacks of September 11, 2001. This most rare and somber session served as a necessary reminder of the human loss and heroism that the City and State of New York, our Nation, and our world experienced on that fateful day. I am proud and blessed—as an American, and as a New Yorker—to have been able to take part.

We met at Federal Hall, the very same site where the First Congress met over two centuries ago. We met just blocks from where the World Trade Center towers once pierced the city's majestic skyline.

Mr. Speaker, most important, we remembered the almost 3,000 innocent civilians who died and their families. We prayed then—and we should pray now—for all of the victims of this most heinous terrorist act. Though a year has passed, the loss of every single person who perished that day is still felt by all those who loved them. The sons and daughters, the brothers and sisters, the mothers and fathers lost that day will never be replaced. We simply hope that the pain will subside, and that the memories will remain strong and vibrant.

Mr. Speaker, we also expressed our deepest gratitude to the firefighters, police officers and emergency personnel who served on that fateful day and in the weeks and months that followed. These brave men and women, and their peers across the country, put their lives on the line—day in and day out—to ensure the safety and well-being of the citizens of our communities. Recognition of the heroism and service of our first responders is overdue and well deserved. We must continue to acknowledge their bravery and sacrifice. And we in Congress must resolve to provide them with the support they need to continue to excel in their chosen duty—to save lives.

Finally, Mr. Speaker, we reaffirmed our commitment, as a Congress and as a Nation, to eradicate the ability of terrorists to ever again carry out such a horrific and offensive act—against us or against other innocent people around the world. We reaffirmed our promise to bring these cowards to justice. A terrorist is, by definition, a coward. It is a person who cannot get what he wants by the power of persuasion, and therefore resorts to the killing of innocent men, women and children.

Mr. Speaker, the American men and women in our Armed Forces are now dispersed throughout the world, seeking out these cowards and introducing them to the might of a Nation that finds its heart bruised but its strength renewed. Freedom is not free. We have paid a tremendous price for it. We must not forget those before us who gave their lives, or those who put their lives on the line today, to allow us the privilege of living in the freest and most open democracy on the face of the Earth. Our patriots fight for the cause of freedom, and we shall support them every step of the way.

The events of September 11, 2001, were basic violations of the fundamental principle that life is to give—not to take. I am proud that this Congress gathered on September 6, in the shadow of Ground Zero, to remember and honor both the victims and the heroes—and to remind the world that the forces of evil shall never prevail.

Monday, September 30, 2002

SEPTEMBER 11 ANNIVERSARY

Hon. David Vitter

OF LOUISIANA

Mr. Speaker, on September 11, 2001, our people, our democracy, and our values were attacked in a cowardly and reprehensible way. I visited the Pentagon 2 days after the attacks, and the sheer devastation viewed in person was beyond the imagination.

When I made it back home to Louisiana, I hugged my wife and kids and could not help but think of the people who never returned home on September 11. That fateful day introduced us to hundreds of heroes. And it reintroduced us to the wonderful spirit of our Nation.

I visited Ground Zero for the commemorative joint meeting of Congress 1 year later and visited the Pentagon on the anniversary of the attacks. These sites—along with the Pennsylvania crash site—stand as reminders of the devastation our country suffered, but they also remind us that America is not devastated.

We are unified in the knowledge that democracy and freedom will prevail. People across the country have, over the last year, demonstrated to the world that terrorism can never destroy our way of life. And I am proud of our country, proud of my fellow citizens for the patriotism, spirit, and strength they have shown over the last year.

It is a great honor to serve in Congress at this time, and I take very seriously my pledge to protect and defend the United States of America.

May God bless us all, and may He continue to bless our great Nation.

Tuesday, October 1, 2002

EXPRESSING SUPPORT FOR GOALS AND IDEAS OF DAY OF TRIBUTE TO ALL FIREFIGHTERS

Hon. Nick Smith

OF MICHIGAN

In 1992, Congress created the National Fallen Firefighters Foundation to lead a nationwide effort to remember our fallen heroes of these first responders. The Foundation has successfully met this challenge, providing a variety of supporting activities to surviving family members, including emotional support, awards, scholarships for surviving spouses and children, and the establishment of the national park in Emmitsburg, MD, to memorialize the fallen firefighters.

Madam Speaker, this Sunday the Foundation will honor 446 firefighters who lost their lives in the line of duty at an annual ceremony usually held in Emmitsburg, but this year will be held here in Washington. The Foundation is expecting over 20,000 people to attend this year's memorial weekend, a record number. Many of these people will be spouses and children of the 343 firefighters that perished September 11.

On that warm September morning in 2001, the firefighters of the New York City Fire Department reported to work as they did every day, ready to respond to whatever emergency situation might occur. But that morning was different. The 110-story World Trade Centers that were both literally and symbolically the center of world commerce were hit by two hijacked 747s.

The firefighters of the New York Fire Department received the most terrifying and overwhelming emergency call that this Nation has ever known. Still, they responded with true bravery, rushing into that burning building without hesitation. They helped evacuate 25,000 people, the largest evacuation in the history of the world, and certainly they struggled until the last possible moments to free those who were trapped. Three hundred forty-three of them lost their lives in doing so. It was on that September day that the American firefighter became the symbol of American freedom and American bravery to not only those of us in the United States, but certainly to millions around the world.

In addition to the heroism displayed on 9/11, we know that first responders all over the country display similar heroism every day, not just when major disaster strikes, but every day, as full-timers and volunteers often risk their lives to protect the lives and property of people around the country.

Fire and emergency service personnel respond to over 16 million calls annually. In addition to the 343 heroes who gave their lives in New York City on 9/11, last year we lost another 99 volunteers and full-time firefighters working in the line of duty to save property and lives.

I think we all agree that it is our job as Americans and as Members of Congress to never forget the sacrifices of those who protect us, and I commend the National Fallen Firefighters Foundation for its exceptional efforts in leading this charge. I offer my strongest support to this resolution, and certainly invite my colleagues to attend this Sunday's tribute.

Hon. Curt Weldon

OF PENNSYLVANIA

Madam Speaker, As my colleagues know, I would not be in public life were it not for the fire service, having grown up in a fire service family. The earliest recollections of spending time with my father was at the firehouse with him and with my six older brothers. It was only natural that when I became 18 I joined the department and eventually worked up to becoming president and chief. I then went back to school while teaching during the day to get a degree in fire protection. So all of my life has been devoted to these outstanding men and women who protect us in 32,000 departments and consist of 1.2 million men and women.

Madam Speaker, this resolution as has been outlined by my colleagues recognizes America's domestic defenders. These are the people who protect our country every day of the year and have been doing so for 350 years, longer than the country has been a country. Because when Ben Franklin started the first fire department in Philadelphia, it was an all-volunteer group, and it was in fact started before America became a Nation. They have been our domestic defenders ever since. In fact, I call them our first responders. Much like our military protects us against threats from outside of our country, our domestic defenders protect us from those threats within America. And in fact they are being asked to do more and more as we face the threats of terrorism on our soil.

Madam Speaker, each year we have in fact acknowledged those who have lost their lives. We average about 100 deaths a year at Emmitsburg, the National Fallen Firefighters Memorial. I have made that trip at least three times since I have been in Congress, and I can tell you there is no more memorable event than to spend time with those families of brave Americans who pay the ultimate price.

Madam Speaker, as founder and original chairman of the Congressional Fire and EMS Caucus, which now has 340 members of the House and Senate, I have traveled to every disaster we have had in the country in the past 16 years. From the wild lands fires in the western part of the country, in California, in the north, in Yellowstone Park, to the hurricanes down in the South, Andrew and Hugo, the Murrah Building bombing down in Oklahoma City, the Northridge and Loma Prieta earthquake. I was up at the World Trade Center in 1993 and back 2 days after September 11.

Madam Speaker, when I went up to the Trade Center in 1993, I was escorted through that building, that bombed-out parking garage, by a rising young emerging chief of the New York City Fire Department. He and I became good friends, and we traveled around the country over the past 9 years talking about the need to understand the first responders and provide support for them.

At 2 o'clock on September 11, I got a frantic phone call on my cell phone from my friends in New York that my good friend who had taken me through the Trade Center in 1993 was killed when the buildings collapsed.

Ray Downing was the chief of all rescue operations for New York City on September 11. He was the guy who was overseeing the bulk of the 343 New York City firefighters who were going up in the buildings when the buildings were coming down. Ray Downing left behind a wife and five kids and grandkids. In fact, 1 month after September 11, I brought them all down to my district. We had a parade with 40,000 people to honor Ray Downing as an American hero.

Perhaps one of the most emotional days I have had during my tenure in Congress was when I went to Ground Zero and spent 12 hours there 2 days after it occurred with my friends of the New York City Fire Department. We went around the back of what used to be one of the huge towers, which is now a seven-story pile of rubble; and I saw two firefighters in their turnout gear among thousands who were collecting rubble. And I looked on the back of their turnout gear and on the bottom were the names Downing and Downing, because Ray Downing's two sons are officers in the New York City Fire Department. One is a captain; one is a lieutenant.

They were looking for the remains of their dad. We did not find the remains of Ray Downing until 3 months ago. It went through DNA sampling. We were able to determine that Ray in fact had been accounted for.

Madam Speaker, this coming Sunday we will honor these brave firefighters, the 343 from New York and the others that combined for a total of 442 brave Americans. Anytime this country has gone to war and lost 442 of our sons and daughters, we have mourned as a Nation. Well, this past year we have lost 442 brave Americans. They were not soldiers on the battlefield overseas. They were our defenders here at home. And it is certainly fitting and proper that we set aside a day to honor them, that we have turnout, as I will be in attendance on Sunday, to pay our respects at the MCI Center here in Washington.

It is my fervent hope that all Americans pause as we begin to celebrate the national week of recognition for the fire service, always the first week in October, and pay tribute to our true American heroes.

As I have said time and time around the country, the firefighter is more than just a person who puts out the fires. It is the person you call upon to handle the hazmat incidents, the fires, and the conflagrations, the floods, the tornados, the earthquakes, the terrorist bombings. They are the people you call when the cat is in the tree, when the cellar has been flooded. They are the first group you call to organize a search party to find your lost child.

In many of our small towns, the firehouse is where you vote on election day. It is the organization that hosts the July 4 and Memorial Day parades. It is the organization where the Boy Scout troops and Girl Scout troops meet. It is the organization that makes our towns what they are. It really is, in my opinion, the heart and soul of America. I can think of no better group that represents what America is all about.

Amazingly, 85 percent of these people are volunteers. What other group in America has their volunteers killed each year in protecting the community while going out and raising the money to buy those fire trucks which cost from $400,000 to $600,000, by having chicken dinners and tag days? Imagine having our police departments or trash departments out raising the money to buy the trash trucks and the police cars.

In every fire department in America, in all of our districts, there are volunteers out there just doing that. In fact, this past Labor Day at Jerry Lewis' annual telethon, as has been the case every year, the IAFF Fire Fighters Union was the largest contributor to the fight against muscular dystrophy, the number one group in the country in reaching out to help other people.

These truly are the heroes of our country. They are the people who time and again have allowed us to understand what America is all about, by offering selfless service to help others. And as our friend, the gentleman from Michigan (Mr. Smith), said, we lost 343 at the Trade Center; but the real story is the success they had in rescuing tens of thousands of people that are today united with their families.

So I ask all of my colleagues to join with us in supporting this resolution and paying tribute to America's heroes.

Hon. Felix J. Grucci, Jr.

OF NEW YORK

Madam Speaker, I would first like to thank my colleague Congressman Weldon for his tireless efforts on behalf of America's firefighters, and for bringing this measure to the floor today.

Few images throughout history more clearly illustrate heroism better than the images of brave firefighters entering the World Trade Center—knowing they very well may never return—with one selfless goal in mind: to save lives.

On that tragic day—September 11, 2001—347 firefighters died in the line of duty, several from my own district on Long Island.

While the heroic efforts of these brave men and women may be more clear on that day there isn't a day that passes when firefighters do not risk their own lives to save others.

Last year alone, 442 firefighters sacrificed their lives in order to save the lives of innocent victims of fire and other emergencies.

Later this week, the National Fallen Firefighters Foundation will honor these firefighters and their families for the sacrifices made over the last year. We will remember the impact these brave firefighters have made in towns and communities throughout America and the heroism that has saved countless lives.

On behalf of the First Congressional District of New York—home to several fallen firefighters—I join my colleagues in support of H. Con. Res. 476.

Hon. George W. Gekas

OF PENNSYLVANIA

Madam Speaker, I rise today in strong support of H. Con. Res. 476 and urge my colleagues to support this important piece of legislation as well.

Our firefighters protect us every day and sometimes give up their own lives to protect and serve their communities and their fellow man. This was never more evident than on September 11, 2001. On that day, as frightened and wounded civilians ran from the World Trade Center, brave firefighters rushed in, in a determined effort to save others. These brave individuals risked everything in an effort to render aid and evacuate the people trapped in the towers. This effort cost many firefighters their lives. The September 11, 2001, attacks highlighted the spirit and courage of firefighters across the Nation. Mostly volunteers, these men and women protect our lives and property, and while they never boast of their heroic deeds, they are truly heroes.

H. Con. Res. 476 reaffirms that Congress supports the goals and ideas of a day of tribute to all firefighters who have died in the line of duty and recognizes the important mission of the National Fallen Firefighters Foundation in assisting family members to overcome the loss of their fallen heroes. I am thankful to be able to rise today and proclaim support of H. Con. Res. 476 on behalf of every firefighter in Pennsylvania's 17th Congressional District.

Madam Speaker, this is the very least we can do. I salute Congressman Weldon for sponsoring this resolution and would like to thank him for his leadership on the Congressional Fire Services Caucus, of which I am a member. The firefighters of the United States should know that the Congressional Fire Services Caucus is continually striving to respond to their needs and to deliver to them the equipment and resources they need to do their job in a safe and effective manner.

Firefighting will never be a safe endeavor but we in Congress must do all we can to help our firefighters. No matter what we provide to our firefighters we will never equal the sacrifices they make for us. Collectively, we in Congress thank you and the passage of H. Con. Res. 476 is just a small token of appreciation. We will never be able to thank you enough.

Madam Speaker, I reaffirm my support of H. Con. Res. 476 and of the firefighters of Pennsylvania's 17th Congressional District.

Hon. Sherwood L. Boehlert

OF NEW YORK

Mr. Speaker, I rise in strong support of H. Con. Res. 476, recognizing the goals and ideas of a day of tribute for fallen firefighters, and supporting the important mission of the National Fallen Firefighters Foundation.

Congress created this non-profit foundation 10 years ago to lead a national effort to honor firefighters who have died in the line of duty and to assist surviving firefighters and family members in rebuilding their lives.

The Foundation has been steadily expanding its activities. In addition to providing emotional support services to survivors and scholarship awards for surviving spouses and children, the Foundation is now creating the first National Park to memorialize fallen firefighters in Emmitsburg, MD. And this Sunday, October 6th, the Foundation will sponsor a memorial weekend to honor the commitment, bravery and sacrifice of the 446 firefighters who died in the line of duty in the past year, 343 whose lives were taken on September 11, 2001.

No one could have anticipated the magnitude of destruction and loss of life that occurred last September. In the wake of those tragic events, the value and contributions of the National Fallen Firefighters Foundation became unmistakably clear.

At the request of the Federal Emergency Management Agency, the National Fallen Firefighters Foundation sent support staff to Ground Zero within days of the attacks, working around the clock to coordinate chaplain support services, survivor support services, as well as logistical and administrative support association with the loss of the firefighters.

The Foundation's efforts in New York City during its time of greatest need were truly invaluable, and I proudly support its cause, as well as this resolution recognizing its importance.

RECOGNITION OF "TOWERS OF LIGHT" BY DOROTHY DIEMER HENDRY

Hon. Robert E. (Bud) Cramer, Jr.

OF ALABAMA

Mr. Speaker, I rise today to share with the House of Representatives a poem written by one of my constituents, Dorothy Diemer Hendry of Huntsville, AL. The poem, "Towers of Light," was written at the time of the 1-year anniversary of September 11 and provides us with hope and encouragement that we can become towers of light in this world. I am pleased to share this poem with you today.

TOWERS OF LIGHT

(By Dorothy Diemer Hendry)

Two shafts of light
stream down through clouds
to bedrock of a crater
cleared of rubble and bones.

We look up, up, up
these twin towers of light
until we must shield our eyes
from their source, more luminous
than sun and moon and stars.

What is the source?
Not firestorm of planes
commandeered and exploded
in misbegotten piety and hate.
Not savage burning of
"heathen"; temples or churches,
mosques or synagogues.
Not merciless holocaust
of "enemy"; fields and forests,
schools and homes and people.

The source of light may
go by different names in
your religion and mine,
yet somehow the twin towers
remind us of two neighbors filled
with the radiance of the Golden Rule.

Neighbor from anywhere,
let us not quarrel about
holy names and ancient cruelties.
Let us fill the crater with loam
and plant a new garden on Earth.
In honor of heroes and loved ones,
let us summon the courage, wisdom,
and kindness to dwell in mutual peace.
Can we not become towers of light?

RECOGNIZING THE WORK OF THE AMERICAN CANINE ASSOCIATION'S SEARCH AND RESCUE TEAMS AT GROUND ZERO IN NEW YORK

Hon. Joseph R. Pitts

OF PENNSYLVANIA

Mr. Speaker, I would like to extend my gratitude and appreciation to the Search and Rescue Teams of the American Canine Association, ACA, who, for 8 weeks following the terrorist attacks of September 11, 2001, tirelessly and selflessly served at Ground Zero in New York City. Following the devastating attacks on our country and our very way of life, handlers Robert Yarnall, Jr., Susan Yarnall, Heather Nothstein, Michael Glass, Avi Thol, Travis Hayden, Sean Hayden, and Amy Dinardi, along with canines Gus, Nela, Nala, Samson, and Duchess, came to the aid of a grieving Nation and endured tremendous personal difficulty to assist in finding survivors of the Twin Towers.

In the year since September 11, we have come to a new appreciation of the heroes who live among us, those who volunteer their time, ability, and talent, without thought of recognition. The actions of these handlers and canines represent the true values of America—generosity, compassion, and service to community—that have made us great. I am proud to have had these extraordinary people and canines represent Chester County and the Commonwealth of Pennsylvania. The ACA deserves our thanks and our commendation for a job well done.

COMMEMORATING SEPTEMBER 11, 2001, AND ON THE OCCASION OF THE SPECIAL JOINT MEETING OF CONGRESSS IN NEW YORK ON SEPTEMBER 6, 2002

Hon. Bob Goodlatte

OF VIRGINIA

Mr. Speaker, 1 year ago, Americans looked on in horror as the events of September 11 unfolded. At the end of the day the skyline of one of our greatest cities was forever changed, the Pentagon, a symbol of America's military might was still smoldering, and a previously indistinguishable field in Western Pennsylvania had suddenly and terribly become an unmarked grave for America's newest heroes.

In the aftermath of the Challenger space disaster, when seeking to comfort a shocked and hurting country, President Reagan told us that "The future doesn't belong to the fainthearted; it belongs to the brave." Overwhelmed by grief, and reeling from a devastating attack, some nations would have been crippled to the point of inaction. Our enemies perceived us as weak, soft, unwilling or unable to respond. It is obviously an understatement to say they miscalculated.

In a tribute to the excellence of our Armed Forces and to the leadership of President Bush, we succeeded in swiftly toppling the Taliban, thereby liberating the people of Afghanistan. In this volatile region of the world, America's national security is directly at stake, for when regimes that tolerate terror and disrespect human life are left to their own devices, they export hatred and murder.

On the home front we moved swiftly to protect against future attacks. Congress gave law enforcement new tools, restructured the beleaguered INS, and took steps to establish a Department of Homeland Security.

Recently, I traveled to New York City for a commemorative joint meeting of the U.S. Congress, which was held at Federal Hall, just blocks from where the Twin Towers once stood. Federal Hall also served as the site of George Washington's swearing in, the location where the Bill of Rights was drafted and the setting of the first meeting of the U.S. Congress. In an era when historical significance is often missed and sentimentality is at times scoffed at, the symbolism of this meeting must not be overlooked.

Democracy is alive and flourishing and despite the best efforts of those who would seek to destroy us, we remain the "shining city on a hill"—envisioned by our Founders—as can be attested to by the resilience, courage and selfless sacrifice, which has characterized our national response.

On the anniversary of this day, which has been eternally seared into our national conscience, our thoughts and prayers go to those Americans whose lives have been forever changed by the loss of a loved one.

Winston Churchill once said, "The price of greatness is responsibility." This September as we mourn the loss and commemorate the lives of our fallen countrymen, we must not forget the raw emotions that marked that day, for they underscore our responsibility and will give us the impetus to continue in the unfinished task before us.

SEPTEMBER 11—FIRST ANNIVERSARY

Hon. Joel Hefley

OF COLORADO

Mr. Speaker, America will never be the same as a result of September 11, 2001. The horrific events of that day dramatically changed the landscape of not only New York City and Washington, DC, but also the entire civilized world. The images of commercial airliners plunging into symbols of American enterprise, economy and security will forever be seared on our individual and national memories. But also vivid are the images of Americans' spirit of community, gratitude and generosity that have been demonstrated these past 12 months.

Through our heartache and sorrow, Americans joined together this year in an unprecedented show of strength and unity. The outpouring of patriotism and pride across the country is displaying itself in every conceivable way. The American flag is flying: large and small; cloth and paper; store-bought and handmade; the red, white and blue is everywhere. Americans opened their homes and wallets to care for the victims. Goods and services were donated to the victims and rescue workers at an almost unmanageable pace. Restaurants in New York and Washington opened their doors to feed the rescuers, people stood on street corners and handed food to passing firemen and companies donated pillows and blankets for weary workers.

The morning after the attack, a column appeared in the *Miami Herald* that spread across the Internet because it captured the thoughts and feelings of our Nation so aptly. In it, the columnist described the "vast and quarrelsome" American family, one "rent by racial, social, political and class division, but a family nonetheless." If the tragedy proved anything, it is that the American family is one that reaches out its hand to help another in need.

The tragedy also redefined the American hero and turned ordinary people into extraordinary Americans. After the first assault on the World Trade Center, New York City firefighters and policemen rushed into the building and began saving lives—even as the buildings were collapsing. Yes, it was the job of firefighters to go into the buildings, but they could have reasoned that the buildings were going to collapse anyway, so why try. When the victims rushed out, they rushed in, and became heroes in the process—343 firefighters sacrificed their lives to save more than 25,000.

Our Nation has had a resurgence of faith and spirituality. The tragedy caused people to reevaluate their core values and cling to their traditions. In one day, everything that we thought was meaningful and important slid to the wayside and we rediscovered fundamental beliefs about faith, family and freedom. If the terrorists had hoped to break the American spirit, they failed spectacularly.

We are now engaged in a war on terrorism and it is a war we will win. This is a struggle that concerns the whole of the democratic and civilized and free world. We will bring to account those responsible, and we will dismantle the apparatus of terror and eradicate the evil of mass terrorism in our world.

The cause that we are fighting is just and it is decent. No citizen, in any country, should live in fear of senseless terrorist attacks. On September 11, 2002, thousands of American civilians gave their lives for a cause they did not know. An attack against civilian targets of women and children, mothers and fathers, peaceful and without prejudice, is beyond comprehension in our modern, civilized world.

America responded to this crisis and emerged from the tragedy stronger and more determined. The course and duration of the conflict is unknown, but its outcome is not. America will prevail and remain the greatest Nation in the world.

EXPRESSING SUPPORT FOR GOALS AND IDEAS OF DAY OF TRIBUTE TO ALL FIREFIGHTERS

Hon. Christopher H. Smith

OF NEW JERSEY

Madam Speaker, I rise in strong support of H. Con. Res. 476 and all of America's firefighters, especially those who have died in the line of duty.

As a proud Member of the Congressional Fire Services Caucus, I think it is fitting and appropriate that we set aside time to pay tribute to our Nation's firefighters, men and women who have dedicated their lives so that the rest of us can sleep in peace. The threat of fire and the calamity an actual fire often creates is a day-to-day concern for all our communities, not to mention the added threats of terrorism now confronting us.

In 1992, on behalf of the more than 1 million firefighters in over 32,000 fire departments nationally, Congress rightly created the National Fallen Firefighters Foundation to lead a nationwide effort to remember our Nation's fallen firefighters and their families. Since its creation, this foundation has assisted many family members, helping them overcome the loss of their fallen champions. Within hours of the September 11 tragedy, the foundation established a process that used resources from across the country to provide the critical support that members of the Fire Department of New York City and their families needed.

This weekend the National Fallen Firefighters Foundation will honor the 442 firefighters who made the ultimate sacrifice in service to their communities last year, including those lost in the World Trade Center's disaster. Also to be honored are five firefighters from my State of New Jersey who served with pride and honor and who dedicated their lives to protect others in their communities. Willie Barns, George "June" Danielson, Jr., James T. Heenan, Alberto Tirado, and Lawrence James Webb are New Jersey's fallen heroes. They will be honored for their ultimate acts of valor this weekend. My prayers and the prayers of New Jerseyans everywhere will be with them and their families.

Madam Speaker, our firefighters and emergency personnel who stand at the ready to protect and help us around the clock deserve our support and dedication.

Thursday, October 3, 2002

HONORING AIR FORCE MAJOR JAMES G. CUSIC III

Hon. Jerry F. Costello

OF ILLINOIS

Mr. Speaker, I rise today to ask my colleagues to join me in recognizing Air Force Maj. James G. Cusic III, a constituent of mine from Fairview Heights, IL.

Major Cusic is receiving a Certificate of Merit from the American Red Cross for his actions on September 11, 2001. This is the highest award the organization gives for someone who saves or sustains a life with skills that were learned in an American Red Cross safety course.

The attacks on the World Trade Center and the Pentagon on September 11, 2001, made this perhaps the most tragic day in our Nation's history. However, the day could have been even more catastrophic if it were not for the efforts of men and women such as Major Cusic.

On the morning of September 11, Major Cusic saw the news of the attacks on the World Trade Center from his Pentagon office. As he watched, he began to feel the floor shake below him, and the television reported that a third plane had been used as a weapon. This time, the target was the Pentagon. A voice came on the Pentagon intercom with a message to evacuate the building.

As the news came that a second hijacked plane might be headed toward Washington, Major Cusic cleared all the rooms in his area of the building to make sure everyone had exited. Next, he assisted five of the approximately 65 patients who were being treated at the Air Force Pararescue triage site.

Major Cusic volunteered to reenter the building as 1 of 5 leaders of a 20-person team to provide medical treatment for survivors in the building. He was responsible for providing treatment for life-threatening injuries. Major Cusic aided one man who had a severe scalp laceration and a spinal injury. He assisted another man who suffered from severe burns on his face and neck and was experiencing difficulty breathing.

Later in the evening, Major Cusic's heroic actions were needed once again. A firefighter who had entered the building as part of the rescue effort collapsed from heat exhaustion and an erratic pulse. Once again, Major Cusic provided the treatment necessary under extreme circumstances.

Major Cusic maintained clarity of mind throughout the day on September 11 and should be commended for his actions in the face of adversity. At the end of the day, he was directly involved in saving three lives and in caring for two more people with severe injuries. In addition, he provided invaluable encouragement to other survivors and those involved with the rescue effort.

Mr. Speaker, I ask my colleagues to join me in honoring Major Cusic and to wish him all the best in the future for him and his family.

Monday, October 7, 2002

EXPRESSING APPRECIATION FOR PRIME MINISTER OF GREAT BRITAIN FOR HIS LOYAL SUPPORT AND LEADERSHIP IN WAR ON TERRORISM AND REAFFIRMING STRONG RELATIONSHIP BETWEEN PEOPLE OF UNITED STATES AND GREAT BRITAIN

Hon. Benjamin A. Gilman

OF NEW YORK

Mr. Speaker, I move to suspend the rules and agree to the resolution (H. Res. 549) expressing appreciation for the Prime Minister of Great Britain for his loyal support and leadership in the war on terrorism and reaffirming the strong relationship between the people of the United States and Great Britain.

The Clerk read as follows:

H. RES. 549

Whereas the people of the United States and Great Britain have a history of shared values and mutual respect for one another;

Whereas the Governments of the United States and Great Britain are close allies and share a deep and abiding friendship based on a shared commitment to democratic values;

Whereas the United States and Great Britain understand the commitment to defend freedom and democracy regardless of the costs involved;

Whereas British Prime Minister Tony Blair has displayed exceptional leadership in the war on terrorism; and

Whereas the United States and Great Britain have been provoked into a war on terrorism that threatens the security of both nations: Now, therefore, be it

Resolved, That the House of Representatives—

(1) expresses sincere appreciation for Prime Minister Tony Blair for his leadership in the war on terrorism;

(2) expresses its deepest sympathy to British victims of terrorism and their families, including the 67 British citizens who were victims of the terrorist attack on September 11, 2001;

(3) commends the efforts of British intelligence and defense agencies for their continued efforts in the war on terrorism; and

(4) reaffirms the strong and special relationship between the people of the United States and Great Britain.

Hon. Eliot L. Engel

OF NEW YORK

Mr. Speaker, I rise in strong support of this resolution. I would like to commend my colleague, the gentleman from Missouri (Mr. Graves), for introducing such a timely resolution.

Mr. Speaker, a Nation discovers its true friends in times of crisis. Since the tragedy of September 11, America has found that it has many friends around the globe. Mr. Speaker, we have seen that the United States has a tremendous friend and ally in the war on terrorism in Great Britain. No head of state has been more supportive of the United States in this battle than British Prime Minister Tony Blair.

Since September 11, British troops fought alongside U.S. forces to liberate Afghanistan and to root out terrorists. Britain acted as the lead Nation for the international security assistance force in Afghanistan until the mission was turned over to Turkey. Humanitarian aid has flowed from Britain to Afghanistan, and the British Government has enacted new counterterrorism legislation.

In short, Mr. Speaker, the United Kingdom has stood shoulder to shoulder with the United States in the war on terrorism. In the horrendous terrorist attacks of September 11, Britain lost 67 of its citizens. The United States has expressed its sympathies to the families of these British victims.

Mr. Speaker, British Prime Minister Tony Blair has shown extraordinary leadership in the war on terrorism. This resolution recognizes his leadership and expresses the appreciation of the Congress and the American people. I urge my colleagues to support this resolution.

Tuesday, October 8, 2002

HONORING MARILYN A. NGUYEN

Hon. Jerry Weller

OF ILLINOIS

Mr. Speaker, I rise today to honor Marilyn A. Nguyen of Bourbonnais, IL. Marilyn was one of over 85,000 secondary school students who participated in a contest through the Veterans of Foreign Wars of the United States and its Ladies Auxiliary (VFW). Each year the VFW conducts a Voice of Democracy audioessay competition designed to give high school students the opportunity to voice their opinion on their responsibility to our county. The contest theme was "Reaching Out to America's Future." Marilyn A. Nguyen was chosen as the 2002 Voice of Democracy broadcast scriptwriting winner this year. Following is Marilyn's winning script.

REACHING OUT TO AMERICA'S FUTURE

(By Marilyn A. Nguyen)

The harmony of an industrious city is disrupted by a deafeningly explosive crash. There is confusion. There are wailing sirens. In another city, the same confusion spreads like wildfire. Lives are forever changed as events unfold and buildings collapse. The horror is almost too much to bear. On September 11, 2001, the gruesome hand of terrorism attempted to reach out and grasp America's future. Fortunately, its grip was too slippery to conquer the heart and soul that is the United States of America.

Over two centuries ago the Founding Fathers of this country left England envisioning better lives for themselves and their posterity. They reached out mentally and physically to find America's future full of promise and patriotism. As this country continues to blossom and mature we must accept the task of reaching out to America's future no matter the cost or hardship.

America is a union for all nationalities. It reaches out to immigrants of all lands. My parents were among these immigrants. As their daughter I especially feel a unique bond to America. I feel that it is my duty to reach out to America's future with my own actions.

But, what does it mean to reach out to America's future? Already, it may seem to some that our future is uncertain because of the terrorist attacks. But, these tragedies only remind us that the time to reach out to America's future is now. We need to rise to the challenge as we have never done before to stand firm as a nation and as human beings to reach forward into the future.

The task at hand is not an easy one. Reaching out to America's future must begin with the individual who believes that America's future is not an abstract idea: it is comprised of neighbors, friends, mothers and fathers, brothers and sisters and especially individuals. America's future depends on what happens today in the lives of ordinary Americans living ordinary lives. It calls for the erasure of color, race and religion. It begins when one person extends respect and acceptance to another person regardless of their background.

Reaching out to America's future as a teenager is not much different from extending a hand as an adult. As a teen, perhaps it may be a difficult step but one which lays the foundations for adulthood. At a time when personal opinions are being formed, it can be easy to declare "it's not my job" to reach out but that is where we are wrong. I am the future of America. It starts with me. I am the voice of influence over my friends and the younger members of my community. Using that influence to promote understanding and cooperation among my peers, family, and community are what I, as a teen individual, can do to reach out to America's future.

It is important to begin with our everyday routines because this is where the impact will be most felt. I must encourage others to talk with friends and family about what it means to be a contribution to America's future. Teach younger children in middle school, neighbors, or even peers in high school that it is wrong to hate and discriminate. I have a responsibility to open my mind to the differences that make us unique and vital components of the future instead of searching for ways to divide. The example I put forth into the world should be one of love and acceptance.

The teenager's job in reaching out to America's future lies in the education of himself and his surroundings. His call to help build America's future is still strong. This Nation has no future without the work of those who believe in its potential for goodness.

Reaching out to America's future can seem like a faraway goal. But, in reality, the future is at our fingertips. We as people of this majestic empire must adopt the task set before us over 200 years ago. Reaching out to the future begins with the person who hears these words. It is he who must first take action. The perfect example of reaching out to America's future is the Statue of Liberty in New York Harbor. She is the example for one and all. Her extended arm holding the torch as a guiding light beckons us to follow her into the future. With her unfailing devotion to the preservation of this land, she reminds us that the future's brightness depends solely on those willing to bear the torch.

Mr. Speaker, I urge this body to identify and recognize others in their own districts whose actions have so greatly benefited and strengthened America's communities.

REMEMBERING SEPTEMBER 11

Hon. Steve Israel

OF NEW YORK

Mr. Speaker, on September 11 I joined with the students, teachers, administrators and parents of the Idle Hour Elementary School in Oakdale for a profound commemoration of those lost in the attacks on America. I know that our colleagues will be as moved as I was to hear the essays of three sixth graders: Emily Pertz, Justin Rigas and April LaValle. I am honored to share them with the entire Congress today:

September 11

(By Emily Pertz)

September 11 was a painful and tragic event. It changed the lives of millions forever.

I don't know anybody who died in my family because of the attack, but knowing that a lot of kids became parentless that day is enough to make my family and I upset. Whenever we go over the bridge I see many buildings and then a big gap where the towers once stood. To me it is very upsetting to see. My family is more cautious than ever.

I think the attacks have changed both our country and our world. The United States became more united. The world together is fighting terrorism. But on the other hand, many people lost loved ones, and the world's tallest towers were destroyed. Many people are still mourning and are still heartbroken.

Our school has done many great things to remember the victims. We raised a lot of money to plant a memorial garden to honor the lives lost from our neighborhood. We made red, white and blue chains that connected every classroom to show we are united. Each student colored in two flags, one to take home and one to hang up in school. The day after the attack our school had a moment of silence. It really made me think and made me a little depressed.

The United States went through a lot, but no matter what we will always be united.

America Changes

(By Justin Rigas)

The terrorist attack made by Osama bin Laden and the Taliban on the Twin Towers, landmarks of our New York City skyline, was a great tragedy. Thousands of innocent people died terribly as the buildings melted and crumbled to the ground. Children are left without their mothers and fathers, families without sisters, brothers, dear friends. Families are left without jobs, without their income, possibly unable to pay their bills and keep their house.

But America has stood together strong. In this moment of sadness and tragedy millions have come together with help and support. People all over our country, not just New York, have sent donations of food, money and clothing to help those families that have lost those dear to them.

The events of September 11, 2001, have changed the attitudes of my family as well as millions of Americans. We all miss those we know and loved that are gone. The Americans' innocence may never again be the same, not able to totally trust the safety we've somehow always felt. Many people hesitate to travel on airplanes which means less people are visiting places where the people there count on them to spend their money. It could hurt business in hotels, restaurants and stores.

We always need to be on guard that something terrible could happen again. Our government cannot sleep, it must always be searching for the next thing to happen.

During the months following September 11, my school painted pictures of the Twin Towers and memories of that day. We made a tree of buttons representing the people that died that day on the wall in our hallway. Collections of food and money were presented to the Red Cross and a garden in the form of our flag was planted at school.

At Dowling College, a memorial garden was planted to be kept forever funded by a dinner our school held.

People everywhere still fly their American flags at their homes and, in their cars.

In the meantime we will rebuild our city and the towers that will again stand, this time as a huge memoriam of 9/11 and those lost. The day that changed America.

September 11, 2001

(By April Lavalle)

9/11 was a day of mixed emotions, sadness, anger and determination. Even though many innocent people were killed, never will the people of America stop the deeds, kind donations and prayers for all who have passed away. Some people were lucky not to know anyone who was in the Twin Towers. But I knew my personal life would never be the same. I took so many things for granted.

I now think about the desperate families of the innocent people who have died. Even though people try to do all they can to make families who lost loved ones feel better, nothing can serve as compensation for those who left us on September 11. America now has to prove to the world that we are a strong Nation and will fight for what we need. The world is no longer a peaceful place for us and no longer united. A gray sky will stay in our minds until we find peace and our sun will again shine through.

Our community hung flags, made donations and I bet you that everyone prayed. We are a proud and patriotic nation. Don't think 9/11 made us a weaker country; it made us a stronger America.

Wednesday, October 9, 2002

EXPRESSING SUPPORT OF OFFICIAL RECOGNITION FOR THE HEROES OF UNITED AIRLINES FLIGHT 93

Hon. Mac Collins

OF GEORGIA

Mr. Speaker, I call to your attention a letter I received from David and Gretchen Nagy and Donald Evans, Jr., of Burke, VA. The letter, addressed to President George W. Bush, urges our government to officially recognize the heroic men and women of United Airlines flight 93 for their actions on the morning of September 11, 2001. These ordinary people aboard flight 93 were thrown into an extraordinary and tragic situation. When their plane was highjacked by Al Qaeda terrorists, these brave souls made a choice to fight back against terror. The citizens on flight 93 became soldiers, and in so doing denied the terrorists of their chosen target, perhaps saving our cherished Capitol from the same fate as the World Trade Center. Mr. Speaker, in support of this letter, I submit it for the *Record*. It reads as follows:

PRESIDENT GEORGE W. BUSH,
The White House,
Washington, DC.

DEAR MR. PRESIDENT: We write as ordinary citizens to ask that you lead our Nation in bestowing some measure of official honor upon a tiny band of extraordinary citizens—the ones who stood up and charged the hijackers of UAL flight 93 over Pennsylvania on 9/11.

There seems little doubt that these heroes spared America another devastating blow with their magnificent stand, possibly even a blow to the Capitol or the White House itself. Thanks to you and others, everyone now knows their rallying cry, "Let's roll!" Surely, everyone with a heart shivered when they heard it, and the story behind it.

And now, Mr. President, how many even remember their names?

According to press reports, they were Todd Beamer, Jeremy Glick, Mark Bingham and Lou Nacke—ordinary yet rare men with the guts to act when most would be paralyzed by fear. Perhaps investigators have identified others who joined their uprising. If so, they remain anonymous and unacknowledged. All the sadder.

In a sense, sir, weren't these men the first combat casualties in our new war against terrorism? The first to go hand-to-hand—and unarmed—against our attackers? They knew they were doomed. ("I'm not going to get out of this," Beamer told a cell phone operator.) They could have curled up and gone passively. But they also knew they could thwart evil and spare many on the ground if they went down fighting.

We respectfully suggest, Mr. President, that valor of this sort is in the grandest traditions of American heroism—something very special, on the order of that which gains our military heroes the Medal of Honor. Yet if anyone has proposed that this Nation extend these men some tangible form of gratitude, something solid their loved ones could touch and treasure, we haven't heard of it. So we are asking you, sir, to consider bestowing such an honor at a fitting, proper ceremony. Perhaps the Presidential Medal of Freedom would be appropriate, perhaps some other award for ultimate service and valor.

We still hope we are merely adding our letter to a growing stack.

God bless you, Mr. President.

DAVID AND GRETCHEN NAGY,
DONALD C. EVANS, JR.

Thursday, October 10, 2002

IN HONOR OF TED MALIARIS

Hon. Jim Davis

OF FLORIDA

Mr. Speaker, I rise in honor of Ted Maliaris, a devoted American who is following his heart and sharing his love for our Nation through his passionate music. Through his "A Tribute to America Tour," Ted is lifting the spirits of Americans across the Nation while teaching children the importance of American values.

Ted was born in South Florida, and thanks to the encouragement of his grandparents, who were both musicians, Ted soon discovered his true love for music. During his years working on the family farm, Ted honed his musical talents and soon decided to follow his dream of sharing his music with others. He recently recorded his first album with the London Symphony Orchestra, where he honored the immigrant farm laborers who worked alongside him during his career on the farm.

After the tragic events of September 11, Ted's mother, Ann S. Miller composed "A Tribute to America—A 21st Century Anthem" to honor the men and women in the Armed Forces. The song inspired Ted to organize the "Tribute to America Tour," which features his performance of his mother's song and performances by various children's groups around the country. Hoping to show children the patriotism and pride that lies in our country, America's Life Line Association is planning a special recording of 50,000 children singing the anthem together.

On behalf of the people of Tampa Bay, I would like to extend my gratitude to Ted for his dedication to our country and this important cause.

IN RECOGNITION OF REV. C.C. CAMPBELL GILLON

Hon. Ken Bentsen

OF TEXAS

Mr. Speaker, I rise to honor Rev. C.C. Campbell Gillon, on the occasion of his retirement from his pastoral duties at the Presbyterian Congregation in Georgetown located in Washington, DC, where he has faithfully served his congregation for more than 23 years. His retirement comes at the end of 50 years in the ministry.

The Presbyterian Congregation in Georgetown has over 200 years of distinguished history, beginning in 1780, under the eminent Stephen Bloomer Balch, pupil of religious leader John Witherspoon and soldier of the Revolution. The church serves as the first Presbyterian Church in what was to become modern-day Washington, DC, and the oldest church of any denomination with an unbroken ministry. A rare charter, still in effect, was granted in 1806 to "the Presbyterian Congregation in George Town" by an act of Congress signed by President Thomas Jefferson. The Presbyterian Congregation in Georgetown, pioneered in both the religious and cultural life of the community, has served as a cornerstone of faith in our Nation's Capital, attracting a wide variety of worshipers from political leaders to those seeking spiritual direction.

Rev. Campbell Gillon was born in Edinburgh, Scotland, into a family immersed in the Christian spirit. Both his father and uncles preceded him as ministers of the Church of Scotland. After 3 years of Army service at the end of World War II, he graduated with a master of arts degree from the University of Glasgow before studying theology at Trinity College, Glasgow, under the tutelage of Professor William Barclay, the noted Scottish New Testament scholar. In 1952, Reverend Gillon began an exceptional career that has spanned 50 years, with his first appointment to the historic Buittle Parish in southwest Scotland. The rest of his 27-year ministry in the Church of Scotland was spent in Glasgow, where he presided over the Milton Saint Stephen's Church. Under his extraordinary leadership, Reverend Gillon's beloved church was united with the noted Renfield Church Center, and was expanded to include a public restaurant, concert hall, and other community oriented facilities. In 1978, Reverend Gillon and his wife Audrey visited the Presbyterian Congregation in Georgetown on a 6-week work exchange, not knowing how their lives would forever be changed. Soon after his short stay, he returned to his 800-year-old parish, the prestigious Cathcart Old Parish, only to receive a call from the Presbyterian Congregation in Georgetown with an offer to join their church family as the senior minister.

Reverend Gillon has earned a reputation of being one of the most thoughtful and provocative interpreters of Christian experience, and has shared his insight and experiences with those who seek knowledge and guidance. He has published *Words of Trust*, a book of sermons produced in both the United States and the United Kingdom. As a testament to his leadership and wisdom, excerpts from Reverend Gillon's sermons have been featured in newspapers, magazines, and Christian publications around the world.

While Reverend Gillon's religious and spiritual obligations to his growing congregation have always been paramount, as a community leader he has shared his faith and free time as chaplain of the Saint Andrew's Society of Washington, DC, a charitable and social organization of men of Scottish birth or ancestry.

Mr. Speaker, at a time when our Nation and many across the world were seeking explanations and direction following the horrific attacks of September 11, terrorism, and war, Reverend Gillon provided comfort after the storm with his prayer before the House of Representatives and a moving sermon before his congregation. He reminded us that suffering is only temporary, and God's love is forever. Deeply rooted in the traditions of Scotland and the Scottish preachers that preceded him, Reverend Gillon has dedicated himself to the principles of the Presbyterian faith, his congregation and his family.

In his own words, Reverend Gillon captured the sentiments of the entire congregation, "never does the heart wish a good relationship to end." I want to thank Campbell for his leadership, spiritual guidance and devotion to the Presbyterian Congregation in Georgetown, the Washington, DC, community, and the many lives he and his wife Audrey have touched throughout his career. He leaves a legacy of good work and grace that will be missed.

TRIBUTE TO CHARLES ROONEY MILLER

Hon. James E. Clyburn

OF SOUTH CAROLINA

Mr. Speaker, I rise today to pay tribute to a fellow South Carolinian and college classmate, Charles Rooney Miller, a good friend, a master teacher, and a survivor of the September 11, 2001, terrorist attacks.

Mr. Miller was born on August 11, 1939, in Clover, SC, in York County. He was the first of two children born to Andrew Charles Miller and Emily Lee (Allison) Miller. After his parents separated, he lived with his grandmother, aunts, uncles and other relatives—he was truly a community-raised child in a Christian environment. From a young age, he attended Clover Chapel Methodist Church Sunday Bible School and later joined Flat Rock Baptist Church where he was baptized. At the age of 11, he became the Sunday School teacher for his age group.

Mr. Miller was an excellent student throughout grammar school and high school. He graduated from Roosevelt High School in May 1956, as valedictorian of his class. From there he went to South Carolina State College (now University), where I had the pleasure of meeting him and beginning a lifelong friendship. We were both active in the civil rights activities on campus and participated in a number of marches and other activities. Rooney graduated from South Carolina State in 1962 and moved to Stamford, CT, where he was later joined by his wife and children.

In Connecticut, Mr. Miller worked two jobs to support his family; at Chemtross, a film developing business, and at Stamford Chemicals, a dry cleaning production business, where his work is associated with the invention of a number of products that are still used in today's dry cleaning industry. Mr. Miller later became a teacher and worked for a short time in South Carolina, Stamford, and in the Norwalk, CT, public school systems.

In 1968, Mr. Miller began a career with the New York City Department of Social Services and worked there until 1994 when he retired as a supervisor with the Bureau of Social Services for Children. His retirement was not long and he returned to work in 1997 as a consultant with PSI International in Fairfax, VA, and was assigned as a conversion specialist for his old office, the New York City Department of Social Services. He worked there until September 11, 2001.

On the morning of September 11, Mr. Miller arrived at work early and spoke to several colleagues on his floor. He thought about how much he enjoyed his post-retirement work as a consultant and his ability to set his own schedule. In the midst of his musings, he heard a loud noise but first thought the sound came from normal truck traffic outside. But this window-rattling occurrence was different. He was astonished when he went to the window and saw the World Trade Center tower on fire and a trail of fluid pouring down the side of the building with fire leaping behind it. He heard other loud explosions and coworkers on his floor began to scream, cry and pray. The radios began broadcasting reports of the fire but no one was sure what was happening. As Mr. Miller and his coworkers continued to watch the building burn he saw people jumping from the windows, some holding hands. They watched as the second plane crashed into the other tower. They knew then they were in the midst of a planned attack, and pandemonium broke out. Finally, they received instructions to leave their building and head down to the South Street Seaport where they thought it would be safer by the water. They were given surgical masks to cover their noses and mouths and instructed to put a moist towel under the mask to help prevent inhalation of smoke, chemicals and other foreign particles. They left the darkened building with smoke and objects flying through the air.

As people were screaming and running out of the building, Miller was knocked to the ground and run over by several people before he could get back to his feet. He thought he would be okay once he caught his breath. He was eventually assisted by a worker from a nearby polling place and taken to a triage location. The medics realized that Miller was suffering from a heart attack and he was then rushed by ambulance to New York Hospital's downtown emergency room. He was hospitalized for 5 days and unable to contact his family. After subsequent angioplasty surgery and treatments for the back injury he received, he is now mending well.

He's still active in his church, Cathedral Baptist, where he serves as chairman of the deacon's board, chairman of the men's department, vice president of the board of directors, a teacher in the Bible Institute and the Adult Sunday School class. He is also a member of South Carolina State University Alumni Chapter of New York.

Mr. Speaker, I ask you and my colleagues to join me in honoring Charles Rooney Miller, a man whose contributions to his community, his friends, and his family will leave lasting impressions on the numerous lives he has touched. As the homecoming celebrations begin at our alma mater, South Carolina State University, I wish him continued success and Godspeed.

SPECIAL JOINT SESSION OF THE U.S. CONGRESS

Hon. Sue Wilkins Myrick

OF NEW YORK

Mr. Speaker, on Friday, September 6 of this year, a special joint session of the U.S. Congress gathered in New York City to remember the terrorist attacks of September 11, 2001. This special session reminded us of the lives that were lost and the heroes that were found all across this country on that terrible day. I am honored to have taken part in this unique session.

We convened at Federal Hall, where the First Congress met over two centuries ago, and a few blocks from where the World Trade Center towers once stood proud and tall.

Mr. Speaker, we met to remember the thousands of lives that ended so abruptly that day. We prayed for the families of those that were lost. We prayed for the families who had to say goodbye before they were ready. The wound that America suffered on that day will always be remembered.

We also expressed our most sincere thanks to the firefighters, police officers, emergency personnel, and all others who risked and gave their lives on that day. These brave men and women, along with their peers across the country, risk their lives every day to protect those around them. Expressing our thanks to them is a long overdue action.

Finally, Mr. Speaker, we recommit ourselves to eradicating terrorism from the world and to making sure that those responsible for this horrible attack on America are brought to justice. American soldiers are now stationed across the globe, helping to create a world where those who live in freedom can also live free of the fear of terrorism. America and the world owe these soldiers a debt of gratitude.

I am proud to have joined Congress on September 6, to remember the lives that were lost and to show those who would harm America that we will not forget, but we will overcome.

DISPOSING OF VARIOUS LEGISLATIVE MEASURES
H. RES. 571, HONORING THE LIFE OF DAVID O. "DOC" COOKE, THE "MAYOR OF THE PENTAGON"
H. RES. 571

Whereas for 44 years, David O. "Doc" Cooke's tireless dedication, skill, and involvement in Department of Defense management issues earned him the respect of his colleagues and distinction as a Pentagon institution;

Whereas as the quintessential civil servant, Doc Cooke rose to become the highest ranking career civil servant within the Department of Defense;

Whereas in his jobs as the Director of Administration and Management for the Office of the Secretary of Defense, and Director of Washington Headquarters Services, Doc Cooke was responsible for maintenance, operation, and security of buildings of the Department of Defense in the Washington, D.C. area, including the Pentagon Reservation;

Whereas because of his propensity to make things happen, Doc Cooke was respectfully known as the "Mayor of the Pentagon";

Whereas Doc Cooke was born in 1920 in Buffalo, New York, and went on to earn a bachelor's degree in education from the State Teachers College at Buffalo in 1941, a master's degree in political science from the New York State College for Teachers in 1942, and a law degree in 1950 from George Washington University, where he was a member of the Law Review;

Whereas Doc Cooke served in the Navy during World War II as an officer on the USS Pennsylvania; returned to active duty during the Korean war, during which time he served as an instructor in the School of Naval Justice; and retired in 1968 as a Navy captain;

Whereas Doc Cooke served on Defense Secretary Neil McElroy's task force on Department of Defense reorganization in 1958; worked for Defense Secretary Robert McNamara, as Director of the Office of Organizational and Management Planning, implementing changes in Department of Defense organization; and worked for every other Secretary of Defense since then;

Whereas during the late 1980s and early 1990s, Doc Cooke was a strong advocate for renovation of the Pentagon;

Whereas many of the construction specifications supported by Doc Cooke helped to save lives during the terrorist attack on the Pentagon on September 11, 2001;

Whereas Doc Cooke could be seen assisting in the response to the terrorist attack on the Pentagon on September 11, 2001;

Whereas throughout the Department of Defense, Doc Cooke was noted for his strong support of equal employment opportunity for minorities, women, and individuals with disabilities;

Whereas Doc Cooke was instrumental in establishing a Public Service Academy at Anacostia High School in the District of Columbia, which has helped to increase the graduation rate of students;

Whereas Doc Cooke served as a member of the seven-member Governance Committee of United Way of the National Capital Area's September 11 Fund, deciding how to distribute disaster relief funds collected after September 11;

Whereas Doc Cooke has been recognized for his extraordinary performance through numerous awards, including the Department of Defense Medal for Distinguished Civilian Service (the Department's highest department career award) seven times; the Department of Defense Medal for Outstanding Public Service; the Department of Defense Medal for Distinguished Public Service twice; the Roger W. Jones Award for Executive Leadership from American University (1983); the NAACP Benjamin L. Hooks Distinguished Service Award (1994); the Presidential Meritorious Rank Award (1994); the Government Executive Leadership Award (1995); a Presidential Distinguished Rank Award (1995); a National Public Service Award (1997); the President's Award for Distinguished Federal Civilian Service (1998), the highest Government service award; the John O. Marsh Public Service Award (2000); the Senior Executives Association Board of Directors Award (2001); the Nelson A. Rockefeller College of Public Affairs and Policy Distinguished Alumnus Award (2001); an award from the University at Albany Alumni Association for "Recognition for Outstanding Service" (2001); and the American Society of Public Administration Elmer B. Staats Lifetime Achievement Award for Distinguished Service (2002); and

Whereas on June 22, 2002, Doc Cooke died as the result of injuries sustained in an automobile accident, after a long and distinguished career in government, in which he became the model for civil servants: Now, therefore, be it:

Resolved, That the House of Representatives—

(1) recognizes David O. "Doc" Cooke's legendary professionalism as a model civil servant;

(2) honors Doc Cooke's life; and

(3) extends its condolences to the Cooke family and the Department of Defense community on the death of an extraordinary human being.

DISCHARGED FROM THE COMMITTEE ON HOUSE ADMINISTRATION, AMENDED, AND AGREED TO

H. Con. Res. 487, authorizing the printing as a House document of a volume consisting of the transcripts of the ceremonial meeting of the House of Representatives and Senate in New York City on September 6, 2002, and a collection of statements by Members of the House of Representatives and Senate from the *Congressional Record* on the terrorist attacks of September 11, 2001.

H. CON. RES. 487

Resolved by the House of Representatives (the Senate concurring),

SECTION 1. AUTHORIZING PRINTING OF VOLUME OF TRANSCRIPTS OF NEW YORK CITY MEETING AND STATEMENTS OF TERRORIST ATTACKS OF SEPTEMBER 11.

(a) IN GENERAL.—A volume consisting of the transcripts of the ceremonial meeting of the House of Representatives and Senate in New York City on September 6, 2002, and a collection of statements by Members of the House of Representatives and Senators on the terrorist attacks of September 11, 2001, shall be printed as a House document under the direction of the Joint Committee on Printing, with suitable binding.

(b) STATEMENTS TO BE INCLUDED IN VOLUME.—A statement by a Member of the House of Representatives or a Senator on the terrorist attacks of September 11, 2001, shall be included in the volume printed under subsection (a) if the statement—

(1) was printed in the *Congressional Record* prior to the most recent date on which the House of Representatives adjourned prior to the date of the regularly scheduled general election in November 2002; and

(2) is approved for inclusion in the volume by the Committee on House Administration of the House of Representatives (in the case of a statement by a Member of the House), or the Committee on Rules and Administration of the Senate (in the case of a statement by a Senator).

SEC. 2. NUMBER OF COPIES.

The number of copies of the document printed under section 1 shall be 15,000 casebound copies, of which—

(1) 15 shall be provided to each Member of the House of Representatives;

(2) 25 shall be provided to each Senator; and

(3) the balance shall be distributed by the Joint Committee on Printing to Members of the House of Representatives and Senators, based on requests submitted to the Joint Committee by Members and Senators.

SEC. 3. MEMBER DEFINED.

In this concurrent resolution, the term "Member of the House of Representatives" includes a Delegate or Resident Commissioner to the Congress.

AMENDED BY COMMITTEE AMENDMENT AND PASSED

Wednesday, October 16, 2002

TRIBUTE TO CPT. KATHY MAZZA OF PORT AUTHORITY POLICE DEPARTMENT

Hon. Peter T. King

OF NEW YORK

Mr. Speaker, I am proud to rise today to pay tribute to Cpt. Kathy Mazza of the Port Authority Police Department who died heroically at the World Trade Center on September 11, 2001.

This past Monday, October 14, 2002, I was privileged to take part in a very moving ceremony which designated North Suffolk Avenue in North Massapequa as Cpt. Kathy Mazza Way. The ceremony, which was held directly across from the home where Captain Mazza grew up, was attended by her husband, Chris Delosh, who is a member of the New York City Police Department; her parents, Rose and John Mazza; her three brothers; and many of her countless friends. The ceremony was conducted by Hon. John Venditto, the supervisor of the town of Oyster Bay.

There were many heroes on September 11 but no one was more heroic than Kathy Mazza. On the morning of September 11, Captain Mazza was in New Jersey, serving as the commanding officer of the Port Authority Police Academy. Immediately upon learning of the attack on the Twin Towers, however, Captain Mazza raced to the World Trade Center in lower Manhattan and entered the North Tower where she proceeded to take a leadership role in the rescue effort—at one point reaching the 22d floor.

What set Captain Mazza apart from all others is that she was personally responsible for evacuating hundreds of people. She did this by having the presence of mind to use her service revolver to shoot out floor-to-ceiling glass walls on the mezzanine level of tower one enabling so many trapped people to escape. Shortly after, at 10:29 a.m., Captain Mazza was killed when tower one collapsed.

This extraordinary heroism and dedication to duty characterized Kathy Mazza's entire life. Prior to becoming a police officer she had been a cardiothoracic operating nurse at St. Francis Hospital in Roslyn, NY. As a police officer she was instrumental in launching the port authority's portable heart defibrillator program at the metropolitan airports. And as commanding officer of the Police Academy she achieved a record of unsurpassed excellence and achievement.

September 11, 2001, was a day of brutality, horror and terror. But it was also a day when brave Americans such as Cpt. Kathy Mazza demonstrated a bravery and courage which will be remembered throughout the history of our Nation. For that and for so much more, we will always be in her debt.

May she rest in peace.

REMEMBERING SEPTEMBER 11

Hon. Robert Menendez

OF NEW JERSEY

Mr. Speaker, September 11, 2001, hit our New Jersey community hard. We lost neighbors and friends, mothers and fathers and children, sisters and brothers—people who left their homes that fateful day, and crossed the Hudson River, never to return.

The Twin Towers that were once visible from our waterfront stand no more. The skyline has forever changed.

But the people who were lost that day, while leaving an unspeakable void in our lives, still live on in our hearts and our minds. They are our heroes: everyday heroes who were providing for their families, contributing to their communities; everyday heroes who lost their lives in their dedication to protect others. Everyday heroes. The most incredible kind of heroes. American heroes. They may not be here, but they do live on, and they will never be forgotten.

We honor them by showing our patriotism; by flying our flag; by fighting terrorism wherever we find its scourge growing; by coming together as one great people and one great Nation; and even by finding the faith and the strength to carry on with our lives, raising our children, building our communities, and moving forward with this wonderful creation of democracy and freedom called America.

Yes, September 11 hit our New Jersey community and indeed our Nation hard. But we remember. We persevere. We move forward. And we are stronger and more united than ever before. God Bless America.

HONORING RON JAMES, MARINE VETERAN AND OUR INTREPID DEFENDER OF THE AMERICAN FLAG

Hon. Benjamin A. Gilman

OF NEW YORK

Mr. Speaker, as the 107th Congress draws to a close, I would like to take this opportunity to recognize a great friend of the American people, a Marine veteran, and our Nation's intrepid defender of the American flag, Mr. Ron James.

Mr. James, who we also know as Mr. Ronald M. Sorenson, a Marine veteran of the Korean war era, and a great friend is a true American patriot. Ron has dedicated his life to preserving the core values of what our great Nation stands for and for more than two decades has educated our Nation on flag etiquette, while paving the way and leading our Nation in seeking a constitutional amendment prohibiting the desecration of our flag, the symbol of our great Nation.

Ron is a familiar face in the Halls of Congress where he regularly visits our offices to seek our support for his noble endeavors. In addition to fighting for our flag, Ron also fights for the rights of our veterans and is active in numerous veterans organizations and assists patients in our VA hospitals. Over the past 20 years, Ron has walked thousands and thousands of miles carrying our flag, to garner support for not only a constitutional amendment protecting it from desecration, but also to raise awareness of its importance to our Nation's youth.

Following the horrific events of barbarity perpetrated against our Nation by forces of true evil on September 11, 2001, Ron met with me to discuss legislation that would benefit the families of our everyday American heroes. On March 14, 2002, I sponsored H.R. 3968, the Fallen Heroes Flag Act of 2002, which provides a flag flown over the U.S. Capitol to the immediate family of our Nation's brave firefighters, law enforcement officers, emergency medical technicians (EMT) and to other relief and rescue workers whose lives are lost in the line of duty. This important legislation ensures that our future generations of public servants who may pay the ultimate price for their service to our Nation and to our communities are accorded the respect and honor that they deserve.

Mr. Speaker, I ask my colleagues to join me in recognizing Ron's hard work and dedication that enabled us to turn an idea into a reality with our "Fallen Heroes Flag Act of 2002." This is yet another selfless act of patriotism by Ron James, a true friend and a great American who lives his life to serve our Nation, our veterans, and our flag.

RECOGNIZING THE WORK OF THE STUDENTS AT VETERANS MEMORIAL ELEMENTARY SCHOOL IN BRICK TOWNSHIP, NJ

Hon. Christopher H. Smith

OF NEW JERSEY

Mr. Speaker, I rise today to honor the hard work and community service exhibited during the past school year by the students of Veterans Memorial Elementary School in Brick Township, NJ. It is my honor and privilege of representing these students, and their parents and teachers in Congress.

During this past year, the students invested many hours of service in projects to help make life better for their school and greater community. For example, in the aftermath of the September 11 terrorist attacks, the students honored local firefighters who participated in rescue and recovery operations at Ground Zero. They also sent thank you notes to New York City police and firefighters, and wrote letters to a local serviceman stationed overseas.

It is important to note, Mr. Speaker, that the students' community service did not just begin, nor will it end, with their outstanding efforts related to September 11.

To highlight the importance of a clean and safe environment, the students commemorated Earth Day by decorating grocery bags with environment-friendly messages. These bags were then used by patrons of a local supermarket so they could take home the students'messages about how we must protect for our environment.

They also implemented a schoolwide paper recycling program, and worked to beautify the school's grounds.

To enhance their own understanding of the challenges that older Americans face, the students visit with senior citizens in their community, exchanging ideas, and striking up new friendships. They make special holiday gifts for the seniors and also put on concerts, including one full of patriotic songs. It's the students' way of thanking America's "greatest generation"; a generation that risked all to secure freedom at home and abroad. As chairman of the House Committee on Veterans' Affairs, I am especially grateful for the outreach our students have initiated with seniors and veterans—the namesakes of their school.

When a peer's house tragically burned to the ground, the students of Veterans Memorial Elementary School responded by holding an emergency fundraiser. They also collected warm winter coats for students in need and helped their school buy new books and playground equipment.

While this is only a small sampling of community service activities performed by the students of Veterans Elementary, it is clear that these children, while learning the subjects and skills they need to succeed in academia, are also learning the generosity, compassion, and service needed to be outstanding members of their community.

I am proud to congratulate the students of Veterans Elementary School. Their leader and my friend, Principal Joe Vicari also deserves our thanks for his many years of hard work and generosity and the dedicated teachers and support staff at Veterans Elementary School also deserve high praise and recognition. I wish them all the very best of success for another year of outstanding community service, and I look forward to working with them in their endeavors in the years to come.

Thursday, October 17, 2002

A TRIBUTE TO MR. IAN GRAY

Hon. Robert L. Ehrlich, Jr.

OF MARYLAND

Mr. Speaker, I would like to take this opportunity to pay tribute to a humanitarian, Mr. Ian Gray, who tragically passed away on September 11, 2001.

Ian Gray worked for Baltimore Medical System (BMS) as part of his personal mission to ensure quality health care access for the underserved in the Baltimore area. He helped to build a health care system which serves over 30,000 patients throughout Maryland. He touched many lives in unseen ways through his commitment to BMS.

Mr. Gray died during the tragic events of September 11, 2001, as he was a passenger on flight 77, which crashed into the Pentagon. His death serves as a reminder to all of us to continue the work he began. While his life was cut short, I know that his many coworkers, friends, and family members continue the noble mission of helping those in need by providing high quality health care.

In recognition of the 1-year anniversary of the attacks on our Nation last month, BMS launched a capital fund drive, named for Ian Gray, to raise money for the health care needs of Baltimore's residents.

Ian Gray's dream of improving health care is something we all share. His work was noble and improved the lives of countless Marylanders. I would like to take this opportunity to extend my best wishes to Ian's wife, Ana, and their children and family members. Ian's commitment to the health care of Marylanders lives on through the dedication of a fund to assist Baltimore Medical System to help those in need.

Over 1 year after the tragic attacks on our Nation, we remember and celebrate the life of Ian Gray and continue his important work.

Wednesday, November 13, 2002

PUBLIC BILLS AND RESOLUTIONS

Under clause 2 of rule XII, public bills and resolutions were introduced and severally referred, as follows:

By Mr. NADLER:

H.R. 5725. A bill to authorize a national memorial at, or proximate to, the World Trade Center site to commemorate the tragic events of September 11, 2001, to establish the World Trade Center Memorial Advisory Board, and for other purposes; to the Committee on Resources.

RECOGNIZING DAVE L. McDONALD

Hon. George Radanovich

OF CALIFORNIA

Mr. Speaker, I rise today in recognition of Dave McDonald upon his receipt of Central Valley Muscular Dystrophy Association's Humanitarian of the Year Award in Fresno, CA, on November 7, 2002. Mr. McDonald is being recognized for his leadership, humanitarian efforts, and dedication to the community.

Mr. McDonald is the president and CEO of PELCO, the world's largest producer of video security systems, as well as Central California's largest manufacturing employer. PELCO produces 3,500 different products which are sold through 5,000 authorized dealers in the United States and abroad. Since its beginning, the company has grown 30-fold under Mr. McDonald's supervision.

Following the disastrous events of September 11, 2001, Mr. McDonald and PELCO assisted the NYPD in its recovery efforts at the World Trade Center site by donating specialized camera equipment and personnel. Additionally, the company's warehouse in Orangeburg, NY, was converted into a distribution center for relief supplies. In November 2001, PELCO created the California Memorial in Clovis as a permanent tribute to the victims and fallen heroes from September 11. The memorial contains hundreds of items donated by the City of New York in remembrance of the events of September 11.

In March 2002, Mr. McDonald was honored by the New York City Fire Department as the Grand Marshall for the St. Patrick's Day Parade. In May 2002, Mr. McDonald also received the Excellence in Business Award from the Fresno Hall of Fame in recognition of his dedication to local businesses and his community.

Mr. Speaker, I want to congratulate Dave McDonald for his contributions to the many people in need, not only locally but nationally. I urge my colleagues to join me in wishing Dave McDonald many more years of continued success.

TRIBUTE TO COMANCHE ELEMENTARY SCHOOL OF SHAWNEE MISSION, KS

Hon. Dennis Moore

OF KANSAS

Mr. Speaker, I want to take this opportunity to recognize and commend the students of Marilyn Tieszen, a kindergarten teacher at Comanche Elementary School of Shawnee Mission, KS, which is located in the Third Congressional District of Kansas.

Following the tragic terrorist attacks of September 11, 2001, Ms. Tieszen had her students create an American flag, using a white sheet, and dipping their hands in red and blue paint to make the stripes and the blue background for the stars. The flag was presented to U.S. Army Capt. John Townsend, who has two children who are students at the school. Captain Townsend is the executive officer of the School of Advanced Military Studies at Fort Leavenworth's U.S. Army Command and General Staff College.

After receiving the flag, Captain Townsend hung it in Eisenhower Hall at Fort Leavenworth, KS, for a few weeks, where, in his words, "it got rave reviews and many people commented that it was unfortunate that all service members would not be able to appreciate it." As a result of the very positive reception it received at the fort, the flag was then mailed to an Army infantry unit conducting a peacekeeping mission in Kosovo. It arrived a few days before Christmas and hung at several of their sites through the new year. From there, it was taken to an airborne unit that "jumped" it into Tunisia, North Africa. Next it was displayed on the aircraft carrier USS *John C. Stennis*, which was supporting Operation Enduring Freedom in the Persian Gulf.

Following the display on the USS *Stennis*, the Comanche Elementary School flag was flown to an Army explosive ordnance unit in Afghanistan which had just lost two soldiers, one being from Kansas. After that, the flag went to the Pentagon where it hung for 3 weeks in the reconstructed part of the building that had been attacked on September 11. As Captain Townsend told me, "few people could pass it without stopping and appreciating the spirit and patriotism it displayed." The flag then was displayed at the U.S. Supreme Court and the U.S. Capitol, before being returned home to Comanche Elementary School, where it is now being displayed.

Mr. Speaker, a book that logged its travels accompanied the flag everywhere it went. At each stop the unit took pictures of its visit and then sent them on with the flag so its travels are well documented. In most cases the unit sent a letter or e-mail praising the students' patriotism. At Comanche Elementary School the students and faculty have created a large display in the foyer just inside the front door so that everyone who comes into the building can see it: a large map with the visited areas highlighted, including pictures from those locations.

As Captain Townsend told me, "I receive e-mails almost daily from soldiers and civilians around the world that have seen this flag and were thankful that they got to see it. In most cases it brought tears to their eyes. In some cases it brought a ray of hope to people that were down from losing friends and comrades and for others it was a sign of support from a community half way around the world."

I am very proud of the patriotism and creativity of Marilyn Tieszen's kindergarten class at Comanche Elementary School, who worked together to create a wonderful symbol of America during our ongoing time of great challenge. Mr. Speaker, I hope that you and all Members of this House will join with me in commending their spirit and thanking them for the inspirational symbol they created for display around the world.

HOMELAND SECURITY ACT OF 2002

Hon. Rob Portman

OF OHIO

Mr. Speaker, I thank the chairman for yielding me this time. His passion and his persistence are the reason that we are here tonight to do this important work, and I appreciate the role he played in moving this legislation through the system as chair of the Select Committee on Homeland Security.

Mr. Speaker, there is an old saying that goes "Times change and we change with them too." Times have changed and it is imperative to the security of our country, security of our families that our government change as well.

On September 11, 2001, the terrorists who struck our homeland killed more civilians than all our foreign enemies combined. We all woke up to the fact that the threats we face now are very different from the ones we faced in the past. During the cold war, we adapted our government structure to better utilize the resources we had to fight then a superpower. Today we face a more unpredictable and a more agile enemy and a very deadly enemy, and today we must reorganize our government again so we can stop that enemy before it strikes again, and we are not ready. There are over 100 departments and agencies with some involvement in homeland security, and when every one is in charge, no one is in charge. There is no accountability in the current system.

Last summer President Bush presented to the Congress a very ambitious and visionary plan to merge and consolidate responsibilities in a new Department of Homeland Security, similar to what Senator Lieberman had proposed and what various commissions had proposed. He laid out three strategic objectives: First, prevention of attacks; second, minimizing our vulnerabilities; and third, minimizing the damage and maximizing recovery should an attack occur. These three pillars provided us with a clear framework to align our resources, people and capital, and to align responsibility and accountability. This single unified structure will make us more efficient and effective in the fight against terrorism. It will not make us immune, but it will make us safer.

I strongly believe in what we are doing tonight, not because we are creating a new department but because we are doing it the right way. We are giving this President and future Presidents the flexibility they will need to make it work. That is budget flexibility; it is organizational flexibility; and, yes, it is personnel flexibility to be sure the right people are in the right place at the right time to protect us. The 21st century threats that we now meet head on cannot be handled by early 20th century civil service rules and bureaucracy. So, yes, the President and the new Secretary of Homeland Security will have the flexibility to design a new human resources management system, but it is one that will preserve fundamental civil service and worker protections while at the same time building a team atmosphere that is absolutely crucial by rewarding and promoting excellence and ensuring that we can do all we can to recruit the best people to this task.

We have before us, Mr. Speaker, a bill that will both protect the homeland and protect workers' rights. It is the right balance.

Mr. Speaker, I urge my colleagues on both sides of the aisle to strongly support this legislation before us tonight. It represents an agreement between the House and the Senate and the White House, and by joining together we will send a strong message to the American people and to the other body that we are committed to doing all we can to protect our families and our country.

Hon. Deborah Pryce

OF OHIO

Mr. Speaker, it has been nearly 5 months since the President called upon Congress to create a new Department of Homeland Security, and nearly 4 months since the House first took up that task. This legislation has been through a long journey, full of procedural and partisan roadblocks, weighed down by special interests, and slowed by a storm of misdirection.

I could not be more pleased that we are here today with this compromise legislation that will finally allow us to move the bill to the President's desk. This is a historic achievement.

In recent days, Members of the House and Senate have been through a thoughtful, thorough, and cooperative process. Every effort was made to address each concern while maintaining a basic framework that creates an effective department.

This legislation will give the new Department of Homeland Security the tools it needs to succeed in its mission. And this, in my mind, is the key, because the new department's most basic and core mission will be to secure America from terrorist attack.

On September 11, 2001, the streets of New York and Northern Virginia were turned to ash, while a grassy field in Pennsylvania played quiet witness to the final act of a heroic group of Americans. Creation of the Department of Homeland Security is the bold and necessary next step we must a take to ensure that this dark day is never repeated.

We are not creating new government, we are creating better government. We are not legislating new bureaucracy, we are streamlining to face a new threat. We are making government smarter, more flexible, and ultimately, better able to secure America.

The perpetrators of terrorism are shadowy and agile, and they target us like predators without distinction between military target and ordinary citizen. They are a 21st century enemy with an agelessly corrupt goal—destruction of life, elimination of liberty, and restriction of human freedom.

Our enemy has recognized that our greatest strength—the open society in which we live—also makes us vulnerable to their attacks. We fight this enemy not just on battlefields abroad, but in our very cities and towns. We must be able to respond at home in a strong, coordinated and agile way.

The new Cabinet-level department is only one part of our national response, but it is an essential part. The new department will consolidate the vital preparedness, intelligence analysis, law enforcement, and emergency response functions that are currently dangerously dispersed among numerous Federal departments and agencies. And in the process, the legislation balances the need to protect America with the need to preserve the American way of life that we are protecting.

Thus far, the government has shown immense resolve and dedication, going to extraordinary lengths to respond to the terrorist threat. We are safer than we were on September 10, 1 year ago. But as the government's efforts reach the limits of their bureaucracies, we must rethink our government structure so that our Nation can be even stronger, smarter, and better prepared.

One of our revolutionary forefathers, George Mason, once said, "Government is, or ought to be, instituted for the common benefit, protection and security of the people, nation or community."

Make no mistake—our work today undertakes this very core function of government to secure the American people. I urge all of my colleagues to take measure of the task before us, and to support this fair rule and the underlying bill.

It has been a long journey, but this legislation, and the American people, are all the better for it.

Hon. Marge Roukema

OF NEW JERSEY

Mr. Speaker, I rise today in strong support of H.R. 5710, the Homeland Security Act of 2002.

At last, Members of both parties and the administration have put their differences aside and agreed on a strong bill that will make America safer by creating a Cabinet-level Department of

Homeland Security to unite essential agencies in our fight against terrorism here at home and abroad.

On the morning of September 11, 2001, a new enemy brought war to our shores. An enemy that considers any innocent man, woman, or child that cherishes freedom a target. An enemy that does not necessarily call any nation home. And an enemy that can hide for years in plain sight and in our own neighborhoods.

This new kind of war, that makes where we live and work a potential battleground, calls for a new response. The United States is a nation at risk of terrorist attacks and it will remain so for the foreseeable future. We need to strengthen our efforts to protect America, and the current governmental structure limits our ability to do so.

When President Bush established the Office of Homeland Security in October 2001, its fundamental mission would be to prevent terrorist attacks within the United States, reduce America's vulnerability to terrorism, and minimize the damage and recovery from attacks that do occur. Mr. Speaker, I believe this new bill will achieve this mission.

The new department will combine 170,000 workers from 22 agencies, including the ATF, Border Patrol, Coast Guard and Customs Service, into a Department of Homeland Security with a $37 billion budget. It balances concerns of Federal workers with the need of the President to make personnel decisions in the interest of national security. It brings all immigration responsibilities under the Secretary of Homeland Security. Immigration services will be kept separate from enforcement functions within the department. This will provide the INS the leadership, direction, and focus that I have been advocating for years.

Now all the necessary functions of government to keep our Nation safe at home will fall under one department—where they should be. And that department will be part of the President's Cabinet—and that is where it should be.

Let me add, Mr. Speaker, that leading a massive new Federal department that is charged with protecting the homeland during such dangerous times is a herculean task. There is no one in the Nation more capable and prepared to provide that leadership than our former colleague Tom Ridge. Governor Ridge was called on by the President shortly after the tragic attacks on our Nation and stepped into the breach to provide leadership on homeland security. This is not the first time he has answered his Nation's call in time of war.

His leadership over the past year has prepared our Nation and our government for the task ahead. Governor Ridge will succeed and I wish him well.

Finally, Mr. Speaker, passage of this bill is the last of the profound accomplishments that this Congress has achieved since September 11, 2001. I am proud of the wise and prudent decisions we have made. Even though many on both sides have disagreed over details and those details have taken longer to work out than I would have liked, we have never disagreed on the goal of our actions. That goal is to protect and defend our Nation in this new and awful era of war.

We may suffer another dastardly attack on our shores—given the diabolic treachery in which our enemy deals, it is probably certain they will attempt to attack us again. But we will endure, care for our own, and stand taller than before. As always, we did not ask for this war, especially one that attacks us at home. But we will fight it. And with the help of this legislation—we will win it.

Mr. Speaker, I urge all Members to support this legislation. God bless America.

Thursday, November 14, 2002

FURTHER MESSAGE FROM THE SENATE

A further message from the Senate by Mr. Monahan, one of its clerks, announced that the Senate has passed without amendment bills and concurrent resolutions of the House of the following titles:

H. Con. Res. 487. Concurrent resolution authorizing the printing as a House document of a volume consisting of the transcripts of the ceremonial meeting of the House of Representatives and Senate in New York City on September 6, 2002, and a collection of statements by Members of the House of Representatives and Senate from the *Congressional Record* on the terrorist attacks of September 11, 2001.

MEMORIALS

Under clause 3 of rule XII, memorials were presented and referred as follows:

422. The SPEAKER presented a memorial of the House of Representatives of the Commonwealth of Pennsylvania, relative to House Resolution No. 685 memorializing the Congress of the United States to declare September 11 as "National Day of Life Appreciation and Freedom"; to the Committee on Government Reform.

429. Also, a memorial of the Legislature of the State of California, relative to Assembly Joint Resolution No. 52 memorializing the Congress and President of the United States to enact H.R. 3917 to designate a National Memorial at the crash site of Flight 93 in Somerset County, Pennsylvania to pay tribute to and honor the true heroes of this nation; to the Committee on Resources.

IN PRAISE OF THE LUTHERAN FELLOWSHIP ASSOCIATION OF THE SAGINAW VALLEY

Hon. James A. Barcia

OF MICHIGAN

Mr. Speaker, I rise today to honor the Lutheran Fellowship Association of the Saginaw Valley as it celebrates 45 years of dedicated service. Over the years, members have made numerous and significant contributions to Lutheran families and to the citizens of many communities in Bay County and elsewhere. They deserve our gratitude and praise.

In Bay County, the Lutheran Fellowship Association began when a handful of faithful Lutherans with a vision of creating a gathering place for families and friends joined together to open a fellowship hall. Since then, the group has grown to include more than 250 members and their families, serving the needs of thousands of Lutherans and others.

The LFA Hall has been a center of Lutheran life in our community for generations. Under the guidance of President Earl Wegener and other leaders past and present, it has continued to fulfill its mission as a venue for individuals, families, groups and organizations to enjoy fraternal, social, educational, recreational and family events of all kinds at a reasonable cost.

It is not a stretch to say that a large percentage of mid-Michigan's Lutheran community has likely attended one or more events at the hall during their lifetime, including wedding receptions, confirmation celebrations, anniversaries and a host of other activities. Many marriages have gotten off to a beautiful start with a reception at the LFA Hall and years later an untold number of couples have returned to the hall to commemorate their anniversary with family and friends.

In addition, LFA members have always put a high premium on charitable donations. Last year, they pooled financial resources to make a contribution to the victims of the September 11, 2001 terrorist attack in New York City. In other years, they have provided assistance to disadvantaged children or those afflicted with serious illnesses.

Finally, Mr. Speaker, I ask my colleagues to join me in praising the members of the Lutheran Fellowship Association for all that they do to meet the needs of the Lutheran community. Fellowship is an integral element in the Lutheran faith and the LFA has served a useful and vital role in fulfilling that need in Bay County. I am confident the LFA will continue to serve its members, their families, our community and our Lord well into the future.

IN CELEBRATION OF NATIONAL BIBLE WEEK

Hon. Bob Etheridge

OF NORTH CAROLINA

Mr. Speaker, I am honored and pleased to serve as congressional cochair for National Bible Week, November 24 to December 1, 2002. National Bible Week has been an annual observance in this country since 1941 when the Nation turned to the Holy Bible for strength, comfort, and guidance. On September 11, 2001, when terrorists destroyed the World Trade Center towers in New York and attacked the Pentagon, another "day of infamy" took place in our Nation's history. President Bush immediately called Americans to prayer, saying, "Our purpose as a Nation is firm, yet our wounds as a people are recent and unhealed and lead us to pray. . . . We ask almighty God to watch over our Nation." I strongly believe that one contribution every American can make in these troubling times is to pray for our Nation, its leaders, and its people.

National Bible Week is celebrated every year from Sunday to Sunday during the week of Thanksgiving. It is a time of prayer, a time to confirm our values and a time to strengthen national resolve. As we gather at our dinner tables in remembrance, let us be thankful to be living in a country where our Constitution guarantees freedom of worship. I commend the National Bible Association for its leadership in promoting this worthy endeavor.

NATIONAL GUARD TROOPS

Hon. Gil Gutknecht

OF MINNESOTA

Mr. Speaker, on September 11, 2001, our generation met its challenge. The attacks against innocent Americans were acts of war. We are still fighting that war. Carl von Clausewitz said that the goal of any military encounter is to destroy the enemy's will to fight. We still have work to do.

But at home we have come far. We have buried our dead. We have comforted our wounded. We have rebuilt the Pentagon. New York is being rebuilt. We have gained a resolve and determination to go on. We will continue to be the shining beacon of liberty. We are willing to bear the price of defending the principles of freedom, justice and honor. We are Americans, and proud to be so.

Generations of Americans have followed the wisdom of President Theodore Roosevelt when he said, "In any moment of decision, the best thing you can do is the right thing. The worst thing you can do is nothing."

From the Barbary Coast to the streets of Kabul, Americans have always sought to do what is right. We have never given way to despots and madmen in the name of artificial peace. More than 48 million men and women have served in our Armed Forces to do the right thing.

The sacrifice of Americans who left their homes and lives for the cause of justice across the globe is a testament to what is good and right about our great Nation. Because of Americans, Europe was liberated from a madman. Because of Americans, communism is left to the ash heap of history. Because of Americans, little girls are going to school in Afghanistan.

Today I honor those Americans who stepped in to secure our domestic defenses during a time of great uncertainty. The brave men and women of the National Guard. As active duty troops were deployed, the men and women of the National Guard dropped what they were doing and answered their call to duty. Careers were put on hold, families parted with loved ones, sacrifices were made to secure our Nation.

Guard members from Minnesota have served in every major conflict since its inception more than 360 years ago. More than 150 Minnesota National Guard soldiers were called to duty following the September 11 attacks.

I am especially grateful to the National Guard soldiers of Company B, Second Battalion of the 135th Infantry. These soldiers performed special duties at the Rochester International Airport. During a time of crisis, they stepped up to join that long grey line. That line that has never failed us.

Thank you First Sergeant Thomas L. Butterfield, Sergeant Samuel M. Adjei, Sergeant First Class Jason R. Schweitzer, Specialist Jason A. Cox, Specialist Benjamin R. Jech, Specialist Jacob R. King, Staff Sergeant Troy D. Landsverk, Sergeant William M. Olson, Sergeant Timothy A. Patterson, Sergeant Daniel J. Prescher, Specialist Brandon L. Riggs, Sergeant Scott J. Saltou, Sergeant Matthew Swiger, and Specialist Benjamin W. Teed.

These soldiers deserve our respect and our gratitude.

As William Jennings Bryan said, "Destiny is not a matter of chance, it is a matter of choice. It is not a thing to be waited for, it is a thing to be achieved." Americans have a history of choosing their destiny. We will continue to do so, because that is who we are.

We must, and we will, continue to achieve this victory for the people of the United States and for all civilized, peace-loving people around the world. There will be a price. The blood and treasure of our Nation will be invested. The leadership, resources and unwavering courage of the United States are critical in this struggle. We shall not falter, we will rise to the challenges. And, in the end, we will leave to future generations a safer planet because we never failed to defend the freedom we cherish. We will continue to practice what we preach.

God Bless America.

In the Senate of the United States

Monday, September 9, 2002

REMEMBERING ALAN BEAVEN

Hon. Dianne Feinstein

OF CALIFORNIA

Madam President, I come to the floor today to honor the heroism of Alan Beaven—a Californian aboard flight 93 who helped prevent the terrorists from crashing another airplane into its intended target on September 11, 2001.

As we approach the 1-year anniversary of that horrible day, our thoughts turn to the heroes like Alan who gave their lives to save others.

To honor the courageous passengers of flight 93, I joined Senator Specter to cosponsor the "Flight 93 National Memorial Act," which I believe the Senate will pass today to establish a memorial at the crash site in Pennsylvania. This legislation will also establish a Flight 93 Advisory Commission to recommend planning, design, construction, and long-term management of the memorial.

I believe it is important to pass this legislation before the anniversary of September 11 to appropriately recognize the heroism of Alan Beaven and the other flight 93 passengers.

I would like to take a few moments to tell the world about Alan and his family.

Alan Beaven wasn't supposed to be on flight 93 that tragic day. On Monday, September 10, Alan and his wife Kimberly were in New York planning for a year-long sabbatical in India to work for a humanitarian foundation. Alan was a top environmental lawyer in San Francisco who planned to volunteer his services in India.

Alan was headed east, not west, but there was one last case involving pollution in the American River near Sacramento and settlement talks had broken down that Monday. Alan had to head back.

Tuesday morning Alan drove to Newark, NJ, to catch a flight to the West Coast. Flight 93 was 40 minutes late that day—giving passengers on board time to learn about the planes that had crashed into the World Trade Center and the Pentagon. A few called home on cell phones to express their love and say that a group of passengers were determined to fight back against the hijackers—Alan Beaven was one of those brave men.

No one knows for sure what happened aboard that airplane, but we do know countless lives were saved when that plane was diverted from its intended target.

Even though Alan's seat was in the back of the airplane, his remains were found in the cockpit at the crash site in Pennsylvania. The Beaven family has also heard Alan on the cockpit voice recorder, so it is clear that Alan, standing 6 feet 3 inches tall and weighing over 200 pounds, fought with the hijackers.

I will enter two letters I have received from the Beaven family into the *Record*. Alan's wife, Kimberly, and his son, Chris, wrote to me about what they heard on the cockpit voice recorder in April when the families of the passengers of flight 93 were allowed to listen to the struggle aboard the aircraft.

My heart goes out to Alan's wife, Kimberly, and his three children, John, Chris, and Sonali. John earned a biology degree at UC San Diego where he was captain of the baseball team and an Olympic torch bearer when the torch went through Sacramento on its way to Salt Lake City this past winter. John's brother Chris attends Loyola Marymount University and sister Sonali is 5 years old.

Alan's great joy was his family. He spent hours reading to Sonali, scuba diving with Chris, and playing catch with John.

In fact, John's early memories of his father were of the two of them playing catch for hours on end. When John was 5, the family moved from London to New York and before they could drop off their luggage, young John made Alan play catch in Central Park.

In a tribute to Alan, the Beaven family decided not to have a funeral, but instead a "Thanksgiving for the life of Alan Anthony Beaven."

And what a life it was.

Alan was born in New Zealand on October 15, 1952. He worked as an attorney in New Zealand, England, New York, and California. As a top environmental lawyer, Alan worked on over 100 clean water cases in just 10 years in California.

Friends and family of Alan say they are not surprised that Alan risked his own life so selflessly to save others.

The day after the terrorist attacks on our Nation, Alan's secretary went into his office and found a single piece of paper tacked up at eye level on the wall in front of his desk. It was a quote he heard that week which summed up how he lived his life, and how he ended it when he joined others to fight back against the terrorists. Alan wrote, "Fear, who cares?" And these words adequately describe his actions aboard flight 93.

I did not know Alan Beaven, but this quote tells me all I need to know about him—that he was a fearless, loving, and devoted man.

One year later, it is clear that our Nation has lost a superstar environmental lawyer, a loving father and husband, and a true hero—Alan Beaven.

I ask unanimous consent to print the two letters to which I referred in the *Record.*

There being no objection, the letters were ordered to be printed in the *Record*, as follows:

AUGUST 9, 2002.

HON. DIANNE FEINSTEIN,
U.S. SENATOR,
Hart Senate Office Building, Washington, DC.

DEAR SENATOR FEINSTEIN: My father, Alan Beaven, was among those 33 passengers of United Airlines flight 93. Their hurried steps toward the cockpit were the first in an international campaign against the threat of fanatical hostility. For this they should be celebrated.

My dad played a central role in the deposing of his flight's assailants. Not only did he cooperate in an organized effort but he commanded it as well. For this effort he should be particularly acknowledged.

The cockpit recorder (C.V.R.) substantiates my claim of his exceptional heroism. At a private listening in Princeton, NJ, I twice heard his accented words. His final phrase, "Turn up!" was shouted at 10:02:17.3 on the official C.V.R transcript. Given the range of sensitivity of the cockpit microphones and my father's seating placement in the rear of the plane I reasonably believe that these findings indicate my dad's extraordinary actions.

Secondly, my father's remains were recovered in the front of the aircraft. Authorities confirmed that D.N.A. testing placed him in the cockpit at the time of impact. Again, given his seating placement, this evidence undoubtedly proves his centrality in the effort to regain custody of United's flight 93.

Though my father did not place a telephone call in his final hour, other such correspondences indicate his exceptional involvement. Reports were made of great men well above the height of six feet leading the passengers toward the captured cockpit. My dad, 6′3″ and 215 lbs., was one of few men who met this description.

Finally, the assumption of his extraordinary bravery in death is founded on the thematic valiance of his life. Whether in his professional or personal activities he met opposition with strength and spirit. It is understood by all who knew him that he continued this trend in passing.

In conclusion, I concede that assumptions based on the thematic valiance of his life do not warrant superlative public recognition. However, his stature and his physical placement at impact beg it. Finally, the cockpit voice recording demands it. I ask you to do all in your power to issue due credit to my father. He led a group that led a nation that led an international campaign against the threat of fanatical hostility. My father is a hero.

Sincerely,

CHRIS BEAVEN.

AUGUST 1, 2002.

HON. DIANNE FEINSTEIN,
U.S. Senate, Hart Senate Office Building, Washington, DC.

DEAR SENATOR FEINSTEIN: On April 18, 2002, in Princeton, NJ, I heard the voice of my husband, Alan Beaven, on the cockpit voice recorder of United Airlines flight 93 that crashed in Shanksville, PA, on September 11, 2001.

I know without a doubt that I heard Alan's voice shout "Turn up!" at the time on the tape's clock of 10:02:17.3. My stepson, Chris Beaven, who was listening to the VCR at the same time, independently made note of the exact same words and time.

There are at least two other occasions that I am very confident that Alan's voice was recorded. These additional times were of shouting and "aargh" noises, familiar to us as Alan often "wrestled" playfully with his sons. The distinct sounds were very similar. The times I noted for these sounds were 9:38:36.3 and 9:40:17.7.

As you know, Alan's physical remains were found in the cockpit area of the plane. Alan was a 6 foot 3 inch, 205 lb powerful man. A brilliant litigator who made his life's work fighting for justice. I, and all who knew Alan, know he was an active participant that fateful day.

Please ensure that Alan Beaven and all the passengers of flight 93 are duly honored for their heroic actions in preventing the terrorists from destroying their intended target in Washington, DC.

Sincerely,

MRS. KIMBERLY BEAVEN.

HONORING NEW YORK CITY'S COURT OFFICERS

Hon. Hillary Rodham Clinton

OF NEW YORK

Madam President, as we approached the 1-year anniversary of 9/11, I rise today to again honor all of the public safety officers whose courageous and heroic acts saved thousands of lives at the World Trade Center. In particular, I want to highlight a group of public safety officers who deserve to be honored for their heroism. The New York City court officers risked their lives and contributed immensely to the rescue and recovery operations at Ground Zero.

I especially would like to honor three court officers who gave the ultimate sacrifice—their lives. Their heroic deeds have earned them the nomination for the Public Safety Officer Medal of Valor—a testament to true American heroes.

I would like to say a little bit on each officer.

Cpt. William "Harry" Thompson, of the Bronx, was widely respected and beloved by all 1,600 court officers in New York City as senior instructor at the New York State Court Officers Academy. A 27-year veteran, he was the father of two adult sons and was the sole supporter for his widowed mother. All who knew Captain Thompson considered him a "spit and polish" type of officer. Captain Thompson was proud of his profession and New York is so very lucky that he devoted his life to public service.

Senior court officer Thomas Jurgens was part of a family who believed in giving back to one's city and country. Senior court officer Jurgens was the son of a firefighter, and was a volunteer fireman from Lawrence, Long Island. He made all of us proud by serving his country in the Persian Gulf war as an Army combat paramedic. Senior court officer Jurgens was a 4-year veteran at the Manhattan Supreme Court, and he was married in June 2001.

Senior court officer Mitchel Wallace, of Mineloa, Long Island, worked at the Manhattan Supreme Court for 2 years. Before September 11, the New York State Court of Appeals Chief Judge Judith Kaye honored him for resuscitating a man who had collapsed from cardiac arrest aboard a Long Island railroad train. Senior Court Officer Wallace planned to marry Noreen McDonough in October, and he called her "Cinderella."

In addition to these brave heroes who were lost, 22 other court officers risked their lives to save others at the World Trade Center. These men and women have been honored for their bravery on September 11. They are: Deputy Chief Joseph Baccellieri, Jr., Officer Tyree Bacon, Sgt. Frances Barry, Cpt. John Civelia, Sgt. Gerard Davis, Officer William Faulkner, Officer Gerard Grant, Officer Edwin Kennedy, Officer Elayne Kittel, Officer William Kuhrt, Officer Theodore Leoutsakos, Officer Craig Lovich, Sgt. Patricia Maiorino, Major Reginald V. Mebane, Sgt. Al Moscola, Sgt. Kathryn Negron, Officer Joseph Ranauro, Sgt. Albert Romanelli, Sgt. Richard Rosenfeld, Officer Andrew Scagnelli, Officer Mahindra Seobarrat, and Sgt. Andrew Wender.

Hundreds of court officers volunteered to work on recovery efforts at Ground Zero. After working full shifts at the courthouse, these officers would then work a full shift at Ground Zero. They would return home, clean the dust and debris from their hands, and return to their jobs at the courthouse. Through valor, duty, and commitment, they did all that they could to assist in the rescue and recovery operations.

On behalf of the American people, I express my thanks and appreciation for these public safety officers whose dedication and patriotism strengthen the resolve of our Nation. These officers went above and beyond the call of duty, sacrificing their lives in order to save others, not because it was their job, but because it was their sense of duty or pride. These officers represent the very best in America.

Tuesday, September 10, 2002

MENTAL ILLNESS PARITY

Hon. Paul D. Wellstone

OF MINNESOTA

As we look back toward September 11, and commemorate this tragic day in America's history, we can be proud of the way in which the American people rallied to support those who suffered such unspeakable losses in their lives. Many of us still feel the shock and the fear of that day, and while we can take great pride in the ways in which our country has recovered, we know that for many, the grief and the trauma is still sharp and constant. We know more about how such events can leave scars on the psyche of a country, as well as individuals. We know that many who had suffered from mental illness prior to September 11 may find they need treatment again. We know that many in New York and other parts of our country are suffering from post-traumatic stress disorder. We show our strength as Americans when we respond not only with our strength and outrage toward the perpetrators of this horror, but also with compassion and support toward the victims.

CONGRESSIONAL MEDALS FOR CREW AND PASSENGERS OF FLIGHT 93

Hon. Arlen Specter

OF PENNSYLVANIA

Mr. President, I have sought recognition to proceed as if in morning business to discuss legislation I have pending, S. 1434, a bill which has 69 cosponsors, which would give the Congressional Medal to all of the crew and passengers on flight 93 that crashed in Shanksville, PA, on September 11, 2001.

As we know from cell phone conversations from passengers on that plane, the passengers took over the plane from the terrorists, at least to the extent of depriving the terrorists control where the terrorists, as was widely suspected, were headed for the Capitol of the United States. And the plane crashed in Shanksville, PA, killing all of those on board.

It seems to me this is a unique place for the Congressional Gold Medal, because those passengers saved the Congress. Had that plane reached the Capitol, this Chamber would not now be in existence, nor the Rotunda, nor the House of Representatives. It is hard to say in the morning, perhaps midmorning, how many Members of the Congress of the United States and staff would not be here today. In seeking this recognition, it is a very unique opportunity to acknowledge those passengers.

This bill has languished because it has gotten tied up, as it is not uncommon for legislation to be tied up for a variety of other reasons. There are some who want to give medals to everyone who died on September 11, which I think is a fine idea. There are some who want to give medals to all of those who were in the rescue squads from the police precincts or fire stations or the port authority. And there, again, I think that is a commendable idea. And all the ideas to recognize other people may be fine, but they can take their turn on legislation.

But this legislation ought to be enacted before sunset tomorrow, before September 11, 2002, expires. I am now working with some of my colleagues in the Senate to accomplish that. If we cannot accomplish that, then I am going to ask unanimous consent to call up S. 1434, which has 69 cosponsors. It should have been discharged from committee a long time ago. With 69 cosponsors, that is 18 more votes than necessary to pass legislation in the Senate.

There is a bill in the House of Representatives which approaches the issue slightly differently. The proposal in the House is to leave the decision up to the Attorney General of the United States. Well, that might be a good idea if there was something for the Attorney General to determine that we do not now know. But all of the knowable facts as to what happened on flight 93 are now known.

The Attorney General cannot conduct an investigation and pinpoint any specific individuals.

And it is doubtless true that some individuals were more responsible for taking control of the plane away from the terrorists than others. But all were present. And all of those who were present were accessories to heroism. They lent their support by their presence. Of course, they could not go anywhere else, but the passengers brought down the plane. And the passengers saved the Capitol of the United States.

Interestingly, just yesterday, the *New York Times* published a release which contains confirmation from key Al Qaeda operatives that flight 93 was, in fact, headed for the Capitol. That has been a fairly accepted conclusion, but this is what the *New York Times* story of yesterday, September 9, says:

Yosri Fouda, correspondent for the satellite station Al-Jazeera, told the Associated Press that he was taken, blindfolded, to a secret location in Pakistan to meet Khalid Shaikh Mohammed and Ramzi Binalshibh in a June interview arranged by Al Qaeda operatives.

The thrust of the story is that the Al Qaeda operatives said that flight 93 was headed for the Capitol. So, in essence, I think we have waited long enough. I think this action ought to be completed before sunset on September 11, 2002. And I hope we can work out an accommodation from the Members who are now with varying points of view. But, as I say, I will ask unanimous consent that the bill be acted upon before sunset tomorrow.

Mr. President, I ask unanimous consent that the full text of this *New York Times* report identifying from Al Qaeda operatives the fact that this plane, flight 93, was headed for the Capitol, be printed in the *Record*.

There being no objection, the article was ordered to be printed in the *Record*, as follows:

[From the *New York Times*, Sept. 9, 2002]

REPORT: CONGRESS WAS ON 9/11 LIST

(By the Associated Press)

Dubai, United Arab Emirates—The U.S. Congress was the fourth American landmark on Al Qaeda's Sept. 11 hit list and the terror group also considered striking U.S. nuclear facilities, according to a purported interview with two Al Qaeda fugitives wanted in the terrorist attack.

Yosri Fouda, correspondent for the satellite station Al-Jazeera, told The Associated Press that he was taken, blindfolded, to a secret location in Pakistan to meet Khalid Shaikh Mohammed and Ramzi Binalshibh in a June interview arranged by Al Qaeda operatives.

Founda said he has waited until now to air the audiotaped interview—it is scheduled to be broadcast Thursday on the pan-Arab satellite station—because he wanted to include it in a documentary marking the first anniversary of the Sept. 11 attacks.

In an article in London's Sunday Times, Fouda wrote that he learned during the interviews that the U.S. Congress had been Al Qaeda's fourth Sept. 11 target. Two hijacked planes slammed into the World Trade Center, another into the Pentagon, and a fourth went down in a Pennsylvanian field.

U.S. counterterrorism officials, speaking on condition of anonymity, said many of Mohammed's statements about the origins of the Sept. 11 plot are plausible, but they have no information that would verify those claims.

The officials could not corroborate Mohammed's statements that the U.S. Capitol was the intended target of the fourth plane or that nuclear power plants had also been considered as potential targets for the Sept. 11 attacks.

Abu Zubaydah, a top Al Qaeda leader in U.S. custody since March, told interrogators that the White House was the fourth plane's target, U.S. officials have said.

U.S. officials regard Mohammed as one of the highest-ranking Al Qaeda leaders still at large and believe he is still planning attacks against U.S. interests. U.S. officials say Binalshibh belonged to a Hamburg-based cell led by Mohammed Atta, an Egyptian suspected of leading the Sept. 11 hijackers.

"I am the head of the Al Qaeda military committee and Ramzi (Binalshibh) is the coordinator of the 'Holy Tuesday' operation," Fouda quoted Mohammed as saying. Sept. 11, 2001 fell on a Tuesday.

Mohammed said planning for the attacks began 2½ years before Sept. 11 and that the first targets considered were nuclear facilities.

We "decided against it for fear it would go out of control," Fouda quoted Mohammed as saying. "You do not need to know more than that at this stage, and anyway it was eventually decided to leave out nuclear targets—for now."

Fouda, an Egyptian reporter and host of al-Jazeera's investigative program "Top Secret," said he flew to Islamabad, the Pakistani capital, and from there to Karachi on Al Qaeda instructions. In Karachi, he was taken blindfolded and via a complicated route to an apartment where he met the two men.

Fouda, speaking by telephone from London, said Al Qaeda operatives told him not to bring any electronic equipment—including a camera or recorder—to the interview. The Al Qaeda members videotaped the interview but instead of sending a copy of the video as promised, sent him only the audiotape, he said.

At one point while being led to the meeting, Fouda said he thought he was going to meet bin Laden. Speculation has been rife that the Al Qaeda leader may be in Pakistan after fleeing U.S. attempts to kill or catch him in neighboring Afghanistan.

Fouda said during the two days he spent talking to the two, Mohammed once referred to bin Laden in the past tense, leading him to believe bin Laden could be dead.

The U.S. officials said they do not consider Mohammed's use of the past tense to refer to bin Laden as any sort of definitive evidence that he is dead.

Fouda said he also learned that Atta, the chief hijacker, had been a sleeper operative in Germany since 1992 and started detailed planning with a 1999 meeting in Afghanistan with other sleepers.

Once in America, Atta communicated with higher ranking Al Qaeda officials via e-mail, Fouda wrote. But when he had determined everything was ready, he telephoned Binalshibh in Germany to tell him the date, using a riddle that referred to the shapes of the numbers 9 and 11.

Al-Jazeera, the Qatar-based satellite broadcaster, has drawn world attention with its broadcast of interviews with and statements by bin Laden and his top lieutenants.

SEPTEMBER 11, 2001, ANNIVERSARY

Hon. Christopher J. Dodd

OF CONNECTICUT

Mr. DODD. Mr. President, I rise this afternoon to share a few thoughts on the eve of tomorrow's anniversary of the terrorist attack, September 11, 2001. Tomorrow, a lot of our colleagues, both in this Chamber and the other body, will be expressing themselves, with many Members attending memorial services at the Pentagon. The Senate, as a body, plans to come together late tomorrow morning to meet as a body and to share our thoughts with the country about the events of a year ago. I take this opportunity to remember and to honor the nearly 3,000 of our fellow citizens and others who had come to this country to work—not all were Americans; the majority were—but lost their lives 1 year ago tomorrow in one of America's darkest of days.

I also join all of America in paying tribute once again to the countless men and women whose acts of bravery and heroism that so inspired us on that day and the days that followed the tragedy of September 11, and continue to serve as a solemn reminder that the American spirit shines as bright as ever despite the events of that day, that horrible day 1 year ago.

Thousands of families across this great country of ours, including families in my home State of Connecticut—families in my State lost some 149 people, most of whom lost their lives in the World Trade Center—these families and their loved ones have endured a year of unimaginable grief at the expense of unimaginable bravery. Every American grieves with them as many of our fellow citizens the world over from around the globe have shared with us the sense of grief and horror of a year ago and have continued to relate to us and to share their thoughts and prayers with all Americans as a result of our commemoration of the events of 12 months ago.

Over the past 12 months, I have heard countless stories, tragedies that were once unthinkable. In Connecticut, I know of a man who lost both his wife and his only child on that day a year ago; of parents who lost their young children in their twenties, just beginning their lives as young adults with professional careers; of wives who had received the last phone calls from their husbands before the Twin Towers fell.

Every American will always remember where they were when the Twin Towers were attacked and collapsed. Every American will always remember where they were when they heard a hijacked plane had crashed into the Pentagon, only a few blocks from where I am sharing these thoughts this afternoon. Every American will always remember how they felt upon learning that a group of passengers fought back against the terrorists who hijacked their plane before it crashed in the field of Pennsylvania. September 11, 2001, is a day that will be etched in all of our memories for the rest of our lives and etched in history forever.

Although all Americans went through that day together, we will always share its memory. Last September 11 was also a deeply personal day for each and every one of us. We each had our own highly personal experiences during those horrid hours that began in the early morning—that wonderful clear, bright, cloudless sky over the eastern part of our country.

For me, the hours and days and weeks following the terrorist attacks were filled with immensely mixed emotions, as most of my colleagues know. I see my friend and colleague from Texas on the floor. We shared the great joy last year of having children come into our lives. My first child, my daughter Grace, was born just 48 hours after the attacks, born on September 13, at a hospital right across the river in Virginia. From the window of the maternity ward, my wife Jackie and I watched the smoke rising from the still-burning Pentagon as we held our newborn child in our hands.

I can still vividly recall trying to balance my feelings of incredible, intense joy with this new beautiful life, mixed with the powerful feelings of horror and trepidation over what kind of a world my daughter Grace would grow up in, in the 21st century.

Something heartened me that day. I have told this story on numerous occasions. In the hospital as my wife held our newborn daughter, many of the doctors and nurses, several of them who held her shortly after she was born, came from places outside of America to become citizens. Three of them came from Afghanistan, Pakistan, and Lebanon. Here we are, 48 hours after the events, those countries had been the places of refuge for those engaged in the attacks on our country, and here were people from that very part of the world, U.S. citizens today, nurturing and caring for my newborn daughter.

That was all the evidence I needed at that particular moment that America was attacked not for who we are, but for what we stand for: freedom, liberty, and community. And we shared something very powerful in common: We were devastated over the attacks, and we were never prouder to be Americans, almost simultaneously.

Word was already out that the terrorist attacks were the work of Al Qaeda, a fanatical group which hijacked planes, but also an otherwise peaceful religion, Islam, to perform their evil deeds.

Word was out that Osama bin Laden and his minions of hate thought that by attacking us, our buildings, our Pentagon, and our planes, they could somehow divide our great Nation and somehow weaken our resolve to be a global power, to be a force for freedom and democracy around the globe.

Word was out that those who hate the United States simply for who we are, for our freedoms, our prosperity, and our diversity, thought that by murdering thousands of innocent Americans and shattering the lives of thousands of families, our Nation would somehow lose its ability to function as a great democracy.

They were wrong. We are today stronger, I argue, than ever.

September 11 changed America forever. At one level, the attacks made us aware of our vulnerabilities and forced us to realize there is no such thing as the unthinkable. Yet at another level, the way in which the entire Nation came together, in the days and weeks and months after the attacks, has served as a profound and inspirational reminder to strengthen the American people and the breadth and depth of the American spirit.

So as we mark this historic day, a day of sadness, we look back and remember September 11, not just for the tragedy it evokes but also in renewing our faith in the greatness of the wonderfulness of our Nation, in which we are charged temporarily to be custodians, as Members of this body, to see that that daughter of mine and the children of our colleague from Texas grow up in a world far safer than what we witnessed a year ago. That becomes our collective responsibility as public officials: To put aside differences and, wherever we can, to work together as one people to make our country stronger and better, to achieve that sense of perfection that the Founders of our Nation envisioned more than 200 years ago.

With those thoughts in mind, I extend my deepest sympathies, my thoughts, and prayers to the families in my State and across our Nation who still grieve terribly for the loss they suffered a year ago.

IN MEMORY OF THE CALIFORNIA VICTIMS OF 9/11

Hon. Barbara Boxer

OF CALIFORNIA

Mr. President, I am here in a very somber mood. We are approximately 15 hours away from the very moment 1 year ago that our Nation was hit, and I want to take just a moment of the Senate's time—maybe 15 minutes—to reflect on that day and, most of all, to remember the Californians we lost that day, numbering 54.

For me, and perhaps for you and many Americans, September has really been a month of excitement and anticipation. I have always loved September. It is the end of the summer, the beginning of a beautiful fall with the changing of the leaves, back to school, and perhaps a little quicker pace, a faster step. September, for most of us, never reminds us of loss, of fear, of shock, of the horrors born of an extreme, unbridled, blind hatred.

In September, we found out about those things. We also found out as a Nation what heroism truly is, how strong and united we can be, how we can set aside differences for the greater good and work together.

The images of September 11 are deep in our minds and deep in our souls. The pain is there, just under the surface. For some of us in America, it is on the surface, and it will always be on the surface for the families who grieve, for the children who will never know a parent—thousands of them—for communities that were decimated.

Today I want to remember those in my State who died on that day. Each was unique. Every one of those planes on that fated day was headed to California. So even though my State was 3,000 miles away from Ground Zero, from the World Trade Center or the Pentagon, we were linked in our sorrow, and we were linked in our outrage.

I am going to read the 54 names, and then I am going to talk a little more about some of the people whose families wanted me to just say a little more about them and show their picture to you.

David Angell; Lynn Angell; David Aoyama; Melissa Barnes; Alan Beaven; Berry Berenson; Dr. Yen Betru; Carol Beug, and her mother Mary Alice Wahlstrom died together on flight 93. Mary Alice is from Utah.

Mark Bingham; Deora Bodley; Touri Bolourchi; Daniel Brandworst, Ronald Gamboa, and their adopted son, David Brandhorst. He was 3 years old.

Charles "Chic" Burlingame, the captain of American Airlines flight No. 77. Technically, he was from McLean, VA, but his family is from California, and they considered him a Californian, and they said he considered himself a Californian.

Thomas Burnett; Suzanne Calley; Jeffrey Collman; Dorothy deAraujo; Lisa Frost; Andrew Garcia; Edmund Glazer; Lauren Grandcolas; Andrew Curry Green; Richard Guadagno; Stanley Hall; Gerald Hardacre; John Hart; John Hofer; Melissa Hughes; Barbara Keating; Chad Keller; Christopher Larrabee; Daniel Lee; Dong Lee; Joe Lopez; Hilda Marcin; Dean Mattson; Dora Menchaca; Nicole Miller; Laurie Neira; Ruben Ornedo; Marie Pappalardo; Jerrold Paskins; Thomas Pecorelli; Robin Penninger; Marie-Rae Sopper; Xavier Suarez; Alicia Titus; Otis Tolbert; Pendyala Vamsikrashna; Timothy Ward; Christopher Wemmers; John Wenckus.

Mr. President, I want these names to be memorialized again today.

There is a beautiful song called "Try to Remember," and one of the lines is: "Try to remember the kind of September when no one wept except the willow."

Sadly, those of us who lived through September 11, 2001, will weep for our lost brothers and sisters, but we will always remember our country, our embrace of freedom, and our democracy. And we will always cling closer to our loved ones. This place, this great democracy, America, will endure.

Now I am going to tell you a little bit more about a few of the people we lost in California. Many people noted that the *New York Times* has run an ongoing biography of the people who were lost on that day. I was talking to Bob Kerrey, the former Senator from Nebraska, and he said this to a group of us: When you read those memorials, what you realize is how wonderful and important each of these people was and what wonderful stories were related from their families, their friends, and their coworkers. What really emerged is why this is such a great country. These people do not get in the news. They get up and go about their lives. That is what you are going to find out as I read about these people and show these pictures in memoriam.

Lauren Grandcolas.—Mrs. Lauren Grandcolas was a 38-year-old advertising sales consultant when the flight she was on, United flight 93, was hijacked by terrorists. As we all know, that plane crashed in a Pennsylvania field killing everyone on board. We also know of the heroism of the passengers on that plane.

Mrs. Grandcolas was born in Bloomington, IN, and attended the University of Texas at Austin where she met her husband, Jack Grandcolas. After graduation, she worked as a marketing director for a law firm and then for PricewaterhouseCoopers.

At the time of her tragic death, Mrs. Grandcolas was working as an advertising sales consultant at *Good Housekeeping* magazine and was researching and writing a nonfiction book to help women boost their self-esteem.

Lauren had enthusiasm and passion for life, loved the outdoors and was devoted to physical fitness. She hiked, jogged, kayaked, and enjoyed inline skating around her neighborhood. Her energy was boundless. She took classes in cooking and gardening, scuba diving, and wine appreciation. Lauren was active with the United Way, March of Dimes, Project Open Hand, Juvenile Diabetes Foundation, Breast Cancer Awareness, and Glide Memorial.

Her husband Jack recalls she had a heart the size of Texas. Knowing her flight had been hijacked, Lauren left her husband a message on their home answering machine and then loaned her cell phone to another passenger to call loved ones.

The joy Lauren felt pursuing new interests and developing new skills was being interwoven in the book she was writing for women. Jack recalls:

> She made a point to do things that were good for her, and she thought she could extend what she had learned to help other adult women gain confidence. Her sister and I will fulfill her dream by completing the book.

Lauren Grandcolas is missed deeply by her family.

I wanted to take a moment to tell you a little bit more about her.

Nicole Carol Miller.—This next picture in memoriam is of Nicole Carol Miller. I want to start out by reading a poem that was dedicated to Nicole that was written by her father, David James Miller. It was written last September 11. If I cannot get through this, I will put it in the *Record.* My daughter's name is Nicole. This is the poem.

> How I love thee My Nicole.
> When the thoughts of you come into my mind
> It's as if a breeze has passed through our rose garden and the sweet savory I smell
> The taste of roses upon my tongue brings the sweetness of your memory to my mind
> It comes upon me as the morning dew weighs the roses down
> Smooth and pleasant are the thoughts of you, as the petals of a rose
> And once again I am nourished with your love.

Nicole Carol was a lovely 21-year-old college student when the flight she was on, United flight 93, was hijacked by the terrorists. That was the plane that was brought down by the passengers in Pennsylvania.

Nicole's memory lives on in the hearts of those she loved. She took great joy in life and exemplified this with her wonderful outlook and her tenacious personality. Nicole's radiant smile, which we can see in this photo, could light up the room as she energized those around her. She knew how to be an outstanding friend. She was blessed with two families, her father and stepmother, David and Catherine Miller of Chico, CA, and her mother and stepfather, Cathy and Wayne Stefani, Sr., of San Jose, CA.

In her father's words:

> She had that sweet baby quality. She could make you smile and forget your troubles for a little bit.

Friend Heidi Barnes describes Nicole as very friendly and welcoming. She had a big heart, and it was open to everyone.

Nicole lived in San Jose, CA, with her mother and stepfather. She attended local schools and graduated from Pioneer High School in 1998. A talented softball player during all 4 years of high school, Nicole won a college softball scholarship during her senior year. Even though she had never been a competitive swimmer, she tried out for the Pioneer High swim team as a freshman and made the team. At the time of her tragic death, she was a dean's list student at West Valley College in Saratoga working part time and weighing whether to transfer to California State University at Chico or San Jose State University.

I offer this tribute to Nicole.

Hilda Marcin.—I would like to take this opportunity to share with the Senate the memory of one of my constituents, Hilda Marcin, who lost her life on September 11, 2001. Mrs. Marcin was 79 years old when the flight she was on, United Airlines flight 93, was hijacked by terrorists. As we all know, that plane crashed in a Pennsylvania field, killing everyone on board.

Mrs. Marcin was born in Schwedelbach, Germany. When she was 7 years old, her family emigrated to the United States to escape oppression. Like many immigrants, her family left all possessions behind and came only with the clothes on their backs.

Her family settled in Irvington, NJ, where she attended local schools. She worked 7 days a week in the payroll department of the New Jersey shipyards during World War II.

A friend arranged a blind date with Edward Marcin and they were married on February 13, 1943. They had two daughters, Elizabeth and Carole. The Marcin family enjoyed participating in school functions, class trips, the PTA, and various church activities. Mr. and Mrs. Marcin were also socially and politically active in Irvington. Mrs. Marcin later worked as a special education teacher's aide.

Hilda Marcin embraced life with enthusiasm and made the most of every minute. She adored her family and her granddaughter, Melissa Kemmerer Lata. She was an inspiration to those she touched, including the special needs children in the school where she worked. Her friends admired her positive attitude and her desire and ability to continue working during the later years of her life. Mrs. Marcin treasured freedom and democracy, and her American citizenship.

At the time of her death, Mrs. Marcin was flying to San Francisco to live with her younger daughter, Carole O'Hare. She is survived by her daughter, Elizabeth Kemmerer and son-in-law Raymond Kemmerer; daughter Carole O'Hare and son-in-law Thomas O'Hare; and granddaughter Melissa Lata and Melissa's husband, Edward Lata. I offer this tribute to her.

Daniel Lee.—Daniel Lee lost his life on September 11, 2001. Mr. Lee was 34 years old when the plane he was on, American Airlines flight 11, was hijacked by terrorists. As we all know, that plane crashed into the World Trade Center, killing everyone on board.

Daniel Lee grew up in Palm Desert, CA. He was a carpenter and a drummer in a local southern California band. He met his wife, Kellie, in 1991 at a rock concert in which he was playing the drums. They were married October 7, 1995, and their first child, Amanda Beth, was born December 11, 1998.

Mr. Lee was a dedicated and successful set carpenter in the music industry, known to work 20-hour days when necessary. He worked with many talented musicians including Neil Diamond, Barbara Streisand, N'Sync, Aerosmith and Yanni. He was touring with the Backstreet Boys when, on September 11, 2001, he left to fly home to be with his wife as she was about to give birth to their second child. Allison Danielle Lee was born September 13, 2001.

Kellie Lee recalls Dan's bright, relaxed and charming smile. "He was caring, loving, funny and romantic. He loved being a Dad and was so excited about having another child on the way," she says. "One of his special joys was getting friends together for barbeques and pool parties," Kellie remembers.

Dan Lee is survived by his wife, Kellie Lee, his daughters, Amanda and Allison, mother and stepfather Elaine and John Sussino, brothers Jack Fleishman and Stuart Lee and sister, Randi Kaye. I offer this tribute to Daniel Lee.

Mari-Rae Sopper.—Mr. President, I take this opportunity to share with the Senate the memory of one of my constituents, Mari-Rae Sopper, who lost her life on September 11, 2001. Ms. Sopper was a 35-year-old lawyer and gymnastics coach when the flight she was on, American Airlines flight 77, was hijacked by terrorists. As we all know, that plane crashed into the Pentagon, killing everyone on board.

Ms. Sopper was a native of Inverness, IL, and attended William Fremd High School in Palatine, IL. At the age of 15 she set the goal of becoming a champion gymnast. She succeeded, becoming all-American in four events, the school's athlete of the year and the State's outstanding senior gymnast of the year.

Larry Petrillo, her high school gymnastics coach, remembers her as brash and committed. "One thing she taught me is, you never settle for less than you are capable of. We should never accept limits. We should always fight the good fight. She was a staunch supporter of gymnastics and what's right," he recalls.

Upon graduating from Iowa State University with a degree in exercise science, Ms. Sopper earned a master's degree in athletics administration from the University of North Texas and a law degree from the University of Denver. Ms. Sopper was an accomplished dancer and choreographer and continued to coach at gymnastics clubs.

Ms. Sopper practiced law as a Lieutenant in the Navy's JAG Corps, focusing on defense and appellate defense. She had left the Navy JAG Corps and was an associate with the law firm Schmeltzer, Aptaker & Sheperd, P.C. when she found her dream job: to coach the women's gymnastics team at the University of California at Santa Barbara.

It was a 1-year appointment and Ms. Sopper was looking forward to the challenge. Her mother, Marion Kminek, says Mari-Rae was excited about the opportunity. "I said go for it. Life is too short. It was something she had always wanted to do and she was so happy and excited," recalls Kminek.

At the time of her death, Ms. Sopper was moving to Santa Barbara to begin her appointment. Her close friend, Mike Jacki, recalls "This was to be a new adventure for Mari-Rae, and an opportunity to get back into the sport she loved. We have lost a very special person. She was prepared to make her dream come true, and in an instant it was gone."

Mari-Rae Sopper is remembered for her loyalty, strong values, excellent work ethic and spirit for life. She is survived by her mother, Marion Kminek, and stepfather, Frank Kminek, her father Bill Sopper, sister Tammy and many loving friends.

Deora Bodley.—Mr. President, the last story I share with the Senate is the memory of one of my young constituents, Deora Bodley, who lost her life on September 11, 2001. Ms. Bodley was a 20-year-old college student when the flight she was on, United Airlines flight 93, was hijacked by terrorists. As we all know, that plane crashed in a Pennsylvania field, killing everyone on board.

Ms. Bodley grew up in San Diego, CA. As a high school student, she visited local high schools to discuss HIV/AIDS with her peers. She volunteered with the Special Olympics and a local animal shelter. Chris Schuck, her English teacher at La Jolla Country Day School, recalls, "Deora was always thinking big and going after big game."

At the time of her death, Ms. Bodley was studying psychology at Santa Clara University. She coordinated volunteers in a literacy program for elementary school students. Kathy Almazol, principal at St. Clare Catholic Elementary, recalls Ms. Bodley had "a phenomenal ability to work with people, including the children she read to, her peer volunteers, the school administrators and teachers. We have 68 kids who had a personal association with Deora."

In the words of her mother, Deborah Borza, "Deora has always been about peace." At the tender age of 11 years, Deora wrote in her journal, "People ask who, what, where, when, why, how. I ask peace." A warm and generous person, Deora was a gifted student and a wonderful friend. Wherever she went, her light shined brightly.

Deora's father, Derrill Bodley, of Stockton, CA, feels her life was about "getting along" and sharing a message of peace. Her 11-year-old sister, Murial, recalls Deora taught her many things and says, "Most of all she taught me to be kind to other people and animals. I cherish the memories of my sister and plan to work hard in school and in everything I do so she can be proud of me like I was of her."

Mr. President, none of us is untouched by the terror of September 11, and many Californians were part of each tragic moment of that tragic day. Some were trapped in the World Trade Center towers. Some were at work in the Pentagon. And the fates of some were sealed as they boarded planes bound for San Francisco or Los Angeles.

So I am honored and very moved to have had this chance to put into the *Record* today the names of these more than 50 Californians, every one now a bright and shining star in the sky. Their memories will live on and their legacies will live on, as will the memories and legacies of every American and every person, every innocent victim, who was cut down in the most hateful way on that tragic day.

FLIGHT 93 NATIONAL MEMORIAL ACT

Hon. Harry Reid

OF NEVADA

Mr. President, I ask unanimous consent that the Energy and Natural Resources Committee be discharged from consideration of H.R. 3917 and the Senate now proceed to its consideration.

The PRESIDING OFFICER. Without objection, it is so ordered.

The clerk will report the bill by title.

The assistant legislative clerk read as follows:

A bill (H.R. 3917) to authorize a national memorial to commemorate the passengers and crew of Flight 39, who, on September 11, 2001, courageously gave their lives thereby thwarting a planned attack on our Nation's Capital, and for other purposes.

There being no objection, the Senate proceeded to consider the bill.

Mr. REID. Mr. President, I ask unanimous consent that the bill be read three times, passed, the motion to reconsider be laid on the table, and that any statements relating thereto be printed in the *Record* as if read, with no intervening action or debate.

The PRESIDING OFFICER. Without objection, it is so ordered.

The bill (H.R. 3917) was read the third time and passed.

Mr. REID. Mr. President, I ask unanimous consent that the Energy Committee be discharged from further consideration of S. 2136, and the Senate now proceed to its consideration.

The PRESIDING OFFICER. Without objection, it is so ordered.

The clerk will state the bill by title.

The legislative clerk read as follows:

A bill (S. 2136) to establish a memorial in the State of Pennsylvania to honor the passengers and crew members of Flight 93, who, on September 11, 2001, gave their lives to prevent a planned attack on the Capital of the United States.

There being no objection, the Senate proceeded to consider the bill.

Mr. REID. Mr. President, I ask unanimous consent that the bill be read the third time and passed, the motion to reconsider be laid upon the table, and that any statements thereon be printed in the *Record*.

The PRESIDING OFFICER. Without objection, it is so ordered.

The bill (S. 2136) was passed, as follows:

S. 2136

Be it enacted by the Senate and House of Representatives of the United States of America in Congress assembled,

SECTION 1. SHORT TITLE.

This Act may be cited as the "Flight 93 National Memorial Act".

SEC. 2. FINDINGS AND PURPOSES.

(a) FINDINGS.—Congress finds that—

(1) on September 11, 2001, passengers and crewmembers of United Airlines Flight 93 courageously gave their lives to prevent a planned attack on the Capital of the United States;

(2) thousands of people have visited the crash site since September 11, 2001, drawn by the heroic action and sacrifice of the passengers and crewmembers aboard Flight 93;

(3) many people in the United States are concerned about the future disposition of the crash site, including—

(A) grieving families of the passengers and crewmembers;

(B) the people of the region where the crash site is located; and

(C) citizens throughout the United States;

(4) many of those people are involved in the formation of the Flight 93 Task Force, a broad, inclusive organization established to provide a voice for all parties interested in and concerned about the crash site;

(5) the crash site commemorates Flight 93 and is a profound symbol of American patriotism and spontaneous leadership by citizens of the United States;

(6) a memorial of the crash site should—

(A) recognize the victims of the crash in an appropriate manner; and

(B) address the interests and concerns of interested parties; and

(7) it is appropriate that the crash site of Flight 93 be designated as a unit of the National Park System.

(b) PURPOSES.—The purposes of this Act are—

(1) to establish a memorial to honor the passengers and crewmembers aboard United Airlines Flight 93 on September 11, 2001;

(2) to establish the Flight 93 Advisory Commission to assist in the formulation of plans for the memorial, including the nature, design, and construction of the memorial; and

(3) to authorize the Secretary of the Interior to administer the memorial, coordinate and facilitate the activities of the Flight 93 Advisory Commission, and provide technical and financial assistance to the Flight 93 Task Force.

SEC. 3. DEFINITIONS.

In this Act:

(1) COMMISSION.—The term "Commission" means the Flight 93 Advisory Commission established by section (4)(b).

(2) CRASH SITE.—The term "crash site" means the site in Stonycreek Township, Somerset County, Pennsylvania, where United Airlines Flight 93 crashed on September 11, 2001.

(3) MEMORIAL.—The term "Memorial" means the memorial to the passengers and crewmembers of United Airlines Flight 93 established by section 4(a).

(4) PASSENGER OR CREWMEMBER.—

(a) IN GENERAL.—The term "passenger or crewmember" means a passenger or crewmember aboard United Airlines Flight 93 on September 11, 2001.

(B) EXCLUSIONS.—The term "passenger or crewmember" does not include a terrorist aboard United Airlines Flight 93 on September 11, 2001.

(5) SECRETARY.—The term "Secretary" means the Secretary of the Interior.

(6) TASK FORCE.—The term "Task Force" means the Flight 93 Task Force.

SEC. 4. MEMORIAL TO HONOR THE PASSENGERS AND CREWMEMBERS OF FLIGHT 93.

(a) ESTABLISHMENT.—There is established as a unit of the National Park System a memorial at the crash site to honor the passengers and crewmembers of Flight 93.

(b) ADVISORY COMMISSION.—

(1) ESTABLISHMENT.—There is established a commission to be known as the "Flight 93 Advisory Commission".

(2) MEMBERSHIP.—The Commission shall be composed of—

(A) the Director of the National Park Service; and

(B) 14 members, appointed by the Secretary, from among persons recommended by the Task Force.

(3) TERM; VACANCIES.—

(A) TERM.—A member of the Commission shall be appointed for the life of the Commission.

(B) VACANCIES.—A vacancy on the Commission—

(i) shall not affect the powers of the Commission; and

(ii) shall be filled in the same manner as the original appointment was made.

(4) MEETINGS.—

(A) IN GENERAL.—The Commission shall meet at the call of the Chairperson or a majority of the members.

(B) FREQUENCY.—The Commission shall meet not less than quarterly.

(C) NOTICE.—Notice of meetings and the agenda for the meetings shall be published in—

(i) newspapers in and around Somerset County, Pennsylvania; and

(ii) the Federal Register.

(D) OPEN MEETINGS.—Meetings of the Commission shall be subject to section 552b of title 5, United States Code.

(5) QUORUM.—A majority of the members of the Commission shall constitute a quorum.

(6) CHAIRPERSON.—The Commission shall select a Chairperson from among the members of the Commission.

(7) DUTIES.—The Commission shall—

(A) not later than 3 years after the date of enactment of this Act, submit to the Secretary and Congress a report that contains recommendations for the planning, design, construction, and long-term management of the memorial;

(B) advise the Secretary on—

(i) the boundaries of the memorial; and

(ii) the development of a management plan for the memorial;

(C) consult with the Task Force, the State of Pennsylvania, and other interested parties, as appropriate;

(D) support the efforts of the Task Force; and

(E) involve the public in the planning and design of the memorial.

(8) POWERS.—The Commission may—

(A) make expenditures for services and materials appropriate to carry out the purposes of this section;

(B) accept donations for use in carrying out this section and for other expenses associated with the memorial, including the construction of the memorial;

(C) hold hearings and enter into contracts, including contracts for personal services;

(D) by a vote of the majority of the Commission, delegate any duties that the Commission determines to be appropriate to employees of the National Park Service; and

(E) conduct any other activities necessary to carry out this Act.

(9) COMPENSATION.—A member of the Commission shall serve without compensation, but may be reimbursed for expenses incurred in carrying out the duties of the Commission.

(10) TERMINATION.—The Commission shall terminate on the dedication of the memorial.

(c) DUTIES OF THE SECRETARY.—The Secretary shall—

(1) administer the memorial as a unit of the National Park Service in accordance with—

(A) this Act; and

(B) the laws generally applicable to units of the National Park System;

(2) provide advice to the Commission on the collection, storage, and archiving of information and materials relating to the crash or the crash site;

(3) consult with and assist the Commission in—

(A) providing information to the public;

(B) interpreting any information relating to the crash or the crash site;

(C) conducting oral history interviews; and

(D) conducting public meetings and forums;

(4) participate in the development of plans for the design and construction of the memorial;

(5) provide to the Commission—

(A) assistance in designing and managing exhibits, collections, or activities at the memorial;

(B) project management assistance for design and construction activities; and

(C) staff and other forms of administrative support;

(6) acquire from willing sellers the land or interests in land for the memorial by donation, purchase with donated or appropriated funds, or exchange; and

(7) provide the Commission any other assistance that the Commission may require to carry out this Act.

Wednesday, September 11, 2002

The Senate met at 11 a.m. and was called to order by the Honorable Tim Johnson, a Senator from the State of South Dakota.

The Chaplain, Dr. Lloyd John Ogilvie, offered the following prayer:

Almighty God, on this day of remembrance and resolve, we praise You for the way You brought us through those dark hours of September 11 a year ago. You were our refuge and strength, a very present help in trouble. We relive the anxious memories of that infamous day of attacks of terrorism on the World Trade Center, the Pentagon, and the airliner crashed in Pennsylvania. Today, as a Nation, we mourn for those who lost their lives as a result of these violent acts of treachery against our Nation. We deepen our ongoing intercession for their loved ones. Continue to comfort them, help them to endure the loneliness of grief, and grant them Your peace. Particularly, we pray for the families of the firefighters, police officers, and military personnel who died seeking to save others. We pray for the thousands of children who lost a parent in these catastrophes.

When we turned over to You our anger, dismay, and grief, you gave us the courage to press on. Thank You for the strong, unified leadership of the President and this Senate in the aftermath of 9/11 and for the decisive engagement of the insidious enemy of terrorism throughout the world. May this be a day of renewed resolve to press on. Protect us from further attacks. Quiet our fears as we reaffirm our trust in You. You are our Lord and Saviour. Amen.

SCHEDULE

Mr. REID. Mr. President, there will be, as the Chair has announced, morning business basically all day. At noon, there will be a moment of silence in recognition of the events of September 11. Both leaders have asked that those Senators who are here and have not gone home to their States try to be in the Chamber for the moment of silence. I hope all Senators will be here.

I also announce that the two leaders are going to speak prior to the noon moment of silence. The minority leader is going to speak at 20 till the hour, and the majority leader will speak at 10 till the hour.

IN REMEMBRANCE OF SEPTEMBER 11, 2001

Hon. Harry Reid

OF NEVADA

Mr. President, we all have been touched by the events at the Pentagon this morning. As I walked in, there was a woman whom I do not know, but she is symbolic of the sacrifices that people have made. Her face had been burned very badly, she had no hands, and her arms had been burned. This is what the terrorist activity is all about.

This innocent woman, who never did anything to anyone, has been subjected to this physical torture. It goes without saying that she has gone through and will go through many skin grafts and other such procedures so that she can learn to use her prosthetic hands, which she does not have yet.

It used to be when a building was constructed, they had a ceremony on every major construction, called the laying of the chief cornerstone. What does that mean? It means that the final stone in the foundation of that building will be laid.

Why did people celebrate that event? They celebrated because they knew if that building had a strong foundation, it would be fine.

In our life in America, that foundation, that chief cornerstone is the Constitution of the United States. That little document that people speak about in this Chamber—discussion led by, more than anyone else, Senator Byrd—is the chief cornerstone of this great democracy.

As we are forced to remember these events of September 11—because it is easy not to put unpleasant thoughts in our minds—as we are forced to remember these events, and rightfully so, we have to remember that this country has a firm foundation because the chief cornerstone of the foundation of this country is our Constitution.

Today, of course, is the first anniversary of the September 11 terrorist attacks on America. On this day we remember, as we will do every year on September 11, those tragic events that our Nation experienced on September 11, 2001.

What happened in New York, at the Pentagon, and in Pennsylvania has left many of us—in fact, all of us—with memories and strong emotions. I know that Nevadans were deeply affected by the terrorist attacks in the aftermath, and I feel good about how people in Nevada have reacted.

We were hurt very badly. Our No. 1 business is tourism, and tourism took a terrible blow. But those business entrepreneurs, people who worked for those large corporations, and the people who worked for the small businesses recognized that time would solve the problems, that time would heal a lot of the tourism problems, and it has. We are not back to where we were, but we are OK. I am proud of how the people of the State of Nevada have reacted.

We also have had from the State of Nevada a pouring out of sympathy, comfort, and consolation for those who were killed and hurt. We lost a teacher in the terrorism attacks, a teacher at Palo Verde High School. We lost two soldiers who were killed in action. So we will always remember what happened.

As individuals and in private, we will often reflect on this national tragedy. We cannot confine our memories to a single day or be guided by the calendar, but September 11 will forever be the day that we collectively, as a Nation, as a people, as America remember. We remember those whose lives were ended so suddenly and violently, not knowing what happened.

We remember the firefighters. We remember the police officers—the firefighters are New York's bravest, the police officers are New York's finest—and all other emergency and rescue workers who accepted the risks in rushing into burning buildings giving their lives, suffering physical and mental injury to help save the lives of people they did not know.

We remember the sacrifice, the selflessness, the heroism, and the courage of all of those who offered aid. We must remember those who survived and the thousands who did not. We must remember the parents, grandparents, children, sisters, brothers, wives, husbands, partners, and friends who have been robbed of not a weekend, not a week, not a month, not a year, but they have been robbed of their loved ones forever.

From the stories they have shared, we remember not only the deaths but the lives of their loved ones, remember their loss, and their struggle to heal. We remember our personal losses, our pain, even our anger, and, of course, our tears.

We remember the shock of seeing massive metal towers collapse as if they were erector sets that our grandchildren constructed. We have seen these massive metal towers reduced to rubble. We all remember the fire and the smoke.

I will never forget leaving room 219, after Senator Daschle told us we had to evacuate the building, looking out the window and seeing the smoke billowing out of the Pentagon where we were this morning. We remember, though, the effort to rebuild the Pentagon. We remember the generosity and spirit of Americans coming together to offer kindness, money, compassion, and consolation. We remember the sympathy expressed by foreign governments. As the President expressed this morning, some 90 foreign governments—I think it was the President; maybe it was Secretary Rumsfeld—are helping us in our battle in Afghanistan.

We remember that individuals all over the world opened their arms and their hearts to America. We remember the gruesome images so vivid that they are etched in our minds, and we remember how the spirit of our Nation was awakened, how Americans demonstrated resilience and resolve. We remember how the country united to support the war on terrorism. We remember the soldiers who were killed as part of our military efforts in Afghanistan. We remember, and we must always remember, the firm foundation of our country. We are a country guided by the Constitution of the United States, which separates us from the rest of the world. That is why we have remained a strong, vibrant democracy for more than 200 years.

Hon. George Allen

OF VIRGINIA

Mr. President, I rise today to offer my thoughts on this very solemn day of remembrance as we all return from a magnificent ceremony at the Pentagon observing all that is strong and good and awesome about our country.

I thank the Senator from Nevada for his very poignant words of empathy, as well as his understanding of the foundations of our country. Nevada, as all States, was hit hard.

We saw the outpouring of compassion all over this country, and I will share some of those stories. I recall in August driving across a lonely two-lane road in South Dakota, which would eventually get to the Badlands. There was a big bale of hay on the side of the road which had painted on it the American flag. It showed the spirit of that farm. We did not see any people, but we knew the sentiment of the folks who lived on that farm and in that region.

September 11, 9/11, just those words evoke sentiments and memories of where we were and what we did on that day of tragedy. As we remember those vile terrorist attacks of 1 year ago, for many of us the emotions and shock, the disbelief and horror that we experienced individually and as a people and a Nation are still fresh. Those memories, however, continue to strengthen our resolve in the same way that our Nation was forged together after those vile attacks a year ago.

Today, we view our Nation in a fundamentally new light. We have a greater understanding of the freedoms we enjoy and how vital it is that they be guarded, preserved, and even fought for, if necessary. We have a greater appreciation for a country that respects people of diverse backgrounds, cultures, and religious beliefs. We have poured out our hearts and our assistance to those who were injured and the families of those who lost a dear one. We view firefighters, police officers, first responders, with much greater appreciation, whether they are the brave men and women of the battalions in New York City or northern Virginia or in communities large and small all across our United States of America. These men and women were transformed on that day into our heroes. We will forever remember the thousands of innocent men, women, and children who were killed at the World Trade Center and in a field in Somerset County, PA.

This Senator will remember the 184 patriots at the Pentagon and on American Airlines flight 77 who lost their lives on Virginia soil. It is indeed the heroes and the innocent patriotic victims we will remember the most. The images of flags raised, the solemn salute of rescuers to their fallen comrades, and people who were rushing into burning buildings on the verge of collapsing hoping to just save one more life.

They and the freedom-loving patriots across our great Nation stand in stark contrast to those who only know hate, destruction, and oppression.

We also see that in a time of trial, ordinary people of all walks of life perform with extraordinary courage and dignity. We remember people such as Ltc. Ted Anderson, who carried two of the injured from the burning Pentagon and reentered through a broken window to drag out two more, one whose clothes were on fire; 1SG Rick Keevill and Virginia State Troopers Mike Middleton and Myrlin Wimbish, who entered the Pentagon three separate times looking for victims; LCDR David Tarantino, who moved a pile of rubble enough to pull a man from the Pentagon just before the roof collapsed; other Pentagon heroes such as SSG Christopher Braman; Ltc. Victor Correa; Sgt. Roxane Cruz-Cortes; Maj. John Grote; Ltc. Robert Grunewald; Col. Philip McNair; Cpt. Darrell Oliver; SP Michael Petrovich; SGM Tony Rose; Ltc. Marilyn Wills; and Cpt. David Thomas.

The Senator from Nevada, Mr. Reid, mentioned a woman who I think was Mrs. Kurtz at the Pentagon. Mrs. Louise Kurtz, though severely burned herself, valiantly tended to the needs of others around her. I am introducing legislation that will change current law so that individuals—such as Mrs. Kurtz, and those in her situation—can contribute to her retirement and so they will be able to afford to return to work after a very lengthy period of recuperation.

We also remember people such as Barbara Olson, a passenger on flight 77 who had the presence of mind to call loved ones on the ground to alert them of the hijacking.

We remember Cpt. "Chic" Burlingame of flight 77 who died fighting off hijackers who commandeered his plane and who is now properly buried at Arlington National Cemetery. These people have all touched our lives.

In talking to Mr. Burlingame's brothers and sister and wife, I find it noteworthy that at the Arlington National Cemetery his grave is on the tour and people in the tradition of those of the Jewish faith will put rocks on his headstone. That is very touching to the family and shows the unity and appreciation of a grateful Nation.

We also remember the survivors, survivors such as Stephen Push, whose wife Lisa Raines perished in the Pentagon and who has become a forceful and articulate spokesman for victims and families.

I will always remember, and thought of it last night while driving home, a young boy, a neighbor, a friend of my children whose name is Nick Jacoby. He lost his father on flight 77.

There are stories all over our communities and Nation. We also, of course, remember the quiet dignity of people such as Lisa Beamer who helped keep their loved ones very much alive for all of us. Her husband Todd, who said, "Let's roll," led an uprising with several other patriots against the hijackers of flight 93 and saved hundreds, if not thousands, of lives at the Capitol and in the Washington, DC, area. Recent reports recognize their likely target was this building.

We will remember countless others whose courageous efforts saved lives and provided comfort. We will remember and we will thank them for their extraordinary, inspirational dignity and their character. We will also remember the construction workers, the hard-hat patriots of the phoenix project who worked around the clock in their inspiring efforts to rebuild the Pentagon in plenty of time for employees to move in before the 1-year anniversary.

We will remember folks from a church that made quilts, the Christ Baptist Church from Prince William in Manassas, a magnificent quilt with the names of all who died. Also, we will remember the International House of Pancakes in Bristol, VA, an IHOP owned by an American who came here from Lebanon. I asked him a few months later how his business was. He said right after the attacks, for a few weeks, there were hardly any customers. But then a Methodist Church in Bristol, on the Virginia-Tennessee line, brought up the situation, and everyone from that church on that Sunday went in with their families and filled up the IHOP. Since then, others were coming back. That is a sign of the decency and the care of communities across the Nation.

Five days ago, in New York City, I had the opportunity to speak to a group of 70 mothers who were pregnant last September 11, and who were made widows on that terrible day. It has been said that suffering makes kinsmen of us all. While those mothers no longer have the physical and emotional support of their husbands, and the fathers of their children, they are now a part of our greater American family. In those babies, all under 1 year, the spirit and blood of their fathers live on. We want the babies to grow up with the optimism of liberty and opportunity and hope that is the spirit of America. These young children represent not just a birth but a rebirth, a rebirth and a rededication of the strength and unity of our Nation and her great, caring people as we move forward. Indeed, our Nation will be changed for generations by the tragic events of a single day and all those that followed September 11. We pray for the souls of all that we lost that day and their surviving families as well.

As a Senator from Virginia, for the permanent *Record* of our Republic, I ask unanimous consent to have printed the names of all the men, women, and children who perished in that attack on Virginia soil.

There being no objection, the material was ordered to be printed in the *Record*, as follows:

THE 184 VICTIMS WHO PERISHED AT THE PENTAGON ON SEPTEMBER 11, 2001

Paul W. Ambrose, Specialist Craig S. Amundson, Yeoman 3d Class Melissa Rose Barnes, Master Sgt. Max J. Beilke, Yeneneh Betru, Information Systems Technician 2d Class Kris Romeo Bishundat, Carrie R. Blagburn, Col. Canfield D. Boone, Mary Jane Booth, Donna M. Bowen

Allen P. Boyle, Bernard C. Brown II, Electronics Technician 3d Class Christopher L. Burford, Capt. Charles F. Burlingame III, Electronic Technician 3d Class Daniel M. Caballero, Sgt. 1st Class Jose O. Calderon-Olmedo, Suzanne M. Calley

Angelene C. Carter, Sharon A. Carver, William E. Caswell, Sgt. 1st Class John J. Chada, Rosa Maria Chapa, David M. Charlebois, Sara M. Clark, Julian T. Cooper, Asia S. Cotton,

Lt. Commander Eric A. Cranford, Ada M. Davis, James D. Debeuneure, Capt. Gerald F. Deconto

Rodney Dickens, Lt. Commander Jerry D. Dickerson, Eddie A. Dillard, Information Systems Technician 1st Class Johnnie Doctor, Jr., Capt. Robert E. Dolan, Jr., Commander William H. Donovan, Lt. Commander Charles A. Droz III, Commander Patrick Dunn, Aerographer's Mate 1st Class Edward T. Earhart, Barbara G. Edwards, Lt. Commander Robert R. Elseth

Charles S. Falkenberg, Leslie A. Whittington, Dana Falkenberg, Zoe Falkenberg, Store Keeper 3d Class Jamie L. Fallon, J. Joseph Ferguson, Amelia V. Fields, Gerald P. Fisher, Darlene E. Flagg, Rear Adm. Wilson F. Flagg, Aerographer's Mate 2d Class Matthew M. Flocco, Sandra N. Foster, 1st Lt. Richard P. Gabriel, Capt. Lawrence D. Getzfred

Cortez Ghee, Brenda C. Gibson, Col. Ronald F. Golinski, Ian J. Gray, Diane Hale-McKinzy, Stanley R. Hall, Carolyn B. Halmon, Michele M. Heidenberger, Sheila M.S. Hein, Electronics Technician 1st Class Ronald J. Hemenway, Maj. Wallace Cole Hogan, Jr., Staff Sgt. Jimmie I. Holley

Angela M. Houtz, Brady Kay Howell, Peggie M. Hurt, Lt. Col. Stephen N. Hyland, Jr., Lt. Col. Robert J. Hymel, Sgt. Maj. Lacey B. Ivory, Bryan C. Jack, Steven D. Jacoby, Lt. Col. Dennis M. Johnson, Judith L. Jones, Ann C. Judge, Brenda Kegler, Chandler R. Keller, Yvonne E. Kennedy, Norma Cruz Khan, Karen Ann Kincaid, Lt. Michael S. Lamana, David W. Laychak

Dong Chul Lee, Jennifer Lewis, Kenneth E. Lewis, Sammantha L. Lightbourn-Allen, Maj. Stephen V. Long, James T. Lynch, Jr., Terrace M. Lynch, Operations Specialist 2d Class Nehamon Lyons IV, Shelley A. Marshall, Teresa M. Martin, Ada L. Mason-Acker, Lt. Col. Dean E. Mattson, Lt. Gen. Timothy J. Maude

Robert J. Maxwell, Renee A. May, Molly L. McKenzie, Dora Marie Menchaca, Patricia E. Mickley, Maj. Ronald D. Milam, Gerald P. Moran, Jr., Odessa V. Morris, Electronics Technician 1st Class Brian A. Moss, Teddington H. Moy, Lt. Commander Patrick J. Murphy, Christopher C. Newton, Khang Ngoc Nguyen, Illustrator-Draftsman 2d Class Michael A. Noeth

Barbara K. Olson, Ruben S. Ornedo, Diana B. Padro, Lt. Jonas M. Panik, Maj. Clifford L. Patterson, Jr., Robert Penninger, Robert R. Ploger III, Zandra F. Ploger, Capt. Jack D. Punches, Aviation Anti-Submarine Warfare Operator 1st Class Joseph J. Pycior, Jr., Lisa J. Raines, Deborah A. Ramsaur

Rhonda Sue Rasmussen, Information Systems Technician 1st Class Marsha D. Ratchford, Martha M. Reszke, Todd H. Reuben, Cecelia E. (Lawson) Richard, Edward V. Rowenhorst, Judy Rowlett, Sgt. Maj. Robert E. Russell, Chief Warrant Officer 4th Class William R. Ruth, Charles E. Sabin, Sr., Majorie C. Salamone, John P. Sammartino, Col. David M. Scales, Commander Robert A. Schlegel

Janice M. Scott, Lt. Col. Michael L. Selves, Marian H. Serva, Commander Dan F. Shanower, Antionette M. Sherman, Diane M. Simmons, George W. Simmons, Donald D. Simmons, Cheryle D. Sincock, Information Systems Technician Chief Gregg H. Smallwood, Lt. Col. Gary F. Smith, Mari-Rae Sopper, Robert Speisman, Lt. Darin H. Pontell, Scott Powell, Patricia J. Statz

Edna L. Stephens, Norma Lang Steuerle, Sgt. Maj. Larry L. Strickland, Hilda E. Taylor, Lt. Col. Kip P. Taylor, Leonard E. Taylor, Sandra C. Taylor, Sandra D. Teague, Lt. Col. Karl W. Teepe, Sgt. Tamara C. Thurman, Lt. Commander Otis V. Tolbert, Staff Sgt. Willie Q. Troy, Lt. Commander Ronald J. Vauk, Lt. Commander Karen J. Wagner, Meta L. (Fuller) Waller, Specialist Chin Sun Pak Wells, Staff Sgt. Maudlyn A. White

Sandra L. White, Ernest M. Willcher, Lt. Commander David L. Williams, Maj. Dwayne Williams, Radioman Chief Marvin Roger Woods, Capt. John D. Yamnicky, Sr., Vicki Yancey, Information Systems Technician 2d Class Kevin W. Yokum, Information Systems Technician Chief Donald M. Young, Edmond G. Young, Jr., Lisa L. Young, Shuyin Yang, Yuguang Zheng

Mr. ALLEN. I add in closing, the Burlingame family, wife and surviving brother and sister, gave me a replica of one of the few things found from Captain Burlingame, other than his wedding ring. He had a picture of his mother and a prayer. They gave this to me a couple hours ago at the ceremony at the Pentagon.

I share it with my colleagues and Americans.

I DID NOT DIE

(By Mary Frye)

Do not stand at my grave and weep;
I am not there, I do not sleep.
I am a thousand winds that blow.
I am the diamond glints on snow.
I am the sunlight on ripened grain.
I am the gentle autumn rain.
When you awaken in the morning's hush
I am the swift uplifting rush
Of quiet birds in circled flight.
I'm the soft stars that shine at night.
Do not stand at my grave and cry;
I am not there, I did not die.

Never forget. We will never forget. We will always remember this day that forged America together. These horrific events have strengthened our unity of purpose and resolve as Americans, that we stand strong together for liberty. I hope and pray that as long as God continues to bless our United States and indeed blesses the entire world with people of such courage, integrity, and character, that liberty and justice will endure and prevail.

Hon. Russell D. Feingold

OF WISCONSIN

Mr. President, today I come to the floor to remember, to reflect, to try to somehow do justice to the memory of those lost to us on September 11. The tremendous grief we felt then, and still feel so sharply today, makes this anniversary a painful one for all of us as a Nation, and as a people.

The anguish of that day will always be with us, but those of us who witnessed those acts of terror on our television screens know that we cannot imagine the suffering of those who perished in the attacks, or those who survived them.

The families and friends of those who died must live with terrible loss, and those who survived must live with searing memories.

No passage of time can ever erase the emotions of that day. But 1 year later, we also know that no passage of time can diminish the legacy left behind by those who perished. They will always be with us, living on in the family and friends who loved them.

No passage of time will allow us to regain what was so tragically lost on that morning. But 1 year later, with the passage of time, we see so clearly what was briefly obscured by smoke and fear and disbelief. We see the strength of the people around us—their everyday heroism, their generosity, and their humanity.

No passage of time can change what happened on September 11. But the last year has shown us that when our Nation was tested by terror, we did not falter, and most of all we did not fail each other. We rose together to meet the challenges before us, and we found that together we were capable of more than we ever imagined.

So today we find strength in each other. We find strength in the acts of heroism, and the acts of simple humanity, that took place on September 11 and in the aftermath of the attacks: the bravery of the first responders at the World Trade Center and the Pentagon, and the acts of kindness of Americans all over the country who donated blood, observed moments of silence, or flew the flag in a show of patriotism and support.

Each of these acts, however large or small, contributed to our growing faith in the Nation's recovery, and in each other.

I was deeply proud of the many Wisconsinites who reached out to the victims of September 11 and their families. Volunteers from around the State flocked to disaster relief organizations to donate money and donate their time to support the victims of the attacks.

A number of Wisconsin volunteers also traveled to the World Trade Center to support the rescue workers. That desire prompted all of us to do something, anything we could to help the victims of the attacks ran deep in my State, as it did everywhere across the country.

Just as the firefighters and police on 9/11 redefined bravery and heroism, in the uncertain days that followed, the Americans who reached out to help the victims and their families redefined generosity and patriotism.

A number of companies in Wisconsin, as so many businesses nationwide, also donated to the rescue efforts. Fire truck manufacturers such as Pierce Manufacturing of Appleton, WI, and Marion Body Works of Marion, WI, donated critical replacement equipment to the New York City Fire Department. Seagrave Fire Apparatus of Clintonville, WI, rallied to complete previously ordered equipment for the New York City Fire Department in the wake of the attacks, and sent staff to New York to help the Department repair damaged equipment.

These efforts reaffirmed our faith that Americans would rise to this challenge, as we have so many times throughout our Nation's history. And we are rising to that challenge.

It has not been easy, and I frankly don't believe that all the choices we have made have been the right ones. But that has never affected the pride I feel to be an American during this extraordinary time in our history. I couldn't be more proud of the way Americans have come together in the wake of this tragedy, and I have been privileged to serve in the Senate during this last year.

What we as a Nation have accomplished over the last year, and what we will accomplish in the years to come to meet the challenge of terrorism, will be our mark on history, not just as a Congress but as a generation.

It is of course impossible to summarize what happened on 9/11 and what it means. There were so many moments—public and private, captured on film and also lost to history—that make up our collective memory of that day.

The *New York Times* section "Portraits of Grief," however, is one laudable effort to pay tribute to the victims as individuals by remembering and celebrating each of their lives. These brief stories of the victims' lives remind us that the people who died that day were from every walk of life, from all over the country, and from all over the world. They remind us of what America truly is—a sea of nationalities and ethnicities never before seen in human history. The bitter irony of Al Qaeda's desire to kill Americans is that people from every corner of the world have become citizens of this Nation. Like places all across America, the World Trade Center, the Pentagon, and the hijacked planes were filled with people with roots in Africa, Europe, the Middle East, Asia, and Central and South America.

These people and their families came here for different reasons, and they likely lived vastly different lives. But all of them had the chance to be a part of this great and free Nation. And all of them were senselessly struck down on September 11.

One such man was Ramzi Doany. I would like to read the story published in the "Portraits of Grief" section of the *New York Times* about this man, who lived for many years in my home State of Wisconsin.

> Ramzi Doany amassed friends. He amassed them with acts of kindness, like tutoring a woman with lupus, two children and no husband, to get her through college, or letting his college roommate and the roommate's wife live in his condo for 2 years so they could save money for a downpayment on a house.
>
> He amassed friends with his sense of humor, which filled a room and flourished at an early age. As a boy of 9 or 10, young Ramzi dug a hole in the backyard for a terrible report card and put a stone on top. "He said it was dead and buried," said his sister, Dina Doany Azzam.
>
> Mr. Doany was born to Palestinian parents in Amman, Jordan, and lived for many years in Milwaukee. At 35, he devoured the novels of Dickens, cooked Thanksgiving turkeys with great pride (even if they were just a bit dry) and had just bought a Harley-Davidson motorcycle. He chose to work as a forensic accountant last March for Marsh & McLennan, the insurance brokerage company, because it would bring him to New York, a city he loved. The job also brought him to the World Trade Center.
>
> It was a funny sort of journey, his sister said.

This man's journey, like so many others, was tragically cut short on September 11.

On this day, the passage of time is bittersweet. Whatever the healing powers of time, no passage of years can change what happened on September 11. But the passage of time brings other gifts.

This last year has brought us resolve—the firm resolve to stop terror, to preserve our liberty, and to do justice to the memory of those who died.

It has also shown us our own resilience—how Americans, even in the initial moments of shock and horror of the attacks, showed so much bravery, so much compassion, and so much generosity.

Finally, time has brought renewal. It has renewed our strength, our hope, and our faith in each other.

So it is with this resolve, this resilience, and this sense of renewal that we move forward, in the name of those who perished, dedicated to fighting terror, and united by our faith in this great and free Nation.

Hon. Ted Stevens

OF ALASKA

Mr. President, in the morning hours of September 11 our Nation endured a terrible tragedy. Though thousands of miles from the crash sites, the response from our "Last Frontier" was overwhelming. Alaskans rushed to aid the victims of the terrorist attacks. They volunteered rescue dogs and handlers. They waited in line for 3 hours to donate blood. Some boarded planes and traveled to Ground Zero to aid in the search and rescue efforts. In December, those Alaskans were still there clearing debris.

Alaskans who could not travel to the crash sites offered support in other ways. Over 10,000 Alaskans signed two 50-foot banners bearing the phrase "Love and Prayers, from the People of Alaska." One banner was presented in New York City by Alaskan firefighters. The other now hangs in the Pentagon. Countless Alaskans donated funds to help victims through the economic hardships brought on by the attacks. In Kenai, the Firefighters Association petitioned our State to name a mountain after St. Florian, the patron saint of firefighters, as a tribute to firefighters killed in the World Trade Center.

This year, Alaskans once again join the Nation in mourning and remembrance. Today, I attended the Pentagon's memorial service, but in my home State. Alaskans will pay tribute to our heroes in their own unique way. Anchorage residents will observe a moment of silence at 8:46 a.m. Emergency responders from across Alaska will gather on Barrow Street in Anchorage and join firefighters and police in a procession. A memorial wall will be erected at Town Square. In Homer, Mozart's "Requiem" will be performed as part of a worldwide sequence of performances beginning at the hour of the attack and moving from one time zone to the next. I hope all Alaskans who cannot participate in these events will attend a memorial and prominently display American flags.

I am proud of Alaska's efforts to honor and remember the victims of this tragedy. On that fateful morning they gave what Lincoln called the "last full measure of devotion." We honor their memory and their sacrifice.

Hon. Craig Thomas

OF WYOMING

Mr. President, I appreciate the opportunity to share some thoughts about this day.

We all have memories, of course, of a year ago. They are so clear after a year. Nothing like that has ever happened in this country. We remember it as we see it again on television today.

We have had a year to react, to recognize and accept the fact that it did happen. The unbelievable thing, shocking as it was, did happen. But we have also had the time to change from the immediate anger that we had, and the disbelief, to a commitment and resolution to do all that is necessary to make certain that it does not happen again.

We have had this year to increase our loyalty to our country and to our flag, to increase our understanding of the values of freedom and democracy, to commit our resolve to help and support those who have lost loved ones, family members, and friends, to accept the reality that here in the Congress we can disagree and have different views on normal, daily issues, but when it comes to protecting our country and to preserving freedom, we all come together.

The events of September 11 have clearly changed the way Americans view the world. We watched the events unfold. No one will ever forget. Everyone around the world has been touched, and we see some of that now. We are embroiled in a struggle against people who do not care about their lives and have set out to ruin ours. Sadly, we lost lives, but we regained a strong commitment to preserving our freedom and our integrity.

So it has been a year of shock, disbelief, anger, followed by commitment, caring, sharing, patriotism, and determination. I think we should be very proud of our fellow Americans for their commitment, their willingness to sacrifice and to give—whether it be on the battlegrounds overseas, whether it be in rescue missions or law enforcement, in charity to the needy, leadership in our country both at the community and national level, or just caring for our friends and neighbors and loving our families. This year has put an emphasis in all these values.

The United States will survive and will strengthen. Freedom will endure, and we thank God for the opportunity to be able to ensure that for our future.

Hon. Trent Lott

OF MISSISSIPPI

I acknowledge the fine statement that was made by Senator Thomas and thank him for his efforts today to make sure that Senators are aware of the opportunity to come to the floor of the Senate and pay appropriate tribute and recognition, and express the condolences that are so appropriate for that occasion. I want to make sure he was aware of our appreciation.

Mr. THOMAS. I thank the Republican leader.

Mr. LOTT. Mr. President, this day, September 11, is its own memorial. A year ago I got a call from my daughter, expressing her horror and her sorrow on this, her birthday—and only 2 weeks after her new daughter was born. She talked about how "I will just change my birth date. I will celebrate it a day earlier." She asked me, "What exactly is this situation in this world I have brought my daughter into?"

It struck me that she would have those questions and those concerns considering what she had seen that day. She worried about what it means for the future.

I talked to her this morning on her birthday. She celebrates her birthday today, as she should—not just because it was the day she was born but because she now realizes that in some ways, in spite of her horror, this is an even more special day—this is Patriot Day.

So my special pen from the Pentagon service will go to my daughter on this day because I think in a way how she felt a year ago and how she feels today reflects what we have all gone through and what we have experienced.

The truth is that this day doesn't really require any speeches or ceremonies, though we certainly will have them all day long. We really need no monument to remind us of the suffering and sorrow that befell our country 1 year ago today.

As we sat there next to the wall of the Pentagon, I kept thinking about the innocent men, women, and children who lost their lives so inexplicably and so mercilessly on that day. But I also think about those who tried so hard that day to save people's lives with danger to themselves. Some of them probably were injured, and some of them maybe were killed—and all that has gone into the work at that building to symbolize the importance of us showing that we are mending our wounds and we are going to be stronger from what we have experienced.

The wound that we had last year hasn't healed, nor should we expect it to be healed so quickly, nor many of the scars. The scars will be there. As a Nation, we lost a great deal—not only these innocent lives in Virginia, Pennsylvania, and New York, but also that sense of security and perhaps even naivete that we have experienced thinking that this is America, we are free and we are accessible, and we go and do what we want without being better in any way.

Well that has changed. I believe we have been hurt deeply—not just those who were directly involved, but all of us who watched it—all Americans and all freedom-loving people all over the world.

I continue to be so pleased and, frankly, thrilled with the reaction I get when I meet with leaders from countries all over the world—and just average people on the streets of other countries. They come up and express their condolences and their support.

Yesterday I met with the President of Bulgaria and the Prime Minister of Portugal. Their comments were so reassuring and satisfying. They have done their part. Bulgaria—yes. Bulgaria has had troops in Afghanistan and, fortunately, has stood with us and will stay with us in the future.

We have been hurt deeply. But our observance of this day is about more than grief, it is about more than anger, and it is about more than appreciation. It is about valor and courage beyond words adequate to describe what has happened and how we feel. It is about compassion and it is about a unity of spirit.

I have felt that I have seen it as I have gone across this country. I do not know how many States I have been in over the past year—but a lot of them, and there is a different feeling. When people sing "God Bless America" and start taking the Pledge of Allegiance to the flag, they sing and speak differently—with a little more gusto. But it is not about a flag, although that is what becomes so much a symbol of what we are experiencing internally. And it is not solely even about individuals. It is about ideas and principles—the values that have made this country what it is.

In this body, we don't take an oath to people or an oath to a person. We take an oath to the Constitution. So that unity of spirit has really been so obvious since I have gone into States in New England and the South and the West and the Far West. It is about faith that looks through death and a consolation beyond all human assurance.

This morning, when we joined President Bush at the Pentagon to formally reopen that section of the building destroyed in the terrorist attack, we all again felt those emotions of a year ago. I was sitting next to Senator Daschle, and we couldn't help but remember a year ago when we flew in a helicopter, along with Senator Reid and Senator Nickles, right over that area. We looked down and saw what was going on—the smoke, the confusion, and the activities in trying to save people's lives, put out the fire, and deal with all that was going on. It is a site that has been burned into my memory forever.

Needless to say, there couldn't be a better symbol than the restoration of the Pentagon for the way America's Armed Forces have responded to the assault on our country. God bless them for what they have done and for what they are doing today.

But those who were responsible for that horror—and all those who shelter them, finance them, abet them, encourage them, or reward them—should understand this: America's most important rebuilding is not the shattered wall of the Pentagon, nor the scar in the Earth in New York City. For what we have restored in the past year cannot be measured in granite and steel, nor even grassy knolls, as in the case of Pennsylvania. We have rebuilt a wall of resolve, of determination, and of steady purpose.

We have renewed trust in our leadership, and in one another, yes. We will disagree on this floor and we will argue about the best way to do the homeland security part and what should be the limitations on terror insurance. That is what democracy is all about. But in the end we have been able to find a way over the past year to come together and get a result. That is through determination and a steady purpose.

We have renewed trust in ways that I hope will stay with us for a long time. We have rediscovered in our shared sorrow the power of a truly free society to overcome the enemies of freedom.

These are our battlements and these are our armaments, and their might is going to be felt both here at home and in lands far away—until the hand of terror is crushed and the work of justice is done.

Again, we extend our heartfelt condolences to those family members who lost loved ones last year. We remind ourselves of how heroes were born on that date out of that horror, and we rededicate ourselves to the purpose of preserving this great young Republic and all the freedoms for which it stands.

Hon. Thomas A. Daschle

OF SOUTH DAKOTA

Mr. President, I begin by complimenting the distinguished Republican leader on his eloquence and his message this morning.

It was 1 year ago today that many of us turned on our television and saw what we could only imagine at the time was a horrible, horrible accident. Soon we realized that it was no accident. Instead, it was the worst terrorist attack on American soil.

Later this morning the wing of the Pentagon that was destroyed is being rededicated. That field in Shanksville is once again green. The debris from the site of the World Trade Center has been removed. The heavy equipment and the workers are now engaged in the act of building—not removing.

Through the physical scars of that day, we see a Nation beginning to be healed. The emotional ones are still raw with our memory. Thousands of families are approaching their second Thanksgiving without a loved one. Children are approaching their second holiday season without a mother—or a father. Empty lockers in firehouses still bear witness to the brave men who are no longer there.

And so, the Pentagon can be restored. New grass can cover the churned earth of a rural field. New towers can begin to rise where others fell. Seasons and years can pass. Through it all, we will never forget.

This day will forever be a part of our national memory. Nine-eleven will forever be our national shorthand for all that we witnessed, all that we have experienced—on that day and the days following.

That is what we remember all across America today.

In my home State of South Dakota, there will be a number of small services, including a memorial ceremony at Mount Rushmore.

In Seattle, WA, citizens will march to a downtown fountain that became the city's unofficial memorial after September 11. Thousands of flowers had been left there. Those flowers were gathered by the city and composted. Each marcher will receive a bulb, in mulch generated by the original memorial flowers, to plant.

Birmingham, AL, is dedicating a new memorial walk. San Francisco is unfurling a 5-mile-long banner along the city's coastline.

From Portland, ME, to Portland, OR, people are pausing, and paying tribute. All told, more than 200 communities are holding events of some kind. In one way or another, all Americans have the opportunity to commemorate our Nation's loss.

And, of course, some people will simply go about their business—and that in itself is a powerful testimonial to our ability to go on.

Today is also a day to remember that our national tragedy is the combination of thousands of individual tragedies.

I think that sentiment was best stated by Janny Scott, a reporter on the Metro desk of the *New York Times*, who was responsible for assembling a number of the "Portraits of Grief" that sought to capture the essence of each of the victims.

She wrote about "the individual humanity swallowed up by the dehumanizing vastness of the toll," and what she called "the preciousness of each life's path."

This morning, in New York, former Mayor Giuliani began the process of reading the names of everyone who perished on that day. If one name is read every 5 seconds, it will take over 4 hours to list every loss.

We also remember the individual acts of heroism: firefighters who rushed up to help others get down; the passengers and flight attendants on flight 93 who showed us that we don't ever have to surrender to evil.

Seeing their selflessness inspired something similar in all of us. In South Dakota, one ranch couple—themselves struggling—sold 40,000 dollars' worth of cattle and donated the proceeds to the victims. Similar acts of selflessness took place all over the country. Millions of hands reached out to those who had lost so much, until, by the act of reaching out and grieving, and remembering, we all came shoulder to shoulder as we understood the extraordinary nature of the loss.

The terrorists who brought down the World Trade Center thought they could shake the foundation of this country. They didn't understand that the foundation isn't concrete and steel; it is our people, it is our commitment—our commitment—to freedom and democracy, and to each other.

So today, we remember those we lost, and we rededicate ourselves to preserving the memory of their lives, and to defeating the terror that took them.

Our military men and women in Afghanistan and those fighting terror around the globe carry with them our pride, and our hopes.

In the most fateful struggles in human history, freedom has triumphed over the worst forms of tyranny, and we will defeat the tyranny of terror as well.

On March 11, 6 months after the attack, Valerie Webb, a 12 year old who had lost her only living parent in the World Trade Center, flipped a switch, sending two towers of light rising into the darkness over lower Manhattan.

Someone compared that memorial to a national votive candle. Others compared it to the lives that were lost: beautiful, powerful, and fleeting. On April 14, as planned, that temporary memorial was extinguished.

At sunset tonight, in Battery Park, New York's mayor will light a flame to commemorate the victims of that day. Unlike the towers of light, that flame will not be extinguished—it will be eternal.

That flame will burn within sight of another eternal flame—the symbolic flame from the torch held by the Statue of Liberty.

Those two eternal flames carry with them two eternal promises.

The torch held by the Statue of Liberty is our Nation's promise that we will never yield in our determination to be a light to all those who seek freedom.

And the flame that will be lit tonight is our promise that though we may be slowly, steadily walking the path from remembrance to recovery—we will never forget.

MOMENT OF SILENCE IN RECOGNITION OF THE EVENTS OF SEPTEMBER 11, 2001

The ACTING PRESIDENT pro tempore. Under the previous order, the hour of 12 noon having arrived, the Senate will now observe a moment of silence in recognition of the events of September 11, 2001.

(Thereupon, the Senate observed a moment of silence.)

IN REMEMBRANCE OF SEPTEMBER 11, 2001

Hon. Ted Stevens

OF ALASKA

Madam President, I was very proud of the efforts of Alaskans in response to the disaster on September 11 of last year. Although we are thousands of miles from New York, they immediately reacted.

Hon. Patrick J. Leahy

OF VERMONT

Madam President, later this afternoon in my home State of Vermont, the chief judge of the Federal district court, Judge William Sessions, will have an immigration ceremony, and I might say that I can't think of anything more fitting. We will have memorials and other events throughout the State of Vermont today, just as we will in other States.

Many of us had been at the Pentagon earlier this morning, heard the moving statements, and saw the resolve of the men and women who protect this Nation. We heard our President and Secretary of Defense and others.

It is right that throughout the country we have different events to mark this occasion.

I want to especially compliment Judge Sessions for what he is doing in Vermont. If there is anything that speaks to the resiliency of this Nation, the greatness of this Nation, it is welcoming immigrants, saying our borders are not sealed, our borders are open.

We want to welcome people who will continue to make this country great, just as did my paternal great-grandparents and my maternal great-grandparents who came to this country not speaking any English but who sought employment and a new life. My grandfathers were stonecutters in Vermont, immigrant stock. My wife was the first generation of her family to be born here in the United States. It is immigrants who have made this Nation strong.

What Judge Sessions is doing is telling us that our borders and our country and our arms are still open to the mix of people from throughout the world who will continue to give us the diversity we need, just as our Constitution gives us diversity and guarantees that diversity in the First Amendment. We now have new Americans who will be here with the same rights and privileges the rest of us have, and the Nation will be a better place for it.

Hon. John W. Warner

OF VIRGINIA

Madam President, we have just returned from a most historic and moving ceremony at the Department of Defense. That building will always occupy in my heart a very special place for I was privileged to serve there during 5 years and 4 months of the period of the war in Vietnam in the Navy Secretariat, including my service as Secretary of the Navy.

On 9/11 I joined colleagues briefly here in the Chamber and then we exited and with other colleagues who were gathered in the park, we chatted a little bit about what we should do. I returned to my office and conducted a brief prayer meeting and recommended to my staff that they proceed to their homes and their loved ones.

In about an hour or two, however, I decided I would like to go to the Department of Defense again because of my very special high regard for the men and women of the Armed Forces and that dastardly act committed by terrorists. I called the Secretary of Defense, whom I had known for many years. We both served in the administration of President Nixon and President Ford. He said: Come right over.

I called my good friend and colleague, Carl Levin, at his home, and Carl immediately said, yes, he would join us, and the two of us then proceeded to the Department of Defense where we joined Secretary Rumsfeld and then-Chairman of the Joint Chiefs, General Shelton. It was a memorable afternoon there at the command post watching the magnificence of our command structure dealing with the many unknowns, and yet taking the proper actions.

The President called in. Both Senator Levin and I spoke with him briefly. Then we went back with the Secretary to where the plane had struck the building and visited with all those who were performing heroic acts right before our eyes in hopes of saving other lives and doing what they could to comfort those wounded.

We then returned with the Secretary. And Secretary Rumsfeld asked Senator Levin and I to accompany him to a press conference. We stood behind the Secretary and the Chairman while they spoke. And then unexpectedly, Secretary Rumsfeld turned to both of us and asked us to make a few remarks.

I have here this morning the remarks I made, with no preparation, just speaking from the heart. And they are as true today, 1 year later, as they were at about 6:30 on the afternoon or the evening of 9/11 when Senator Levin and I joined the Secretary. I will just repeat these remarks.

I stated that I was joined by my distinguished chairman, Carl Levin, and I said, speaking to the Nation:

> I can assure you that the Congress stands behind our President and the President speaks with one voice for this entire Nation. This is, indeed, the most tragic hour in America's history, and yet I think it can be its finest hour, as our President and those with him, most notably our Secretary of Defense, our Chairman {of the Joint Chiefs} and the men and women of the Armed Forces all over this world stand ready not only to defend this Nation and our allies against further attack, but to take such actions as are directed in the future in retaliation for this terrorist act—one of the most unprecedented in the history of the world.
>
> We call upon the entire world to step up and help, because terrorism is a common enemy to all, and we're in this together. The United States has borne the brunt, but {which nation} can be next? Step forward and let us hold accountable and punish those that have perpetrated this attack.

Under the leadership of our President and the courage of the men and women of the Armed Forces and the strength of the citizens of this Nation, that has been done, is being done, and will be done.

Hon. Dianne Feinstein

OF CALIFORNIA

Madam President, I rise today to share some of my thoughts on this very special day, a day that commemorates one of the darkest days in our Nation's history.

Those of us who listened this morning to the recitation of the names of those killed in the World Trade Center and the Pentagon found in those names both a message of grief and one deep in sorrow. Also in those names was a profound message of how deeply the world is interwoven. The reading of these names was, for me, an unforgettable message of our diversity.

My sorrow, my sympathy, my condolences go to those who have lost so much. For many, they have lost everything; yet they still have their spirit, their hope, and their determination, and they still have the love of a very sympathetic Nation.

On September 11, we all felt as if the loss was too much to bear, as if it would be impossible to go on. But out of the ashes of the World Trade Center and the Pentagon, we in Congress returned to work. We tried in our legislative ways to address the terrorist threat. Within a week of the attack, we approved a resolution authorizing the President to use force against those who would perpetuate or harbor the terrorists.

Within a month, we approved the USA Patriot Act, which authorized our law enforcement and intelligence agencies to take the necessary steps to root out the terrorist threat and to protect the Nation.

In May of this year, we approved the border security and visa reform legislation, which overhauled the way this Nation allows immigrants and visitors into the country.

In June, we approved a bioterrorism bill that included strict certification requirements for laboratories that handle anthrax, smallpox, and more than 30 other deadly pathogens.

At the same time, the United States launched a war against terror. In Afghanistan, the U.S. forces, working with the Northern Alliance, ousted the Taliban, fought Al Qaeda troops, and made it possible for Hamid Karzai to be elected

President—Afghanistan's first democratic election.

U.S. Special Forces were also sent to the Philippines, to Yemen, and Georgia to train local troops on how to fight the war against terror. We have broken up Al Qaeda cells in Spain, France, Morocco, and Singapore, thereby preventing planned attacks.

In the financial world, the Treasury Department began examining the financing of terrorist organizations, freezing more than $34 million in terrorist assets.

Now the Senate is considering two additional steps to defend our Nation: a bill to create a new Department of Homeland Defense and a comprehensive review of the intelligence failures that led to 9/11.

I would expect the Senate to approve the homeland defense bill in the coming weeks, and, hopefully, it will be signed into law by the end of the year.

On September 17, the Intelligence Committees of both the House and the Senate will open their first hearings on our intelligence review, which has been going on now for 6 months.

One year has now passed. The Nation has shown its resolve and resiliency. Now we must show our staying power.

For me, what emerged from 9/11 were four specific points:

First, we must stay the course on the war on terror. We must ferret out, bring to justice, one by one, group by group, those Al Qaeda, or others, who would simply kill because they hate.

Second, we must make this country as safe as possible: eliminate loopholes in laws, prevent fraudulent entry into our country, ensure that deadly chemicals and biological agents are properly handled, and see that the national security is protected, wherever possible.

Third, we have to reinforce the hallmarks of America: liberty, justice, freedom. Despite this crisis, the Constitution and the Bill of Rights remain strong and central to our way of life.

Finally, we need to celebrate our democracy, and the way we do that is simple: We make it work. We produce for our people. We pass good legislation. We administer the programs. We show that democracy offers solutions to the real problems of our society.

Let me say one thing about remaining vigilant in the war against terror. Much of the Al Qaeda organization remains intact, including two-thirds of the leadership, and possibly Osama bin Laden himself. Afghanistan is our beachhead in the war on terror. We cannot lose it or we lose the war on terror. Yet Afghanistan's leadership is fragile. Just last week there was an attack on President Karzai's life.

We have an obligation to provide for the security of Afghanistan and its leaders and ensure that the nation does not fall under the control of regional warlords. We must ensure that the Afghan economy becomes upwardly mobile.

We have work to do to find those in hiding, whether in Pakistan, Yemen, Saudi Arabia, the Sudan, Southeast Asia, and, yes, in our own country. Al Qaeda remains poised to do their dirty deeds.

America learned on September 11 a very simple and sobering lesson—that there are people in the world who would destroy us if they could. We must remember this fact and do all we can to stop them. This means staying the course and winning the war against terror. This means keeping focused on the immediate threat from Al Qaeda, and this means looking for new ways to strengthen our Nation's homeland defense.

As we all consider the past year, let us remember all of those who perished in the attacks and in their memory rededicate ourselves to doing all we can to making our Nation strong and preventing a similar attack in the future.

Hon. Kay Bailey Hutchison

OF TEXAS

Madam President, 1 year ago today, 3,000 people woke up, kissed their loved ones goodbye, went to work, and never returned. In the blink of an eye, their lives were brutally taken by the violent acts of terrorists.

Together, we grieved and mourned for those who lost their lives. We marveled at the heroism and bravery of the first responders—the firefighters, emergency personnel, and police officers—who rushed into the devastation to help others, many sacrificing their own lives in the process.

But the American spirit of resilience rose from the ashes of Ground Zero, the Pentagon, and that quiet field in Pennsylvania. Our collective anguish became our national resolve. We focused our energies on destroying the cowardly instigators of this tragedy so they could not do it again to us or any other nation on Earth.

We will forever recall this day, but we are not a vengeful people. As Americans, we value peace, freedom, and liberty. We know our diversity and tolerance of other views, religions, and ways of life are what make our Nation great. We do not perpetuate hatred or violence. We teach our children to love one another and treat others with respect.

America was born out of a great struggle. The words of our Founding Fathers ring as true today as they did more than 200 years ago. In 1771, Samuel Adams said:

> The liberties of our country, the freedom of our civil constitution, are worth defending at all hazards; and it is our duty to defend them against all attacks. We have received them as a fair inheritance from our worthy ancestors: they purchased them for us with toil and danger and expense of treasure and blood, and transmitted to us with care and diligence.

It is our duty to carry on the crusade for freedom that generations of Americans have fought and died to keep. The heroes of September 11 did not lose their lives in vain. The protection of our liberty and freedom remains resolute.

It is the words of a civilian hero that remain with us, a young man with a pregnant wife at home. He saw the horrors on his airplane that morning on September 11 as they were flying over Pennsylvania. He realized from telephone reports that this airplane, too, was part of a terrible plan headed for one of our treasured symbols of freedom in Washington, DC. Though he had little time to prepare, he and other brave passengers decided to fight. And Todd Beamer's last words in his valiant effort are our battle cry in this war on terrorism: "Let's roll."

America is ready to roll, Madam President, and we will never forget those who gave their lives for our freedom on September 11, 2001.

Hon. Phil Gramm

OF TEXAS

Madam President, I thank my dear colleague and tell her I am very happy to have my remarks appear next to hers.

A year ago today, terrorism struck at the very symbols of American democracy and capitalism, as if by destroying those symbols, as if by destroying the buildings, as if by killing innocent people, they could destroy those institutions.

They failed.

Like millions of Americans, a year ago today I watched the horror of the terrorist attacks. But then I watched the triumph of the human spirit. I watched ordinary Americans, people pretty much like us, who on that day did extraordinary things.

A year ago today, our Nation's leaders watched, and we were helpless, like everybody else, to do anything about the problem. I am proud to say today that we are not helpless, that we have started to fight back.

Our homeland is more secure today than it was a year ago, but it is not as secure as it has to be. We are fighting a war, but the Congress has to give to our military and to our law enforcement officials the tools they need to finish the job.

When in doubt, I believe we must act. What is at issue is the safety of the American people, and I am not willing to turn that safety over to our allies, to the United Nations, or to anybody else. Where terror hides, it must be rooted out and it must be destroyed, and if we have to do that alone, then America is willing and capable of doing that alone.

In my 24 years of public service in Congress, I have always been proud of my country and my countrymen, but I have never been prouder than I have been in the last 12 months. It has always been a privilege to serve, but in the last 12 months it has been my great privilege to serve the greatest country in the history of the world.

Hon. Carl Levin

OF MICHIGAN

Madam President, 1 year ago today, the openness and freedom of American society were used against us when terrorists hijacked civilian passenger jets and used them as missiles to demolish the Twin Towers of the World Trade Center and destroy a large section of the Pentagon. The fourth jet hijacked that infamous morning, United flight 93, may well have been headed for this Capitol Building before brave crew members and passengers fought back against their captors.

One year ago today, over 3,000 people had their lives snatched away from them. The emotional trauma of those losses has affected each and every American. No State, no town, no community, no person has been left untouched.

The despicable actions of the terrorists last September 11 have changed the world, not only because of what they have destroyed, but also because of what they have kindled in the American people.

In New York City, at the Pentagon, in the skies over Pennsylvania and across America, 1 year ago today and in the days since then, we have seen the bravery, compassion, determination, and shared sense of purpose of Americans from all walks of life. As one writer put it, "September 11 did not alter the American character, it merely revealed it."

I would add that it did not weaken our spirit, it strengthened it immeasurably.

We have, astoundingly, already rebuilt the mangled section of the Pentagon, and we have cleared Ground Zero in New York City. We have consecrated time and place and commemorated the heroic individuals who faced 9/11 head on.

We are now engaged in a war on terrorism. It is unlike any war we have ever fought. It has no boundaries. It has no clear end. Our enemies target civilians. They are not soldiers. They are not warriors. They are murderers.

We have taken the battle to our enemy. We have destroyed the Taliban and disrupted the Al Qaeda network. Those who have not been killed or captured we have driven into hiding. We have liberated Afghanistan from the clutches of terrorists, and we have put the rest of the world on notice that to harbor terrorists is to invite disaster.

In these sterner times, we have rediscovered that we are made of sterner stuff.

Yesterday, I had the honor of helping to plant a memorial red ash tree at the Pentagon. That tree, and eight others like it planted at the site over the weekend, were propagated from parts of a champion red ash tree in Dowagiac, MI, named as such because it is the largest example known of its species. That champion is 450 years old and 21 feet around at its trunk. It spans the history of America. And, like the American spirit, it is indomitable.

At yesterday's ceremony, I remarked that we Americans are as well-rooted as that champion red ash in Dowagiac, and like its crown, our Nation's aspirations reach high into the skies above. The tendrils of democracy root us; our aspiration is an unquenchable desire for freedom—for ourselves and for all people everywhere.

Archibald MacLeish wrote,

> There are those who will say that the liberation of humanity, the freedom of man and mind, is nothing but a dream. They are right. It is the American dream.

We have shared that dream with the rest of the world.

For the better part of the last century, the United States and our allies fought a successful battle against the genocidal forces of fascism and totalitarianism. We defeated the Nazis. We won the cold war. In the bloody struggle between ideologies, democratic governments triumphed over repressive regimes.

This democracy of ours and our allies will prevail against the likes of Al Qaeda because the overwhelming majority of people in the world want freedom and justice and dignity and opportunity. America remains a beacon of hope to the oppressed everywhere. Our current generation of servicemen and women, and the American people generally, will meet the new challenges and threats that we face as a Nation as successfully as we met the challenges and threats of the last century.

The people who perished 1 year ago did not do so in vain. We will always remember them and, most important, we will honor them by carrying on that noble struggle for what has been called the American dream but what is actually humankind's dream.

Madam President, I ask unanimous consent that an article from the Detroit Free Press titled "Michigan's 16 Legacies" be printed in the *Record.*

[From the Detroit Free Press, Sept. 5, 2002]

MICHIGAN'S 16 LEGACIES

(By Sheryl James)

They were among America's best, brightest and happiest.

Many had attended some of the finest schools—Cranbrook, Detroit Country Day, the University of Michigan, Harvard, Yale, Princeton and Vanderbilt.

As teens, they walked the halls of schools from Cass Tech in Detroit to Traverse City High, all of them contributing, achieving, giving back. They were young scholars and financial wizards, technology gurus, National Honor Society members, athletes, musicians, champions of theater, contributors to their communities.

Most of them were well traveled—and well on their way to the kind of success that defines the American Dream. A few already had achieved that dream, with homes in Manhattan, book credits, TV appearances.

One of them survived the 1993 terrorist bombing of the World Trade Center.

They are gone now, these 16 terrorism victims who had significant Michigan ties. But their legacies live on—in their accomplishments and through their loved ones left behind.

FINANCIAL WHIZ KID ON FAST TRACK TO SUCCESS

Terence Adderley Jr., 22, had a head and a heart for finance by the time he was a teenager. Before even graduating from Detroit Country Day in 1997, he had started an investment club. His grandfather, William Russell Kelly, founded Kelly Services of Troy in 1946, and his father, Terence E. Adderley, is its president and chief executive officer.

Adderley, who grew up in Bloomfield Hills, took his love of finance to Vanderbilt University in Nashville, Tenn., where he earned a bachelor's degree in economics in May 2001. Soon after, he landed a job with Fred Alger Management Inc., which had offices on the 93rd floor of the trade center's north tower.

Joseph White, a University of Michigan business professor and former interim president, said Adderley had tremendous potential.

White knew Adderley through U–M alumnus David Alger, president of Fred Alger Management. As a young analyst for the company, "Ted was thriving," White said. "He loved what he was doing."

Besides his father, survivors include his mother, Mary Elizabeth; five sisters, and a grandmother.

EXECUTIVE HAD EYE FOR BEST, BRIGHTEST

David Alger, 57, president of Fred Alger Management Inc., was a familiar fixture on CNN, MSNBC and CNBC—and at many U–M events. He was a prominent alumnus and supporter of the U–M business school, where, White said, he was the spring commencement speaker in 1997 and served on the school's senior advisory board.

Alger loved grooming young people for business and often returned to his alma mater.

"David was a terrifically talented man," White said. "A colleague of mine said, 'I loved David's rational exuberance,' and she got it just right. That was David: very analytical and very optimistic . . .

"David encouraged people to participate in what he thought would be the biggest bull market in American history—in 1991. It was an incredible prediction."

Alger was born Dec. 15, 1943, in California but grew up in Grosse Pointe. He received his undergraduate degree from Harvard University and a master's in business administration from U–M in 1968. He joined his brother Fred's company, Fred Alger Management, in 1972 as an analyst. He eventually owned 20 percent of the business.

In 1995, Alger's brother moved to Geneva, Switzerland, and left him in control of the company's daily operations. At the time of Alger's death, the firm's assets had grown from $3 billion to $15 billion and its workforce from 82 to 220.

Alger, who owned homes in Manhattan and Tuxedo Park, N.Y., loved technology stocks and managed mutual funds that ranked near the top of the 1990s bull market. He often appeared on financial TV programs and wrote "Raging Bull: How to Invest in the Growth Stocks of the '90s."

On Sept. 11, Alger was working in company offices on the 93rd floor of the trade center's north tower. "There were 1,300 people at his funeral," said White, who attended. Alger is survived by his wife, Josephine; two daughters; his brother, and a sister.

UNFETTERED SPIRIT LOVED THE CITY LIFE

Eric Bennett, 29, a Flint native, caught the travel bug early when he took a trip overseas with a high school foreign language club. He traveled often afterward, said his mother, Kathy Bennett of Flint.

He visited Brazil, Puerto Rico, Rome, London and Paris, but Bennett also loved the Brooklyn, N.Y., brownstone where he lived, the big-city life in New York and his job as area vice president for Alliance Consulting Group. His office was on the 102nd floor of the trade center's north tower.

"From his home, he could see the towers, and from his desk at work, he could see Brooklyn," his mother said. "He just loved life."

In 1989, Bennett graduated from Flint's Kearsley High School, where he had been co-captain of the football team the year before. In 1993, he received a bachelor of science degree in computer information systems from Ferris State University. He also played football at Ferris State and earned an All-Midwest Intercollegiate Football Conference honorable mention in 1992.

In addition to his mother, he is survived by his father, Terry Bennett, and a sister.

WINGS FAN HELD FAMILY CLOSE TO HIS HEART

Frank Doyle, 39, formerly of New Boston and Bloomfield Hills, was a loyal Detroit Red Wings fan. He grew up playing hockey and was the varsity goalie from his first year on at Bowdoin College in Brunswick, ME. Later, he played on three mens hockey teams near his Englewood, N.J., home.

Doyle attended Huron High School in New Boston and graduated from the Cranbrook Schools in Bloomfield Hills in 1980. He earned a bachelor of arts degree from Bowdoin with majors in economics and government in 1985. He also earned a master's in business administration from New York University's Stern School of Business in 1993.

Doyle was senior vice president of the Keefe Bryuette & Woods brokerage in the trade center's south tower. He directed its equity and trading department and was on the company's board of directors.

Just before his death, Doyle was training for triathlons. "He was probably in the best shape of his life" and planned to run a triathlon the weekend after the terrorist attacks, said his wife, Kim Chedel. But, she said, Doyle mostly "loved being a dad" to their children, Zoe and Garrett, who were 3 and 16 months when their father died.

Doyle and Chedel had both escaped harm when the trade center was bombed by terrorists in 1993. Chedel, who then worked at a brokerage in a different part of the complex than her husband, escaped within 2 hours. She said she cried for hours while waiting for Doyle to emerge.

On Sept. 11, Doyle called Chedel after the first plane hit. He was on the 87th floor of the south tower—the second hit but the first to fall.

"He said, 'If you think we got rocked in '93, this was 10 times worse,' " Chedel said.

"I said, 'Frank, get out of there.' He said to me, 'The PA system said it was . . . more secure to stay in the building.' "

Doyle called his wife again at 9:22 a.m. "He said, 'Sweetie, we've gone up to the roof. The doors are locked, and we can't go down. I know you know this, but I love you . . . and you need to tell Zoe and Garrett every day for the rest of their lives how much Daddy loves them.' "

His survivors also include his mother, Maureen Doyle of New Boston, and three siblings.

TEACHER KEPT HER FRIENDS FOR A LIFETIME

Barbara Edwards, 58, who grew up in Wyoming, near Grand Rapids, was a woman of character and warmth. She kept friends for a lifetime and, as a high school teacher, showed up at her students' soccer games on her own time.

"Barb was a people person," said her sister Jane Gollan of Seattle. "If you met her 30 years ago, she would still be friends with you. She had a knack for keeping in touch."

Edwards also was a fan of Bette Midler and of personal mementos like the 40-year-old accordion she had as a child. She never threw anything out, and her garage never had room for a car, family members said.

Edwards, who lived in Las Vegas, graduated from Kelloggsville High School in Grand Rapids in 1961 and from Western Michigan University in 1965. She worked for a time at Catholic Central High School in Grand Rapids. She also married, had three children and lived in various states before divorcing in the early 1990s.

She was a high school French and German teacher in Las Vegas when she went to a friend's wedding in Connecticut the weekend before the Sept. 11 attacks. She was supposed to return home, but friends convinced her to stay for a couple of days. She wound up on American Airlines Flight 77, which left Dulles International Airport in suburban Washington, and crashed into the Pentagon.

Just before the attacks, one of Edwards' sons had left a job as a broker in the World Trade Center. In addition to her children, Edwards' survivors include her parents, Jack and Liss Vander Baan who live south of Grand Rapids in Allegan County; a sister, and two grandchildren.

AVID READER RELISHED HIS MICHIGAN ROOTS

Brad Hoorn, 22, originally from Richland, near Kalamazoo, never lost his affection for his favorite childhood book, "Charlotte's Web." He learned to read using that book and he reread it periodically, said his mother, Kathy Hoorn of Richland. A voracious reader, Hoorn often read an entire book to relax before an important college exam, she said.

Bright, energetic and outgoing, Hoorn played several musical instruments; had been president of the National Honor Society at Gull Lake High School in Richland, from which he graduated in 1997, and cocaptain of the tennis team.

He was a computer whiz, his mother said, and loved coming back to Michigan from his New York City apartment to golf with his father, Dennis; play with the family's two Labrador retrievers; visit friends and enjoy boating on lakes near the family home.

In May 2001, Hoorn received a bachelor's degree in economics from Yale University. On Sept. 11, he was working at Fred Alger Management Inc. on the 93rd floor of the north tower.

In addition to his parents, he is survived by a sister.

CONSULTANT MADE HER OWN WAY IN THE WORLD

Suzanne Kondratenko, 27, formerly of Romeo, had such zeal and spark, her sister called her a spitfire. Patricia Kondratenko said Suzanne was creative, independent and had a daring sense of humor.

"Things she would say, other people wouldn't get away with," Patricia Kondratenko of Rochester said. She especially remembers Suzanne's beautiful skin and how she always smelled like flowers.

Kondratenko and her sisters—all seven of them—attended the Academy of the Sacred Heart in Bloomfield Hills. Suzanne graduated in 1992. In 1996, she earned a bachelor's degree in English literature and humanistic studies from St. Mary's College in Notre Dame, Ind.

An employee of Keane Consulting in Chicago, Kondratenko was in New York on Sept. 11 to do consulting work for Aon Corp. on the 92nd floor of the trade center's south tower.

"Suzanne committed herself, entirely, to her every endeavor,' said her sister Aimee Kondratenko of Chicago. "She was capable of so much."

She is survived by her sisters and her parents, Eric and Patricia Kondratenko, of Romeo.

ACTRESS CAPTIVATED BY THE ALLURE OF THEATER

Margaret Mattic, 51, knew by her senior year at Cass Tech High in Detroit that she wanted to be an actress and live in New York. She accomplished that goal, and more. Before she died, she was writing plays and planning to produce them, her sister, Jean Neal of Detroit, said.

"My earliest memories of Margaret are of her performing in school plays," said Peggy Robinson, who grew up with Mattic on Detroit's east side. "She was always the lead. I was a narrator. When we did 'Snow White,' she was Snow White. And she was Gretel when we did 'Hansel and Gretel.' "

Mattic also was adventurous, Neal said. "She did more traveling and living away from home. The rest of us remained in Detroit. Margaret lived in New Orleans and New York."

While Mattic was a student at Wayne State University, where she received a bachelor's degree in liberal arts in 1973, "she traveled to Europe for 8 weeks, all by herself," Neal said. "She had more nerve than all of us."

Mattic, the youngest of five children, always loved to read and write, said her mother, Katie Mattic of Detroit. As an adult, she bought dozens of books for herself and for her mother.

After graduating from Cass Tech in the late 1960s, she appeared in several plays in Detroit and New York, particularly ones with African-American themes. Mattic worked as a customer service representative for General Telecom in the trade center's north tower.

She was single and had no children.

HE WAS ON WAY TO A HAWAII HONEYMOON

Robert R. Ploger III, 59, of Annandale, Va., approached his life's work with a sense of adventure, said his father, Maj. Gen. Robert Ploger of Ann Arbor.

He studied philosophy in college but wound up working with computers. He worked for major corporations, established his own successful company and finally became a computer architect for Lockheed-Martin in Washington.

Ploger's parents—his father is retired and his mother, Marguerite, is deceased—were originally from Owosso. Their son graduated in 1959 from Paris American High School in France and attended Michigan State University in 1959–60. He served in the U.S. Army from 1960–62. Ploger then earned a bachelor's degree in philosophy from the University of Denver in 1965, married and raised two children. He and his first wife, Sheila, later divorced.

Ploger had lived in California, Virginia and Maryland, working as a computer specialist.

He married his second wife, Zandra, in May 2001. On Sept. 11, they were on their way to Hawaii for a honeymoon. Both were aboard American Flight 77 when it crashed into the Pentagon. A memorial service was held at the same hotel in Annandale where the couple celebrated their wedding.

GENTLE GIANT WON PEOPLE OVER EASILY

David Pruim, 52, was "the kindest, nicest, most gentle, 6-foot-4 person there ever was," his wife of 28 years, Kate, told the *New York Times* shortly after his death. "He made everyone he came into contact with feel good about himself, from children to adults."

Pruim was senior vice president of risk services for Aon Corp., on the 103rd floor of the trade center's south tower.

The Pruims, both originally from Michigan, lived in Upper Montclair, N.J., with their 10-year-old daughter, Carrington. David was a 1966 graduate of Western Michigan Christian High School in Muskegon. He received a bachelor's degree in political science from Hope College in Holland in 1970.

Last October, the college dedicated its homecoming football game to Pruim.

He is survived by three brothers and his stepmother, Louise Pruim, who lives in Norton Shores. His late father, James Pruim, was mayor of Muskegon from 1994 to 1996.

BRILLIANT MAN HAD AN EMPATHETIC EAR

Josh Rosenthal, 44, was a brilliant guy with a big heart. He doted on his nieces and, despite a terrible allergy to cats, kept his Manhattan apartment window open to provide a scratching post for his neighbor's cats, his sister Helen Rosenthal recalled.

"He had this ability to reach in and really touch people and make them feel like he was truly listening and understanding what they were saying," she said.

Rosenthal was raised in Livonia and graduated from Stevenson High School in 1975. His mother, Marilynn Rosenthal of Ann Arbor, is a professor of behavioral sciences at the University of Michigan-Dearbon. His father, Avram (Skip) Rosenthal of Southfield, owns Books Abound in Farmington and is a former director of the Henry Ford Community College Library in Dearborn.

In 1979, Rosenthal earned a bachelor of arts degree in political science from the U–M in Ann Arbor. In 1977, he was named a Truman Scholar, a merit-based scholarship award given to outstanding college juniors.

In 1981, Rosenthal received a master's degree in public affairs from the Woodrow Wilson School of Public and International Affairs at Princeton University. He moved to New York afterward, and on Sept. 11, was at work as senior vice president of Fiduciary Trust Co. on the 90th floor of the trade center's south tower.

Rosenthal lived in Manhattan, but stayed involved with family and friends in Michigan. He was godfather to several children of friends his sister said.

"Josh had a large and wonderful family that he was very close to," his mother said.

ONETIME STAR PITCHER HAD NEW LIFE WITH WIFE

Brock Safronoff, 26, originally from Traverse City, worked as a computer systems analyst for Marsh & McLennan Cos. Inc. on the 96th floor of the trade center's north tower. He was a 1993 graduate of Traverse City Central High School, where he had been a star pitcher for the baseball team.

In 1997, he earned a bachelor's degree in chemistry from Amherst College in Massachusetts, where he also was a starting pitcher on the baseball team. Later, he finished course work toward a master's degree from Columbia University in New York.

Safronoff and his wife, Tara, were married in August 2001 on Staten Island. They had just moved to a Manhattan apartment before the Sept. 11 attacks. In addition to his wife, his survivors include his parents Joel and Debra Safronoff of Traverse City; a brother, and a sister.

MILITARY MAN HAD 2D CHILD ON THE WAY

Lt. Col. Kip Taylor, 38, originally from Marquette, came from a military family. He loved his job as an assistant to three-star Gen. Tim Maude at the Pentagon.

But at home, Taylor loved the gentler art of cooking. He especially loved trying out new recipes on unsuspecting visiting friends, his wife, Nancy, said.

Taylor also loved working with wood. He build a deck and worked on other projects at his McLean, VA., home. He had a 22-month-old son Dean. On Oct. 25, his wife gave birth to his second son, John Luke who will be called Luke.

"He considered his most important role as that of father," his wife said. "We were both so excited about the baby."

In 1985, Taylor graduated from Northern Michigan University with a bachelor's degree in management. He had two scholarships, one for basketball and one for the ROTC program.

He died when American Airlines Flight 77 crashed into the Pentagon. He was promoted to lieutenant colonel from major posthumously, his wife said.

His survivors also include his mother, Kay Taylor, who is executive director of a child care referral agency in Marquette; a brother, also in the Army, and a sister. Taylor's late father, Don, was a retired Army lieutenant colonel who gave his son his commission in 1985 and later taught military science at Northern Michigan University.

HORSE ENTHUSIAST COORDINATED SHOW

Lisa Marie Terry, 42, of Oakland Township found time every summer, no matter how busy she was, to hold her Summer Sizzler Horse Show in Mt. Pleasant.

It was not an easy task for Terry to put on a quarter horse event while working full time with Marsh Inc., a national construction company with offices in Detroit and in the trade center's north tower. But horses were her passion.

"She did it all by herself," said Sarah Tupper of Metamora, Terry's best friend and fellow horse enthusiast. "She worked on it all year, getting sponsorships, making it nice for exhibitors. She made a huge effort to put stuff in the show people wanted."

Terry especially encouraged young exhibitors, Tupper said.

On Sept. 11, Terry, a vice president in charge of construction for the New York-based Marsh, was going to Hawaii for a project. She stopped at the World Trade Center for a business meeting and was among some 300 Marsh employees lost in the attacks, company officials said.

A couple of years before her death. Terry, who was single, celebrated her rise to vice president—a rare position for women in her field—by buying a red BMW convertible, said her aunt, Olga Stevens of Troy. Soon after, she bought her Oakland Township home.

Terry had one brother and grew up in Troy, graduating from Troy High School in 1977. She studied a social services program at Ferris State University from 1977 to 1980 and was a member of the Theta Tau Alpha sorority. She studied insurance at Michigan State University in the mid-1980s.

She loved skiing, flowers and her two cats. An accomplished horsewoman, she showed for the American and Michigan Quarter Horse associations. The latter named her Sportswoman of the Year in 1993.

Terry also was a member of the American Business Women's Association and the National Association of Women in Construction.

FLIGHT ATTENDANT HAD THE HEART OF A CHILD

Alicia Nicole Titus, 28, whose parents live in Dexter, was a positive, peace-loving, let's-make-the-world-a-better-place kind of person.

"She was very much into acceptance of world cultures and . . . very embracing of people with different belief systems," said her father, John Titus. "So it is ironic, sadly so," he said, that she was a flight attendant on United Airlines Flight 175, which crashed into the trade center's south tower Sept. 11.

Titus had just become a flight attendant. Disillusioned with corporate life as a marketing director for a firm in San Francisco, where she lived, she switched careers, said her father, who is director of student advisement services at Schoolcraft College in Livonia. Alicia's mother, Beverly, teaches part-time at the college's Women's Resource Center and Continuing Education Department.

Titus grew up in St. Paris, Ohio. She graduated from Graham High School there in 1991 and earned a bachelor's degree in international business from Miami University in Oxford, OH, in 1995.

"She had lead roles in musicals in high school and was into track, cheerleading and National Honor Society," her father said.

She also loved outdoor sports such as snowboarding, mountain climbing and sky diving. She had traveled to Spain, Morocco, England and France.

Titus' roommate in San Francisco told her parents that the Sunday before the attacks, the two went to a local park, where they twirled hula hoops and played on the swings and slides—typical of Titus' childlike zest for life, her father said.

Titus, who was single, also is survived by two brothers; a sister, and a nephew.

RECENT GRADUATE HAD ENERGY, POTENTIAL

Meredith Lynn Whalen, 23, who was originally from Canton Township, loved animals, particularly horses. She always wanted to own a horse, said her mother, Pat Whalen of Canton.

But Whalen valued friends most of all. Her mother was comforted by her daughter's friends after the Sept. 11 attacks.

"Her friends have all described her as a very energetic, caring person with a lot of compassion for others," Pat Whalen said.

Whalen was just as energetic in high school. She was a varsity swimmer and in the National Honor Society at Plymouth Salem High School, from which she graduated in 1996. She earned a bachelor's degree in business administration with honors from the U–M in Ann Arbor in 2000.

"Meredith was an outstanding graduate of our 2000 BBA program," said White, the U–M business professor and former interim president.

White said David Alger, another U–M graduate and World Trade Center victim, spotted Whalen as a young person with great potential and convinced her to work for his company.

She became a research assistant for Fred Alger Management on the 93rd floor of the trade center's north tower and lived in Hoboken, N.J.

She is survived by her mother; three sisters, and a brother. Her late father, Henry (Hank) Whalen, had been a Canton Township trustee.

Hon. Olympia J. Snowe

OF MAINE

A few short hours ago, at the Pentagon, we bore witness to the essence of this solemn anniversary. It was a message of restoration and renewal from a grateful Nation.

Today is the commemoration of both incalculable loss and limitless courage, of enduring sorrow and indomitable spirit.

We seek to honor the bravery and heroism displayed by so many for so long on this day and the days following. We are moved to grieve for what and whom we lost—such as Maine victims Anna Allison, Carol Flyzik, Robert Jalbert, James Roux, Stephen Ward, Robert Norton, Jacqueline Norton, and Robert Schlegel.

Today, we embrace all that we have retained as a Nation—our strength, our sense of purpose, our unity, and our veneration of the principles of freedom and justice—for today, the hearts of Americans and freedom-loving people across the globe are beckoned at once by sorrow and resolve, and we should heed the call of both.

The snapshots of insanity etched in our minds, the indelible stain of unfathomable inhumanity, these must remain if we are to triumph over the tyranny of terror, and triumph we must.

In a horrific irony, the forces of darkness had their way on an especially bright and beautiful morning, much like today, and the evil that fueled their horrible deeds lives on in the shadows of the world. The struggle before us will be constant, and therefore our vigilance must be unflagging.

So on this first anniversary of a new era, let us continue to brace ourselves to perpetuating what is good and just, as we and our allies did in the 20th century's great struggles against evil. And let us remember how that one day in September not only changed America and the world but also reminded us of what really matters, of the principles and the people we value and certainly should appreciate—our firefighters, police officers, rescue workers, our troops, and seemingly ordinary Americans who, when faced with the horrible certainty of their circumstances, knowingly bring down a plane to save the lives of others, not to mention the very symbols of our democracy, the Capitol and the White House.

On this solemn occasion, we celebrate those heroes who walk among us today, while the legacy of those who made the ultimate sacrifice reverberates throughout New York, Washington, Virginia, Pennsylvania, and every town and city in this land. Indeed, if it is true that a nation is defined by its response to adversity, then America redefined its own greatness.

Men and women searching and clearing the World Trade Center site worked day and night, while volunteers brought them food and water. Their labor will stand as a memorial beside the hallowed site's eternal flame near the hole in the Earth that mirrors the hole in our heart that will never fully mend. And just across the Potomac, engineers and construction workers poured forth every last ounce to rebuild the Pentagon within 1 year in a gesture comprised of equal parts defiance and pride. At the building's D-ring, a father literally helped repair the broken stone and mortar near the very spot where his son perished that fateful morning.

What is lost can never be recovered, but with this first anniversary of September 11, it is as though life has reclaimed its rightful place where destruction dared intrude. At the Pentagon there is a single blackened stone set within the new wall to symbolize what cannot and must not ever be forgotten. We have witnessed an almost incomprehensible transformation from the blackened devastation we saw a year ago, just as America itself has been transformed.

An unparalleled sense of unity and compassion swept across America, proving once again that the true strength of our Nation has always flowed from the fortitude of our people.

As we lifted up the hearts of those grieving for loved ones, we moved toward a swift and just defense of our freedom, and the President worked vigorously and mightily to build an international coalition. And while the war on terror will unquestionably be long and dangerous, our heroic men and women in uniform struck quickly and decisively at the heart of the Taliban.

In February, I had the privilege of visiting our troops and meeting with President Karzai as part of a congressional trip to Afghanistan. What left the most profound impression on me, one I will never forget, was the unflinching commitment, the indefatigable resolve, and highest level of professionalism, not to mention bravery, of our troops.

Indeed, much was revealed to us on the morning of September 11, 2001: The extent of the threat against us, the image of the devil incarnate, but also the face of a resilient and passionate and a united Nation that would not allow this travesty to stand.

We have learned that we can continue the process of healing, even knowing we will never fully be healed. We have learned we can move forward, without moving away from the anger we justifiably feel. Indeed, if we are to properly memorialize those whom we lost on that day and the days since, then we must maintain a boundless resolve in perpetuity that is so essential to keeping America secure and eradicate the roots of terrorism and the bloody instruments of fear.

At Gettysburg, President Lincoln said:

> It is for us, the living, rather to be dedicated here to the unfinished work which they have, thus far, so nobly carried on. It is rather for us to be here dedicated to the great task remaining before us—

That is our call yet again today. That is the destiny to which we must rise. Now, like then, we are equal to the challenge. God bless America.

Hon. Byron L. Dorgan

OF NORTH DAKOTA

Madam President, I was not sure if I was going to come and say a few words today. I am almost fatigued by the coverage of 9/11, and yet there is something so important about this moment that silence somehow is not an appropriate response.

The horror of the moment of September 11 last year remains with us even as we ache in our heart for those who lost their lives on that day. I think all of us understand the target was not buildings. It was not buildings in New York or Washington, DC. The target was the spirit of our country. The target was democracy. The target was Americans and what Americans represent.

With the 1-year anniversary of that event, it is important for our country again to take stock of where we are, who we are, and what our citizenship responsibilities are as Americans.

One year ago, I left the Capitol Building late at night to drive home and drove past the Pentagon. It was still burning, with smoke billowing out of the Pentagon that was then bathed in floodlights. It was an eerie sight to see the fire at the Pentagon even late at night and to hear and see the F–15 and F–16 fighter planes flying combat missions over our city and the Capitol that day and that evening.

We went back to the Pentagon a few days later, and we were, as Members of Congress, meeting with Pentagon officials and viewing the damage. As we were there, one young Marine was hanging by a crane in a bucket up near what had been the fourth floor of the Pentagon in what now was an open wound and gash in that concrete building. He had been hoisted up in the bucket by a crane that was moving toward this open gash. As we watched him, he reached around into this open area where this airplane hit and he pulled out a flag he had spotted up in an open area that had miraculously burned, and he brought this red and gold flag, which was the Marine flag, a brilliant red and gold color. He had the crane lower him to the ground. He marched over to where we were, walked past us and said: I am going to give this flag to the Marine commandant. He said: Terrorists could not destroy this flag, and they cannot destroy this country.

I think the spirit of that young Marine and the spirit of people at Ground Zero, where we visited a week following the attacks, is something I will always remember.

The visit to Ground Zero that many in Congress conducted was a very sad visit, showing the carnage and destruction of the World Trade Center where so many thousands died. The event I recall from that day, among many, was a firefighter who came to me with a several-day growth of beard and bloodshot eyes. He had worked around the clock for many days. He told me of the friends he had lost, those who were his fellow firefighters who had died in the tragedy. Here was a man who obviously had very little rest, had worked day and night. Through his bloodshot eyes and with a uniform that was quite dirty, having worked around the clock, he looked at me and said: Senator, you must promise me to do one thing.

I asked: What is that?

He said: Get them. Get them. If you do not get them, they will do this to this country again.

He represented the feeling of all Americans. We must make certain that terrorists are not able to do this again in our country. Our country is, in my judgment, as united as ever, united to battle terrorism wherever it exists in the world. We have come to understand as a country that a battle against terrorism is not quick. It is not easy. But it is something to which all America is committed. Every fanatic anywhere in the world who thinks terrorism is an acceptable means to an end needs to hear and know that America is united.

My State is half a continent away from Washington, DC., and New York City where the attacks took place. Let me speak for a moment about my rural State, so distant from the urban areas where the attacks took place.

First, tragically, we too experienced the loss of life. A young North Dakotan, Ann Nelson of Stanley, ND, died when the World Trade Center collapsed. I knew Ann and her family. She was a very special young woman. Her father has been a good friend of mine for many years. She had a bright future ahead of her, and she was a joy to all who knew her. Ann Nelson was a young North Dakota woman seeking a career, pursuing a job in the World Trade Center in New York City. She died because she was an American. She was one of thousands of innocent Americans who lost their lives because of these heinous, unspeakable acts of horror committed by terrorists.

The day of the attack in Washington, DC, I looked up in the sky to see fighter jets flying overhead. I found out later that day they were pilots from Fargo, ND, members of the Air National Guard called the Happy Hooligans, some of the best fighter pilots in the world. Over the years, they have won three William Tell Awards which is the award for the best fighter pilots. They are stationed on a rotating basis at Langley Air Force Base.

Part of their mission is to protect the Nation's Capital. They were the first scrambled and the first in the air and the first over the Nation's Capital on that very day, 9/11. We are so proud of them. They are the ones we saw in the air almost immediately after these attacks.

Their mission, I told them, reminds me of something I read some long while ago. I don't know where it comes from, a verse that said: When the night is full of knives, and the lightning is seen, and the drums are heard, the patriots are always there, ready to fight and ready to die, if necessary, for freedom.

A lot of patriots last September 11 said: I'm here and I'm ready to fight for freedom.

In the weeks and months that followed the attacks of September 11, our country has come together like never before. We understand that we face a very special and unusual challenge. We are a big, wide open, free democracy. It is very hard to provide absolute security in every way, every day all around our country. We do not want any of us to diminish the basic freedoms that represent America, the basic freedoms in our Constitution. We do not want to diminish those freedoms in order to fight terrorism. We want to fight terrorism and eradicate terrorism wherever it exists. We want to preserve that which makes America unique, the most wonderful country on the face of the Earth.

A year ago when I spoke in this Chamber about the events of September 11, I recalled the words that Shakespeare wrote: "Grief hath changed me since you saw me last."

It continues to change us as we go forward carrying the heavy burden that grief imposes. But part of that change is much more than grief. Part of it is a steely resolve to respond forcefully and strongly and with precision to those who carried out the attacks 1 year ago. Our resolve in this year, in my judgment, has grown even stronger.

This will be a day that Americans will think about for many decades to come, the September 11 anniversary of 2001. My hope is we rededicate ourselves today to the mission ahead and the challenge ahead to preserve our democracy. My hope is that today we also pay honor to the memory of those innocent Americans who lost their lives, and then say thank you to all of the heroes who, on September 11, extended forward and said: Let it be me to reach out and help. And especially we say thank you to the men and women of the Armed Forces who serve in harm's way all across the world.

Finally, months after September 11, I was in Afghanistan, Baghram, Kabul, and that region of the world. As you fly into Afghanistan and look down through an airplane window to the hills and the mountains and understand that somewhere in caves deep in the mountains there were people plotting the murder of innocent Americans, you understand we cannot ever be oblivious to what is happening in the rest of the world. It can have a profound impact on the lives of those who cherish freedom.

I say to the young men and women I met in Kabul and Baghram and elsewhere, thanks for your service to America. Thanks for helping us wage the fight against terrorism, a fight this country is determined and destined to win.

Hon. Richard C. Shelby

OF ALABAMA

Madam President, 1 year ago, America awoke to a tragic and devastating incarnation of hatred and evil. Incomprehensible to most citizens only a day before, the terrorist attacks of September 11 dramatically changed our people, our country, and the entire world.

Insulated for over 50 years from foreign attacks on our soil, Americans in an instant grasped the magnitude of the threat we face from terrorism. In the days after the attacks, the dangerous world in which we live never seemed more precarious.

The immediate aftermath brought a tremendous outpouring of grief and sorrow.

Our Nation mourned as the realities of the events of September 11 penetrated our collective psyche. Candles were lit in remembrance and flags were flown in patriotic displays of unity. Stories of bravery and courage emerged in the wake of the attacks which helped to inspire and remind us of all that is great about the American spirit.

This foundation of strength which was built in the days after the attacks prepared us for the challenges ahead, and helped harden our national resolve to deliver our enemies to justice.

We live in a far different world than the one we occupied just 12 months ago.

With a clear sense of purpose our country has engaged the war on terrorism on every conceivable front. The vision outlined by President Bush in his September 20 address to the Nation has been undertaken with extraordinary success. We have been vigilant in our fight to hunt down those responsible for the attacks, as well as those who might do us harm in the future. We have fortified our defenses and reorganized our government. Americans everywhere are more aware of their surroundings and remain defiant of those wishing to do us harm. We as a Nation have grown stronger and more united than ever.

We have been blessed with enormous freedoms and prosperity in this country. Over the course of our history, many Americans have made the ultimate sacrifice by giving their lives to protect our freedoms. Although we have enjoyed many years of peace, the events of September 11 showed us that this fight is far from over.

We must continue to build on the successes of the past year, and never become complacent with our victories. The burden cast upon our great Nation was one we neither asked for nor deserved, but we carry it on our shoulders consoled by the memories of those who went before us who sacrificed in the name of freedom.

Hon. Arlen Specter

OF PENNSYLVANIA

Madam President, Senator Santorum, Senator Ensign, and I have just returned from memorial services in Shanksville, PA, commemorating the downing of flight 93. It was truly an inspirational and emotional occurrence. The families of the victims of flight 93, the crew and passengers, were seated front and center, and then a large crowd was assembled, estimated in advance to be in the range of 20,000 to 30,000 people.

Gov. Mark Schweiker, Governor of Pennsylvania, spoke, as did former Gov. Tom Ridge, now the Homeland Security Director. There was not a dry eye in the entire assemblage. The message delivered by Governor Schweiker and Governor Ridge was a moment of remembrance, a moment of commemoration, and a moment of hope for the future, with a determination that a united America will repel terrorists wherever terrorists exist and that the struggle for freedom will be maintained and will be won.

Governor Schweiker went to the Shanksville Elementary School in advance of the ceremony and brought to the assemblage, especially the families of victims of flight 93, this message from the Shanksville Elementary School: If God brought you to it, God will bring you through it.

That brought quite a response.

Churchill was quoted, I think, so appropriately: "Never was so much owed by so many to so few."

I think that is especially applicable to the Members of the Senate and the Members of the House of Representatives because flight 93 was headed to the U.S. Capitol. That had long been the speculation, and it was confirmed 2 days ago in an article in the *New York Times*, quoting members of Al Qaeda.

Ms. Sandy Dahl, wife of pilot Jason Dahl, made an emotional speech and later came down and sat right next to where I was sitting and was holding her infant daughter, Michaela, who will be 2 at the end of September. It was quite a poignant picture of the widow, grieving for what happened a year ago today, but holding her child and looking forward to the future. The child was smiling, and so was Mrs. Dahl, looking at her infant daughter.

It is my hope that the Congress will yet act on legislation which has been introduced to grant Congressional Gold Medals to the 40 who were crew and passengers of flight 93. As I moved through the assembled ladies and gentlemen who were families of the victims and spoke to them and heard of their grief, the common thread was: Please move ahead. Thank you for the legislation—thanking the Congress for the legislation authorizing the creation of a memorial at Shanksville, a national memorial site, but also asking that our legislation for the Congressional Gold Medals be completed.

The family of Georgine Rose Corrigan presented me with this photo and the ribbons, red, white, and blue. These photos were worn by so many—virtually all of the families of the victims who were in attendance.

Yesterday, I spoke on the floor of the Senate and said that sometime before dusk today I would ask unanimous consent for the consideration of the bill S. 1434, which has 69 co sponsors, which would grant the Congressional Gold Medal posthumously to the victims of flight 93. This bill should have been moved a long time ago. I have taken it up with the appropriate Senators to get it moved, and it has not moved because of the interest of some in expanding it to cover other victims—the firefighters, the police, and others.

I certainly think it would be appropriate to grant recognition to all of those people. However, I think the victims of flight 93 are in a special category because they saved the Capitol.

In order to avoid the complications of having a bill discharged from committee, I have consulted with the Parliamentarian as to the procedures for having a bill held at the desk.

MEASURE READ THE FIRST TIME—S. 1434

Mr. SPECTER. At this time, I submit on behalf of myself and 69 cosponsors, a bill to authorize the President to award posthumously the Congressional Gold Medal to the passengers and crew of United Airlines flight 93 in the aftermath of the terrorist attack on the United States on September 11, 2001.

The PRESIDING OFFICER. The bill will be read for the first time.

The assistant legislative clerk read as follows:

> A bill (S. 1434) to authorize the President to award the Congressional Gold Medal to the passengers and crew of United Airlines flight 93 in the aftermath of the terrorist attack on the United States on September 11, 2001.

Mr. SPECTER. Madam President, for purposes of completing the procedure, I intend to object after asking the bill be read the second time.

The PRESIDING OFFICER. Objection has been heard.

Mr. SPECTER. Madam President, I ask unanimous consent that at the conclusion of my remarks the program for the commemorative ceremony in Shanksville be printed in the *Congressional Record*, and I yield the floor.

There being no objection, the material was ordered to be printed in the *Record*, as follows:

"A Time For Honor and Hope"—One Year Anniversary Memorial Service, Wednesday, September 11, 2002

The County of Somerset wishes to express heartfelt thanks to all who have come forward to assist, contribute and participate in the One Year Anniversary Memorial Service. We would also like to extend a special thank you to these sponsors:

DIAMOND LEVEL

United Airlines Corporation
Deitrick & Associates Interiors, Inc.

UNDERWRITER LEVEL

Davis Bus Lines/STA
Fedex Ground
Meyer Sound Labs
Northern Sound & Light, Inc.

GOLD LEVEL

BMW of North America
Under the Sun Productions, Inc.

SILVER LEVEL

Benack Sound Productions, Inc.
Hershey Foods
Thoratec
Daily American Newspaper
Merrill Lynch-Southwestern Pennsylvania Offices

BRONZE LEVEL

Aurora Casket Company
Community Foundation for the Alleghenies
Don & Becky Kelley
Ohio Gratings
Seven Springs Mountain Resort
Staley Communications

PEWTER LEVEL

Mississippi Association of Supervisors
Somerset Trust Company
Pennsylvania Funeral Directors Association
Kendall, Inc.
Roth Brothers
Color ID
Baw Plastics, Inc.

HEARTFELT SUPPORT LEVEL

Ironworkers Local Union 46
Radio Shack Corporation
Rockwood Area School

One Year Anniversary Memorial Service—9:30 AM

MUSICAL SELECTIONS

Johnstown Symphony Orchestra and the 2d Marine Aircraft Wings Band, 2d Marine Aircraft Wing, Cherry Point, NC

"OLD GLORY" FLAG PRESENTATION

101st Airborne Division (Air Assault), Fort Campbell, KY
Remarks by Mr. Dave Pawlewicz, Century Link America

PRESENTATION OF THE COLORS

U.S. Marine Corp Honor Guard, 2d Marine Aircraft Wing, Cherry Point, NC

PLEDGE OF ALLEGIANCE

Miss Priscilla Gordeuk and Mr. Elwood Brant, Top Honor Senior Students, Shanksville-Stonycreek School District

NATIONAL ANTHEM

Ms. Jeanne Wentworth

FLYOVER

C–130s—911th Airlift Wing, Pittsburgh International Airport Air Reserve Station

REMARKS

Sandy Dahl, Wife of Flight 93 Pilot, Jason Dahl

"ONE MINUTE OF SILENCE FOR WORLDWIDE PEACE"

Murial Borza, Sister of Flight 93 Passenger, Deora Bodley

10:06 AM—"TOLL THE BELLS"

Br. David W. Schlatter, O.F.M., Bells of Remembrance toll forty times as names of passengers and crew are read.
Presentation of Names: Mr. Tony Mowod

INTRODUCTION OF GOVERNOR MARK SCHWEIKER

REMARKS

Governor Mark Schweiker

INTRODUCTION OF KEYNOTE SPEAKER

Governor Tom Ridge, Homeland Security Advisor

REMARKS

Governor Tom Ridge, Homeland Security Advisor

TWENTY-ONE GUN SALUTE

Marines of the 4th Marine Aircraft Wing, Site Support Element, Johnstown, PA

ECHO TAPS

2d Marine Aircraft Wing Band, 2d Marine Aircraft Wing, Cherry Point, NC

MUSICAL SELECTION

"God Bless America"—2d Marine Aircraft Wing Band, 2d Marine Aircraft Wing, Cherry Point, NC and public

AIR FORCE FLYOVER WITH MISSING MAN FORMATION

AT–38's, 80th Flying Training Wing (80FTGW), 88th Flying Training Squadron (88FTS) Sheppard AFB, TX

HEROES MEMORIAL RELEASE

Release of forty white birds representing the fallen heroes of flight 93

"GOD BLESS AMERICA"

Ms. Jeanne Wentworth, leading assembled public and Johnstown Symphony Orchestra

FORMAL RETIREMENT OF THE COLORS

Honor Guard 2d Marine Aircraft Wing, Cherry Point, NC

DEPARTURE OF FLIGHT 93 FAMILIES

DEPARTURE OF DIGNITARIES

PUBLIC DEPARTURE

Hon. Tim Johnson

OF SOUTH DAKOTA

Madam President, I rise today to commemorate the vicious terrorist attacks that occurred on our Nation a year ago today, September 11, 2001.

Over the past few days, in large cities and small towns, in New York and the Pentagon today, in my home State of South Dakota, people have gathered to remember and to reflect upon what happened this last year.

In many ways it still does not seem possible that a year has passed since the attacks of September 11, 2001. It seems too soon to look back and reflect on the meaning of September 11 because the events of that day still echo in our daily lives. The wound is still too fresh, the loss too great. Every American will remember exactly where they were when they realized that our Nation was under attack. The images of that day will be with us forever: The burning buildings, the endless television footage of airplanes crashing into the Twin Towers, the images of New York covered in rubble.

I will always remember the smoke of the Pentagon as it appeared through a too-perfect blue sky here in Washington, DC. But other thoughts—candlelight vigils, American flags adorning buildings, fences, and barns, the heroism of the passengers of American flight 93, and the lasting bravery of firefighters and other first responders—will also remain in our hearts as constant reminders of that day in September.

Across the Nation and around the world, people came together to do what they could for the victims and their families. I was proud, but not at all surprised, when South Dakotans overwhelmingly responded with offers of blood and other assistance. Millions of dollars were donated by South Dakotans to relief organizations, and thousands of hours of time were volunteered in efforts to aid survivors and the family members of victims. Employees at Luverne Fire Apparatus in Brandon put in 2,000 hours of their own time to build a fire truck that was donated to New York City. Red Cross volunteers from Rapid City assisted in recovery efforts at the Pentagon. Farmers and ranchers throughout the State sold cattle and grain at auctions and livestock drives and donated that money to relief efforts. In one extraordinary example, Don and Adeline Hight of Murdo sold 100 calves and donated the proceeds, about $40,000, to help victims of the terrorist attacks. In Brown County, the Rural American Patriot Fund used the money they collected from fellow farmers and ranchers to buy thousands of dollars in patriot bonds. The idea of patriot bonds began with a call from a South Dakotan to my office, and were approved by the Treasury Department last December. Patriot bonds, similar to World War II war bonds, allow Americans to support the relief and recovery efforts at the World Trade Center and the Pentagon and to help fund the war on terrorism.

South Dakotans also helped to ease the emotional strain that the attack had on survivors and the victims' families. Police officers specializing in stress management from Mitchell and Yankton went to New York to help the police officers there deal with the emotional aftermath of the terrorist attacks. Lance Fillspipe, Junior Rodriguez, and eight other police officers from Pine Ridge Indian Reservation traveled to New York to help the police there handle security. Bonnie Riggenbach and Bob Holmes of Rapid City, both therapists, traveled to New York to do what they could to help people mend their lives. The Disaster Mental Health Institute at the University of South Dakota went to New York City in the wake of the September 11 attacks and played a key role in helping the recovery process. Students at Mount Marty College put together a banner signed by members of the community with words of sympathy and support for the City of New York. That banner is being considered by the *Guinness Book of World Records* as being the largest handmade banner ever made. Through gestures large and small, South Dakotans united with their neighbors and worked to bring something positive from all of the terrible destruction.

A lot has changed in our country, and in our world, since September 11. Our Nation has learned, to our vast sorrow, that we were not as untouchable as we had believed ourselves to be. Our country is involved in a war against terror that has taken our courageous military men and women, including my son Brooks, to Afghanistan as well as other far-off corners of the world. Our military effort in Afghanistan has helped to free people who were oppressed by a dictatorial regime that, in addition to the atrocities that the government inflicted upon its own people, harbored a terrorist group representing the worst humanity can become.

Here on the home front, things have changed as well. Barricades have been erected around national monuments, the Capitol, and the White House. Lines are longer and security more thorough at airports. Despite the longer lines and tighter security, our Nation still moves and functions much as it has for the last 225 years. We remain a beacon of democracy and justice for much of the world, and I work very hard as a Senator to make certain that new regulations, however necessary they may be in our post-September 11 world, do not infringe upon the basic rights of our citizens that we seek to secure.

So as we take this day to reflect upon the many lives lost last year, we are to reflect on the courage and heroism of those who did so much to save lives and defend our liberty today. We take comfort that the terrorists' goals were not realized—that there were attacks on buildings, but there were also attacks on everything America stands for—on individual liberty, on religious tolerance, on democracy, on free speech, and all the rights of our Constitution. These forces of hate, these forces of intolerance tried to destroy the very things that make this Nation strong. Buildings are being repaired but, more important, the light of democracy that holds this Nation and our fundamental values together burns just as brightly as it ever has.

The United States took a hit, but we have responded aggressively, and America will remain a beacon of liberty and freedom for the world forever after.

Hon. Robert F. Bennett

OF UTAH

Madam President, as I contemplated what I might say here today, I went back to my computer and pulled up a letter I wrote on September 22, 2001, in response to a friend who corresponded with me at that difficult time. He is a history professor. I have known him since high school. I have now decided to share that letter with the Senate and with the country as I look back on it after a year's time. I think it reflects better than anything I could create now not only my feelings at the time but my concern for where we should be and what we should be doing.

I read the letter dated September 22, 2001:

I have your letter, dear friend, and am moved to reply in depth using you as my audience as I make an attempt to speak to all the issues I see.

I start with the President's address to the Congress last Thursday. It has been called historic, one of the greatest Presidential addresses ever given. It certainly had that character and, listening to it in a packed chamber more hushed than I can ever remember, I was struck by the power, eloquence and directness of his words. From a distance of a day or two, however, as I analyze it and discuss it with my colleagues, I realize that the speech was more than moving rhetoric. For the first time since the end of the cold war, a U.S. President has laid out a clear statement of what the world is now like and what the U.S. role must be in that world.

The fact that this has not been done before is less a knock on previous Presidents than a realization that, until September 11, Americans in general were probably not yet free of all our cold war illusions. We are the world's only remaining superpower, we told ourselves. We are a just compassionate nation, we said. Ergo, we reasoned, it follows that, under our vigorous stewardship, the world itself will become a just and compassionate place, albeit little by little.

With Hitler and Stalin and Mao all dead, we thought, with the Soviet Union gone, evil—true malevolence—has gone from the world stage. It only pops up here and there in the form of an isolated Serb or Somali warlord.

No more.

Now we know that evil is alive and thriving, still threatening the peace everywhere in the world. Irrational hatred has not disappeared. The same mindset of fanaticism that built gas chambers 60 years ago is now hijacking airplanes and flying them into buildings, overseeing the preparation of chemical and biological weapons of mass destruction. As the President made so starkly clear, the world's new enemies hate freedom as much as Hitler did, and are prepared to kill millions as much as Stalin did. Evil has not gone away; it has simply changed its political language and its physical address.

Our cold war mentality told us that the trouble in the Middle East was about Israel, about power politics between established nation states, about borders and economics and markets. I readily agree that Israel has real problems with her neighbors, and they with her, but this is not about those problems. It is not about Israel. It is about defending the helpless against evil.

President Bush told us that America is feeding the poor in Afghanistan. I didn't know that before. Since his speech I have been in briefings from those familiar with the region who tell us that the Taliban uses food as a weapon, denying it to those that oppose them. They say they hate us for our support of Israel, but they also hate us because we are trying to feed the starving in their own country, and thus undermine their effort to starve everyone into submission.

They hate us because we profane their world with our notions of freedom—we "pollute the holy places" with business people and diplomats who let women drive and appear in public with bare faces. They hate us because we take the youth of all countries, including theirs, into our universities and teach them about science and economics and democracy, as well as about blue jeans and movies and freedom to travel and open debate. President Bush said it better than I can, and it was necessary for him to lay all that out if the Country is to "get" what we are facing.

The President spoke of the diplomatic front in this war, of our need for partners . . . . He reported good progress there, citing Pakistan as an example. In the same briefings that told us about food shipments to Afghanistan I learned that the current leaders of Pakistan really don't have much of a choice in this fight because they are a target themselves. They hardly qualify as democrats by our definition, but the radicals still hate them for even their tenuous ties to us. By some estimates, the radicals are close to bringing the government down and turning Pakistan into another Afghanistan. The same is true, in terms of the radical's end goal, in Egypt, Jordan, Saudi Arabia and elsewhere.

The leaders of those countries know they are at risk, and have been for some time. Sadat was murdered because he was willing to go to Jerusalem and say, "No more war." Those leaders need our help and are willing to help us in return because, long term, they know that the only nation with the capacity to lead a worldwide campaign to eliminate this evil is ours. The success we are having in building a coalition of partners in the first days of this conflict is one of the most encouraging signs that things are, for the moment, going somewhat well.

You are a historian; you know that the Second World War didn't begin on December 7. Neither did this one begin on September 11. As was the case with Europeans in the 30s, Americans have been in foreign policy denial in the 90s. Thrilled with the demise of our four-decade Soviet enemy, we read articles about the "end of history" and ignored the signs that were there to be seen. Now we have to go back and examine those signs . . .

We must realize that we are truly at war, and, as was the case in 1941, really have been for some time. The Embassy bombings, the attack on the USS *Cole*, the intelligence warnings—all these should have told us that this is a war and not, as some of the commentators have described it, a law enforcement problem. What's the difference? . . . In a war you focus on prevention of attack, not punishment. You . . . go after the enemy's assets to destroy them before they can be used to destroy you, gather the best intelligence you can and then play hunches and probabilities. You don't give out Miranda warnings.

Please accept my assurance that our leaders know how different and difficult this war will be. They know that we have to have partners, and that many of these partners have internal problems that will prevent them from being the kind of "allies" on which we could traditionally count . . . The team that President Bush has assembled is experienced, intellectually nuanced in its understanding, and deep. Down below the level of Cheney, Powell, Rumsfeld and Rice is a significant bench of very solid players who understand what we are up against . . .

So there we are. It has fallen our lot . . . to be the leader of the free world in a struggle that is global and against an enemy that is fanatic, decentralized, persistent, completely fearless and very, very patient . . . bin Laden and his fellow fanatics have decided that they can defeat [us] . . . by keeping intact their capacity to visit horror on us at unexpected times. We will not have won until that capacity is destroyed. This will be a very long, tricky and difficult fight.

But, as the President said, we will win it. And it will be worth it. The stakes are nothing less than they were in 1941 and through the cold war years, for us and for all the rest of those who want to live in freedom.

Madam President, reading that a year later, I still feel the same way.

Hon. Joseph I. Lieberman

OF CONNECTICUT

Madam President, it is truly an honor to have the opportunity to come and speak on the floor of the Senate today. This is one of those days when the Senate Chamber is really the people's forum, when the partisan or ideological or regional or whatever differences that sometimes separate us in votes fall aside and we stand here before the Chair, under the slogan that describes us—"E pluribus unum," "Out of many, one"—and truly represent the common and shared values, hopes, and aspirations of our people.

I am proud of what my colleagues have said thus far in this discussion and very grateful to be a part of it.

In New York City today, they are reading the names of the victims, the names of 3,000 of God's children, magnificent in their characteristic American diversity, whose lives were savagely taken on September 11 of last year simply because they were American. None of us here can say anything as powerful or profound as the recitation of those names today.

The Pentagon, the World Trade Center, and that field outside Shanksville, PA, will forever be hallowed battlegrounds, places where we will take our children to stand silently and contemplate their meaning, places of worship, really, where we will revere the lives lost and honor their place in our history.

This morning, as I left the very moving and unifying commemorative rededication service at the Pentagon, I came across a family, and I said hello and shook their hands, and I realized these were survivors of a man killed in the Pentagon on September 11 of last year: A young boy about 10 or 12 years old, full of innocence and youth, a great looking kid, carrying an American flag in a case—I presume the flag that was either placed over his father's coffin or given to him in memory of his father—a woman, who was the wife of the deceased and his parents—strong American stock—a man wearing the cap of a veteran, tears under his eyes. And there it was: A son without a father, a woman without her husband, parents without their child. I was speechless. There was nothing I could say except to shake their hands and put my hands over my heart.

In some ways, silence is a more appropriate response to the dreadful losses that were suffered on September 11. Silence, somehow, speaks more loudly to the horror and the complicated feelings that we all had on that day. Nevertheless, we must speak, to reflect on what happened that day, in the year that has passed, and to try to learn from that day and chart our way forward.

Madam President, our enemies hoped that September 11, 2001, would be the first page of a new chapter in world history: The end of the American century; the end of America as we know it; the beginning of a civilizational conflict, based on theological differences, taken to an inhumane extreme, which would end in the victory of radical extreme Islam.

As a distinguished Muslim citizen of Connecticut said yesterday at a public ceremony, Al Qaeda hijacked his religion.

In this the terrorists betrayed their ignorance, not just about Islam but about America; not just about the American people but American democracy and its values.

I wish to speak for a moment about this conflict that September 11 has put us into and the differences between us and our enemies, which is what this is all about. This is not a simple struggle for power. This is a global conflict for values, for ideals. We are idealists. We and our many allies around the world, including so many millions in the Muslim world, believe in the inalienable and inviolable rights of every individual. Our enemies are craven cynics who desire raw power for themselves and seek to crush those who look or act or think differently. They claim to be religious, but how can they be religious and faithful in any way in which any of the world's religions understand it, if they are prepared to kill thousands of God's children allegedly in the name of God?

We are different. We are optimists. We grant people liberty, not as the gift of politicians but as our Declaration of Independence says: As the endowment of our Creator. We have confidence that a society governed by its people will progress, and that is why we seek to open the world and broaden the community of nations living under democracy, as we have so magnificently since the fall of the Berlin Wall.

Our enemies are not just pessimists; they are fatalists. They fear the voices of the people. They want to bring down a theological Iron Curtain to divide the world into acceptable and unacceptable people and nations and faiths, to those worthy of living and those targeted for death and domination.

Third, we are skeptics in a very healthy way. We question one another and ourselves. We are proud of who we are but not so proud that we pretend to be without fault. Our enemies proceed with a chilling sense of certainty and an unwillingness to look at themselves in the mirror.

It is those values that have guided us through our history and distinguish us now from our enemies. The men and women of our military performed brilliantly in unfamiliar territory against an unprecedented foe. Our police officers, firefighters, and other first responders have had reason to despair, but they have risen to the immense challenge and reminded us of what heroism they display every day. Every day Americans in our communities have had reason to lose faith and to turn from hope to fear, but they have not faltered. They have come together, finding our strength, not losing our optimism and our courage.

Here in Congress, though we still have work to do, we have faced the new reality of the post-September 11 world. We have asked tough questions of ourselves. We have supported our President as Commander in Chief. We have realized that we have not been as prepared as we should have been on September 11 last year, and we are taking steps to close our vulnerabilities.

As we do, we must remember that September 11 was not just a tragedy that happened. It was not just a natural disaster. It was an unnatural disaster, carried out as an intentional act by people who were evil.

That is why, as Charles Krauthammer wrote in the *Washington Post* a while ago, we must understand this anniversary as more than a day of mourning and solemn remembrance. It must be not just a day of commemoration but a day of rededication. Charles Krauthammer wrote:

> We would pay such homage had the World Trade Center and the Pentagon collapsed in an earthquake. They did not. And because they did not, more is required than mere homage and respect. Not just sorrow, but renewed anger. Not just consolation, but renewed determination . . . .

We will build beautiful memorials to those killed on September 11, but there are other memorials that we here in Congress can and must build: a Department of Homeland Security that does everything humanly possible to prevent anything such as September 11 from recurring, and it need not recur. We must support and encourage our military to search out and destroy or capture Al Qaeda wherever they exist. We must reach out to the Muslim world, the great majority who are not fanatics or extremists, who suffer from a lack of freedom and a lack of material resources and hope, and offer them the support and the freedom that they desire and that is ultimately the best defense against the evil terrorism of the minority in the Islamic world that Al Qaeda represents.

As we approach the great debate in this Chamber on the questions around Saddam Hussein and Iraq, we must remember the lessons of September 11. As we look back, having heard the warnings of Osama bin Laden, having experienced the attack against the World Trade Center in 1993, against the two embassies in Africa, against the USS *Cole*, as we look back, don't we wish we had taken the kind of action we are taking today to destroy Al Qaeda?

In her foreword to "At Home In The World," a collection of Daniel Pearl's writings in the *Wall Street Journal*, his widow Mariane Pearl wrote:

> The terrorists who killed Danny stood at the other extreme of what Danny represents. They could only wield their knife and cowardice against Danny's intellectual courage and bold spirit. Danny died holding only a pen. They stole his life but were unable to seize his soul. By killing Danny, terrorists took my life as well but could not lay claim to my spirit. We will never let them win.

So, too, the terrorists may have killed 3,000 innocent Americans on September 11 last year, but they will never lay claim to America's living spirit. We will never let them win.

Hon. Sam Brownback

OF KANSAS

Madam President, I rise to speak about the events of 1 year ago. I am delighted to follow my colleague, Senator Lieberman of Connecticut, with whom I have worked and will continue to do so. I add my name to his comments.

Today we are gathered to remember those who lost their lives on September 11, to honor those who sacrificed everything for the concept of and belief in freedom. That day and every day since then are stark and simple reminders that freedom is not free and that it is never secure.

This is something the people of my home State of Kansas have long understood. Our very motto is "ad astra per aspera"—"To the stars through difficulties." We have lived this every day, every year since before we were a State.

It is also a theme our entire Nation embraces today. We were not bowed by last September's attack. In fact, we arose from the ashes stronger and more resilient than ever.

"The greatest victories come when people dare to be great," Ronald Reagan said, "when they summon their spirits to brave the unknown and go forward together to reach a greater good."

In the days immediately following that fateful day, we summoned our spirits and went forward together. We dared to be great.

As you look at the memorials, and as you listen to the speeches of remembrance, think of the sacrifice of all those involved, and of the lives cut short, the promises broken, the happiness destroyed. It is all too easy to cloak these sacrifices in mere platitudes. We must remember exactly what it means to sacrifice and what exactly was sacrificed.

These were not nameless, faceless people who just simply acted out some role. These were sons, daughters, husbands, wives, brothers, sisters, mothers, and fathers. They were scared men and women who had thrust upon them the part of hero, and they lived up to the billing.

It is also too easy to just call them heroes and walk away. It is not easy to recognize the fear and the strength and the courage they exhibited on that particular day.

I have been particularly taken now, reading stories of the heroes of 9/11 and the miracles that happened on that day—stories that we are all familiar with now—Todd Beamer and "let's roll," and the flight that went into the field in Pennsylvania, which was the very flight headed for this building. It probably would have reached its target had they not been heroes on that day. Would this place even be here now? It may have been rebuilt, but would we be back here yet? How many lives would have been lost here?

I read last Friday in *USA Today* about miracles of 9/11. Some police officers, one a rookie, went into the South Tower; they were buried in 20 feet of rubble. Three of them were together. The first was killed in the first crushing, but two survived and they were able to crawl around. The second tower came down and they were pinned underneath the rubble and stayed there almost 24 hours. They could see a light about 20 feet up, and they knew there was a possibility they would get out. As they faded in and out of consciousness during the night, one of them had a vision, it said in USA Today. The vision he saw was Jesus coming toward him, bringing him a bottle of water. It gave him strength. He wasn't fearful of death. He was able to reach out with strength and yell for help. They were eventually found by a Marine and were dug out from the rubble. That is one of the miracles of 9/11.

I think of the heroes that were going up the tower, instead of coming down, on 9/11. It was an amazing day, a tragic day, one we should not and we won't forget.

Also, sometimes it is easy to think that perhaps life does not change that much when actually life has irrevocably changed. It is not that life doesn't go on, it certainly does. We must never forget.

As author Elie Wiesel said in his Nobel lecture:

> For me, hope without memory is like memory without hope. Just as a man cannot live without dreams, he cannot live without hope. If dreams reflect the past, hope summons the future. Does this mean that our future can be built on a rejection of the past? Surely, such a choice is not necessary. The two are incompatible. The opposite of the past is not the future, but the absence of the future; the opposite of the future is not the past, but the absence of the past. The loss of one is equivalent to the sacrifice of the other.

We must not forget our past or the attacks or the outpouring of generosity and patriotism and simple kindnesses in the days following the attacks. All of this must continue. We cannot return to the safety of our homes and pretend the storms buffeting the lives of people hundreds and thousands of miles away does not affect us.

September 11 was a wake-up call that we cannot and will not forget. It has changed us. It has changed us in substantial ways that we can see and feel, and in ways that I don't think we have wrestled with yet.

One simple thing: "God Bless America" has become a national song—not the national anthem but the national song. We gathered again today as Members of the Congress on the steps and sang it as we did on September 11. I hope we can officially continue to do that. Even though it was unofficial today, I hope our national song will become official.

We are a Nation founded by men and women who are willing to stake their lives upon the conviction of universal rights and freedoms. This was larger than their own lives and small roles that they felt they would play. Their actions were just a shot across history's bow on behalf of all people who both desired to be free and honored the sense of duty that liberty engendered.

On September 11 we saw a number of people step forward to recognize and fight for those universal rights and freedoms, each of us in our own way in our own actions. Today, we still have a torch to carry—for all those who died on September 11, all those who have died in the war against terrorism, and all peoples across the world who desire freedom.

These may seem to be the worst of times, but we are resilient and, most important, we are a hopeful people and we will prevail. There is a Biblical verse that says:

> And not only so, but we glory in tribulation also: knowing that tribulation worketh patience; and patience, experience; and experience, hope: and hope maketh not ashamed . . . .

We are a hopeful people. God bless America.

Hon. Richard J. Durbin

OF ILLINOIS

Madam President, I joined my colleagues in the Senate this morning on a trip to the Pentagon. The project there is known as the phoenix project, and those familiar with mythology know that the phoenix is the great bird that rose from the ashes and flew again. Well, the spirit of America was flying again this morning at the Pentagon as we looked at a restored building—more important, a restored spirit.

The President, the Secretary of Defense, and others spoke. We all gathered—thousands of us—to pay tribute to those who lost their lives on September 11 there, as well as the victims in Pennsylvania and in New York.

As our buses came back, there were a number of people gathered in Washington on the Mall. Many of them were—in their own way, with their families and friends—commemorating September 11. As I passed, I saw one man standing there with a handwritten sign that said "United in Memory." I thought that really captures what we are doing today. We have stood united since September 11, but today we reflect. We are united in memory. We grieve for the victims. We mourn those who died.

But we also stand in praise of those heroes of September 11. Each one of us carries certain images in our minds of where we were when we heard it, what we did. For most of us, the first calls were to our families, and maybe it was indeed proper that we would turn to those we love the most to make certain they were safe.

I still remember images of that day, and the days following, just as real as the moment when I experienced it. One was a photograph from the *New York Times*, which showed a New York firefighter racing up the stairs, as you saw a long line of people racing down the stairs of the World Trade Center. A young, handsome face—probably a man with a family himself, thrusting himself into the jaws of that disaster in the hope that he could save someone's life.

Many like him—firefighters, policemen, first responders, medics, and others—gave their lives on September 11. They got up that morning and put their badges and uniforms on and probably never thought twice about whether they would return to their families. Sadly, many never did. They are truly American heroes.

I can recall a few weeks later being out on Lake Michigan, near Chicago, in a Coast Guard vessel with the young men who were keeping 12-hour shifts, patrolling the shores to keep them safe, checking every vessel that came on the Great Lakes. Most people in Chicago didn't even know they were there. But they were—every single day.

I can remember, as well, the Capitol Police just outside this door and around this complex, who worked 12-hour shifts day after day, week after week, to protect us, to protect the visitors, to protect the staff, to protect this great building after September 11. They are truly American heroes.

In January, as part of the first congressional delegation to visit Afghanistan in the daylight hours, we went to Bagram Air Force Base. It was an old Soviet base, and we were using it as part of our efforts to liberate Afghanistan from the Taliban.

We sat down with those young men and women in uniform who had missed Christmas with their families, did not know how long they were going to be there, and just talked with them and ate with them and shared some stories about home. One young man came up to me and said: Senator, I am from Illinois. Can I ask you a favor? When I come back after this, could you give me a helping hand?

I said: Sure, what is it?

He said: I would like to become an American citizen.

I said: Wait a minute, you're a soldier here.

He said: Yes, I am. I was born in Panama, and I am not an American citizen. Will you help me become an American citizen?

I said: You got it, buddy; whatever you want, I will be there.

I also remember another incident in the middle of December. I flew into O'Hare, and I went down to get in the line for a taxicab. I drew a taxicab, and the driver was wearing a black turban and a beard. As we started to move along, I said: Excuse me, sir, would you happen to be a member of the Sikh religion?

He said: Yes, I am.

I remember I had been visited by Muslims, Sikhs, and others worried about people who would discriminate against them, and I knew a little bit about some of the terrible things that happened to them—they were isolated, but that did happen.

I said to the taxicab driver: How have things been for you over the last several months, wearing your turban, trying to be a regular taxicab driver?

He said: Most people couldn't be nicer. There were bad ones, too. Some cussed me out; some wouldn't get in my cab. They think I am a terrorist, too.

He said most people could not be nicer. He said I have been in the United States for 33 years. I wish they would get in my cab because I would like to show them something.

I said: What is that?

He reached over and pulled down the visor, and there was a photograph of a young man in a U.S. Army uniform.

He said: I want to show them a picture of my son Michael.

I said: Michael is in the Army?

He said: Oh, yes; he was in Kosovo.

I said: Where is he now?

He said: He is with the Special Forces in Afghanistan, and I haven't heard from him in 6 weeks.

I thought to myself: Boy, does that tell the American story. Here we have a man who some people, with little education or learning, in their ignorance, would say is an enemy of America. No, that man is a loyal American who was offering his greatest treasure on Earth, his son, to our Nation to serve and who was in harm's way at that very moment.

Just a few weeks ago, four widows from the World Trade Center came to see me. They want a public investigation of what happened leading up to September 11. I completely support them. I think it is now overdue. We should do it.

They talked about their experiences with their families. They told their stories over and over in all the Senate offices. Some of them carried around their necks photographs of husbands and families.

I remember one saying: I am lucky. My three friends here do not have any evidence of their husbands they lost, but I was a lucky one because they found a hand, and on that hand was my husband's wedding ring which I now have on my hand. That is all that survived.

She was grateful for that one memento of his life and how much it meant to her, and what a reminder it is to all of us of the true grief and loss that so many families have endured.

I suppose the lesson from September 11 should be clear: Let all those around the world who would attack the United States know that they will pay a heavy price. We approve of that. But also let everyone around the world know that we are not an aggressive, angry people. We are a caring and compassionate nation, and if others will reach out with a hand of peace, we will extend ours as well, no matter where you are from, no matter what your religion or ethnic or cultural background. Osama bin Laden and Al Qaeda did not understand that, but we in America understand it well.

When I reach back in history for words that bring inspiration, I so often turn to one of our favorite sons, Abraham Lincoln from Illinois, and his second inaugural address right outside this building in which he said:

> With malice towards none, with charity for all, with firmness in the right as God gives us to see the right, let us strive on to finish the work we are in, to bind up the Nation's wounds, to care for him who shall have borne the battle and for his widow and his orphan, to do all which may achieve and cherish a just and lasting peace among ourselves and with all nations.

Lincoln said those words as we came to the close of the most devastating war in our history. He reached out to try to find common ground, even with the enemy, to bind this Nation. So, too, should we reach out in this world to tell the story of America, to help build a more peaceful world, a world where our children and grandchildren never have to fear another September 11.

After September 11, we were not just united in anger, not just united in sympathy. We were united in memory and united in hope—hope for a world of peace, hope that our children and children around the world will be spared the horror, the disaster, and the tragedy of September 11.

Hon. Mitch McConnell

OF KENTUCKY

Madam President, as we observe the 1-year passing of the day Al Qaeda attacked America, we have in our hearts, our thoughts, and our prayers the victims and their families. On this painful anniversary, they do not stand alone in their grief. All Americans of all faiths, colors, and creeds mourn the senseless loss of life on September 11, 2001.

The war on terrorism is a fight against evil forces bent on destroying America and our many freedoms. President Bush said this war will be unlike any we have ever seen, and he is certainly right. This is a war without borders and one in which the battle must be brought to the enemy, lest terrorists strike again on our own soil.

In the long proud arc of our Republic, America's courage has been too great, its values too strong, and its freedoms too dear to ever be turned back by an enemy. As we stand at the beginning of a new century, I am certain in the knowledge that we will prevail again.

Madam President, the watchwords for Kentuckians and all Americans on this day must be: Never again.

I think we can safely say that 1 year after September 11, 2001, we have ended the first chapter in the book about the war on terrorism. But the second chapter is going to be, in many ways, much more challenging.

The President and many of us in this Chamber are haunted by the notion that a year ago today, had those planes been full of something other than gasoline—a chemical weapon, a biological weapon or, worse still, a nuclear weapon—all of the destruction that we remember so vividly today would have been dramatically worse.

We will have before us in the Senate in the next few weeks a resolution giving the President the authority he will be seeking, and we will need to give him, to begin to launch the second chapter in this war, which is to target weapons of mass destruction, wherever they may be in the world, in the hands of leaders or gangs who wish to use them against our own people here at home.

This is a new doctrine the President will be laying down. In the past, we have turned the other cheek, if you will; we have waited to be attacked, and then we have responded. But that approach, when one considers the devastation of weapons of mass destruction, is simply unacceptable. The American people will not accept a strategy based upon responding after the next attack on our own soil using weapons of mass destruction.

This will be one of the most important debates we will have in the history of this body, and it will come up in the next few weeks. It will be an appropriate memorial and remembrance to those who lost their lives a year ago today as a result of a conventional attack. Were they alive today, I am sure they would applaud our efforts to prevent another attack with weapons even more devastating on other Americans here at home.

Make no mistake about it, this is the new challenge of the 21st century: Weapons of mass destruction in the hands of gangs such as Al Qaeda or regimes such as the one in Baghdad used on Americans here at home by people who really are against modernity, who want to roll the clock back to the Middle Ages where women had no rights, where people had no opportunity to speak or to worship as they chose.

This is a war between modernity and the Middle Ages. Our enemies are quite intelligent and resourceful, and this challenge is going to go on for quite some time.

In conclusion, this would be a fitting memorial to those who died a year ago today, that America in a very proactive way seeks to prevent the next attack in the United States using weapons of mass destruction.

Hon. Larry E. Craig

OF IDAHO

Madam President, I think all Americans today are pausing, if not for a moment, for a longer time just to think, to kneel and pray. On their mind is a historic incident that occurred a year ago today in this country.

Many of my colleagues and I went to the Pentagon this morning to recognize that great tragedy once again and to be there to honor those 184 civilian and military men and women who tragically died in the Pentagon when it was struck by terrorists.

There is no doubt in my mind, and my guess is there is no doubt in any American's mind, they again relived the events of the phenomenal and tragic incident that occurred a year ago today, both mentally and visually on television or in ceremonies or prayers or moments of silence around this country.

In rethinking that day myself, I thought of my own emotions; that I grew angry at first to realize we were being attacked by terrorists. Then I experienced for a moment on Capitol Hill that day a sense of fear that maybe the Capitol itself would be struck, or our office buildings, and that my staff might be in some way injured or my family may not be safe. Out of that fear, though, grew in my mind, and grew in most Americans' minds, a phenomenal sense of resolve.

Since that tragic day, we have seen that resolve take shape in so many forms, whether it is the celebrating of a complete reconstruction of the Pentagon today in almost unbelievable time, or whether it was citizens across this Nation reaching deep in their pocket to give a little or a lot of their personal wealth to help the remaining citizens whose husbands, wives, sons, or daughters were the victims of the 9/11 incidents.

As I was listening to our Secretary of Defense and our President today, I thought of two Idahoans who died a year ago today at the Pentagon, one Ltc. Ron Vauk of Nampa, ID, and one Brady Howell of Sugar City, ID.

I have known Lieutenant Colonel Vauk only by a piece of paper. When I was a young Member in the House of Representatives, I had looked at his resume. I had studied his grades and I, along with the rest of my colleagues, had decided he was eligible for and ought to be nominated to the Naval Academy at Annapolis. We did nominate him, and he served with honor.

He had retired from the military and was serving in the Navy Reserve, teaching in this area. He was serving only as a reservist at the Pentagon in a temporary status for a few days, having been called from his job to do so when that plane struck. I will never forget the time I spent with his bride and their small son in Maryland. I watched the unity of that family coming around the widow and that small son of Ltc. Ron Vauk. That was the kind of resolve we have seen repeated time and time again out of the tear-stained faces of Americans as they recognized that they had to commit themselves, as our President and as this Congress has committed itself, to never letting this happen again.

I remembered Jennifer Vauk today, and I can only say to that brave widow that I thank her for her courageousness at this tremendously difficult time for her. Resolve and resilience flowing from the veins of Americans into the energy of their souls clearly speaks so well in this country today. It is not just a 9/11 feeling, it is a sense of patriotism and resolve that has grown out of nearly every crisis this great country has experienced down through the years. It comes in all different forms.

At this instance, in Idaho, it was the Red Cross sending volunteers all the way across the country to Ground Zero in New York, or it was the numerous churches or memorial services held across the State of Idaho, or it was a marvelous little gal in Pocatello who had saved $1,000 of her own money to buy a horse, and she gave all of it to the 9/11 charities so some other child could have a little bit because that child had lost so much, a mother or a father.

It was not just an Idahoan doing it. It was thousands of Americans speaking out from the smallest, almost the poorest, to the tallest and the most wealthy in our country who found the capacity in their heart to experience this resolve and dedicate themselves, as did Leah Wright in Pocatello with her $1,000.

I suspect every generation has a defining moment. My guess is that September 11 is the defining moment for America's current generation. Our President, in speaking today, has given a name for all of us who would call it 9/11. It will be a Patriots Day, and I hope that every year we stop to remember Patriots Day and why we now recognize it in that capacity.

Congress is now debating legislation to create homeland security as a department, hopefully to bring our country together more cohesively, to allow our law enforcement communities to do so in a way that will give us greater intelligence and therefore greater resolve. In doing so, we must not allow terrorism, or our commitment to stopping it, deny us our own personal freedoms. We should never select security over freedom because it is the very freedom of our country that gives us the resolve we have today. Tragically enough, it was the very freedom of our country that caused terrorists to strike at us because we do not speak of freedom for Americans only, we speak of it for all citizens of the world and citizens of all countries as a right of humankind to be as free as possible, and for this great country to be dedicated to that freedom.

In our search for security, let us not deny ourselves the very freedom that is the strength of our country.

Many more will speak today, and at the end of the day many tears will be reshed in memory of the men and women who died on 9/11. I am so proud of my country and so proud to be but a small part of its leadership because I have sensed in the Senate that while we may have our differences politically, a resolve all Americans have at this moment is to never allow this to happen again, never allow our citizens to be the target of an enemy that would choose to strike them down for political expression.

So be it 9/11 or be it Patriots Day, I hope on September 11 next year we will once again be speaking out about that day on September 11, 2001, when thousands of Americans lost their lives, but America found once again a revitalized reason for being what we are and striving to allow the rest of the world to have the same kind of human freedoms we have and cherish.

Hon. Bill Frist

OF TENNESSEE

Madam President, though we would never wish to relive the horror of September 11, 2001, we must dedicate ourselves to appropriately remembering it. That is the task we begin with this first anniversary of that darkest of days, to properly and lastingly honor the sacrifice of the more than 3,000 women, men, and children who perished at the Pentagon, at the World Trade

Center and at the crash site of flight 93 in Shanksville, PA.

September 11 will be a day of mourning for many years to come. And it should be, for the grief of those who lost loved ones on that day will pass only with their passing. Nothing can wipe away the memory of a friend or a family member taken before their time. The victims of September 11—those who died and the friends and family who survived them—deserve our enduring respect.

Though the attacks were carried out in New York, Washington, and Pennsylvania, no American was left untouched by this tragedy. That includes the men and women of my home State of Tennessee. I think of John and Pat Lenoir of Knoxville who lost their son, Rob, when the World Trade Center collapsed. Francis Hall of Knoxville lost her sister-in-law. And Otis and Nancy Tolbert of Brentwood, TN, lost their son when flight 77 crashed into the Pentagon. We keep those Tennesseans in our thoughts and prayers today.

It is entirely appropriate that the President and First Lady began their September 11 by attending a church service. I hope Americans all across this country follow their example by spending some part of their day in a house of worship or on bended knee in prayer. Regardless of the God we may worship, faith in a higher power can help heal and explain and console and reassure us today, just as it did a year ago.

Though September 11 attacks did bring one of the darkest days in our history, a few rays of light did shine through. Americans rallied to help those in need by waiting hours to give blood, by donating supplies to the rescue effort, by digging deep in their own pockets for the September 11 charities. I am especially proud of the Tennessee Baptist Convention that sent 30 volunteers to prepare food for the rescue workers at the World Trade Center.

I am still moved to this day, as we were at the Pentagon a few hours ago, by the presence of members of the Tennessee Task Force One who helped search for survivors and recover the fallen at the Pentagon.

America will always remember the men and women who risked their lives to save the lives of others on September 11: Those on the front line, the medical personnel, the firemen, the police officers, all who rushed into harm's way, who forever touched our hearts with their heroism. Their example exists, survives, as an inspiration to us all. It will remain so for generations to come.

Britt Brewster, a 12-year-old Tennessee girl, who came up yesterday from Tennessee to participate in the remembrance services said earlier this week: The one good thing [about September 11] was that America started coming together as one.

I remember visiting Ground Zero with about 40 of my colleagues from this body a couple days after the attacks. Smoke was still rising from the debris. Almost everything was covered with the fine ash. The only color, other than the workmen's bright yellow hats, was the American flags that hung so proudly posted on the buildings around that World Trade Center site. We should fly our flags on this anniversary and show our common love for country and our fellow countrymen.

There has been much debate about what we should teach our children on this first anniversary of the September 11 attacks. I believe they need to know the truth. I had the opportunity to take my wife and my three teenage boys to Ground Zero about 2 months after the attacks. I wanted them to see first-hand the destruction with their very own eyes and remember, long after I am gone and my generation is gone, what evil once did—and, I should add, can do again—to our country. I will take them back to New York. We were just there 5 days ago and saw the rebirth, the vitality of that remarkable city. I also want my sons to see what good can be done, and can always be done, in our country.

The Gettysburg Address is considered one of the most powerful pieces of funeral oratory ever delivered on American soil. As Lincoln himself admitted, even he could not dedicate the battlefield beyond what those who fell there had already done. Instead, he urged his audience at the time to dedicate themselves, "that from these honored dead we take increased devotion to that cause for which they here gave the last full measure of devotion."

The terrorists attacked on September 11 and continue to make deadly threats because they hate our country and everything we represent. The 3,000 women, men, and children who died on that tragic day did so for the same cause as those who fell on the battle green of Lexington, and the forests of Argonne, and on the beaches of Normandy—justice, equality, liberty, democracy.

I urge every American to offer their respects to families who lost loved ones, to put those who perished in their prayers, and to show their patriotism by unfurling the American flag. But above all, I hope we will rededicate ourselves to those values, to the values that have been the core of the greatness of our country for more than two and a quarter centuries. Those values may be threatened sooner than we may think. If they are, we will find strength and hope and resolve in remembering, properly and lastingly, September 11, 2001.

Hon. Paul S. Sarbanes

OF MARYLAND

Madam President, this past year has been one of tragedy and challenge for the American people. Just a year ago, on September 11, 2001, we experienced a dawning national tragedy.

Just as the workday was beginning—8:46 a.m. to be precise—terrorists struck this country in a series of savage attacks. Over 3,000 were killed and many more were injured. Those attacks struck a vicious blow at every American everywhere.

Over the past year we have labored with the highest degree of human spirit to address our grievous losses—as individuals, in our families, in our communities, and as a nation. At the same time, we have worked hard to deal with the challenges that confront us now and into the future. We are resolved to put an end to the scourge of terrorism and to bring its perpetrators to justice. Our response to terrorism must be committed and complex, for no simple solution or single action can accomplish our goal. We must engage in the broadest possible international effort, for we know that terrorists are not contained by national borders. As we move forward, we take our inspiration from the calm determination and steely resolve of the firefighters, police, emergency personnel, and airline passengers who responded to the attacks, and from the resilience of those who are rebuilding lives and families and communities.

And we shall move forward, for we have families to care for, neighbors to look after, jobs that must be done, and civic obligations that must be met. The events of September 11, 2001, were tragic beyond measure, but our response to those events demonstrates the great strength of America and provides a new sense of what it means to be an American. The future of our Nation is ours to make.

Hon. Don Nickles

OF OKLAHOMA

Madam President, a year ago today our Nation was savagely attacked in maybe the most evil attack ever on American soil. Over 3,000 innocent people were killed. I call it an evil attack because how can it be more evil than to kill people who are totally innocent—men, women, and children?

The attack on the World Trade Center was an attack on the United States, on our economic beliefs and foundations. The attack on the Pentagon was an attack on our national security and defense. Flight 93 was, we now find out, headed for the Capitol, an attack on our democracy. I thank God for the heroes, the passengers on the plane.

A lot of heroes came out as a result of these savage attacks a year ago—men and women who were running into the buildings, not away from the buildings; into the buildings to save lives.

It is amazing. If you look at the pictures we have seen in the last few days of the World Trade Center and Pentagon—it is amazing that there are only 3,000 that were lost. That number could have been significantly higher. If you look at the devastation in New York City alone, it would not have been hard to imagine 20,000 lost, not 3,000. It probably would have been 20,000 lives lost had it not been for the courageous acts of firemen and policemen and fellow workers putting their own safety at risk to save other lives, not to mention the passengers on flight 93 who kept that plane from running into our Capitol, from hitting our Capitol. I cannot imagine the loss that would have happened, not just the loss of life—of Senators and Congressmen, our staffs, our employees, our security officers—but also the effect it would have on democracy. I shudder to think what would have happened if they would have hit our Nation's Capitol Building.

Today I joined with the President and many others in rededicating the Pentagon. It is great to see the Pentagon rebuilt, and my compliments go to the workers and others who rebuilt such a wonderful building in such a short period of time. But we also remember the loss of life in each of these instances.

In the Pentagon, a former employee and personal friend of mine, Barbara Olson, was killed. She was a passenger on that airplane. My heart still aches for Ted Olson and their family. What a terrible loss that was, taking the life of a person who was so bright, had so much life, was so engaged in the political life of our country. To have that life taken is just a very sad tragedy. To think that is multiplied by 3,000 times all across the country, it is a very sad reminder of the enormous tragedy we have suffered.

It reminds me of the Oklahoma City bombing we suffered on April 19, 1995. In Oklahoma City, we lost 168 lives. I knew some of those people as well. When you know somebody it makes it more personal. It is not just 3,000 lives. You realize it is individual families and some of those families were totally devastated and their futures enormously changed, if not destroyed because of this senseless, cowardly, evil attack that happened a year ago.

Like Oklahoma City, we had a lot of heroes. The heroes, the firemen who raced into the building, the heroes on flight 93, the heroes who were saving lives in the Pentagon, the medical personnel and others who saved countless lives, in some cases they gave up their life in order to save lives. The Bible says: Greater love hath no man than this, that a man lay down his life for his friends.

We had a lot of American citizens who laid down their lives to save other lives. What an enormous gift they have given. They did this to pay the ultimate price or make the ultimate sacrifice—to save the lives of other Americans.

Thousands of people killed 1 year ago today. Why? Because they were Americans, because they happened to be citizens of the United States, because they stood for freedom, they happened to share freedom.

Our country was attacked economically and militarily and politically. However, we survived that attack. The American economy is fine. Our American military stands strong. Our American democracy remains steadfast.

My compliments to the men and women in the military who are protecting our freedom daily and who have done a fantastic job going after the culprits, those who are responsible for this attack, in Afghanistan and other places.

My compliments to the administration, President Bush, Secretary Powell and Secretary Rumsfeld and others, who are going after the perpetrators of this crime—not just in Afghanistan, but in countries all across the world. My compliments to them for building up an international coalition of over 90 countries who are joining us in this attack, fighting the battle against terrorism throughout the world.

There is a lot of work that has been done and a lot of work that yet needs to be done. This Congress needs to join with the administration, both legislatively but also in support in continuing this attack and this battle on terrorism. We are not finished. There are still a lot of trained terrorists who threaten our country. Unfortunately, maybe they have been brainwashed into thinking it is good to try to kill innocent people if they happen to be Americans, or maybe if they happen to be friends of Israel. There is a lot of hatred that has been fomented for a long time, and that is very regrettable, but it is important that we band together—people all across the world—to condemn and combat terrorism.

I think the President has done an outstanding job, leading this country and leading the free world in that battle. I compliment him for it. We have a lot of work ahead, but I am absolutely confident that freedom will prevail. We are a great country because we are a free country. We have greater freedoms—political freedom, economic freedom, religious freedom—than any other country in the history of mankind. I am absolutely confident, though, in 10 years from now or 20 years from now, we will still be able to say that we live in the greatest and most free country in the history of mankind. However, these freedoms have been attacked. Frankly, these freedoms have been under attack for several years. Now we are responding and we are responding strongly. Yet we still have a lot to do. I am confident that the people who challenge us will not be successful. Freedom will prevail.

Hon. James M. Inhofe

OF OKLAHOMA

Mr. President, I think my colleague from Oklahoma is exactly right. We can carry it one step further. We have these freedoms and that is why they hate us so much; they don't have these freedoms. The idea that individuals can have the freedoms we in this country have is totally alien to everything they believe in.

I sat there as others did—I am sure my colleague from Oklahoma did this morning—and looked at the Pentagon, and I know what went through his mind and what went through my mind was the Murrah Federal Office Building in Oklahoma City in 1995. We lost about the same number of lives back then as we did in the Pentagon. There are a lot of other similarities there.

The appearance after the airplane struck was so similar to that which we experienced in Oklahoma City. That brought back those very sad memories.

I sometimes look at things and ask, How can anything good come from something as bad as all that? Yet I can see—it is obvious, as I saw the changes in attitudes of people here in this body, and also the body down the hall—they are reflecting the interests of the American people.

Hon. Richard C. Lugar

OF INDIANA

Mr. President, it is a privilege to welcome a distinguished delegation from the Norwegian Parliament. Nine members of the Committee on Defense are with us today in the Senate Chamber. They have come to the United States this week as a part of an ambitious series of events which will include meetings at the Pentagon, the State Department, the National Security Council, and Central Command in Florida.

They had been scheduled to leave Washington this morning, but they have changed their itinerary deliberately because they wanted to be with us here, the U.S. Senate, in the Capitol Building on this solemn day.

As fellow legislators and close NATO allies, the Norwegian Defense Committee wanted to express its solidarity with Congress and with the American people on the first anniversary of September 11.

I would like to read a letter into the *Record* from the Defense Committee of Norway.

They have written:

To the Senate of the United States:

The Standing Committee on Defense of the Norwegian Parliament wishes to express its deepest sympathy and solidarity with the American people on this day of remembrance—one year after the horrible terror attack on the United States that occurred September 11, 2001.

Let us never forget all those individuals who lost their lives in New York, Virginia, and Pennsylvania, including firemen, police officers, and volunteers who tried to rescue people from the flames.

September 11 changed the world and international politics. Norway is proud to participate in the broad coalition against terrorism and does so by taking part in "Operation Enduring Freedom" under U.S. command. The fight against terrorism is a fight for democracy, for an open and free society, and for human rights.

Sincerely,

The Standing Committee on Defense [of Norway]:

Ms. Marit Nybakk, chairman DC, Ms. Aase Wisloeff Nilssen, Member DC, Mr. Bjoern Hernaes, Member DC, Mr. Kjetil Bjoerklund, Member DC, Mr. Per Roar Bredvold, Member DC, Mr. Gunnar Halvorsen, Member DC, Mr. Aage Konradsen, Member DC, Mr. Leif Lund, Member DC, Mr. Per Ove Width, Member DC, Mr. Joern Olsen, Secretary DC.

I know that I speak for all Members of the Senate when I say that we deeply appreciate your support today. Your presence here reminds us of the importance of allies and the enduring bond between the United States and Norway.

During the long decades of the cold war, Norway was the only NATO member to border directly on the Russian Republic. This front line position imposed a special burden on Norway, and its value as a member of the Alliance far exceeded the size of its population.

The border between Norway and Russia is now peaceful and cooperative. Yet Norway still bears burdens from its history as a front line state. In particular, it must contend with the environmental dangers created by the nuclear-powered Soviet-era fleet that is deteriorating on the nearby Kola Peninsula.

In June of this year, I had the pleasure to visit Norway following an extensive trip to Russia. There I met with many members of the Norwegian defense establishment, including members of the Defense Committee. We talked a great deal about nuclear cleanup issues on the Kola Peninsula. Norway has been an invaluable partner in addressing this nuclear threat through its support for the Nunn-Lugar program and its participation in the trilateral Arctic Military Environmental Cooperation Program or AMEC. Under AMEC, our country has been working with the Russians and Norwegians to safely dispose of the nuclear material from decommissioned vessels.

We have had great success so far, but the challenges of safeguarding weapons and materials of mass destruction are immense. I am hopeful that our efforts can be expanded and accelerated, and I know that Norway will work closely with us to address these dangers.

So we welcome the Norwegian Defense Committee and draw encouragement from their presence here on this day of remembrance. We look forward to all that we can accomplish together, as we strive to make the world safe from terrorism and weapons of mass destruction.

Hon. Jim Bunning

OF KENTUCKY

Mr. President, I rise today in honor of those in New York, at the Pentagon, and in the skies over Pennsylvania who lost their lives 1 year ago today.

None of us in the Senate will ever forget the events of that terrible morning, nor will we ever forget the courage and compassion displayed by Americans everywhere in response to the attacks.

Today, my thoughts and prayers are with the victims, their families, and all of those who risked their lives to save others on that awful day The despicable acts that we witnessed were meant to create havoc and to test our resolve as Americans. America passed that test. We are stronger and more united as a Nation today than we ever were. Despite the fact that our buildings were damaged, America's foundations remain unshaken. And despite the fact that 3,000 of our friends, relatives, and neighbors were murdered, America's sense of community is stronger than ever.

This morning, many of us took part in a memorial service across the river at the newly restored Pentagon. One year ago today, that building was on fire. One of its five sides essentially lay in ruin.

Last September, I visited the Pentagon just days after the attack. It was a terrible scene of devastation. But today we saw a building that has been completely rebuilt. The Pentagon, both on the inside and on the outside, is better than before. The offices within are busy now with the activity of military men and women who are hard at work in the war against terror.

The Pentagon today stands as a reminder of the American spirit and a warning to those who want to terrorize us: America will triumph, and those who want nothing less than to destroy our way of life will fail. They will fail because of the American spirit. They will fail because of our faith in freedom and democracy. They will fail because of the strength and character of the American people.

I believe Americans have emerged from the attacks even stronger and more dedicated to our beliefs and to our Nation. But we cannot let our guard down again. We cannot forget that evil is lying in wait for another opportunity to attack. So far, we have been able to anticipate, with intelligence, any future attacks. But we know the enemy will try again.

It has been said many times—but it bears repeating—it might not seem that we are at war, but we are at war. It is a different kind of a struggle than we have ever fought before.

On the surface, it might not seem like World War II, Korea, Vietnam, or any other conflict of the past. Make no mistake about it, danger still lurks, and we must remain vigilant.

Americans have made many sacrifices, big and small, over the last year. They gave blood and contributed to relief efforts. They became more vigilant in their communities. They volunteered to help those in need. We have come a long way since the attacks a year ago, and I could not be prouder of our people.

In that time, I believe we have found new national unity, not only from the heroism of firefighters, police, and our military, but also from the everyday efforts of regular everyday Americans.

I am proud of the way we in Congress responded to the attacks. By putting aside politics and working together with President Bush for the greater good, we have shown that, while we all wear political labels as Republicans and Democrats, we are Americans first.

The President has done a superb job leading our country in the war on terror, and we in Congress have done our best to provide him with the resources necessary to persecute and win that war.

Much has been done, but we must continue to remain focused on the task at hand—protecting our homeland. And that job continues tomorrow in the Senate. We will finish it, and we will finish it successfully.

People often ask me how things have changed in Washington since 9/11. Some things on the surface certainly have changed. There are more concrete barriers, roadblocks, and security precautions, but looking beneath the surface, I think the better question to ask is, What did 9/11 reveal about us? It showed that we are still a good and compassionate Nation and people. It showed that, under the worst of circumstances, we will come to the aid not only of our friends and neighbors, but to complete strangers. It showed that America is still the greatest nation on Earth, and it showed that, in the war on terror, we will prevail.

Hon. Mary L. Landrieu

OF LOUISIANA

Mr. President, I will spend a few minutes remembering the day, as we all have taken some time to express our individual thoughts and honor this day in the best way we can, to express what it means for us, for all Americans, and actually for millions of people around the world.

It is truly a somber day for all of us. In a nation that has known unparalleled success, coping with a tragedy of this enormity is all that more difficult. We love our liberty, we love our freedom, and we want nothing from the world or for the world but peace and prosperity. Yet today we find our liberty and our personal freedoms restrained. We are fighting one war and are poised on the brink of another. It is no wonder the anxiety of the American people is palpable.

As we search for certainty and leadership in these uncertain times, it is only natural we turn toward one of our greatest leaders, Abraham Lincoln. As our leader during our greatest crisis, his words carry a resonance and wisdom that ring true today.

President Lincoln's second inaugural address, delivered at the twilight of the Civil War, reads like a prayer. It is a request to God to show us how to be just, and to grant this Nation peace. Yet, while it is a prayer, it is also a plan. President Lincoln wrote:

> Fondly do we hope, fervently do we pray, that this mighty scourge of war may speedily pass away. Yet, if God wills that it continue until all the wealth piled by the bondsman's 250 years of unrequited toil shall be sunk, and until every drop of blood drawn with the lash shall be paid by another drawn with the sword, as was said 3,000 years ago, so still must it be said, the judgments of the Lord are true and righteous altogether.
>
> With malice towards none, with charity for all, with firmness in the right as God gives us to see the right, let us strive on to finish the work we are in, to bind up the Nation's wounds, to care for him who shall have borne the battle and for his widow and his orphan, to do all which may achieve and cherish a just and lasting peace among ourselves and with all nations.

President Lincoln asked for God's assistance, but also to shine a light on the righteous path to victory. We must finish the work we are in, but we must also care for the victims of war and do all we can to achieve a lasting peace.

That is how President Lincoln wanted the Civil War to end. Not with recriminations that would tear the country apart forever and sow seeds for the next war but, rather, to approach victory with charity. For a powerful Nation such as the United States, it is only in demonstrating our commitments to a world order that we can attain a lasting peace.

It is only in our generosity to the struggling nations of the world we can enjoy the full fruits of our labors and the great bounties of our democracy. It is part of the twin ironies of being the sole superpower in the world. The first is that to preserve the peace, we must prepare for war. The second is that to attain any real victory, we must show mercy to those we have vanquished.

Louisiana lost three sons on September 11: Petty Officer Second Class Kevin Yokum of Lake Charles, Navy LT Scott Lamana of Baton Rouge, and Louis Williams of Mandeville. The quick and expedient thing for our country would be to cry for their revenge and the revenge of the thousands of other Americans who died for freedom that day. Yet that is not America. It is not what these three men would have wanted. Rather, they would want to know that their deaths had helped to sow a lasting and just peace among ourselves and with all nations.

So I join my colleagues today in saluting the heroism of these men, of the men and women who died and were wounded at the Pentagon, the heroism of the men and women of our great metropolis, New York, and the surrounding States and regions—of course, New Jersey lost many people—and the men and women aboard flight 93. Let their heroism be our inspiration to finish the work that we have at hand.

We have a great amount of work ahead. We can be proud of the work we have accomplished in the last 12 months, any number of initiatives and bills and legislative proposals and endeavors that have really made this country much stronger, more secure than we were on this day, this hour, a year ago. But there is no doubt there is a great deal of work to be done. Let us remember that we will show leadership in our might and power. We will also show leadership in our mercy, in our willingness to leave this world to a much more just and fair place, where democracies rule the day and people can enjoy freedoms unheard of, really, and not yet experienced in the world. That is America's greatest challenge.

Hon. Orrin G. Hatch

OF UTAH

Mr. President, 1 year ago today, September 11, 2001, terrorists attacked our country, killing almost 3,000 people Each of us has, in the year since the attack, had our lives touched by the horrific events of September 11. Each of us has, in the year since the attack, been shocked by the terrible images of destruction and suffering at the sites of these attacks—at the World Trade Center, in Pennsylvania, and at the Pentagon. Paradoxically, each of us has also been uplifted by the stories of heroism and self-sacrifice that have emerged from around the country in the wake of these terrible events.

From this act of war and hatred against all Americans, our country has demonstrated once again the resilient strength in the fabric of our people. Make no mistake about it: Our country is at war. But unlike past wars, we face a new challenge, a new type of enemy, one that is loosely organized throughout the world, with tentacles stretching into every corner of the globe, and one which is steadfast in its determination to defeat America.

We cherish our freedoms, our opportunities, and our tolerance. But we remain vigilant in our determination to meet and defeat our enemy—terrorists who threaten our security and our freedoms. Throughout our country's history, our people, its leaders, and Congress have demonstrated time and time again that when we work together—when we harness the full energy and commitment of our country—we can overcome any adversity or any enemy to our people. The tragic events of September 11 have united this country and challenged our country once again to face down the terrible threat of terrorism.

While we have accomplished much, there is still much to do. Since September 11, we all recognize that we live in a different and more dangerous world. We must unite in our continuing support for our country's war against terrorism; we live with an ongoing and serious threat to our society. We must remain vigilant in protecting our way of life and meeting the challenges ahead.

I want to take a moment and offer my prayers and condolences for those families and friends who lost loved ones on September 11. Today we all join together, hand in hand, heart in heart, tear in tear, to share as a country all of the pain of September 11. We recognize your terrible loss, we offer you our support and we give you our love. We will never forget the terrible tragedy of September 11. We will do all we can in our prayers and in our deeds to make sure that such an attack never occurs again.

It is a new era in America and I ask for your prayers and support as we face many difficult challenges ahead. We do so with a steely resolve to never, ever let this horrible event ever occur again.

Hon. Ernest F. Hollings

OF SOUTH CAROLINA

Mr. President, in the year since last September 11, there has been much debate on the Senate floor on how to stop terrorists; and there will be plenty of time in the coming months to debate what we do insofar as organizing homeland security, and how we deal with Iraq and the fanatics who want to blow us up and the like.

But on this September 11, this Senator wants to remember the 3,000 lives lost in New York, at the Pentagon, and in Pennsylvania. Obviously their loss has been felt by their families in ways none of us can imagine. But their loss has also been felt by all Americans. As a Senator, in the last year, not a day has gone by when I haven't thought about what happened last September 11, and what actions this Congress can take to prevent such horrifying events. September 11 has moved this Nation to respond and to defend ourselves in ways that has made America stronger, I have no doubt.

In addition, I want to honor the men and women in the Armed Forces who have put their lives on the line in the last year to track down terrorists in caves and everywhere else they are hiding. I honor the law enforcement officials all over this country who protect our homeland every day. And I have great respect for the newly hired men and women of the Transportation Security Administration, who are wearing the newest American uniform to ensure the safety of our airports.

September 11 made us address our security vulnerabilities, but there is more work to be done. In the coming months, on days that are less emotional than this anniversary, I hope we remain as strong and determined to win the war on terrorism.

Hon. Pete V. Domenici

OF NEW MEXICO

Mr. President, many of us will spend a good deal of this day reflecting on what happened 1 year ago in New York, Washington, DC, and Pennsylvania. The images of horror, confusion and bravery that dominated our television screens last September will, once again, be refreshed in our minds. The raw emotion that we felt then will also be revisited. And, as is our nature, we will, again, try to rationalize why such tragedy befell our Nation. But this effort will be futile, because those who attacked us are filled with a hatred that is incomprehensible to American logic. Simply put, for us, life is a precious gift of God; to our enemy, it is utterly dispensable. So how can we constructively approach September 11, 2002? I believe that Americans can do three things today to accomplish a sense of healing.

The first step focuses on our children. No group was scarred as much by the terrorist attacks as were they. Their innocent view of the world did not contemplate the kind of evil that was perpetrated on September 11, 2001. So for them, the images of crashing planes, burning buildings, and crying adults shattered their belief in a world that was good and safe. Not only were they frightened, they were also confused about why others wanted to hurt us. Today, many children may experience the same anxiety about terrorism that they did 1 year ago; let us recognize that and take a moment to reaffirm to them that they are loved, that they are protected, and that the good people in the world far outnumber the bad.

Second, be a patriot. This can be accomplished in many ways. Flying the flag is the most recognized. But telling a service-veteran that you appreciate his or her sacrifice is equally valuable. The civic heroes of September 11, firefighters and police officers, also deserve our recognition for selflessly responding to the needs of the country. And acts such as giving blood, helping a neighbor in need or giving to a charity are just as patriotic. All these acts have the effect of uniting us behind a common purpose and remind us that no enemy can weaken our moral fabric.

Last, reflect back upon these words spoken by President Bush last September 20:

> We will direct every resource at our command—every means of diplomacy, every tool of intelligence, every instrument of law enforcement, every financial influence, and every necessary weapon of war—to the disruption and defeat of the global terror network.

This bold commitment by the President signaled to all that the United States was entering a long struggle that would require our desire for action to be checked by patience.

Patience, of course, remains necessary, but we have achieved much in our initial response to last year's attack. U.S. military action has unseated the Taliban government that once protected Al Qaeda in Afghanistan, while terrorist training camps in that country have been rendered inoperable. At home, Congress and the President are working to establish a new Department of Homeland Security to enhance coordination of our government's antiterror effort, both Chambers of Congress have passed the largest defense budgets in our Nation's history, and extraordinary effort has been made to improve air safety, intelligence gathering and counterterrorism methods. To be sure, our war on terror is just beginning, but we should remember that American action since last year's attack has been strong and has yielded positive results.

Contemplating what happened to our country 1 year ago is difficult for all of us. It is difficult for the friends and family of flight attendant Al Marchand, a New Mexico native who was one of the victims aboard United Airlines flight 175. It is difficult for the urban rescue team that traveled from New Mexico to New York in hopes of finding survivors. But remembering those lost is a duty. Today, if we focus on our children, our communities and the progress we have made in the last year, we will honor the fallen as well as reenergize ourselves for the struggle ahead.

Before I end, I want to make note of a poem I received from a young girl from Los Alamos, NM. The title of the poem is "Who Am I," and it reflects some of the very serious thoughts that the reality of terrorism has forced upon our young people, thoughts about humanity, and thoughts about whether peace can prevail. I ask unanimous consent to print this poem in the *Record*.

WHO AM I?

(By Noopar Goyal, May 2002)

A face, different from others around me.
A name, unusual to outsiders, yet beautiful in meaning.
A voice, bold but not abrasive when spoken to the world.
To my parents, I am their pride, their courage.
To my teachers, I am a word of exquisiteness.
Me: lucky and fortunate to be here and to have what I have.
In my family I am the listener and the speaker.
To my friends, I am the fun and happiness.
To my enemies, I am ignored like dust swept away.
To many strangers, I am another face smiling in the crowd.
My mind is mature, but there is much I don't know.
I am a child in every way.
Successes come and go, and I'm sure there will be disappointments.
I dream about the future and what it brings.
I always remember the good things and seldom the bad.
I forget the days when I was little, and they disappear into vast space.
People don't understand my thoughts, my culture, or sometimes, just me.
My frustration makes me want to be alone.
Who am I?
I am a voice with laughter, thoughts and opinion.
A name with pride and courage.
But most of all, a person waiting to fulfill a life of wonders, dreams, and the happiness that comes with it.

Hon. Jeff Bingaman

OF NEW MEXICO

Mr. President, I rise today in remembrance of the events of September 11, 2001. It hardly seems that it was 1 year ago that New York City and Washington, DC, awoke to an astonishingly sunny late summer day whose calmness belied the tragic events that would come that morning. While the passage of time has continued as it did before that day, our lives, our Nation, and the entire world have since been profoundly transformed.

As that day unhinged 1 year ago, we all struggled to answer the questions that raced through our heads: "Who is capable of such monstrous violence?" "Why would they do this to us?" And like the families of the victims, many of us sat at the end of that long day and wondered: "How do we go on from here?"

A deep sense of loss and uncertainty permeated the Nation in the weeks that followed the attacks. But in spite of the somber mood, we did what Americans do best. United as never before, we found our resolve to forge ahead. We found strength by turning to family and by turning to our neighbors. We rededicated ourselves to the civic responsibility that is the cornerstone of freedom.

While memory remains scarred by the worst act of terrorism on American soil, the past year has been a time for healing. Today, we continue to heal by remembering those who lost their lives on September 11. We remember the men and women who worked at the World Trade Center, the military and civilian personnel at the Pentagon, the firefighters who did what they could to ease the tragedy, and the heroes of United Airlines flight 93 who gave their lives to spare the Nation an even larger loss.

Each day that passes will bring us a bit closer to becoming whole again. But we must all take time on this day to mourn those who lost their lives on September 11 and honor the heroes who saved so many lives. We must also keep in our thoughts the troops who are fighting overseas in defense of our Nation.

Today is about remembrance, but tomorrow is always about the future. Once again, we have to ask ourselves, "How do we go on from here?" And answer with certainty and strength. America will not be deterred by terrorism. Instead we will celebrate what it means to live as citizens of this country and honor our continued responsibility to advancing the freedoms that are the hallmark of this country.

Hon. Strom Thurmond

OF SOUTH CAROLINA

Mr. President, I rise today, on this somber occasion, to remember and honor the thousands of lives that were lost 1 year ago today in the tragic terrorist attacks on our Nation. The United States will never forget the horrific events that occurred on what began as a peaceful morning of Tuesday, September 11, 2001. The tranquility of that morning was shattered by the evil acts of terrorists, filled with hatred for our Nation and opposed to the ideals we treasure. With their terror, our Nation was plunged into one of the darkest days of our history as thousands of Americans lost mothers and fathers, brothers and sisters, sons and daughters, friends and associates.

However, during this tragedy, the American spirit shined through that darkness and continues today, as a beacon to the world. It is that spirit, the willingness to expend the last full measure of one's life in service to others, that is the strength of our Nation. It is a power that grows as we are challenged. It is a force which has helped this Nation through difficult times and will see us through these times as well.

On that dreadful day, the terrorists failed miserably in their attempt to weaken our Nation. Their goal was destined for failure, for America has faced adversity numerous times before and has always emerged stronger. The acts of heroism and charity by Americans in New York City, at the Pentagon, over the skies of Pennsylvania, here in Washington and across the Nation were extraordinary but not surprising.

Today, I also honor the men and women of our Armed Forces. They serve around the world defending the freedoms we enjoy and securing the liberty we cherish. I have stated many times that the highest obligation of American citizenship is to defend this country in time of need. Our citizens have accepted that obligation, some giving their all. Whether serving abroad or at home, the men and women in uniform are performing in an outstanding manner and deserve the appreciation and respect of all Americans.

History will not forget the events of this day. Likewise, we must never forget the thousands of Americans to whom we pay tribute today.

Hon. Kent Conrad

OF NORTH DAKOTA

Mr. President, anniversaries are a time for reflection, and I wanted to take a few moments to share what emotions this North Dakotan is experiencing.

First, I feel sorrow. I feel sorrow for the thousands of innocent victims and the scores and scores of families and friends most directly and tragically affected. Those murdered were fathers and brothers; daughters and spouses; they were coworkers earning a living and supporting their families; they were best men in weddings and mothers flying home to see their children.

They were also honorable Americans, as well as citizens from 86 countries around the world. The victims included North Dakotan Ann Nelson. Ann was the kind of daughter every parent wishes for. She was intelligent, caring, adventurous, and had a real zest for life. Ann was simply doing her job that day in the World Trade Center. Ann, the pride and joy of Stanley, ND, didn't make it home from work that fateful day.

When I think about Ann and her family and the other victims and their loved ones, I also feel anger. One of my home State newspapers, the *Grand Forks Herald* put it best,

> Americans should accept that at the core of their grief is a white-hot fury and a sense of being outrageously wronged .... The World Trade Center didn't collapse in a strong wind. The Pentagon didn't fall into a sinkhole from an abandoned mine, and United flight 93 didn't crash in a Pennsylvania field because it ran out of gas. No, those things happened because a band of terrorist fanatics slit the pilots' throats and then flew the planes, passengers and all, into the buildings for the mad glory of killing infidels by the score.

This North Dakotan's anger and sorrow also fuels my resolve. I feel resolve to continue working with Members from both sides of the aisle and with our President to make sure we are doing everything in our power to protect North Dakotans and all Americans.

Over the past year, we have enacted vital antiterrorism legislation, including provisions I authored to shore up our visa and border security laws. And while we have paid more attention to the challenges of protecting our 4,000-mile northern border from terrorist infiltration, I continue in my resolve to focus the necessary attention and resources to get the job done right.

So I feel sorrow, anger and resolve, but I also feel pride, pride in how our heroes performed that day in response to the attacks; pride in our police officers and firemen; pride in those risking their lives to save coworkers; and pride in members of flight 93 who lost their lives to save countless others.

I also feel pride at North Dakota's own, the Happy Hooligans, who minutes after the terrorist attacks took to the skies over Washington, protecting our Nation's Capital at this most critical time.

I feel pride at the men and women in uniform, who have served and continue to serve in Afghanistan and across the world, with the unfailing twin goals of eradicating global terrorism and protecting their fellow citizens. I feel pride in the dedication of those reservists who have put their lives on hold to serve our country.

I also feel pride in our country, and it is certainly appropriate that Congress and the President agreed to designate September 11 as "Patriot Day." September 11 brought out the best in our fellow citizens and showed us again why we are all so fortunate to be a part of the greatest Nation on Earth. Rather than the disillusionment that the terrorists hoped for, our country responded with renewed patriotism.

And finally, on this first anniversary of one of the darkest days in our country's history, this North Dakotan feels optimism. In times of challenge, moments of great opportunity also present themselves. In this instance, we stand at a critical time in our Nation's and the world's history, and the decisions we now make will influence the shape of our world in the 21st century and beyond.

Will we live in a world of freedom or fear? Will democracy reign or will fanaticism retain its lure? Will our country try to build a wall around ourselves or will we continue to be a beacon of freedom, democracy, and tolerance around the world?

Some fifth grade North Dakota students are planning to commemorate the September 11 attacks by planting trees at the International Peace Garden, just north of Dunseith, ND. The name of the program is "Seeds of Peace." I have the optimism to believe that this is the perfect symbol to commemorate the first September 11 anniversary.

Hon. E. Benjamin Nelson

OF NEBRASKA

Mr. President, I am here on this day of remembrance to express on behalf of the National Funeral Directors Association and all Nebraskans our heartfelt sorrow for those who lost their loved ones on September 11, 2001. We Americans can be proud to declare the continued strength of our Union 1 year after this horrific act.

The United States of America continues to stand as a beacon of freedom and opportunity for everyone, regardless of race, creed, or religious belief.

The United States of America was founded on the fundamental principle that all citizens have the inalienable right of life, liberty, and the pursuit of happiness. The vitality of the United States of America is in the diversity of ideas, the freedom to express those ideas, and the opportunity to achieve one's potential and direct one's destiny.

These principles are absolute and will not be surrendered or weakened by the cowardly acts of terrorists who are afraid of the sunshine of freedom and the responsibility it brings.

On this day we must continue our unity, which reaffirms the principles for which this country was founded and that on this day freedom shall ring from every community in this great land and the voice of America will be heard around the world.

Hon. Robert G. Torricelli

OF NEW JERSEY

Mr. President, we will always remember where we were on this day, 1 year ago. As on other moments of tragedy in American history, September 11 will forever be in our hearts and mind. For those who were lost, for those who gave of themselves to save others on that day, we were all one. We were all Americans.

When we reflect upon a tragedy such as this, there are many who come into our minds. We reflect upon the honored dead who we remember today, including the friends and family we lost. We think of our loved ones who are still with us today. We remember all of the firefighters, police and ordinary citizens who risked their lives to save people they often did not even know. We reflect upon the members of our Armed Forces who diligently work to protect us from any future tragedy.

On this solemn occasion, I would like to take a brief moment to recognize the efforts of the members of the AFL–CIO on September 11 and its aftermath. Indeed, there were few others as affected by September 11 than the labor community. The firefighters and police who bravely sacrificed and risked their lives were union members. The laborers, ironworkers, and operating engineers who helped dig for survivors while the fires still burned were union members. The nurses, doctors and EMTs who cared for the injured and dying were union members. Those who manned the ferries and fireboats that transported both the survivors and the bodies of the victims across the harbor were union members. Their efforts greatly affected the lives of many.

In the aftermath, unions across America started up blood drives and the AFL–CIO Union Community Fund along with dozens of local and international unions raised relief funds for the families and children who have been left behind.

On this day of solemn remembrance, I want to recognize all of the sacrifices of these valiant men and women. Their response to this tragedy was truly heroic.

Hon. Herb Kohl

OF WISCONSIN

Mr. President, today we remember the terrible events of September 11. A year has passed but for those who lost loved ones or sustained serious injuries in the terrorist attacks on the World Trade Center or at the Pentagon, the pain is still fresh and the loss is still palpable. Losing a loved one is always difficult but to experience loss as a result of a senseless act of terror can only compound the pain.

For Americans in general, the sheer number of lives lost on September 11 was a national tragedy. Those of us who did not lose friends and family also experienced loss on September 11, albeit a loss of a different kind.

On September 11, we lost our sense of personal safety. The idea that terrorism could strike Americans going about their business, working in their offices or taking a simple plane ride for business or pleasure, has changed us forever. Travel by air will never be the same post-September 11. And on a less tangible level, we are now cognizant that in a free society like ours terrorists cannot truly be contained. The threat of terrorism may subside but it will never disappear.

On September 11, we lost our sense of trust. We have become more suspicious of those who want to enter our country. The Federal officials who protect our borders and control access to our country continue on heightened alert, on the lookout for aspiring terrorists. Our first responders, our local police and fire officers, have been tasked to survey our towns with a new eye and have entrusted all of us with the unsettling job of reporting suspicious activity in our neighborhoods.

Most of all, on September 11, we lost our sense of national security. The attacks of September 11 brought with them the realization that our robust defenses, the biggest and best in the world, cannot protect us from terrorists. Our sophisticated planes, submarines and missiles cannot deter a terrorist attack, and cannot protect us from the unconventional attacks we now know the Al Qaeda terrorists were contemplating.

Today, however, is not just a day to reflect on loss. Just as the stories of those who experienced personal loss on September 11 have evolved into stories of determination to carry on, our losses are tempered by resolve.

We are resolved to uproot the terror cells which may now be lurking in as many as 60 countries, waiting for us to let down our guard so they can attack us at home or abroad. Working with our allies around the world we are determined to disrupt these cells by stopping their funding and prosecuting their members. We will also remain vigilant. To the best of our abilities, we will take all precautions to deny these terrorists the opportunity to strike again.

We are resolved not to succumb to hate and to stereotyping of those who share ethnic or religious backgrounds with the terrorists. One of the biggest fears after September 11 was that there would be a backlash in this country against those of Middle Eastern descent or against adherents of Islam.

While there were reports of hate crimes, many Americans reached out to their Muslim or Middle Eastern neighbors to reassure them—whether they were American citizens or just residents—that they should not feel at risk. While the heinous acts of September 11 elicited many emotions, I was proud that most Americans recognized that taking our revenge against those who had nothing to do with Osama bin Laden would have sunk us to the level of the terrorists themselves.

The threat of terrorism does have the potential to change the character of our Nation. Just as we are vigilant about our physical security, we are resolved not to let terrorism curtail our freedoms. We must not allow the war on terror to infringe on the rights and liberties we hold dear. Terrorism will not go away, but it will have succeeded if we use it as an excuse to trample on the Constitution. The wonderful outpouring of patriotism which occurred this past year was not just an expression of national unity, it was a strong statement that we cannot and will not allow terrorism to undermine our democratic way of life.

In the days after September 11, many Americans wondered how we would carry on. And yet we have carried on. We have danced at weddings, rejoiced in new babies, and it is the brave family members of those who perished on September 11 who have led the way: the mothers who gave birth without the presence of their husbands and the brides who walked down the aisle without their fathers. This has been a difficult year, a year of inconceivable loss, but a year which has been marked by resolve and a rededication to the ideals and principles upon which our Nation was founded.

Hon. Charles E. Grassley

OF IOWA

Mr. President, I rise today on this 1-year anniversary of September 11 to join my colleagues, as well as the rest of the country, in a day of reflection and remembrance. It is often said that time heals all wounds. Obviously, it is going to take an eternity for the grief to subside for the families of the 3,000 people mass-murdered last year, with the mental image of commercial jetliners searing into the national landmarks remaining fresh in our mind. If they do not, we are reminded of it by watching television this very day and maybe all week.

Such horrific acts that happened a year ago today seemed impossible at that time. But, of course, all that has changed. Now it is hard to understand how such hate and extreme acts of horror against humanity can take place. While the Federal Government is working to beef up the military to protect our borders and to improve its intelligence gathering, it is virtually impossible to guarantee a risk-free environment in a free and open society that we proudly claim as America.

One year after the attacks, I am not so sure the American people really remember that we are in a war on terrorism. But the fact is, we are, and we will be for some time to come. I pray that we do not end up with a situation in the Middle East that we see too often on television where there are random suicide bombings. The risk exists and Americans are not realistic if they do not think of those things happening here like they might happen in Jerusalem. Terrorism has changed our way of life. We might be complacent about it and not want to realize it, but it is here. And every one of us, then, has a responsibility to remain vigilant.

The 1-year anniversary of September 11 is an important reminder that the war against terror has not yet been won. Considering the loss of human life, imminent terrorist threats to our security, and even our ailing economy, it is not easy to look for the silver lining on the 1-year anniversary of the September 11 attacks. But remember what the President said in the week after those attacks: The terrorists succeeded in tearing down bricks and mortar, but they failed to rock the foundation that has kept America strong.

A year later, America's resilience can be seen from sea to shining sea.

September 11, 2001, ushered in a new era. Notions of invincibility have been shattered. But the uncommon courage of first responders called to duty on that day reflects the steadfast spirit of our great Nation and our people called Americans. Our resolve to pull together and to stand united against evil immediately resurrected the principles on which this country was founded some 226 years ago. Despite the attacks, attacks defined to pit fear against freedom, the United States of America is yet stronger than ever.

Like the rest of the country, lawmakers in Washington, DC, dropped partisan pretense, worked quickly to assist survivors, backed recovery efforts, ensured the safety of the flying public, and got the economy rolling again. The 107th Congress threw its support behind the President to root out the terrorist networks responsible for the attacks, realizing the war in Afghanistan is probably only one of many battles to be fought and hopefully won.

Thanks to courageous servicemen and women, the Al Qaeda network has been largely dismantled from its base in Afghanistan but not elsewhere. That evil continues to lurk in other regions of our world. And with the security of the American people first and foremost in our mind, the President has worked to leave no stone unturned. That includes creating a new Cabinet-level Department of Homeland Security and keeping Iraqi dictator Saddam Hussein from unleashing weapons of mass destruction against the United States or other places within our world.

The President needs to make the case to the American people, to Congress, and our allies abroad, and he will do that hopefully within 48 hours, and do it in a way that says freedom and peace will remain at risk, as he explains it, until rogue dictators and others who harbor terrorists and finance their evil acts are no longer able to do those things.

This month, the Senate is debating the proposal to realign the Federal Government's infrastructure and operations charged with thwarting acts of terrorism. I am working to make sure the new department helps to solve the shortcomings exposed by September 11 and not create new ones.

Many recall the patriotism displayed by native Iowan Coleen Rowley, who blew the whistle on bureaucratic bungling at the FBI. I will work in this bill to see that new department employees are guaranteed strong whistleblower protections and to strengthen accountability within the intelligence community. These protections for whistleblowers are very important to make sure our intelligence community and the homeland security is working for the good of the American people and to see the statutory requirements are carried out.

When the Department of Homeland Security bill is up, I will make sure that hard-working taxpayers' money is not wasted with this new Federal agency.

One year later, after September 11, life goes on in America. It is not the same as it was a year ago. Life is not as secure or risk free as we once thought it was—and maybe we should not have thought that it was, but we did. Air travelers deal with tightened security measures at the Nation's airports. People are staying closer to home, flying less. Ordinary Americans and law enforcement officials do not hesitate to report suspicious activity. For many, it has enhanced common courtesies and boosted greater appreciation for the simpler things of life.

Iowans deserve a lot of credit for their outpouring of support in the last year. From a remarkable quiltmaking project for the victims, particularly in New York City, to generous charitable cash donations, and to those serving in our military, Iowans are proud, compassionate Americans. Many agree that the tragedy a year ago has renewed a sense of civic duty, patriotism, and appreciation for the U.S. military. There is a spirit of all-for-one and one-for-all, as we wear, display, and decorate with all things red, white, and blue.

With this 1-year anniversary, I join my colleagues in reflection and remembrance. While we go about our daily business, we can consider the tragic loss in human life and the acts of heroism by brave defenders. In the weeks and years ahead, we can continue to work for the betterment of our communities. We can donate blood, pray for the victims and their families, support emergency workers, and give thanks for the precious freedoms we enjoy every day.

We cannot erase the sorrow and suffering brought by September 11, but with our actions each one of us can make America stronger.

I remember this day especially Miss Kincade, from Waverly, IA, who was on the plane that hit the Pentagon. She was an intern in my office in 1984.

I remember Mr. Edward V. Rowenhorst, whom I did not know but I know his brother who goes to my church in Cedar Falls, IA. He was in the Pentagon working.

I remember traveling to a ceremony last Veterans Day in Anamosa, IA, where they honored one of their own who was also killed in the Pentagon 1 year ago today.

So Iowans, as most people in most States, have victims to remember. I remember them.

Hon. James M. Jeffords

OF VERMONT

Mr. President, today we as a Nation commemorate a most unfortunate milestone, the 1-year anniversary of the tragic attack of September 11, a day that will sadly live in infamy. Since that time, much has been said and written about the terrible events of that day.

As we reflect on the events of the past year, I would like to commend the thousands of rescue workers, volunteers, and countless others who helped rebuild our Nation in the months following the attacks. In particular, I would like to especially commend the work of the Federal Emergency Management Agency, fondly known as FEMA.

To give some background, on August 15, 2001, I became chairman of the Environment and Public Works Committee, which has jurisdiction over FEMA. A short month later, terrorists attacked the Pentagon and the World Trade Center. Later that day, following the attacks, I visited the Pentagon. I was amazed that literally overnight FEMA had established a well-coordinated Federal response at the Pentagon. Virginia, Maryland, and other first responders came—from as far away as Nebraska. As soon as possible, they were all working, as coordinated as I have ever seen.

As soon as possible after that, I traveled to the World Trade Center—again to take a look at FEMA's response. Again, I was overwhelmed by the organizational capacity of FEMA and the fine work being done by that agency's men and women, under the guidance of FEMA Director Joe Allbaugh. There, on the piers of Manhattan, FEMA had quickly created a disaster field office that was a small city unto itself. Thousands of workers from around the country came together to bring calm and order to an otherwise chaotic situation.

Visiting the Pentagon this morning brought back a flood of memories for me about my own visits to the Pentagon and the World Trade Center. Just days after those attacks, I vividly remember the sights and sounds and smells. The devastation I witnessed was incredible, and difficult to put into words. Thousands of people lost their lives due to the cruel and cunning acts of evil perpetrated by a few. The victims of these attacks were men, women, and children, people with well-laid plans for their pleasant futures.

Although I left both the Pentagon and the World Trade Center with a heavy heart, I also left with a profound sense of gratitude for the gallant efforts of these rescuers and volunteers who tirelessly, and mostly anonymously, worked in places reserved only for the Ground Zero heroes.

In the year following the attacks, I have spoken with many people, Vermonters and others, about the attack. We have all expressed profound sadness for our Nation's great loss. They have also left me with the confidence that freedom will prevail, that good will triumph over evil, that these horrible attacks cannot break our resolve to stand together as free Americans.

Abraham Lincoln once said: Freedom is the last best hope on Earth.

Time is a great healer. The passage of time has brought thoughtful recollection. The passage of time has not dulled my recollection of what I saw and felt in those days following September 11. For me, this healing process has brought a renewed commitment to move forward with the hope that freedom prevails.

In closing, I would also like to pay tribute to the work of the Vermont Air and National Guard, which did an outstanding job of protecting our skies and our borders and our airports in the days after September 11. When our Nation was most in need, we pulled together successfully to bring this Nation into a position where it feels secure and with hope for the future.

MESSAGES FROM THE HOUSE

At 3:15 p.m., a message from the House of Representatives, delivered by Ms. Niland, one of its reading clerks, announced that the House has passed the following bill, in which it requests the concurrence of the Senate:

H.R. 3880. An act to provide a temporary waiver from certain transportation conformity requirements and metropolitan transportation planning requirements under the Clean Air Act and under other laws for certain areas in New York where the planning offices and resources have been destroyed by acts of terrorism, and for other purposes.

Under the authority of the Senate of January 3, 2001, the Secretary of the Senate, on September 6, 2002, during the recess of the Senate, received a message from the House of Representatives announcing that the House has passed the following concurrent resolution, in which it requests the concurrence of the Senate:

H. Con. Res. 464. Concurrent resolution expressing the sense of the Congress on the anniversary of the terrorist attacks launched against the United States on September 11, 2001.

INTRODUCTION OF BILLS AND JOINT RESOLUTIONS

The following bills and joint resolutions were introduced, read the first and second times by unanimous consent, and referred as indicated:

By Mr. SPECTER (for himself, Mr. Allard, Mr. Allen, Mr. Bayh, Mr. Bennett, Mr. Biden, Mr. Bond, Mrs. Boxer, Mr. Breaux, Mr. Bunning, Mr. Burns, Ms. Cantwell, Mrs. Carnahan, Mr. Carper, Mr. Chafee, Mr. Cleland, Mrs. Clinton, Mr. Cochran, Ms. Collins, Mr. Conrad, Mr. Corzine, Mr. Crapo, Mr. Dayton, Mr. Domenici, Mr. Dorgan, Mr. Durbin, Mr. Edwards, Mr. Ensign, Mr. Enzi, Mr. Feingold, Mrs. Feinstein, Mr. Fitzgerald, Mr. Grassley, Mr. Harkin, Mr. Helms, Mr. Hollings, Mr. Hutchinson, Mrs. Hutchison, Mr. Inhofe, Mr. Inouye, Mr. Jeffords, Mr. Johnson, Mr. Kennedy, Mr. Kerry, Mr. Kohl, Ms. Landrieu, Mr. Levin, Mrs. Lincoln, Mr. McCain, Mr. McConnell, Ms. Mikulski, Mr. Miller, Mr. Murkowski, Mr. Nelson of Florida, Mr. Nelson of Nebraska, Mr. Nickles, Mr. Reed, Mr. Reid, Mr. Roberts, Mr. Rockefeller, Mr. Santorum, Mr. Schumer, Mr. Sessions, Ms. Snowe, Mr.

Thomas, Mr. Thompson, Mr. Torricelli, Mr. Warner, Mr. Wellstone, and Mr. Wyden):

S. 2924. A bill to authorize the President to award posthumously the Congressional Gold Medal to the passengers and crew of United Airlines Flight 93 in the aftermath of the terrorist attack on the United States on September 11, 2001; read the first time.

Hon. Thomas A. Daschle

OF SOUTH DAKOTA

Mr. President, what we saw happen to our country 1 year ago today will be forever etched in our memories. Several of our colleagues have taken time here on the floor today to reflect on that horrible day.

Though our Nation was wounded deeply that day, we learned a great deal about ourselves—and that has made our country stronger. The courage of the first responders, the valor of the passengers on flight 93, the strength of the families of the victims, the character of our Armed Forces, and the generosity of Americans from each and every State in the Union have shown to terrorists, and to the world, that America is strong and will not bow to terror.

H. Con. Res. 464 is a small tribute to each of these heroes. It spells out, in broad bipartisan fashion, Congress' memory for lost loved ones, our deep admiration for the families of these innocent victims, our respect for the work of our first responders and Armed Forces, and our resolve to find and bring to justice those responsible for the attacks.

That resolve was made clear on September 14, 2001, when we overwhelmingly passed S.J. Res. 23 In that resolution, we granted the President the authority to pursue the nations, people or organizations who perpetrated the September 11, 2001, terrorist attacks on the United States in order to prevent any future acts of international terrorism against the United States by such nations, people, or organizations.

With today's resolution, we look back at the horror and the hope we saw on a day we will never forget. There may come a day when we must again look ahead to threats to our Nation that lie on or beyond the horizon. I am confident that when that time comes, Congress will again act in a bipartisan fashion to take the steps needed to keep America strong, and Americans safe.

But today, as we walk the path from remembrance to recovery, this resolution says what we all know in our hearts: We will never forget.

Thursday, September 12, 2002

SEPTEMBER 11 REMEMBRANCE

Hon. Michael B. Enzi

OF WYOMING

Mr. President, yesterday we marked the anniversary of one of the most horrific events in our Nation's history. On September 11 of last year, without provocation or warning, extremists took control of four of our planes and used them as weapons of destruction against us to cowardly take from our lives our friends and neighbors, our mothers and fathers, and our sons and daughters.

As we watched those events unfold, during the subsequent rescue attempts, we saw more of our Nation's brave men and women lose their lives in the support and defense of others. It was not only a terrible loss of life. It was a loss of our most vital and valuable resource, our Nation's people and the potential they carried within them for greatness in so many different fields and endeavors of importance to them and to us.

As we watched the images broadcast around the world, we all made a decision in our hearts to do everything we could to respond to the attack on our Nation, our freedom, our liberty and our way of life. For each of us it meant something different, but for all of us, it helped to know there was something we could all do to help.

For Congress, that meant expressing our strongest support for the President and his ambitious and necessary plan to end the global network of terror that has sown the seeds of despair and hatred wherever it has found fertile ground. The President's plan is to do more than defeat the forces of terror. It is to replace those seeds of anger and hatred with seeds of hope and peace.

For our Nation's Armed Forces, it meant answering the call to duty and taking arms against an enemy who placed no value on human life.

The rules of war are not many, but one unavoidable one is that it takes the lives of our young men and women. One of those we lost in the early stages of the war was one of Wyoming's own, Jonn Edmunds, an Army Ranger from Cheyenne, who gave his life in Afghanistan as he fought and died for a cause that he believed in.

For all Americans, it meant an awakening of our sense of patriotism and our love of country, as we put aside our differences and unfurled our flags and proudly displayed them on our porches and windows. We came together as one, united, in support of our leaders and our President.

We know from past experience that the effort to respond to challenges like this is not a quick or easy one. It takes a lengthy and determined commitment to principle if we are to succeed.

I have no doubt our resolve will remain strong and we will be united in purpose, as we have done before when called to respond to a threat to our way of life.

A little over 50 years ago, on a day that has been compared to this one, those who opposed us were heard to say after their attack that they may have done nothing more than awaken a sleeping giant. On that day in December and this one in September, we may have been a sleeping giant, but when the time came to respond, we did, and by so doing, we changed the world.

We have to respond with strength and determination because those who attacked us chose their targets with such clear and evil intent. They attacked the World Trade Center because of its symbolic representation of our economic power. They attacked the Pentagon because of its symbolic representation of the power of our military. And they sought to attack our Nation's Capital because it is the heart of our government and it represents our democracy and our way of life.

No one will ever forget where they were or what they were doing as they first heard the news of the terrorist attack on our Nation. We all sat and watched in stunned silence as events unfolded that are now forever etched in our mind.

In the days that have passed since then, we have kept alive the memory of those we lost, repaired and restored what we could, and made plans to recreate what could not be saved. It has been a difficult and daunting task.

Through it all the President has led a united Nation, committed to ending the threat of terrorism, not just for us, but for our children, and for all the children of the world who deserve to grow up and pursue a dream of peace, hope and opportunity.

When the terrorists struck at the heart of our Nation that day they took something more precious than our buildings, and the symbols of American pride and ingenuity we all hold dear. When they took our loved ones from us, they also took the innocence of our children who had to learn quickly, and at a young and tender age, that there are bad people in the world who do bad things. And that all too often, bad things happen to good people.

But, when they looked at us with questioning eyes, did any of us have a good answer to the question they wanted answered the most, "Why?"

Fortunately, the President's leadership has enabled him to put together an international coalition dedicated to dismantling the network of terror and to bringing those responsible to justice, wherever they may try to hide.

The conspiracy of terrorism can only survive in the darkness of hatred. It can not long survive when we bring the light of peace to bear on all the nations of the world. That light is the symbol of freedom that our Statue of Liberty holds proudly and with purpose in the harbor of New York, not far from where the Twin Towers once stood. It is a light that will someday shine for everyone in every country in the world, and we will all live in peace and freedom.

We are, and always will be, a Nation of individuals. We all have our own stories, our own goals and ambitions, and our own plans for our lives. But, when faced with a crisis, as we were last year, we come together as one united in our commitment that no one will ever have to endure a tragedy as terrible as the events that unfolded last year.

Yesterday was a day of remembrance. It will always be so. May it serve as a constant reminder that we are one Nation, under God, with liberty and justice for all.

The lives of all those who were lost are like an unfinished symphony that has been left to us to continue and complete. We carry their dreams, their hopes, their ambitions, their challenges and their plans for the future with us. With God's strength and the support of each of us we will complete the work they started and ensure the safety and security of all people, of all countries, and of all regions of the world for generations to come.

MESSAGE FROM THE HOUSE
ENROLLED BILLS SIGNED

Under the authority of the Senate of January 3, 2001, the Secretary of the Senate, on September 11, 2002, during the recess of the Senate, received a message from the House of Representatives announcing that the Speaker has signed the following enrolled bills:

H.R. 3917. An act to authorize a national memorial to commemorate the passengers and crew of Flight 93 who, on September 11, 2001, courageously gave their lives thereby thwarting a planned attack on our Nation's Capital, and for other purposes.

H.R. 5207. An act to designate the facility of the United States Postal Service located at 6101 West Old Shakopee Road in Bloomington, Minnesota, as the "Thomas E. Burnett, Jr. Post Office Building".

MEASURE PLACED ON THE CALENDAR

The following bill was read the second time, and placed on the calendar:

S. 2924. A bill to authorize the President to award posthumously the Congressional Gold Medal to the passengers and crew of United Airlines Flight 93 in the aftermath of the terrorist attack on the United States on September 11, 2001.

Friday, September 13, 2002

Hon. Don Nickles

OF OKLAHOMA

Mr. President, I ask unanimous consent to have printed in the *Record* an article from the *Wall Street Journal* dated September 11, 2002.

There being no objection, the material was ordered to be printed in the *Record*, as follows:

[From the Wall Street Journal, Sept. 11, 2002]
WE WILL PREVAIL

(By Theodore Olson)

From a speech by Solicitor General Theodore Olson to the Federalist Society on Nov. 16, 2001, Mr. Olson's wife, Barbara, was one of the airplane passengers murdered on Sept. 11, 2001.

September 11, 2001 was unprecedented in our Nation's history. Our country has been attacked before. Our soldiers and innocent citizens have been the victims of terrorism before. But never before in our history have so many civilian citizens, engaged in the routines of their daily lives, who neither individually nor collectively had done anything to provoke the savage attack that they were to experience that day, been brutally murdered for the simple reason that they were Americans, and because they stood, in their countless individual lives, for all the things that America symbolizes.

As President Bush immediately recognized, Sept. 11 was an act of war. But it was much more than that. It was also a crime, an act of pure hatred and unmitigated evil.

The victims were of all races, backgrounds, religions, ages and qualities. They had one thing in common. They were nearly all Americans. Their lives were extinguished because they were the embodiment of the aspirations of most of the world's peoples. The people who killed them hate the beacon that America holds out to people who are impoverished, enslaved, persecuted and subjugated everywhere in the world.

The men who planned the savage acts of Sept. 11 cannot prevail as long as American ideals continue to inspire the people they hope to tyrannize and enslave.

It is a cynical lie that the animals that killed our loved ones were motivated by Islam, or because this nation of ours is anti-Islamic. Enshrined in the First Amendment to our Constitution is freedom of expression and the free exercise of religion. This continent was populated by people who crossed a terrifying ocean to reach a rugged and inhospitable frontier to escape religious persecution.

From its birth, this nation and the American people have offered sanctuary and shelter to all faiths. Our Constitution—always with the support of our people—has extended its embrace to the unpopular, the unusual, the unconventional and the unorthodox. We protect not only those who will not salute our flag, but those who would spit upon it or burn it. We pledge our allegiance to a Constitution that shelters those who refuse to pledge their allegiance to it.

It is true, I suppose, that there are many in the Middle East who hate this country for its support of Israel. But how tragic and misguided to despise us for extending comfort and defense to a people who have so long, and so recently, been the victims of indescribable ethnic persecution. Nor has America's support for Israel ever been rooted in or manifested by hostility to the Muslim faith or those who practice it. The terrorists and their apologists have lied about these things, but what is another lie when their goals and tactics are so vastly more evil?

The terrorists can succeed only through corruption and brutality. Thus they must tear down America and its system of laws which shields its people from those malevolent acts. They can enslave the people they wish to subjugate only by keeping them poor and destitute, so they must undermine and discredit the one place in all the world that stands the most for the rule of law and allows its people the opportunity to rise above all those conditions.

Abraham Lincoln was paraphrasing our Declaration of Independence when he characterized our Nation as having been "conceived in liberty and dedicated to the proposition that all men are created equal." That revolutionary document set down our collective belief in inalienable human rights, the proposition that governments derive their powers from the consent of the governed, the principle that tyrants who would oppress their people are unfit to be rulers of a free people, and the right to the pursuit of happiness.

The terrorists of Sept. 11 cannot prevail in a world occupied by the Declaration of Independence, the Constitution and its Bill of Rights, the Emancipation Proclamation, the Gettysburg Address, the Statue of Liberty, the World Trade Center, the Pentagon, the Capitol, the Supreme Court and the White House. They cannot coexist with these ideals, these principles, these institutions and these symbols. So they cannot survive, much less prevail, in the same world as America.

America is not today, or ever, without imperfections and shortcomings. Implementation of our lofty ideals has never been without error, and some of our mistakes have been shameful. But the course of our history has been constant, if occasionally erratic, progress from the articulation of those lofty ideals to the extension of their reality to all our people—those who were born here and those, from hundreds of diverse cultures, who flock here.

There is no segment or class of the world's peoples who have exclusive claim on the term "American," and no segment of the world's population to whom that claim has been denied. We welcome 100,000 refugees per year into this country. Over 650,000 people immigrated legally to America in the most recent year for which we have reliable statistics. Over five million people are in this country today who were so desperate to come here that they did so illegally.

There are more Jews in New York City than in Israel. More Poles in Chicago then any city in the world except Warsaw. America is home to 39 million Irish-Americans, 58 million German-Americans, 39 million Hispanic-Americans and nearly a million Japanese-Americans. And there are seven million Muslims in America, nearly the population of New York City.

How tragic it is that the agents of the Sept. 11 terrorist acts were people whom we welcome to this country, and to whom we extended all of our freedoms, the protections of all of our laws, and the opportunities this country affords to everyone to travel, work and live. But we welcome immigrants because nearly all of us are immigrants or descendants of immigrants who came here to enjoy freedoms, rights, liberties, and the opportunity, denied elsewhere, to pursue happiness and prosperity.

Ronald Reagan often said that "every once in a while, each of us native-born Americans should make it a point to have a conversation with someone who is an American by choice." Mr. Reagan was fond of quoting from a letter he received from a man who wrote, "you can go to live in Turkey, but you can't become a Turk. You can't go to live in Japan and become Japanese, {and so on for Germany, France, etc.}. But . . . anyone from any corner of the world can come to America and be an American."

So it is particularly sad and a bitter irony that the 19 savages who took the lives of thousands of Americans were able to come here because we welcomed them, and trusted them, and allowed them to learn to fly our airplanes and gave them the freedom to travel. They took these precious gifts and turned them into instruments of hatred and death.

It has, I suppose, always caused some resentment that we believe so passionately and unquestioningly that the freedoms we value should belong to all people. But we know that these are enduring values. We can debate nearly everything else, but we don't need to debate that. We know that these principles lift everyone up.

We have now been reminded, in the most horrible way, that there are those who not only hate our principles, but who would dedicate their lives—and surrender their lives—to banish those ideals and the incentives they provide for tyrannized and impoverished people everywhere to do what Americans did in 1776. We have tragically learned again, in the most unthinkable fashion, that our values and our principles are neither self-executing nor self-sustaining, and that we must sacrifice and fight to maintain what our forebears sacrificed and fought to bequeath to us.

And now the rest of the world is learning again that Americans will not flinch from that fight or tire of it. Americans will fight, they will sacrifice, and they will not give up or leave the job unfinished. This war is for all living Americans. It is for the parents, grandparents and great-grandparents that fought and sacrificed to come here. And it is for our children and generations to come. And it is for those who choose to become Americans in the future.

America will not lose this war because we cannot even consider that we will lose what centuries of Americans fought to create, improve and maintain. We cannot, and we will not, betray the people who gave us this glorious heritage. We cannot and will not, dishonor or wash away the memories of those who somehow clawed their way out of poverty, tyranny and persecution to come to this country because it was America, and because they were willing to risk death to become Americans, and to give their children and grandchildren the opportunity and freedom and inspiration that makes this place America. Americans could no longer call themselves Americans if they could walk away from that legacy.

People who write for newspapers and who offer opinions on television, or who send advice to us from other parts of the world, sometimes say that America is too rich, lazy, complacent, frightened, soft and enervated to fight this fight. That we have no stamina, strength, will, patience, or steel. That we will collapse.

They are so wrong. We will prevail for the very reason that we have been attacked. Because we are Americans. Because the values that made us free, make us strong; because the principles that made us prosperous, make us creative, resourceful, innovative, determined and fiercely protective of our freedoms, our liberties and our rights to be individuals and to aspire to whatever we choose to be. Those values and those characteristics will lift us and will defeat the black forces who have assaulted our ideals, our country and our people.

The very qualities that bring immigrants and refugees to this country in the thousands every day, made us vulnerable to the attack of Sept. 11, but those are also the qualities that will make us victorious and unvanquished in the end.

Tuesday, September 17, 2002

ENROLLED BILLS SIGNED

The following enrolled bills, previously signed by the Speaker of the House, were signed on September 12, 2002, by the President pro tempore (Mr. Byrd).

H.R. 3917. An act to authorize a national memorial to commemorate the passengers and crew of Flight 93 who, on September 11, 2001, courageously gave their lives thereby thwarting a planned attack on our Nation's Capital, and for other purposes.

H.R. 5207. An act to designate the facility of the United States Postal Service located at 6101 West Old Shakopee Road in Bloomington, Minnesota, as the "Thomas E. Burnett, Jr. Post Office Building".

Wednesday, September 18, 2002

TRIBUTE TO BRADY HOWELL

Hon. Larry E. Craig

OF IDAHO

Mr. President, the terrorism of September 11 changed America forever, and it profoundly changed Americans, as well. The people we lost left behind legacies, the compilation of the meaningful things they accomplished throughout their lives, actions and words that still touch their friends and families after their deaths. Those legacies inspire all of us with the bravery and courage of the human spirit, and also remind us of the precious frailty of life.

Brady Howell lost his life in the attack on the Pentagon. This letter, written by Brady's brother, Carson Howell, to commemorate the 1-year anniversary of that terrible event, articulates the legacy Brady left behind. I would like to enter this letter into the *Congressional Record* so all my colleagues can remember the great example these Americans are to us. In the words of Carson Howell, "The men and women who perished that day are not heroes because of how they died; they are heroes because of how they lived."

Let me read the letter in its entirety:

Today is a sad day for our family. Not just our family, but also families just like the Vauk family, the Conaty family, the Andrews family, and thousands of others. It's a sad day for our American family as we all remember and pay tribute to the thousands of friends, family, and fellow Americans that lost their lives 1 year ago today. It's a day that many will remember as the day we learned that heroes aren't found only in comic books. No, there are heroes greater than Superman and my brother is one of them.

Brady Kay Howell loved this country. He was an Eagle Scout. He loved children and taught the youth in Sunday School classes while living in New York and later Virginia. He loved his family and actually had plans to return to Idaho that following weekend for a welcome home party for my parents and for my wedding reception. He loved his wife, Liz, to whom he'd been married for only five short years.

Brady was working in naval intelligence as an intern. Shortly before his death, he and I had a telephone conversation. In it he told me that one of his goals in his life was to have top-secret clearance. I'm proud to say that he accomplished that goal.

I could go on and on about how great my brother was. But, if it were he speaking here today, he wouldn't use this opportunity to speak of his accomplishments. I believe that he would talk about service. He would talk about what a great country this is that we live in and how proud he was to serve and protect all of us.

The work that Brady and many others did that died that day was for all of us. Brady prepared briefings for the Chief of Naval Intelligence and other high-ranking officials so that they could best be informed of how to protect us—the American public. Every day he was protecting our country. Every day he was fighting for our freedoms that we enjoy. To Brady, it didn't matter how much money you had, it didn't matter what the color of your skin was, it didn't matter which religion you believed. To Brady, what mattered were the people.

Ongoing community service initiatives to commemorate Brady's commitment to public service are being conducted in the Washington, D.C., area and there are plans for at least one such initiative in Utah. Generous contributions from all over the country have allowed us to create an endowed memory in Brady's name to continue the influence of his story. These contributions will also support an endowed lecture series in Brady's name that has been established and now approved by the BYU-Idaho Board of Trustees.

I miss Brady very much. I remember with fondness building bases and battling with our G.I. Joe action figures, waking up early Saturday morning to watch the "Bugs Bunny and Tweety Show" together, and climbing trees together. I always looked up to Brady and for me, he was always a hero. As his story is told, others are hearing about the hero whom I was privileged enough to call "brother."

September 11 wasn't the first day that this country has known heroes, nor will it be the last. We should take this time to pay tribute not only to the heroes of September 11, but all of the heroes that have fought for freedom. Thousands of men and women are working today to protect us from evil. The men and women who perished that day are not heroes because of how they died; they are heroes because of how they lived. Heroes are the men and women who have put themselves in harms way for the cause of democracy and freedom since long before September 11, 2001. Heroes are the men and women who serve each day to protect people they will never know. Heroes are the men and women who spend more waking hours caring for and about others than they do for themselves. Let us remember the heroes of September 11, 2001, along with the heroes who stood before, who stand now, and who are preparing to stand against evil. Because it is to all of you who have served this country, have given your children for the service of America, and are currently serving that we, the American people, pay tribute this day; the firefighters, the police officers, the emergency medical crews, and the soldiers of freedom.

If the mark of a hero is one that cares about and fights for others, I hope that the destruction of September 11 has facilitated the construction of tomorrow's heroes. Wouldn't the greatest honor that we could pay to those that perished be if we could follow their example and give of ourselves as they did? We may not be called upon to die for this country, but we are all called upon to live for it. This country doesn't need more martyrs, but this country could use more doers.

Tens of thousands have given their time and tens of thousands have given their lives for America; this "one nation, under God, indivisible, with liberty and justice for all." To be "one nation," we need to be one state, one neighborhood, one home. Let us rededicate ourselves as we did after September 11, to being Americans. Never in my life before September 11, had I seen such a display and attitude of patriotism. We were friendlier, we were more patient, and we looked out for each other. I wish that those who died that day could have seen the America that we became. We became strong and united. We showed forth the America that we always should have been; the America that those men and women sacrificed their lives for. Let us honor all of the heroes of America by not letting their sacrifices be in vain. Let us continue their legacies. Let us live for what they died for—The United States of America.

Thursday, September 19, 2002

THE POEM AMERICAN PRIDE

Hon. Mary L. Landrieu

OF LOUISIANA

Mr. President, the events of September 11 were very tragic and very traumatic for our Nation, especially our children. This poem, written by 10-year-old James Dillon Hughes of Bourg, LA, demonstrates, very simply, what is great about America. In these few lines James captures the spirit of his country, stronger now than ever before. It is our job to ensure that the freedoms we enjoy now will still ring true for our children and future generations to follow. James wrote this on September 13, 2001, only 2 days after the terrible events of September 11. Even after those tragic events, James was still able to show his American pride. I was so moved upon reading this poem that I ask that it be printed in the *Record*.

The poem follows:

AMERICAN PRIDE

(By James Dillon Hughes)

I am proud to be an American
I am proud to be free
I'm proud to be able to choose anything I want to be.

I can be a doctor, a lawyer or a priest
Because I live in a country
That allows me to be free.

Our country was somewhat divided
Now it has united
Let's keep it strong and free.

Where leaders teach and guide us
Always stand beside us
And show us the way to be.

Our country is rich
Our army is strong
Living in America
Could never be wrong.

Wednesday, September 25, 2002

NEW JERSEY TASK FORCE ONE

Hon. Robert G. Torricelli

OF NEW JERSEY

Mr. President, I rise today to bring to the attention of the Senate the outstanding efforts and contributions of the New Jersey Task Force One Urban Search and Rescue Team.

On September 11, 2001, at approximately 9 a.m., New Jersey Task Force One was activated and deployed to the World Trade Center attack. Requested by New York City's Office of Emergency Management, New Jersey Task Force One had an advance team in New York City by 11:15 a.m. The team's entire equipment cache, along with 140 members was set up at the Jacob Javits Center by 2 p.m. that same day. They deployed search assets to search for survivors during the afternoon and evening of the 11th and for the next 10 days they worked continuously as part of the urban search and rescue effort.

In a short period of time, New Jersey Task Force One became an integral part of the rescue efforts. The team was integrated into the FEMA Urban Search and Rescue System and was included in the FEMA Urban Search and Rescue Action Planning. The well-equipped New Jersey Task Force One had the ability to lend critical equipment to some of the responding FEMA teams and developed an excellent working relationship with many of these teams. Structural engineers and planners from New Jersey Task Force One were used by the FEMA Incident Support Team to brief the incoming FEMA teams as they rotated into New York City.

FEMA gave an operational performance evaluation to New Jersey Task Force One for their participation related to the World Trade Center operations. New Jersey Task Force One was rated as superior in six out of nine categories. Those categories included attitude, initiative, consideration for personnel welfare, obtain necessary equipment and supplies, physical ability for the job, and safety.

With valor and honor, the members of New Jersey Task Force One have proven themselves as one of the finest urban search and rescue teams in the country. It is for these reasons that my colleague, Senator Corzine and I introduced S. 2618 to designate New Jersey Task Force One as the 29th national urban and search rescue team.

New Jersey Task Force One was formed in 1997 and they have used the FEMA urban search and rescue guidelines in its development. The team is organized under the New Jersey State Police Office of Emergency Management Domestic Preparedness and Hazardous Material Planning and Emergency Response Unit. The State provides an annual budget of $1.5 million to support New Jersey Task Force One.

The team consists of 180 members from over 100 different fire, police and EMS agencies throughout the State. Over the past few years, New Jersey Task Force One has purchased the necessary cache of equipment, based on the FEMA equipment list, and has trained its members to FEMA urban search and rescue standards. New Jersey Task Force One is in the process of purchasing weapons of mass destruction response equipment.

The 180 courageous men and women of New Jersey Task Force One have dedicated their lives to saving others. They have proven themselves to be highly qualified to join FEMA's urban search and rescue system. I strongly believe that New Jersey Task Force One would enhance FEMA's urban search and rescue program. I urge the FEMA Director to include New Jersey Task Force One as the 29th FEMA urban search and rescue team.

Thursday, September 26, 2002

Hon. Hillary Rodham Clinton

OF NEW YORK

Mr. President, on September 12, 2001, hundreds of thousands of New Yorkers woke up to a changed world. Thousands had lost family, friends and coworkers to the terrorist attacks of September 11 and hundreds of thousands more New Yorkers had lost their jobs. America watched the scenes of New York and felt pride in the firefighters, the police officers, the emergency workers, and the construction workers who had all fled to Ground Zero to help with recovery.

Monday, September 30, 2002

HOMELAND SECURITY

Hon. Joseph I. Lieberman

OF CONNECTICUT

As the Senator from Utah said, we never want to give the impression we do not think the employees who will move to this Department are as concerned about homeland security as we and the rest of the American people are. In fact, the evidence before us is quite ample that Federal employees are concerned.

The stories are legion and numerous of Federal employees—I think of FEMA employees—they were somewhere else and they rushed to the Pentagon to be of help; they flew to New York; they worked hours and hours of overtime. Of course, the most vivid demonstration of the way in which union membership is not inconsistent with national service or sacrifice is the firefighters in New York, several hundred of whom were off duty on September 11. When they heard what had happened, they just rushed to the scene. Nobody was thinking about whether this was supposed to be a day off under the collective bargaining agreement, what risks they were assuming, or they were going to be asked to do things that were not quite in their job description. Needless to say, a lot of them not only rushed into the building, but they never came out.

Tuesday, October 1, 2002

EMERGENCY UNEMPLOYMENT COMPENSATION ACT OF 2002

Hon. Edward M. Kennedy

OF MASSACHUSETTS

I want to speak today about America's workers. We live in a Nation forever changed by the tragic events of September 11. The dreadful images seared into our memories on that fateful day were grim proof to every American that we are vulnerable to grave new threats. We must take the necessary steps to protect America from these new dangers. We must act wisely as we create a new Department of Homeland Security. We must ensure that our actions truly enhance, rather than diminish, our Nation's security. And we must meet our security needs in ways that reflect the values that make America the envy of the world.

As we debate the formation of this new agency, we should remember the events of September 11 and the heroism of our Nation's union workers in the cause of homeland security. Union members risked and lost their lives and saved countless others through their actions on September 11. We will never forget the example that firefighters, construction workers and many government workers set that day.

Union workers have also shown great bravery and extraordinary sacrifice in the service of homeland security since September 11. The postal workers and the hospital worker killed as a result of bioterrorism were all union members. The brave flight attendant, whom the President recognized in the State of the Union Address for preventing terrorism, is a member of a union.

The dedication and resolve of these union members truly represents the best of America. Over 43,000 of the Federal workers affected by the proposed government reorganization are currently union members. These are the workers who risk their lives each day to protect our Nation's borders. They are the workers from the Federal Emergency Management Authority who coordinated the Federal emergency response on September 11. These workers are out every day on the high seas to rescue those in need and to prevent dangerous cargo from reaching our shores. They are also the workers dedicated to making our Nation safer from the threat of bioterrorism.

Among the ranks of unionized Federal workers are true heroes who have served their Nation with distinction in battle and are now contributing to our Nation as civilian employees and as active members of their community. I am talking about Federal workers like Robert J. Patterson, who was awarded the Purple Heart Medal and the Bronze Star and many other honors for his service in Vietnam. He was ambushed and shot in the legs, the stomach and the shoulder while on patrol in Vietnam, but he still managed to call for backup and save the lives of many other members of his squad. For nearly 20 years now, Mr. Patterson has worked as a civilian employee for the Federal Government, and he now serves as vice commander of his local VFW post and is active with the Boy Scouts and as a mentor for troubled youth.

Dedicated Federal workers like Mr. Patterson take pride in their work, love their country, and have served it with distinction for decades. Nearly half a million Federal workers are veterans of our Nation's armed services. Veterans are represented at twice the rate in the Federal workforce as in the private sector. Disabled veterans, those who have paid a great price for serving this Nation, are five times more likely to work in the Federal Government as the private sector.

On September 11, unionized Federal workers were on the scene and played critical roles at both the World Trade Center and the Pentagon as they worked round the clock to make our homeland secure. Denise Dukes, of the Federal Emergency Management Agency, worked a 24-hour shift in Washington on September 11 to ensure that food and water was reaching the rescue personnel at Ground Zero. Afterward, she left her two children to go to New York and coordinate the response and recovery effort on the ground. As Ms. Dukes explains of her fellow Federal workers: "We were proud and eager to serve our fellow Americans, and we would never allow anything to stand in the way of that mission."

Michael Brescio, who works for the Environmental Protection Agency's Response Team, got tens of thousands of urgently needed respirators to the rescue workers at Ground Zero immediately after the attack. Far away in Kodiak, AK, Mark Andrew Jamison went on high security alert in order to protect our Nation's coastline. Mr. Jamison, a veteran of our Nation's armed services who was entrusted with a top secret security clearance, loves his job because, as he put it: "Above all . . . I'm a patriot like the hundreds of thousands of other Federal employees who keep our country secure and safe day-in and day-out."

We must protect the rights of these dedicated Federal workers to remain union members and we must allow other workers in the new department to exercise their fundamental right to form a union.

REFLECTING ON THE ANNIVERSARY OF SEPTEMBER 11

Hon. John Ensign

OF NEVADA

Mr. President, 1 year ago, this Nation stood united. Together we mourned, prayed, and hoped. We hugged our loved ones a little bit longer and a little bit tighter. Our hearts wept for the thousands of families who unexpectedly and unbelievingly lost a husband or wife, a mother or father, a son or daughter at the hands of evil.

It's hard to believe that an entire year has passed since that surreal day. While we have observed holidays, celebrated milestones, and continued with life, there are still daily reminders of the horrific events of 1 year ago. Flags still fly more frequently than before, security precautions still cause delay, and our hearts still weigh heavy when we think about the dreams that were cut short that tragic day.

As we remember September 11, I encourage you to make today a day of introspection and compassion.

Remember where you were last year when you heard the news. Remember the footage you watched in disbelief. Remember the pain you felt in your heart. Take those images with you throughout the day. Make it a point to leave work on time, have dinner with your family, talk to each other about what today means, and hug your loved ones a little bit longer and a little bit tighter.

On your own or as a family, consider doing something for your community in honor of the victims of 9/11. It can be donating blood, making a financial contribution to a needy cause, or giving your time and energy to a worthwhile organization.

I hope that we can all make today a positive and meaningful opportunity to unite our communities in helping others and honoring the victims of 9/11. Together we will send a strong message to the world that Americans remain united. Time will not steal our memory of the victims and attacks of September 11.

Friday, October 4, 2002

ADDITIONAL COSPONSORS

S. 1434

At the request of Mr. Specter, the name of the Senator from Michigan (Ms. Stabenow) was added as a cosponsor of S. 1434, a bill to authorize the President to award posthumously the Congressional Gold Medal to the passengers and crew of United Airlines Flight 93 in the aftermath of the terrorist attack on the United States on September 11, 2001.

Tuesday, October 15, 2002

MASSACHUSETTS MEMORIAL SERVICE

Hon. Edward M. Kennedy

OF MASSACHUSETTS

Mr. President, I am honored to join all of you, the families of loved ones from across our Commonwealth who lost their lives last September 11.

We come to this birthplace of liberty to remember, to give honor, and to express our resolve.

All around us in this historic place are the images of famous leaders who brought life and nationhood to the ideals that were attacked a year ago, on a day whose dawn had seemed almost uniquely American in its sunny optimism.

Etched on the wall around this stage are the names of heroes who gave their lives for our country on September 11, 2001. The list is heartbreaking, and it goes on and on. These heroes were famous in a different way, famous to their friends for their fabled jumpshot in a neighborhood park, or prized in their firms for a brilliance tempered by laughter, or celebrated by their young children as superheroes, able to launch them into the air with an easy toss, and always there to catch them. They expected to pass the ball again, to make another trade or tell another joke, to come home that night and read a bedtime story.

Then they were gone, in the darkness at midmorning which succeeded that sunny dawn. We mourn them for the years that were too few and the hopes that were unfulfilled. We praise them for the way they lived, and in so many cases for the bravery in the way they died. And we as a country, as a community, as friends and neighbors and family, hold them in our hearts.

I spoke with a member of almost every family in Massachusetts who lost a loved one on the planes, or at Ground Zero in New York, or at the Pentagon. To those left behind, I say on this sad day: I know something of what you feel. To lose someone you love, and to lose them so suddenly, so unexpectedly, so terribly, to see them torn out of the fabric of life, is almost more than one can bear.

And then, although we know the passage of a year cannot heal that memory, we move on, because we have to, because they would want us to, and because there is still light left in the world, including the love they left us.

In a different time of grief, my brother Robert Kennedy quoted the ancient poet Aeschylus:

> In our sleep, pain, which cannot forget, falls drop by drop upon the heart until, in our own despair, against our will, comes wisdom through the awful grace of God.

May God, this year and every year and every day, grant that grace to you the families.

And for all of us, there is something else that comes from last September 11. From the pain that day have come both wisdom and will.

We have learned anew the wisdom that as Americans, we are many, but we are also one. On flight 93, there was a unity of purpose and a fierce pride. Passengers who had never met before became a band of brothers and sisters, sacrificing their lives so that others might live. Many other individual acts of courage saved more lives than we can know or count at Ground Zero and the Pentagon.

People all across the country and of all ages asked what they could do, from giving their blood, to clearing rubble at the World Trade Center, to giving their dollars, to lending a shoulder to their neighbor to cry on. In countless ways, we came together, and founded a new American spirit of service to others.

The terrorists taught us a lesson different from the one they expected. They acted with hate, but we reached out to comfort and support one another with love. No one asked whether the rescuer leading them down the packed stairwell of the World Trade Center was rich or poor, Anglo or African-American or Hispanic, gay or straight. We gained a new determination as Americans to reject discrimination in all its hateful forms.

Out of the pain that day, Americans understood more powerfully than perhaps ever before the pledge of "liberty and justice for all."

To help those in need;

To give hope;

To share what we have;

To see suffering and try to heal it—

That is our lesson from this tragedy, and it is wisdom that must guide us over time. The new American spirit of service can and must become a new era of commitment to the ideals of compassion, equality, opportunity, and concern for one another. We as a society seek to save a life when a terrorist strikes, and we as a society must do as much when the terror or a dread disease strikes, or the terror of poverty steals opportunity.

May that legacy of 9/11, that legacy of love and compassion and caring, become our enduring tribute to all those who were lost.

Out of that day also came a new sense of national resolve and will. We are at war today, with a terrorism that has plagued too many places for too many years, and that has finally struck at the heart of America.

This is a conflict we did not seek, but must win, not alone for ourselves, but for the cause of freedom, tolerance and human rights around the world.

The ideas and ideals created long ago in this great hall have shaped the dreams of countless millions yearning to be free.

Now, as the greatest power on Earth, we have a responsibility. Our gifts of strength and wealth and values can decide that the future will belong to the forces of hope and not to hate.

This brighter future depends on victory against terrorism. It demands that we then continue in a long, tireless endeavor to make the world not only safer for us, but better for all. In our determination to defeat those who have attacked our people and our principles, we truly are "one Nation under God, indivisible."

How true that was, how deeply we felt it, a year ago today. Together that day, we hurt and feared and hoped and prayed. And together now, we will prevail.

God bless all who were lost and all who lost them. God give us strength, and the wisdom to use it well. God bless America.

MESSAGES FROM THE HOUSE

At 1:05 p.m., a message from the House of Representatives, delivered by Ms. Niland, one of its reading clerks, announced that the House has agreed to the following concurrent resolution, in which it requests the concurrence of the Senate:

H. Con. Res. 487. Concurrent resolution authorizing the printing as a House document of a volume consisting of the transcripts of the ceremonial meeting of the House of Representatives and Senate in New York City on September 6, 2002, and a collection of statements by Members of the House of Representatives and Senate from the *Congressional Record* on the terrorist attacks of September 11, 2001.

Thursday, October 17, 2002

AMERICA'S STRENGTHENED RESOLVE

Hon. Rick Santorum

OF PENNSYLVANIA

Mr. President, this year, we did not wait passively for September to arrive; we began preparing weeks ago to greet this month with offerings of memorial in hand. At services across the Commonwealth and in remembrances around the country, last fall's attacks have again drawn the focus of our Nation. There is a new sentiment this time around, though, one that is hopeful, grateful, more determined, and less confused.

For all of us, it has been a week of reflection on the losses and lessons of the attack that changed our history and our lives. The destruction wrought by a hateful few was intended to unravel America's strength, but it has only made us stronger. And from this strength, we have come to understand that the tragedy of last September 11 has in fact blessed us with an opportunity. The attacks are still tangible in Pennsylvania, and so we take this opportunity very seriously, proud to have a part in creating a positive legacy for 9/11. It was aboard the plane that crashed in Shanksville that America's response to terrorism first began.

Somerset County, for this reason, will be a symbol of the heroism and sacrifice that a few brave, ordinary citizens chose to exhibit when faced with the most difficult and dangerous situation of their lives. Shanksville, the World Trade towers, and the Pentagon can all be reminders of what the American spirit is capable of overcoming, of what Todd Beamer meant when he said, "Let's roll," if we as a Nation choose to make it so. The anniversary of September 11 should, therefore, be about the resolve to honor the memories of all those lost to the terrorist attacks by living to make ourselves, our communities, and our country better.

Looking back over the past 12 months, the most inspiring aspect of the national recovery effort was the compassion, cooperation, and concern that citizens across the country shared with one another. Through the charity of time, prayer, blood, consolation, money, and other expressions of support, Americans exhibited a good will that is rarely seen so universally, but comes so naturally to us all at times of crisis. As we settle back into our normal, peaceful lives, however, this good will tends to steal away from us. As a result, our collective awareness of a common humanity and a world view larger than our own backyards also begins to fade. In the aftermath of 9/11 and the years to follow the shock of terrorism on our soil, we must renew the commitment we have to our neighbors, our communities, and our Nation. Across the country, we can make the courage and responsibility displayed by the heroes at Ground Zero endure. In this way, we will triumph over evil and devastation, and we can try to make sense out of all that we have suffered.

When I first visited the cratered field in Shanksville, and when I returned to that crash site this week, I was struck by the importance of our continued hope. I was also inspired by the strength of those flight 93 family members, now carrying the torches of their loved ones who gave their last measure of bravery for our Nation. I have resolved to make every day a memorial to September 11 by working to keep the bigger picture in mind and a better world in sight. I hope you will find your own way to keep and exhibit this renewed American spirit in your lives. May God bless you and our great country.

IN CELEBRATION OF THE WOMEN AT GROUND ZERO

Hon. Barbara Boxer

OF CALIFORNIA

Mr. President, I would like to take this opportunity to share with the Senate my thoughts on 33 women who courageously served as rescue and medical workers, firefighters and police officers in New York City on September 11, 2001.

It is my great honor to recognize the extraordinary contributions made by these rescue workers who bravely worked to save lives at Ground Zero in New York City during the horror of September 11, 2001. The selfless actions of these women helped heal our country during a time of national tragedy. On September 11, we found out as a Nation what heroism truly is, how strong and united we can be, how we can set aside differences for the greater good and work together. And these women helped show us the way.

Some wonderful people in my home State of California are bringing these women to Sonoma County for an all-expense-paid week in the wine country to pay tribute to their heroism. I want to send my warmest thanks to Susan Hagen and Mary Carouba, authors of "Women at Ground Zero," who wanted to make sure that the contributions of women rescue workers were recognized and honored along with their male counterparts.

In honor of their incredible efforts on September 11 and the important work they do every day, I am going to read the names of 30 women who worked at Ground Zero and then I will remember 3 women rescue workers who lost their lives on September 11, 2001: Detective Jennifer Abramowitz; Rose Arce, who is not a rescue worker but who was doing a live broadcast next to Ground Zero on September 11 in order to get vital escape and rescue information out; Lt. Doreen Ascatigno; Cpt. Brenda Berkman; Maureen Brown; Tracy Donahoo; Major Kally Eastman; Bonnie Giebfried; Lt. Kathleen Gonczi; Sarah Hallett, PhD; Cpt. Rochelle "Rocky" Jones; Sue Keane; Tracy Lewis; Patty Lucci; Christine Mazzola; Lt. Ella McNair; Cpt. Marianne Monahan; Lt. Amy Monroe; Lois Mungay; Cpt. Janice Olszewski; Carol Paukner; Sgt. Carey Policastro; Mercedes Rivera; Lt. Kim Royster; Maureen McArdle-Schulman; Major Molly Shotzberger; JoAnn Spreen; Cpt. Terri Tobin; Nancy Ramos-Williams; and Regina Wilson.

I also want the following names to be memorialized today: Yamel Merino, emergency medical technician; Cpt. Kathy Mazza, commanding officer of the Police Academy at the Port Authority Police Department; and Moira Smith, police officer with the New York Police Department. All three of these women sacrificed their lives on September 11, 2001, in their heroic efforts to save the lives of others.

None of us is untouched by the terror of September 11, and many Californians were part of each tragic moment of that tragic day. I offer today this tribute to the heroic women who worked tirelessly and selflessly at Ground Zero. I want to assure the families of Yamel Merino, Cpt. Kathy Mazza, and Officer Moira Smith that their mothers, daughters, aunts, and sisters will not be forgotten. And we will always be grateful to the brave men and women who worked tirelessly and selflessly at Ground Zero.

Wednesday, November 13, 2002

INDEPENDENT COMMISSION TO INVESTIGATE SEPTEMBER 11 ATTACKS

Hon. John McCain

OF ARIZONA

Mr. President, the legislation Senator Lieberman and I introduced last year to create an independent commission to investigate the September 11 attacks passed the Senate as an amendment to the homeland security bill by a vote of 90 to 8 in September. Days before the vote, the administration issued a letter supporting the creation of an independent commission. But Congress is about to adjourn without having done so.

The agreement that was reached on the homeland security bill is a welcome development and will make our Nation more secure. But the agreed text does not include our independent commission proposal, despite an overwhelming Senate vote in September and despite its previous inclusion in both the Lieberman and Gramm-Miller bills.

I believe President Bush and his team have responded admirably and with a sense of purpose to the terrorist attacks, and the joint intelligence committee investigation into the associated intelligence failings has added to our understanding of what went wrong. But neither the administration nor Congress is alone capable of conducting a thorough, nonpartisan, independent inquiry into what happened on September 11, or to propose far-reaching measures to protect our people and our institutions against such assault in the future.

To this day, we have little information on how 19 men armed with boxcutters could have so effectively struck America. After every other such tragedy in our Nation's history, like Pearl Harbor and President Kennedy's assassination, independent investigations were immediately appointed to examine what went wrong and recommend needed reforms to prevent such tragedies from happening again. There has been no such review since September 11.

This is what our proposed commission would do. Its goal would be to make a full accounting of the circumstances surrounding the attacks, including how prepared we were, and how well we responded to this unprecedented assault. The commission would also make comprehensive recommendations on how to protect our homeland in the future. It would examine not just intelligence but the range of government agencies and policies, from border control to aviation security to diplomacy.

Learning the lessons of September 11 will require asking hard questions. It will require digging deep into the resources of the full range of government agencies. It will demand objective judgment into what went wrong, what we did right, and what else we need to do to deter and defeat depraved assaults by our enemies in the future.

No such review has occurred to date. Passage of the homeland security legislation is a good start to making needed reforms, but to some extent we are flying blind in our efforts to reform our approach to homeland defense because we still do not know what parts and policies of the government failed the American people last September 11.

We do know, thanks to press leaks and the work of the joint intelligence committee, that significant failures occurred.

The chairman and ranking member of the Senate Intelligence Committee have suggested we might have prevented the September 11 attacks had we properly analyzed available information. They strongly support our independent commission legislation to carry on the work their joint intelligence investigation started. Together with Senators Bob Graham and Dick Shelby, we have been negotiating intensively with the White House and remain hopeful we might reach an agreement with them to create a commission, but we believe Congress must speak on this issue.

The families of September 11 will not rest until they have answers about how their government let them down and what we can do to make sure such tragedy never strikes America again. This is not a witch hunt. It is a search for the answers that will enable us to better protect our Nation against future attack by terrorists. It is about the future, not the past. It is worthy of the strong bipartisan support it has already received. I urge my colleagues to support this amendment.

I want to thank my friend from Connecticut for his efforts on behalf of this commission. I want to thank him for his efforts on behalf of the families, and I want to thank the White House for their continued negotiations. It is time we wrapped up these negotiations so this commission can be part of the homeland security bill.

Hon. Joseph I. Lieberman

OF CONNECTICUT

Mr. President, I thank my dear friend from Arizona for his strong statement, for joining me in introducing this amendment, and for his characteristic steadfastness in pursuit of an important cause regardless of the opposition and where it comes from.

He and I introduced legislation last December. We are approaching a year ago. It was a few months after the tragic attacks of September 11. We felt there should be an independent citizen commission, nonpartisan, with full powers of subpoena and adequate resources to investigate how could September 11 have happened, because clearly the fact it did happen said we were not adequately protecting the American people. We were insistent that this kind of investigation occur so we could learn how to prevent it from ever happening again.

There have been roadblocks along the way, but we have continued to state, and we state again, we are not going to give up this fight until such an independent commission is created because we cannot rest until the truth and the whole truth, so help us God, as best as anyone is able to find it, is determined about September 11. Because without that unlimited, unvarnished, uninhibited truth, we are not going to be able to inform this new Department of Homeland Security adequately.

This measure of ours passed the Senate earlier this year when we were considering the homeland security measure. It passed overwhelmingly with bipartisan support. In fact, the so-called Gramm-Miller substitute incorporated this provision, which I was very grateful to Senator Gramm and Senator Miller for doing, and Senator McCain was a great advocate for that cause.

In the substitute introduced by Senator Thompson, in coordination with the White House and the House, the commission proposal is not in it, and that is not acceptable. Senator McCain has said happily we continue to negotiate with the White House up until this moment, hopeful that an agreement can be secured that will create the aggressive, independent, nonpolitical commission this tragedy requires. But if it is not, and we have not reached an agreement yet, we are going to do everything we can to reinsert this commission into this homeland security bill where it belongs.

I think I can say for my friend from Arizona and myself if for some reason that does not work, we are going to keep introducing it wherever and whenever we think we can get a vote that will make it law. We owe this to the families of the September 11 victims.

I have met with them, as Senator McCain has, several times. Their desire for this commission is in some ways the strongest and most compelling argument anyone can make on its behalf, because they asked us and they asked America, having lost loved ones, how could September 11 have happened? We owe them an answer to that question, and we have not given it to them yet.

As Senator McCain said, the work by the Joint Intelligence Committee has revealed information, media investigations have revealed information, that only increases our understanding of how much more we need to know. The Senate coleaders of the Intelligence Committee, Senator Graham and Senator Shelby, are now strong supporters of this commission idea.

Going back to the families of the September 11 victims, I do want to say the persistent advocacy of these families, led by Steve Push, Kristen Breitweiser, Mary Fetchet, Beverly Eckert, and so many others, despite their great personal loss, has inspired not only my deep admiration but our continuing commitment to fight for this commission until it comes to fruition. We are not interested in pointing fingers. This is all about our common security, and improving it is our common responsibility.

I hope our colleagues will join us in supporting this amendment to the homeland security bill and restoring this provision to create an independent commission on September 11.

It is the clear desire of the families of the victims of September 11 that this commission be created by Congress. We ought to create it. This was a national catastrophe.

As we create a Department of Homeland Security to protect the American people from that ever happening again—as the representatives of all the people of this country, all of them in this terrible new era we have entered, potentially victims of terrorism—ought to say loudly and together, hopefully together with the administration, we can never know too much about how September 11 happened. We do not know enough now how September 11 happened. The one best way to know as much as we can of the truth about September 11 is to create a strong, nonpolitical commission with full resources and powers of subpoena to get to the truth.

The day for this commission will come. The arguments for it are irresistible. Let us hope that day is sooner than later. I thank my friend from Arizona for his persistence and advocacy. Also, it is an honor to work with him. We will stand shoulder to shoulder with a lot of other Members, of both parties, of this body to get this commission created.

AMERICA'S POWER

Hon. Jeff Sessions

OF ALABAMA

Madam President, I join my colleagues today to commemorate September 11, 2002, and the special joint session of Congress held in New York City. Americans are a generous people, with both our time and our money. We are a people committed to our religious beliefs. We are a people who place great value on education and the individual. We seek opportunity at every turn for our children, and we strive to take care of those who are elderly, infirm, and less fortunate.

We are also a people who take great pride in the protections we collectively offer one another through a common defense. We wear most humbly the mantle of "super power."

Last September 11, our collective vulnerability in securing the borders of our Nation was made known to all. In those initial chaotic hours, we watched the opening battle of what is now called the "war on terror."

Like many States, Alabama lost sons and daughters that day. Indeed, five Alabamians died in the Pentagon. Families were broken and great symbols of our might and entrepreneurial achievement made waste. We felt, and I believe we still feel, a collective pain in our hearts that will never heal. But the world has witnessed the development of a new resolve among Americans. A resolve too quiet for too long.

In the past year, we have taken a long and hard look at our defense posture. We have found great problems that must be fixed. We have found strength. We have committed our uniform services to battle, and we must give our President tools and the authority to get the job done.

Nearly a year has elapsed. Our emotions still run high.

America is demanding much from itself and its governmental leaders. The creation of the Department of Homeland Security has resulted in a vigorous and healthy debate and a strong interest in making our homeland safer and improving our intelligence gathering capabilities.

Fundamentally, the homeland defense debate is about change. Specifically, it is about protection of American citizens.

I am proud of this country and how we have reacted. Everything has not been perfect, but great progress has been made. Noted columnist Mr. Charles Krauthammer recently wrote, "National character does not change in a day. September 11 did not alter the American character, it merely revealed it." I could not agree more.

The American character displayed "courage, resolve, resourcefulness and above all resilience" Krauthammer wrote and I agree. We are a great power and indeed a super power.

We are a nation that believes in freedom and progress and are forgiving and slow to anger, but when aroused we have proven once again we can be a terrible force.

Our President is leading us with strength and resolve. Homeland defense is but a part—an important part—of that resolve. Foreign policy initiatives, social policy changes and prosecution of the war on terror are other aspects as well. Of the latter, winning is no simple matter. Patience, superior planning, and the support of the military are all required to complete the tasks which lie ahead.

The Nation has met the challenge this year. Now we must work hard as the memories of the horror of September 11 fade, to finish the job of making our homeland safe and ensuring that our magnificent military continues to expand its capabilities and world leadership. We must not sleep.

CHANDLER RAYMOND KELLER: IN MEMORIAM

Hon. Barbara Boxer

OF CALIFORNIA

Mr. President, I take this opportunity to share with my colleagues the memory of one of my constituents, Chandler Keller, of Manhattan Beach, CA, who lost his life on September 11, 2001. He was a passenger on American Airlines flight 77. As we all know, that plane crashed into the Pentagon, killing everyone on board.

Mr. Keller was a 29 year-old lead propulsion engineer and project manager with Boeing Satellite Systems in El Segundo, CA.

Chandler Keller was known to his family and friends as "Chad." He was born in Manhattan Beach, CA. Chad mostly grew up there, with the exception of some time spent in Hong Kong, New York and Sydney, Australia, due to his father's work assignments with Security Pacific Corporation.

As a child, Chad enjoyed a great love of rocketry and an avid interest in space. As a young boy he had an innate ability to understand machines and how to make them work. In 1993, Chad graduated from the University of Colorado's aerospace engineering program and pursued his career at HughesBoeing, working in their satellite launching program.

Chad and his wife, Lisa Hurley Keller, were married on July 22, 2000, at the Old Mission in Santa Barbara. During their brief time together Chad and Lisa enjoyed travel, outdoor activities, and most of all, being with one another.

Chad Keller enjoyed surfing, skiing and snowboarding. He loved to cook and possessed a wonderful sense of humor. "He had the ability to bond with people, and he touched many lives during his short life. He lived his life to its fullest," says his father, Dick Keller.

In celebration of his life, the Keller family established the Chandler Keller Memorial Scholarship at the University of Colorado. It is to be awarded to well-rounded aerospace engineering students. Chad was posthumously awarded the Defense of Freedom Medal for his work with the Department of Defense in conjunction with Boeing Satellite Systems.

Chad Keller is survived by his wife, Lisa Hurley Keller; parents Kathy and Dick Keller; brothers Brandon and Gavin; mother-in-law and father-in-law Shirley Ann and Jim Hurley; and brother-in-law James Hurley.

Mr. President, none of us is untouched by the terror of September 11, and many Californians were part of each tragic moment of that tragic day. Some were trapped in the World Trade Center towers. Some were at work in the Pentagon. And the fates of some were sealed as they boarded planes bound for San Francisco or Los Angeles.

I offer this tribute to one of the 54 Californians who perished on that awful morning. I want to assure the family of Chad Keller, and the families of all the victims, that their fathers and mothers, sons and daughters, aunts, uncles, brothers and sisters will not be forgotten.

DINO XAVIER SUAREZ RAMIREZ: IN MEMORIAM

Hon. Barbara Boxer

OF CALIFORNIA

Mr. President, I take this opportunity to share with my colleagues the memory of one of my constituents, Dino Xavier Suarez Ramirez, who lost his life on September 11, 2001. Mr. Ramirez was a 41-year-old civil engineer returning to Los Angeles from vacation when the flight he was on, American Airlines flight 11, was hijacked by terrorists. As we all know, that plane crashed into the World Trade Center, killing everyone on board.

Xavier Ramirez was born in Guayaquil, Ecuador and completed his primary and secondary education there, receiving certification in mathematics, physics, chemistry and biology. Mr. Ramirez achieved his goal of becoming a civil engineer by working during the day and completing his university studies at night. He majored in civil engineering at the University Laica Vicenete Rocafuerte de Guayaquil. "He was very intelligent, and his hobby was reading. He knew very much about the history of nations because of his reading," recalls his mother, Blanca Vilma Ramirez.

Upon coming to this country, Mr. Ramirez worked to have his degree recognized here. In his native country of Ecuador he worked in the construction of roads. His mother further recalls that, "He worked very hard and was not afraid of any kind of job, wanting only to go beyond himself in what he did." Xavier Ramirez is survived by his mother, Blanca and his brother, Klinger David Suarez Ramirez.

Mr. President, none of us is untouched by the terror of September 11, and many Californians were part of each tragic moment of that tragic day. Some were trapped in the World Trade Center towers. Some were at work in the Pentagon. And the fates of some were sealed as they boarded planes bound for San Francisco or Los Angeles.

I offer today this tribute to one of the 54 Californians who perished on that awful morning. I want to assure the family of Dino Ramirez, and the families of all the victims, that their fathers and mothers, sons and daughters, aunts, uncles, brothers and sisters will not be forgotten.

JOHN D. YAMNICKY, SR.: IN MEMORIAM

Hon. Barbara Boxer

OF CALIFORNIA

Mr. President, I take this opportunity to share with the Senate the memory of Capt. John D. Yamnicky, Sr., of Waldorf, MD, who lost his life on September 11, 2001. He was a passenger on American Airlines flight 77. As we all know, that flight was hijacked by terrorists and crashed into the Pentagon, killing everyone on board. Captain Yamnicky was a 71-year-old retired naval aviator who, since his retirement from the U.S. Navy in 1979, continued to work as a defense contractor for Veridian Engineering.

Captain Yamnicky was a gentleman and a scholar. He graduated from the Naval Academy in 1952, and devoted a 26-year career in service to this country. He served a combat tour in Korea and served two tours in Southeast Asia flying from aircraft carriers. He earned several military honors, including the Distinguished Flying Cross.

After graduating from the Naval Test Pilot School at Patuxent River, MD, in 1961, one of his first assignments was to determine the minimum acceptable airspeed for the A–4 aircraft after a catapult launch from an aircraft carrier. He was inducted into the Society of Experimental Test Pilots in recognition of his contributions. In 1963, after that honor, Captain Yamnicky reported to VA–146 at NAS in Lemoore, CA.

Captain Yamnicky met his wife, Jann, while she was working as a nurse at Jacksonville Naval Hospital. They married in 1959 and had four children. Their son John David, of California, said of his father, "This guy was the head of the family, he made everyone feel safe. If he ever talked about accomplishing something, it was as a group or a team. He was a modest man."

Friend and colleague, Dennis Plautz, commented that, "John Yamnicky emphasized teamwork. His style was never to leave a teammate straggling, rather work with them, help them, encourage them to maximize their potential."

He applied this attitude in all areas of his life, including his community contributions. Captain Yamnicky served on the Board of Directors at his daughter Lorraine's high school, St. Mary's Academy, was a member of the Knights of Columbus, and the Elks Lodge. He was proud of his volunteer contributions to the De La Brooke Foxhounds Hunt Club, where he and Jann were members for 25 years.

His best times were spent away from the office, riding on a tractor through the fields of his Waldorf horse farm. "He loved being out there. His nature was not to stand around. He was always out in the fields, always working on something," remembers his son, John.

Captain Yamnicky is survived by his wife Jann and their four children, John, Jr., Lorraine, Mark and Jennifer.

Mr. President, none of us is untouched by the terror of September 11, and many Californians were part of each tragic moment of that tragic day. Some were trapped in the World Trade Center towers. Some were at work in the Pentagon. And the fates of some were sealed as they boarded planes bound for San Francisco or Los Angeles.

I offer today this tribute to one American who perished on that awful morning. I want to assure the family of Captain John D. Yamnicky, Sr., and the families of all the victims, that their fathers and mothers, sons and daughters, aunts, uncles, brothers and sisters will not be forgotten.

DOROTHY A. DEARAUJO: IN MEMORIAM

Hon. Barbara Boxer

OF CALIFORNIA

Mr. President, I take this opportunity to share with my colleagues the memory of one of my constituents, Dorothy A. deAraujo, of Long Beach, CA, who lost her life on September 11, 2001. Mrs. deAraujo was an 80-year-old retiree when the flight she was on, United Airlines flight 175, was hijacked by terrorists. As we all know, that flight crashed into the World Trade Center killing everyone on board.

Mrs. deAraujo was returning to her home in Long Beach after enjoying a lengthy visit with her son, Joaquim ("Tim"), his wife Rita and their two sons, Jonathan and Jason in Bedford, MA.

Dorothy worked as an executive administrative assistant in the business office of California State University, Long Beach for 20 years. She retired in 1979 and returned to the university as a student, earning her bachelor's degree in fine arts. During her retirement, Mrs. deAraujo traveled extensively and pursued her passion for watercolor painting. She was a talented artist, and her paintings won several prizes in various competitions.

During the 1970s Dorothy successfully overcame breast cancer. She was active in her community and devoted her spare time to the American Cancer Society. She was especially involved in operating the American Cancer Society's Discovery Shop in Belmont Shore.

Mr. President, none of us is untouched by the terror of September 11, and many Californians were part of each tragic moment of that tragic day. Some were trapped in the World Trade Center towers. Some were at work in the Pentagon. And the fates of some were sealed as they boarded planes bound for San Francisco or Los Angeles.

I offer today this tribute to one of the 54 Californians who perished on that awful morning. I want to assure the family of Dorothy deAraujo, and the families of all the victims, that their fathers and mothers, sons and daughters, aunts, uncles, brothers and sisters will not be forgotten.

ALAN BEAVEN: IN MEMORIAM

Hon. Barbara Boxer

OF CALIFORNIA

Mr. President, I take this opportunity to share with my colleagues the memory of one of my constituents, Alan Beaven, who lost his life on September 11, 2001. Mr. Beaven was a 48-year-old environmental lawyer when the flight he was on, United Airlines flight 93, was hijacked by terrorists. As we all know, that plane crashed in a Pennsylvania field, killing everyone on board.

Mr. Beaven was born in New Zealand and was educated at the University of Auckland, New Zealand where he was a recipient of the

Butterworth Prize. He taught law and practiced in the areas of securities, class actions and environmental law in New Zealand, England, New York and California.

Considered one of our Nation's leading environmental lawyers, over the past 9 years Mr. Beaven prosecuted nearly 100 clean water cases. His law firm partner, Joe Tabbacco, observes, "This is an absolutely remarkable record. Alan's efforts had almost single-handedly cleaned up the waters in Northern California through his aggressive prosecutions."

California lost an environmental champion, and Mr. Beaven's family lost a loving and devoted husband and father. His proudest achievement was his family. His wife, Kimi Beaven, recalls, "He would do anything for his children and spent hour after hour reading to Sonali, playing ball with John and scuba diving with Chris."

Mr. Beaven was flying back to California to prosecute one more water pollution case before taking a sabbatical in India where he was to volunteer his services as an environmental lawyer. Alan Beaven was one of many heroes on flight 93 who, aware of the terrorist attacks on the World Trade Center and the Pentagon, chose to fight back against the hijacking terrorists. His voice was recognized by his family on the cockpit voice recorder, and his remains were found in the wreckage of the cockpit.

His son John perhaps describes Alan Beaven best when he writes, "His love for simplicity and genuine appreciation for the happiness he held within was not lost on others; friends would always leave his company with uplifted spirits."

Mr. President, none of us is untouched by the terror of September 11, and many Californians were part of each tragic moment of that tragic day. Some were trapped in the World Trade Center towers. Some were at work in the Pentagon. And the fates of some were sealed as they boarded planes bound for San Francisco or Los Angeles.

I offer today this tribute to one of the 54 Californians who perished on that awful morning. I want to assure the family of Alan Beaven, and the families of all the victims, that their fathers and mothers, sons and daughters, aunts, uncles, brothers and sisters will not be forgotten.

Thursday, November 14, 2002

TRIBUTE TO GUIDE DOGS

Hon. Hillary Rodham Clinton

OF NEW YORK

Madam President, after reading an article in the *Saturday Evening Post* about the Germans training dogs to aid veterans blinded during World War I, Morris Frank, a blind man living in Tennessee, wrote to the author, "Thousands of blind like me abhor being dependent on others. Help me and I will help them. Train me and I will bring back my dog and show people here how a blind man can be absolutely on his own." The author, Dorothy Harrison Eustis, agreed to Mr. Frank's request, and Mr. Frank's dog Buddy became the first guide dog in America.

That was in 1929 and today there are more than 7,000 guide dogs serving in America, and 2 performed miracles in New York on September 11.

That morning, Roselle, a yellow labrador retriever and her owner, Michael Hingson, went to the office on the 78th floor of the World Trade Center. While Mr. Hingson worked, Roselle slept underneath his desk. Then the plane hit the South Tower, and what she did next was nothing short of heroic. She guided Mr. Hingson through the smoke and to the stairwell. Not only did Roselle help Mr. Hingson down 78 flights of stairs, but another woman who had been blinded by debris clutched Roselle's harness until they reached safety.

There was another yellow lab in the World Trade Center named Salty. His owner, Omar Rivera, worked on the 71st floor of the Port Authority. After the planes hit, Salty refused to leave Omar's side and walked through the smoke-filled stairway, broken glass and debris to get Mr. Rivera and a coworker to safety. Even as the North Tower collapsed and the debris cloud filled the streets, Salty remained calm, loyal, and focused on guiding Mr. Rivera to a place free from danger.

These two guide dogs performed their jobs under the most extreme circumstances. But what they did that day reinforced what guide dogs do every day—they provide independence to individuals who are blind and visually impaired so that they can live their life free from constraints. To serve as another's set of eyes, to navigate busy city streets, and to keep their owners from harm's way is a responsibility that only a loyal dog would welcome with no questions asked.

Throughout the United States and around the world, guide dog schools have given more than 100,000 people the chance to move about the world with freedom and dignity. Each school offers their guide dogs at no cost to the owners. All they have to do is apply, attend training, and promise to care for their dog for the rest of his or her life. The success of each school is dependent upon thousands of staff, volunteers, and generous supporters. Many people volunteer to raise puppies, socialize them and then give them up at the end of the year. And we see these dogs every day sitting patiently on the subway, stopping at walk lights, and maneuvering people around hazards that prevent a safe, straight path. They wear bright colored vests that read "Guide Dog in Training."

Not only did Morris Frank bring the first guide dog to America, he opened the first school in 1929, the Seeing Eye. Now in every State, guide dog schools provide an invaluable service. In California, The Guide Dog School just celebrated its 60th anniversary, and in New York, the Guide Dog Foundation in Smithtown has assisted New Yorkers and others from around the world since 1946. And Guiding Eyes for the Blind in Yorktown Heights has graduated more than 5,000 dogs and owners since 1954.

Each success story is testament that one good idea can transform the lives of many. But the success of the guide dog schools would not have occurred without two key components: those who believed that the blind and visually impaired could lead more independent lives with the right kind of help, and the dogs, the labrador retrievers, the golden retrievers, the German shepherds, and other breeds that are ready, willing, and able to guide their owners through the world.

Every day, thousands of people grab on to the harness and place their trust in their companion. Some have acted with remarkable heroism like those on September 11, and we have all heard the stories about guide dogs waking their owners in the event of a fire and blocking them from the path of a speeding car. But most go through their days with quiet dignity and they deserve our utmost respect. Whether they are named Roselle or Salty or Buddy, they all respond in the same way. That harness goes on, their eyes open, and they show us that it is possible to walk through this world with a profound desire to help another so that life is limitless.

AUTHORIZING PRINTING OF HOUSE DOCUMENT

Mr. REID. Madam President, I ask unanimous consent that the Senate proceed to the consideration of H. Con. Res. 487 received from the House and which is now at the desk.

The PRESIDING OFFICER. The clerk will report the concurrent resolution by title.

The legislative clerk read as follows:

A concurrent resolution (H. Con. Res. 47) authorizing the printing as a House document of a volume consisting of the transcripts of the ceremonial meeting of the House of Representatives and Senate in New York City on September 6, 2002, and a collection of statements by Members of the House of Representatives and Senate from the *Congressional Record* on the terrorist attacks of September 11, 2001.

There being no objection, the Senate proceeded to consider the concurrent resolution.

Mr. REID. Madam President, I ask unanimous consent that the concurrent resolution be agreed to; that the motion to reconsider be laid upon the table; and that any statements relating to the concurrent resolution be printed in the *Record*.

The PRESIDING OFFICER. Without objection, it is so ordered.

The concurrent resolution (H. Con. Res. 487) was agreed to.

AMERICAN AIRLINES FLIGHT 11

Anna Williams Allison • David Lawrence Angell • Lynn Edwards Angell • Seima Aoyama • Barbara Jean Ares Tegui • Myra Joy Aronson • Christine Barbuto • Carolyn Mayer Beug • Kelly Ann Booms • Carol Marie Bouchard • Neilie Anne Heffernan Casey • Jeffrey Dwayne Collman • Jeffrey W. Coombs • Tara Kathleen Creamer • Thelma Cuccinello • Patrick Currivan • Brian P. Dale • David DiMeglio • Donald Americo DiTullio • Alberto Dominguez • Paige Farley-Hackel • Alexander Milan Filipov • Carol Flyzik • Paul Friedman • Karleton D.B. Fyfe • Peter Alan Gay • Linda M. George • Edmund Glazer • Lisa Reinhart Fenn Gordenstein • Andrew Peter Charles Curry Green • Peter Paul Hashem • Robert Jay Hayes • Edward (Ted) R. Hennessy • John A. Hofer • Cora Hidalgo Holland • John Nicholas Humber • Waleed Iskandar • John Charles Jenkins • Charles Edward Jones • Robin Kaplan • Barbara Keating • David Kovalcin • Judy Larocque • Natalie Janis Lasden • Daniel John Lee • Daniel C. Lewin • Sara Elizabeth Low • Susan A. MacKay • Karen A. Martin • Thomas F. McGuinness • Christopher D. Mello • Jeffrey Peter Mladenik • Antonio Jesus Montoya Valdes • Carlos Alberto Montoya • Laura Lee Morabito • Mildred Naiman • Laurie Ann Neira • Renee Lucille Newell • Kathleen Ann Nicosia • Jacqueline J. Norton • Robert Grant Norton • John Ogonowski • Betty Ann Ong • Jane M. Orth • Thomas Nicholas Pecorelli • Berinthia Berenson Perkins • Sonia Morales Puopolo • David E. Retik • Jean Destrehan Roger • Philip M. Rosenzweig • Richard Barry Ross • Jessica Leigh Sachs • Rahma Salie • Heather Lee Smith • Dianne Bullis Snyder • Douglas J. Stone • Xavier Suarez • Madeline Amy Sweeney • Michael Theodoridis • James Anthony Trentini • Mary Barbara Trentini • Pendyala Vamsikrishna • Mary Alice Wahlstrom • Kenneth E. Waldie • John Wenckus • Candace Lee Williams • Christopher Rudolph Zarba

UNITED FLIGHT 175

Alona Avraham • Garnet Edward (Ace) Bailey • Mark Lawrence Bavis • Graham Andrew Berkeley • Touri Bolourchi • Klaus Bothe • Daniel R. Brandhorst • David Reed Gamboa Brandhorst • John Brett Cahill • Christoffer Mikael Carstanjen • John (Jay) J. Corcoran • Ana Gloria Pocasangre de Barrera • Dorothy Alma DeAraujo • Robert John Fangman • Lisa Frost • Ronald Gamboa • Lynn Catherine Goodchild • Peter Morgan Goodrich • Douglas A. Gowell • Francis E. Grogan • Carl Max Hammond • Christine Lee Hanson • Peter Hanson • Gerald F. Hardacre • Eric Samadikan Hartono • James E. Hayden • Herbert W. Homer • Michael R. Horrocks • Robert Adrien Jalbert • Amy N. Jarret • Ralph Francis Kershaw • Sue Jue Kim-Hanson • Heinrich Kimmig • Amy R. King • Brian Kinney • Kathryn L. LaBorie • Robert George LeBlanc • Maclovio Lopez • Marianne MacFarlane • Alfred Gilles Padre Joseph Marchand • Louis Neil Mariani • Juliana Valentine McCourt • Ruth Magdaline McCourt • Wolfgang Peter Menzel • Shawn M. Nassaney • Marie Pappalardo • Patrick J. Quigley • Frederick Charles Rimmele • James M. Roux • Jesus Sanchez • Victor J. Saracini • Mary Kathleen Shearer • Robert Michael Shearer • Jane Louise Simpkin • Brian D. Sweeney • Michael C. Tarrou • Alicia Nicole Titus • Timothy Ray Ward • William M. Weems

THE TOWERS

Gordon McCannel Aamoth • Edelmiro (Ed) Abad • Maria Rose Abad • Andrew Anthony Abate • Vincent Abate • Laurence Christopher Abel • William F. Abrahamson • Richard Anthony Aceto • Alicia Acevedo Carranza • Heinrich B. Ackermann • Paul Andrew Acquaviva • Donald L. Adams • Patrick Adams • Shannon Lewis Adams • Stephen Adams • Ignatius Adanga • Christy A. Addamo • Terence E. Adderley • Sophia B. Addo • Lee Adler • Daniel Thomas Afflitto • Emmanuel Afuakwah • Alok Agarwal • Mukul Agarwala • Joseph Agnello • David Scott Agnes • Joao A.D. Aguiar • Brian G. Ahearn • Jeremiah J. Ahern • Joanne Ahladiotis • Shabbir Ahmed • Terrance Andre Aiken • Godwin Ajala • Gertrude M. Alagero • Andrew Alameno • Margaret Ann (Peggy) Jezycki Alario • Gary Albero • Jon L. Albert • Peter Craig Alderman • Jacquelyn Delaine Aldridge • Grace Alegre-Cua • David D. Alger • Boutros al-Hashim • Ernest Alikakos • Edward L. Allegretto • Eric Allen • Joseph Ryan Allen • Richard Dennis Allen • Richard Lanard Allen • Christopher Edward Allingham • Janet M. Alonso • Anthony Alvarado • Antonio Javier Alvarez • Telmo Alvear • Cesar A. Alviar • Tariq Amanullah • Angelo Amaranto • James Amato • Joseph Amatuccio • Christopher Charles Amoroso • Kazuhiro Anai • Calixto Anaya • Joseph Peter Anchundia • Kermit Charles Anderson • Yvette Anderson • John Andreacchio • Michael Rourke Andrews • Jean A. Andrucki • Siew-Nya Ang • Joseph Angelini • Joseph Angelini • Laura Angilletta • Doreen J. Angrisani • Lorraine D. Antigua • Peter Paul Apollo • Faustino Apostol • Frank Thomas Aquilino • Patrick Michael Aranyos • David Gregory Arce • Michael G. Arczynski • Louis Arena • Adam Arias • Michael J. Armstrong • Jack Charles Aron • Joshua Aron • Richard Avery Aronow • Japhet J. Aryee • Carl Asaro • Michael A. Asciak • Michael Edward Asher • Janice Ashley • Thomas J. Ashton • Manuel O. Asitimbay • Gregg Arthur Atlas • Gerald Atwood • James Audiffred • Louis Frank Aversano • Ezra Aviles • Samuel (Sandy) Ayala Arlene T. Babakitis • Eustace (Rudy) Bacchus • John James Badagliacca • Jane Ellen Baeszler • Robert J. Baierwalter • Andrew J. Bailey • Brett T. Bailey • Tatyana Bakalinskaya • Michael S. Baksh • Sharon Balkcom • Michael Andrew Bane • Kathy Bantis • Gerard Jean Baptiste • Walter Baran • Gerard A. Barbara • Paul V. Barbaro • James W. Barbella • Ivan Kyrillos Fairbanks Barbosa • Victor Daniel Barbosa • Colleen Ann Barkow • David Michael Barkway • Matthew Barnes • Sheila Patricia Barnes • Evan J. Baron • Renee Barrett-Arjune • Arthur T. Barry • Diane G. Barry • Maurice Vincent Barry • Scott D. Bart • Carlton W. Bartels • Guy Barzvi • Inna Basina • Alysia Basmajian • Kenneth William Basnicki • Steven J. Bates • Paul James Battaglia • W. David Bauer • Ivhan Luis Carpio Bautista • Marlyn C. Bautista • Jasper Baxter • Michele (Du Berry) Beale • Paul F. Beatini • Jane S. Beatty • Larry I. Beck • Manette Marie Beckles • Carl John Bedigian • Michael Beekman • Maria Behr • Yelena Belilovsky • Nina Patrice Bell • Andrea Della Bella • Debbie S. Bellows • Stephen Elliot Belson • Paul Michael Benedetti • Denise Lenore Benedetto • Bryan Craig Bennett • Eric L. Bennett • Oliver Duncan Bennett • Margaret L. Benson • Dominick J. Berardi • James Patrick Berger • Steven Howard Berger • John P. Bergin • Alvin Bergsohn • Daniel D. Bergstein • Michael J. Berkeley • Donna Bernaerts-Kearns • Dave Bernard • William Bernstein • David M. Berray • David S. Berry • Joseph J. Berry • William Reed Bethke • Timothy D. Betterly • Edward F. Beyea • Paul Michael Beyer • Anil T. Bharvaney • Bella Bhukhan • Shimmy D. Biegeleisen • Peter Alexander Bielfeld • William Biggart • Brian Bilcher • Carl Vincent Bini • Gary Bird • Joshua David Birnbaum • George Bishop • Jeffrey D. Bittner • Balewa Albert Blackman • Christopher Joseph Blackwell • Susan L. Blair • Harry Blanding • Janice L. Blaney • Craig Michael Blass • Rita Blau • Richard M. Blood

THE TOWERS

• Michael A. Boccardi • John Paul Bocchi • Michael L. Bocchino • Susan Mary Bochino • Bruce Douglas (Chappy) Boehm • Mary Katherine Boffa • Nicholas A. Bogdan • Darren C. Bohan • Lawrence Francis Boisseau • Vincent M. Boland • Alan Bondarenko • Andre Bonheur • Colin Arthur Bonnett • Frank Bonomo • Yvonne L. Bonomo • Sean Booker • Juan Jose Borda Leyva • Sherry Ann Bordeaux • Krystine C. Bordenabe • Martin Boryczewski • Richard E. Bosco • John Howard Boulton • Francisco Bourdier • Thomas H. Bowden • Kimberly S. Bowers • Veronique (Bonnie) Nicole Bowers • Larry Bowman • Shawn Edward Bowman • Kevin L. Bowser • Gary R. Box • Gennady Boyarsky • Pamela Boyce • Michael Boyle • Alfred Braca • Sandra Conaty Brace • Kevin H. Bracken • David Brian Brady • Alexander Braginsky • Nicholas W. Brandemarti • Michelle Renee Bratton • Patrice Braut • Lydia Estelle Bravo • Ronald Michael Breitweiser • Edward A. Brennan • Frank H. Brennan • Michael Emmett Brennan • Peter Brennan • Thomas M. Brennan • Daniel Brethel • Gary L. Bright • Jonathan Eric Briley • Mark A. Brisman • Paul Gary Bristow • Victoria Alvarez Brito • Mark Francis Broderick • Herman C. Broghammer • Keith Broomfield • Janice J. Brown • Lloyd Brown • Patrick J Brown • Bettina Browne • Mark Bruce • Richard Bruehert • Andrew Brunn • Vincent Brunton • Ronald Paul Bucca • Brandon J. Buchanan • Greg Joseph Buck • Dennis Buckley • Nancy Bueche • Patrick Joseph Buhse • John E. Bulaga • Stephen Bunin • Matthew J. Burke • Thomas Daniel Burke • William F. Burke • Donald James Burns • Kathleen A. Burns • Keith James Burns • John Patrick Burnside • Irina Buslo • Milton Bustillo • Thomas M. Butler • Patrick Byrne • Timothy G. Byrne • Jesus Cabezas • Lillian Caceres • Brian Joseph Cachia • Steven Cafiero • Richard M. Caggiano • Cecile M. Caguicla • Michael John Cahill • Scott W. Cahill • Thomas J. Cahill • George Cain • Salvatore B. Calabro • Joseph Calandrillo • Philip V. Calcagno • Edward Calderon • Kenneth Marcus Caldwell • Dominick E. Calia • Felix (Bobby) Calixte • Frank Callahan • Liam Callahan • Luigi Calvi • Roko Camaj • Michael Cammarata • David Otey Campbell • Geoffrey Thomas Campbell • Jill Marie Campbell • Robert Arthur Campbell • Sandra Patricia Campbell • Juan Ortega Campos • Sean Canavan • John A. Candela • Vincent Cangelosi • Stephen J. Cangialosi • Lisa B. Cannava • Brian Cannizzaro • Michael R. Canty • Louis A. Caporicci • Jonathan N. Cappello • James Christopher Cappers • Richard M. Caproni • Jose Cardona • Dennis M Carey • Edward Carlino • Michael Scott Carlo • David G. Carlone • Rosemarie C. Carlson • Mark Stephen Carney • Joyce Ann Carpeneto • Jeremy M. Carrington • Michael T. Carroll • Peter Carroll • James J. Carson • Christopher Newton Carter • James Marcel Cartier • Vivian Casalduc • John F. Casazza • Paul Cascio • Margarito Casillas • Thomas Anthony Casoria • William Otto Caspar • Alejandro Castano • German Castillo Galicia • Arcelia Castillo • Leonard M. Castrianno • Jose Ramon Castro • Richard G. Catarelli • Christopher Sean Caton • Robert J. Caufield • Mary Teresa Caulfield • Judson Cavalier • Michael Joseph Cawley • Jason D. Cayne • Juan Armando Ceballos • Marcia G. Cecil-Carter • Jason Cefalu • Thomas J. Celic • Ana M. Centeno • Joni Cesta • Jeffrey M. Chairnoff • Swarna Chalasini • William Chalcoff • Eli Chalouh • Charles Lawrence (Chip) Chan • Mandy Chang • Mark L. Charette • Gregorio Manuel Chavez • Pedro Francisco Checo • Douglas MacMillan Cherry • Stephen Patrick Cherry • Vernon Paul Cherry • Nestor Chevalier • Swede Joseph Chevalier • Alexander H. Chiang • Dorothy J. Chiarchiaro • Luis Alfonso Chimbo • Robert Chin • Wing Wai (Eddie) Ching • Nicholas P. Chiofalo • John Chipura • Peter A. Chirchirillo • Catherine E. Chirls • Kyung (Kaccy) Cho • Abdul K. Chowdhury • Mohammed Salahuddin Chowdhury • Kirsten L. Christophe • Pamela Chu • Steven Paul Chucknick • Wai-ching Chung • Christopher Ciafardini • Alex F. Ciccone • Frances Ann Cilente • Elaine Cillo • Edna Cintron • Nestor Andre Cintron • Robert Dominick Cirri • Juan Pablo Alvarez Cisneros • Benjamin Keefe Clark • Eugene Clark • Gregory A. Clark • Mannie Leroy Clark • Thomas R. Clark • Christopher Robert Clarke • Donna Clarke • Michael Clarke • Suria R.E. Clarke • Kevin Francis Cleary • James D. Cleere • Geoffrey W. Cloud • Susan M. Clyne • Steven Coakley • Jeffrey Coale • Patricia A. Cody • Daniel Michael Coffey • Jason Matthew Coffey • Florence Cohen • Kevin Sanford Cohen • Anthony Joseph Coladonato • Mark J. Colaio • Stephen J. Colaio • Christopher M. Colasanti • Kevin Nathaniel Colbert • Michel Paris Colbert • Keith Eugene Coleman • Scott Thomas Coleman • Tarel Coleman • Liam Joseph Colhoun • Robert D. Colin • Robert J. Coll • Jean Marie Collin • John Michael Collins • Michael L. Collins • Thomas J. Collins • Joseph Collison • Patricia Malia Colodner • Linda M. Colon • Soledi Colon • Ronald Comer • Jaime Concepcion • Albert Conde • Denease Conley • Susan Clancy Conlon • Margaret Mary Conner • Cynthia L. Connolly • John E. Connolly • James Lee Connor • Jonathan (J.C.) Connors • Kevin P. Connors • Kevin Francis Conroy • Brenda E. Conway • Dennis Michael Cook • Helen D. Cook • John A. Cooper • Joseph J. Coppo • Gerard J. Coppola • Joseph Albert Corbett • Alejandro Cordero • Robert Cordice • Danny A. Correa-Gutierrez • Ruben D. Correa • James Corrigan • Carlos Cortes • Kevin M. Cosgrove • Dolores Marie Costa • Digna Alexandra Rivera Costanza • Charles Gregory Costello • Michael S. Costello • Conrod K.H. Cottoy • Martin Coughlan • John Gerard Coughlin • Timothy John Coughlin • James E. Cove • Andre Cox • Frederick John Cox • Michelle Coyle-Eulau • James Raymond Coyle • Anne M. Cramer • Christopher Seton Cramer • Denise Crant • James L. Crawford • Robert James Crawford • Joanne Mary Cregan • Lucia Crifasi • John Crisci • Daniel Hal Crisman • Dennis A. Cross • Helen Crossin-Kittle • Kevin Raymond Crotty • Thomas G. Crotty • John Crowe • Welles Remy Crowther • Robert L. Cruikshank • Francisco Cruz • John Robert Cruz • Kenneth John Cubas • Francisco C. Cubero • Richard Joseph Cudina • Neil James Cudmore • Thomas Patrick Cullen • Joan McConnell Cullinan • Joyce Cummings • Brian Thomas Cummins • Nilton Albuquerque Fernao Cunha • Michael Joseph Cunningham • Robert Curatolo • Laurence Curia • Paul Dario Curioli • Beverly Curry • Michael Curtin • Gavin Cushny • John D'Allara • Vincent D'Amadeo • Jack L. D'Ambrosi • Mary D'Antonio • Edward Alexander D'Atri • Michael D. D'Auria • Michael Jude D'Esposito • Manuel Da Mota • Carlos S. DaCosta • Caleb Arron Dack • Thomas A. Damaskinos • Jeannine Marie Damiani-Jones • Patrick W. Danahy • Nana Kwuku Danso • Vincent G. Danz • Dwight Donald Darcy • Elizabeth Ann Darling • Annette Andrea Dataram • Lawrence Davidson • Michael Allen Davidson • Scott Matthew Davidson • Titus Davidson • Niurka Davila • Clinton Davis • Wayne Terrial Davis • Anthony Richard Dawson • Calvin Dawson • Edward James Day • Jayceryll M. de Chavez • Emerita (Emy) De La Pena • Azucena de la Torre • Cristina de Laura • Oscar de Laura • Francis (Frank) Albert De Martini • Robert J. DeAngelis •

THE TOWERS

James V. DeBlase • Paul DeCola • Jason Christopher DeFazio • Jennifer DeJesus • Monique E. DeJesus • Nereida DeJesus • Martin DeMeo • Jean C. DePalma • Michael DeRienzo • David Paul DeRubbio • Jemal Legesse DeSantis • Christian D. DeSimone • Edward DeSimone • Melanie Louise DeVere • Jerry DeVito • William T. Dean • Thomas P. Deangelis • Tara Debek • Anna Debin • Simon Dedvukaj • David A. Defeo • Manuel Del Valle • Donald A. Delapenha • Vito Joseph Deleo • Danielle Delie • Joseph A. Della Pietra • Palmina Delli Gatti • Colleen Ann Deloughery • Anthony Demas • Francis X. Deming • Carol K. Demitz • Kevin Dennis • Thomas F. Dennis • Jose Nicholas Depena • Robert J. Deraney • Andrew Desperito • Cindy Ann Deuel • Robert P. Devitt • Dennis Lawrence Devlin • Gerard Dewan • Simon Suleman Ali Kassamali Dhanani • Michael L. DiAgostino • Patricia F. DiChiaro • John DiFato • Vincent F. DiFazio • Carl DiFranco • Donald J. DiFranco • Debra Ann DiMartino • Anthony DiOnisio • George DiPasquale • Joseph DiPilato • Douglas Frank DiStefano • Michael Diaz-Piedra • Judith Belguese Diaz-Sierra • Lourdes Galletti Diaz • Matthew Diaz • Nancy Diaz • Obdulio Ruiz Diaz • Joseph Dermot Dickey • Lawrence Patrick Dickinson • Michael David Diehl • Stephen P. Dimino • William J. Dimmling • Christopher Dincuff • Jeffrey M. Dingle • Ramzi A. Doany • John J. Doherty • Melissa C. Doi • Brendan Dolan • Neil Dollard • James Joseph Domanico • Benilda Pascua Domingo • Charles (Carlos) Dominguez • Geronimo (Jerome) Mark Patrick Dominguez • Kevin W. Donnelly • Jacqueline Donovan • Stephen Dorf • Thomas Dowd • Kevin Christopher Dowdell • Mary Yolanda Dowling • Raymond M. Downey • Frank Joseph Doyle • Joseph M. Doyle • Randy Drake • Stephen Patrick Driscoll • Mirna A. Duarte • Luke A. Dudek • Christopher Michael Duffy • Gerard Duffy • Michael Joseph Duffy • Thomas W. Duffy • Antoinette Duger • Sareve Dukat • Christopher Joseph Dunne • Richard A. Dunstan • Patrick Thomas Dwyer • Joseph Anthony Eacobacci • John Bruce Eagleson • Robert D. Eaton • Dean P. Eberling • Margaret Ruth Echtermann • Paul Robert Eckna • Constantine (Gus) Economos • Dennis Michael Edwards • Michael Hardy Edwards • Christine Egan • Lisa Egan • Martin Egan • Michael Egan • Samantha Egan • Carole Eggert • Lisa Caren Weinstein Ehrlich • John Ernst (Jack) Eichler • Eric Adam Eisenberg • Daphne F. Elder • Michael J. Elferis • Mark J. Ellis • Valerie Silver Ellis • Albert Alfy William Elmarry • Edgar H. Emery • Doris Suk-Yuen Eng • Christopher S. Epps • Ulf Ramm Ericson • Erwin L. Erker • William J. Erwin • Sarah (Ali) Escarcega • Jose Espinal • Fanny M. Espinoza • Brigette Ann Esposito • Francis Esposito • Michael Esposito • William Esposito • Ruben Esquilin • Sadie Ette • Barbara G. Etzold • Eric Brian Evans • Robert Edward Evans • Meredith Emily June Ewart • Catherine K. Fagan • Patricia M. Fagan • Keith G. Fairben • William F. Fallon • William Fallon • Anthony J. Fallone • Dolores B. Fanelli • John Joseph Fanning • Kathleen (Kit) Faragher • Thomas Farino • Nancy Carole Farley • Elizabeth Ann (Betty) Farmer • Douglas Farnum • John G. Farrell • John W. Farrell • Terrence Patrick Farrell • Joseph Farrelly • Thomas P. Farrelly • Syed Abdul Fatha • Christopher Faughnan • Wendy R. Faulkner • Shannon M. Fava • Bernard D. Favuzza • Robert Fazio • Ronald C. Fazio • William Feehan • Francis J. (Frank) Feely • Garth E. Feeney • Sean B. Fegan • Lee S. Fehling • Peter Feidelberg • Alan D. Feinberg • Rosa Maria Feliciano • Edward T. Fergus • George Ferguson • Henry Fernandez • Jose Manuel Contreras Fernandez • Judy H. Fernandez • Elisa Giselle Ferraina • Anne Marie Sallerin Ferreira • Robert John Ferris • David Francis Ferrugio • Louis V. Fersini • Michael David Ferugio • Bradley James Fetchet • Jennifer Louise Fialko • Kristen Fiedel • Samuel Fields • Michael Bradley Finnegan • Timothy J. Finnerty • Michael Curtis Fiore • Stephen J. Fiorelli • Paul M. Fiori • John Fiorito • John R. Fischer • Andrew Fisher • Bennett Lawson Fisher • John Roger Fisher • Thomas J. Fisher • Lucy Fishman • Ryan D. Fitzgerald • Thomas Fitzpatrick • Richard P. Fitzsimons • Salvatore A. Fiumefreddo • Christina Donovan Flannery • Eileen Flecha • Andre G. Fletcher • Carl Flickinger • John Joseph Florio • Joseph W. Flounders • David Fodor • Michael N. Fodor • Steven Mark Fogel • Thomas Foley • David Fontana • Chih Min (Dennis) Foo • Del Rose Forbes-Cheatham • Godwin Forde • Donald A. Foreman • Christopher Hugh Forsythe • Claudia Alicia Martinez Foster • Noel J. Foster • Ana Fosteris • Robert J. Foti • Jeffrey L Fox • Virginia Fox • Joan Francis • Pauline Francis • Virgin (Lucy) Francis • Gary J. Frank • Morton Frank • Peter Christopher Frank • Richard K. Fraser • Kevin Joseph Frawley • Clyde Frazier • Lillian I. Frederick • Andrew Fredericks • Jamitha Freemen • Brett O. Freiman • Peter L. Freund • Arlene E. Fried • Alan Wayne Friedlander • Andrew K. Friedman • Gregg J. Froehner • Peter Christian Fry • Clement Fumando • Steven Elliot Furman • Paul James Furmato • Fredric Gabler • Richard S. Gabrielle • James Andrew Gadiel • Pamela Gaff • Ervin Vincent Gailliard • Deanna L. Galante • Grace Galante • Anthony Edward Gallagher • Daniel James Gallagher • John Patrick Gallagher • Tomas Gallegos Linares • Cono E. Gallo • Vincenzo Gallucci • Thomas Edward Galvin • Giovanna (Genni) Gambale • Thomas Gambino • Giann F. Gamboa • Peter J. Ganci • Claude Michael Gann • Charles William Garbarini • Cesar Garcia • David Garcia • Jorge Luis Morron Garcia • Juan Garcia • Marlyn C. Garcia • Christopher Gardner • Douglas B. Gardner • Harvey J. Gardner • Jeffrey B. Gardner • Thomas A. Gardner • William Arthur Gardner • Francesco Garfi • Rocco Gargano • James M. Gartenberg • Matthew David Garvey • Bruce Gary • Boyd A. Gatton • Donald Richard Gavagan • Terence D. Gazzani • Gary Geidel • Paul Hamilton Geier • Julie M. Geis • Peter Gelinas • Steven Paul Geller • Howard G. Gelling • Peter Victor Genco • Steven Gregory Genovese • Alayne F. Gentul • Edward F. Geraghty • Suzanne Geraty • Ralph Gerhardt • Robert J. Gerlich • Denis P. Germain • Marina R. Gertsberg • Susan M. Getzendanner • James Gerard Geyer • Joseph M. Giaccone • Vincent Francis Giammona • Debra L. Gibbon • James A. Giberson • Craig Neil Gibson • Ronnie Gies • Laura A. Giglio • Andrew Clive Gilbert • Timothy Paul Gilbert • Paul Stuart Gilbey • Paul John Gill • Mark Y. Gilles • Evan H. Gillette • Ronald Gilligan • Rodney C. Gillis • Laura Gilly • John F. Ginley • Donna Marie Giordano • Jeffrey Giordano • John Giordano • Steven A. Giorgetti • Martin Giovinazzo • Jinny Lady Giraldo • Kum-Kum Girolamo • Salvatore Gitto • Cynthia Giugliano • Mon Gjonbalaj • Dianne Gladstone • Keith Alexander Glascoe • Thomas I. Glasser • Harry Glenn • Barry H. Glick • Steven Lawrence Glick • John T. Gnazzo • William (Bill) Robert Godshalk • Michael Gogliormella • Brian Fredric Goldberg • Jeffrey Grant Goldflam • Michelle Herman Goldstein • Monica Goldstein • Steven Goldstein • Andrew H. Golkin • Dennis James Gomes • Enrique Antonio Gomez • Jose Bienvenido Gomez • Manuel Gomez • Wilder Gomez • Jenine Gonzalez • Joel Guevara Gonzalez • Mauricio Gonzalez • Rosa J. Gonzalez • Calvin J. Gooding • Harry Goody • Kiran Reddy Gopu • Catherine Carmen Gorayeb • Kerene Gordon • Sebastian Gorki • Kieran Gorman • Thomas E. Gorman • Michael Edward Gould • Yugi Goya • Jon Richard Grabowski • Christopher Michael Grady • Edwin John Graf • David M. Graifman • Gilbert Granados • Elvira Granitto • Winston Arthur Grant • Christopher Stewart Gray • James Michael Gray • Linda Mair Grayling • John Michael Grazioso • Timothy Grazioso • Derrick Arthur Green

THE TOWERS

• Wade Brian Green • Elaine Myra Greenberg • Gayle R. Greene • James Arthur Greenleaf • Eileen Marsha Greenstein • Elizabeth (Lisa) Martin Gregg • Denise Gregory • Donald H. Gregory • Florence M. Gregory • Pedro (David) Grehan • John M. Griffin • Tawanna Griffin • Joan D. Griffith • Warren Grifka • Ramon Grijalvo • Joseph F. Grillo • David Grimner • Kenneth Grouzalis • Joseph Grzelak • Matthew J. Grzymalski • Robert Joseph Gschaar • Liming (Michael) Gu • Jose A. Guadalupe • Yan Zhu (Cindy) Guan • Geoffrey E. Guja • Joseph Gullickson • Babita Guman • Douglas B. Gurian • Janet H. Gustafson • Philip T. Guza • Barbara Guzzardo • Peter Gyulavary • Gary Robert Haag • Andrea Lyn Haberman • Barbara M. Habib • Philip Haentzler • Nizam A. Hafiz • Karen Hagerty • Steven Hagis • Mary Lou Hague • David Halderman • Maile Rachel Hale • Richard Hall • Vaswald George Hall • Robert John Halligan • Vincent Gerard Halloran • James D. Halvorson • Mohammed Salman Hamdani • Felicia Hamilton • Robert Hamilton • Frederic Kim Han • Christopher James Hanley • Sean Hanley • Valerie Joan Hanna • Thomas Hannafin • Kevin James Hannaford • Michael L. Hannan • Dana Hannon • Vassilios G. Haramis • James A. Haran • Jeffrey P. Hardy • Timothy John Hargrave • Daniel Harlin • Frances Haros • Harvey L. Harrell • Stephen Gary Harrell • Aisha Harris • Stewart D. Harris • John Patrick Hart • John Clinton Hartz • Emeric J. Harvey • Thomas Theodore Haskell • Timothy Haskell • Joseph John Hasson • Leonard William Hatton • Terence S. Hatton • Michael Helmut Haub • Timothy Aaron Haviland • Donald G. Havlish • Anthony Hawkins • Nobuhiro Hayatsu • Philip Hayes • William Ward Haynes • Scott Hazelcorn • Michael K. Healey • Roberta Bernstein Heber • Charles Francis Xavier Heeran • John Heffernan • Howard Joseph Heller • JoAnn L. Heltibridle • Mark F. Hemschoot • Ronnie Lee Henderson • Janet Hendricks • Brian Hennessey • Michelle Marie Henrique • Joseph P. Henry • William Henry • John Henwood • Robert Allan Hepburn • Mary (Molly) Herencia • Lindsay Coates Herkness • Harvey Robert Hermer • Claribel Hernandez • Norberto Hernandez • Raul Hernandez • Gary Herold • Jeffrey A. Hersch • Thomas Hetzel • Brian Hickey • Ysidro Hidalgo-Tejada • Timothy Higgins • Robert D. Higley • Todd Russell Hill • Clara Victorine Hinds • Neal Hinds • Mark D. Hindy • Katsuyuki Hirai • Heather Malia Ho • Tara Yvette Hobbs • Thomas A. Hobbs • James L. Hobin • Robert Wayne Hobson • DaJuan Hodges • Ronald George Hoerner • Patrick Aloysius Hoey • Frederick J. Hoffman • Joseph Hoffman • Marcia Hoffman • Michele L. Hoffman • Stephen G. Hoffman • Judith Florence Hofmiller • Thomas Warren Hohlweck • Jonathan R. Hohmann • John Holland • Joseph Francis Holland • Elizabeth Holmes • Thomas P. Holohan • Bradley Hoorn • James P. Hopper • Montgomery McCullough Hord • Michael Horn • Matthew D. Horning • Robert L. Horohoe • Aaron Horwitz • Charles J. Houston • Uhuru G. Houston • George Howard • Michael C. Howell • Steven L. Howell • Jennifer L. Howley • Milagros Hromada • Marian Hrycak • Stephen Huczko • Kris R. Hughes • Melissa Harrington Hughes • Paul R. Hughes • Robert T. "Bobby" Hughes • Thomas F. Hughes • Timothy Robert Hughes • Susan Huie • Mychal Lamar Hulse • Kathleen (Casey) Hunt • William C. Hunt • Joseph G. Hunter • Robert Hussa • Thomas E. Hynes • Walter Hynes • Joseph Anthony Ianelli • Zuhtu Ibis • Jonathan Lee Ielpi • Michael Patrick Iken • Daniel Ilkanayev • Frederick Ill • Abraham Nethanel Ilowitz • Anthony P. Infante • Louis S. Inghilterra • Christopher N. Ingrassia • Paul Innella • Stephanie V. Irby • Douglas Irgang • Kristin A. Irvine-Ryan • Todd A. Isaac • Erik Hans Isbrandtsen • Taizo Ishikawa • Aram Iskenderian • John Iskyan • Kazushige Ito • Aleksandr Valeryerich Ivantsov • Virginia Jablonski • Brooke Alexandra Jackman • Aaron Jacobs • Ariel Louis Jacobs • Jason Kyle Jacobs • Michael Grady Jacobs • Steven A. Jacobson • Ricknauth Jaggernauth • Jake Denis Jagoda • Yudh V.S. Jain • Maria Jakubiak • Ernest James • Gricelda E. James • Mark Jardim • Mohammed Jawara • Francois Jean-Pierre • Maxima Jean-Pierre • Paul E. Jeffers • Joseph Jenkins • Alan K. Jensen • Prem N. Jerath • Farah Jeudy • Hweidar Jian • Fernando Jimenez Molina • Eliezer Jimenez • Luis Jimenez • Charles Gregory John • Nicholas John • LaShawana Johnson • Scott M. Johnson • William Johnston • Allison Horstmann Jones • Arthur Joseph Jones • Brian L. Jones • Christopher D. Jones • Donald T. Jones • Donald W. Jones • Linda Jones • Mary S. Jones • Andrew Jordan • Robert Thomas Jordan • Albert Joseph • Ingeborg Joseph • Karl Henri Joseph • Stephen Joseph • Jane Eileen Josiah • Anthony Jovic • Angel Luis Juarbe • Karen Susan Juday • Mychal Judge • Paul W. Jurgens • Thomas Edward Jurgens • Shashi Kiran Lakshmikantha Kadaba • Gavkharoy Mukhometovna Kamardinova • Shari Kandell • Howard Lee Kane • Jennifer Lynn Kane • Vincent D. Kane • Joon Koo Kang • Sheldon R. Kanter • Deborah H. Kaplan • Alvin Peter Kappelmann • Charles Karczewski • William A. Karnes • Douglas G. Karpiloff • Charles L. Kasper • Andrew Kates • John Katsimatides • Robert Kaulfers • Don Jerome Kauth • Hideya Kawauchi • Edward T. Keane • Richard M. Keane • Lisa Kearney-Griffin • Karol Ann Keasler • Paul Hanlon Keating • Leo Russell Keene • Joseph J. Keller • Peter Rodney Kellerman • Joseph P. Kellett • Frederick H. Kelley • James Joseph Kelly • Joseph A. Kelly • Maurice Patrick Kelly • Richard John Kelly • Thomas Michael Kelly • Thomas Richard Kelly • Thomas W. Kelly • Timothy C. Kelly • William Hill Kelly • Robert C. Kennedy • Thomas J. Kennedy • John Keohane • Ronald T. Kerwin • Howard L. Kestenbaum • Douglas D. Ketcham • Ruth E. Ketler • Boris Khalif • Sarah Khan • Taimour Firaz Khan • Rajesh Khandelwal • Bhowanie Devi Khemraj • SeiLai Khoo • Michael Kiefer • Satoshi Kikuchihara • Andrew Jay-Hoon Kim • Lawrence Don Kim • Mary Jo Kimelman • Lisa M. King-Johnson • Andrew Marshall King • Lucille T. King • Robert King • Takashi Kinoshita • Chris Michael Kirby • Howard (Barry) Kirschbaum • Glenn Davis Kirwin • Richard J. Klares • Peter A. Klein • Alan D. Kleinberg • Karen J. Klitzman • Ronald Philip Kloepfer • Andrew Knox • Thomas Patrick Knox • Yevgeny Knyazev • Rebecca Lee Koborie • Deborah Kobus • Gary Edward Koecheler • Frank J. Koestner • Ryan Kohart • Vanessa Lynn Kolpak • Irina Kolpakova • Suzanne Kondratenko • Abdoulaye Kone • Bon-seok Koo • Dorota Kopiczko • Scott Kopytko • Bojan Kostic • Danielle Kousoulis • John J. Kren • William Krukowski • Lyudmila Ksido • Shekhar Kumar • Kenneth Kumpel • Frederick Kuo • Patricia Kuras • Nauka Kushitani • Thomas Joseph Kuveikis • Victor Kwarkye • Kui Fai Kwok • Angela R. Kyte • Andrew LaCorte • Jeanette LaFond-Menichino • David LaForge • Michael Patrick LaForte • Stephen LaMantia • Carol Ann LaPlante • Jeannine M. LaVerde • Amarnauth Lachhman • Ganesh Ladkat • James P. Ladley • Joseph A. Lafalce • Alan Lafranco • Juan Lafuente • Neil K. Lai • Vincent A. Laieta • William David Lake • Franco Lalama • Chow Kwan Lam • Amy Hope Lamonsoff • Robert T. Lane • Brendan M. Lang • Rosanne P. Lang • Vanessa Langer • Mary Lou Langley • Peter J. Langone • Thomas Langone • Michele B. Lanza • Ruth Sheila Lapin • Ingeborg Astrid Desiree Lariby • Robin Larkey • Christopher Randall Larrabee • Hamidou S. Larry • Scott Larsen • John Adam Larson • Gary E. Lasko • Nicholas C. Lassman • Paul Laszczynski • Jeffrey Latouche • Charles Laurencin • Stephen James Lauria • Maria Lavache • Denis F. Lavelle • Anna A. Laverty • Steven Lawn • Robert A. Lawrence • Nathaniel Lawson • Eugen Lazar • David Prudencio LeMagne • Jeffrey Earle LeVeen • James Patrick Leahy • Joseph Gerard Leavey • Neil Leavy • Leon Lebor • Kenneth Charles Ledee • Alan J. Lederman • Elena Ledesma • Alexis Leduc • David S. Lee • Gary H. Lee • Hyun-joon (Paul) Lee • Jong-min Lee • Juanita Lee • Kathryn Blair Lee • Linda C. Lee • Lorraine Lee • Myung-woo Lee • Richard Y.C. Lee

THE TOWERS

• Stuart (Soo-Jin) Lee • Yang Der Lee • Stephen Lefkowitz • Adriana Legro • Edward J. Lehman • Eric Andrew Lehrfeld • David Ralph Leistman • Joseph A. Lenihan • John J. Lennon • John Robinson Lenoir • Jorge Luis Leon • Matthew Gerard Leonard • Michael Lepore • Charles Antoine Lesperance • John D. Levi • Alisha Caren Levin • Neil D. Levin • Robert M. Levine • Robert Levine • Shai Levinhar • Adam J. Lewis • Margaret Susan Lewis • Ye Wei Liang • Orasri Liangthanasarn • Daniel F. Libretti • Ralph M. Licciardi • Edward Lichtschein • Steven B. Lillianthal • Carlos R. Lillo • Craig Damian Lilore • Arnold A. Lim • Darya Lin • Wei Rong Lin • Nickie L. Lindo • Thomas V. Linehan • Robert Thomas Linnane • Alan Linton • Diane Theresa Lipari • Kenneth P. Lira • Francisco Alberto Liriano • Lorraine Lisi • Paul Lisson • Vincent Litto • Ming-Hao Liu • Joseph Livera • Nancy Liz • Harold Lizcano • Martin Lizzul • George A. Llanes • Elizabeth Claire Logler • Catherine Lisa Loguidice • Jerome Robert Lohez • Michael W. Lomax • Laura M. Longing • Salvatore P. Lopes • Daniel Lopez • George Lopez • Luis Lopez • Manuel L. Lopez • Joseph Lostrangio • Chet Louie • Stuart Seid Louis • Joseph Lovero • Michael W. Lowe • Garry Lozier • John Peter Lozowsky • Charles Peter Lucania • Edward (Ted) H. Luckett • Mark G. Ludvigsen • Lee Charles Ludwig • Sean Thomas Lugano • Daniel Lugo • Marie Lukas • William Lum • Michael P. Lunden • Christopher Lunder • Anthony Luparello • Gary Lutnick • Linda Luzzicone • Alexander Lygin • Farrell Peter Lynch • James Francis Lynch • Louise A. Lynch • Michael F. Lynch • Michael Francis Lynch • Michael Lynch • Richard Dennis Lynch • Robert H. Lynch • Sean Patrick Lynch • Sean Lynch • Michael J. Lyons • Monica Lyons • Patrick Lyons • Catherine Fairfax MacRae • Robert Francis Mace • Jan Maciejewski • Richard B. Madden • Simon Maddison • Noell Maerz • Jeannieann Maffeo • Joseph Maffeo • Jay Robert Magazine • Brian Magee • Charles Wilson Magee • Joseph Maggitti • Ronald E. Magnuson • Daniel L. Maher • Thomas Anthony Mahon • William Mahoney • Joseph Maio • Takashi Makimoto • Abdu Malahi • Myrna T. Maldonado-Agosto • Debora Maldonado • Alfred R. Maler • Gregory James Malone • Edward Francis (Teddy) Maloney • Joseph E. Maloney • Gene E. Maloy • Christian Maltby • Francisco Miguel (Frank) Mancini • Joseph Mangano • Sara Elizabeth Manley • Debra M. Mannetta • Marion Victoria (vickie) Manning • Terence J. Manning • James Maounis • Joseph Ross Marchbanks • Peter Edward Mardikian • Edward Joseph Mardovich • Charles Joseph Margiotta • Kenneth Joseph Marino • Lester Vincent Marino • Vita Marino • Kevin D. Marlo • Jose J. Marrero • John Marshall • James Martello • Michael A. Marti • Peter Martin • William J. Martin • Brian E. Martineau • Lizie Martinez-Calderon • Betsy Martinez • Edward J. Martinez • Jose Martinez • Robert Gabriel Martinez • Paul Richard Martini • Joseph A. Mascali • Bernard Mascarenhas • Stephen F. Masi • Nicholas G. Massa • Patricia A. Massari • Michael Massaroli • Philip W. Mastrandrea • Rudolph Mastrocinque • Joseph Mathai • Charles William Mathers • William A. Mathesen • Marcello Matricciano • Margaret Elaine Mattic • Robert D. Mattson • Walter Matuza • Charles A. (Chuck) Mauro • Charles J. Mauro • Dorothy Mauro • Nancy T. Mauro • Tyrone May • Keithroy Maynard • Robert J. Mayo • Kathy Nancy Mazza-Delosh • Edward Mazzella • Jennifer Mazzotta • Kaaria Mbaya • James J. McAlary • Brian McAleese • Patricia A. McAneney • Colin Richard McArthur • John McAvoy • Kenneth M. McBrayer • Brendan McCabe • Michael J. McCabe • Thomas McCann • Justin McCarthy • Kevin M. McCarthy • Michael Desmond McCarthy • Robert Garvin McCarthy • Stanley McCaskill • Katie Marie McCloskey • Tara McCloud-Gray • Charles Austin McCrann • Tonyell McDay • Matthew T. McDermott • Joseph P. McDonald • Brian G. McDonnell • Michael McDonnell • John F. McDowell • Eamon J. McEneaney • John Thomas McErlean • Katherine (Katie) McGarry-Noack • Daniel F. McGinley • Mark Ryan McGinly • William E. McGinn • Thomas H. McGinnis • Michael Gregory McGinty • Ann McGovern • Scott Martin McGovern • William J. McGovern • Stacey S. McGowan • Francis Noel McGuinn • Patrick J. McGuire • Thomas M. McHale • Keith McHeffey • Ann M. McHugh • Denis J. McHugh • Dennis P. McHugh • Michael Edward McHugh • Robert G. McIlvaine • Donald James McIntyre • Stephanie McKenna • Barry J. McKeon • Evelyn C. McKinnedy • Darryl Leron McKinney • George Patrick McLaughlin • Robert C. McLaughlin • Gavin McMahon •rt Dismas McMahon • Edmund M. McNally • Daniel McNeal • Walter Arthur McNeil • Christine Sheila McNulty • Sean Peter McNulty • Robert William McPadden • Terence A. McShane • Timothy Patrick McSweeney • Martin E. McWilliams • Rocco A. Medaglia • Abigail Medina • Anna Iris Medina • Deborah Medwig • Damian Meehan • William J. Meehan • Alok Kumar Mehta • Raymond Meisenheimer • Manuel Emilio Mejia • Eskedar Melaku • Antonio Melendez • Mary Melendez • Yelena Melnichenko • Stuart Todd Meltzer • Diarelia Jovannah Mena • Charles Mendez • Lizette Mendoza • Shevonne Mentis • Steve Mercado • Wesley Mercer • Ralph Joseph Mercurio • Alan H. Merdinger • George C. Merino • Yamel Merino • George Merkouris • Deborah Merrick • Raymond J. Metz • Jill A. Metzler • David Robert Meyer • Nurul Huq Miah • William Edward Micciulli • Martin Paul Michelstein • Luis Clodoaldo Revilla Mier • Peter T. Milano • Gregory Milanowycz • Lukasz T. Milewski • Corey Peter Miller • Craig James Miller • Douglas C. Miller • Henry Miller • Joel Miller • Michael Matthew Miller • Phillip D. Miller • Robert Alan Miller • Robert C. Miller • Benjamin Millman • Charles M. Mills • Ronald Keith Milstein • Robert Minara • William G. Minardi • Louis Joseph Minervino • Thomas Mingione • Nana Akwasi Minkah • Wilbert Miraille • Domenick Mircovich • Rajesh A. Mirpuri • Joseph Mistrulli • Susan Miszkowicz • Paul Thomas Mitchell • Richard Miuccio • Frank V. Moccia • Louis Joseph Modafferi • Boyie Mohammed • Dennis Mojica • Manuel Mojica • Kleber Rolando Molina • Manuel Dejesus Molina • Carl Molinaro • Justin J. Molisani • Brian Patrick Monaghan

THE TOWERS

• Franklin Monahan • John Gerard Monahan • Kristen Montanaro • Craig D. Montano • Michael Montesi • Cheryl Ann Monyak • Thomas Moody • Sharon Moore • Krishna Moorthy • Abner Morales • Carlos Morales • Paula Morales • John Christopher Moran • John Moran • Kathleen Moran • Lindsay S. Morehouse • George Morell • Steven P. Morello • Vincent S. Morello • Arturo Alva Moreno • Yvette Nicole Moreno • Dorothy Morgan • Richard Morgan • Nancy Morgenstern • Sanae Mori • Blanca Morocho • Leonel Morocho • Dennis G. Moroney • Lynne Irene Morris • Seth A. Morris • Stephen Philip Morris • Christopher M. Morrison • Ferdinand V. Morrone • William David Moskal • Marco Motroni • Chung Mou • Iouri A. Mouchinski • Jude J. Moussa • Peter C. Moutos • Damion Mowatt • Christopher Mozzillo • Stephen V. Mulderry • Richard Muldowney • Michael D. Mullan • Dennis Michael Mulligan • Peter James Mulligan • Michael Joseph Mullin • James Donald Munhall • Nancy Muniz • Carlos Mario Munoz • Theresa (Terry) Munson • Robert M. Murach • Cesar Augusto Murillo • Marc A. Murolo • Brian Joseph Murphy • Charles Murphy • Christopher W. Murphy • Edward C. Murphy • James F. Murphy • James Thomas Murphy • Kevin James Murphy • Patrick Sean Murphy • Raymond E. Murphy • Robert Eddie Murphy • John Joseph Murray • John Joseph Murray • Susan D. Murray • Valerie Victoria Murray • Richard Todd Myhre • Robert B. Nagel • Takuya Nakamura • Alexander J.R. Napier • Frank Joseph Naples • John Napolitano • Catherine A. Nardella • Mario Nardone • Manika Narula • Narender Nath • Karen S. Navarro • Joseph M. Navas • Francis J. Nazario • Glenroy Neblett • Marcus R. Neblett • Jerome O. Nedd • Laurence Nedell • Luke G. Nee • Pete Negron • Ann Nicole Nelson • David William Nelson • James Nelson • Michele Ann Nelson • Peter Allen Nelson • Oscar Nesbitt • Gerard Terence Nevins • Nancy Yuen Ngo • Jody Tepedino Nichilo • Martin Niederer • Alfonse J. Niedermeyer • Frank John Niestadt • Gloria Nieves • Juan Nieves • Troy Edward Nilsen • Paul R. Nimbley • John Ballantine Niven • Curtis Terrence Noel • Daniel R. Nolan • Robert Walter Noonan • Daniela R. Notaro • Brian Novotny • Soichi Numata • Brian Felix Nunez • Jose R. Nunez • Jeffrey Nussbaum • Dennis O'Berg • James P. O'Brien • Michael O'Brien • Scott J. O'Brien • Timothy Michael O'Brien • Daniel O'Callaghan • Dennis J. O'Connor • Diana J. O'Connor • Keith K. O'Connor • Richard J. O'Connor • Amy O'Doherty • Marni Pont O'Doherty • James Andrew O'Grady • Thomas O'Hagan • Patrick O'Keefe • William O'Keefe • Gerald O'Leary • Matthew Timothy O'Mahoney • Seamus L. O'Neal • John P. O'Neill • Peter J. O'Neill • Sean Gordon Corbett O'Neill • Kevin O'Rourke • Patrick J. O'Shea • Robert W. O'Shea • Timothy O'Sullivan • James A. Oakley • Jefferson Ocampo • Douglas Oelschlager • Takashi Ogawa • Albert Ogletree • Philip Paul Ognibene • Joseph J. Ogren • Samuel Oitice • Gerald Michael Olcott • Christine Anne Olender • Elsy Carolina Osorio Oliva • Linda Mary Oliva • Edward K. Oliver • Leah E. Oliver • Eric T. Olsen • Jeffrey James Olsen • Maureen L. Olson • Steven John Olson • Toshihiro Onda • Michael C. Opperman • Christopher Orgielewicz • Margaret Orloske • Virginia A. Ormiston-Kenworthy • Juan Romero Orozco • Ronald Orsini • Peter K. Ortale • Alexander Ortiz • David Ortiz • Emilio (Peter) Ortiz • Pablo Ortiz • Paul Ortiz • Sonia Ortiz • Masaru Ose • James Robert Ostrowski • Jason Douglas Oswald • Michael Otten • Isidro Ottenwalder • Michael Ou • Todd Joseph Ouida • Jesus Ovalles • Peter J. Owens • Adianes Oyola • Angel M. Pabon • Israel Pabon • Roland Pacheco • Michael Benjamin Packer • Deepa K. Pakkala • Jeffrey Matthew Palazzo • Thomas Anthony Palazzo • Richard (Rico) Palazzolo • Orio Joseph Palmer • Frank A. Palombo • Alan N. Palumbo • Christopher M. Panatier • Dominique Pandolfo • Paul Pansini • John M. Paolillo • Edward J. Papa • Salvatore Papasso • James N. Pappageorge • Vinod K. Parakat • Vijayashanker Paramsothy • Nitin Parandkar • Hardai (Casey) Parbhu • James Wendell Parham • Debra (Debbie) Paris • George Paris • Gye-Hyong Park • Philip L. Parker • Michael A. Parkes • Robert Emmett Parks • Hasmukhrai Chuckulal Parmar • Robert Parro • Diane Marie Moore Parsons • Leobardo Lopez Pascual • Michael J. Pascuma • Jerrold H. Paskins • Horace Robert Passananti • Suzanne H. Passaro • Victor Antonio Martinez Pastrana • Avnish Ramanbhai Patel • Dipti Patel • Manish K. Patel • Steven B. Paterson • James Matthew Patrick • Manuel Patrocino • Bernard E. Patterson • Cira Marie Patti • Robert Edward Pattison • James R. Paul • Victor Paz-Gutierrez • Patrice Paz • Sharon Cristina Millan Paz • Stacey L. Peak • Richard Allen Pearlman • Durrell Pearsall • Thomas E. Pedicini • Todd D. Pelino • Michel Adrian Pelletier • Anthony Peluso • Angel Ramon Pena • Jose D. Pena • Richard Al Penny • Salvatore F. Pepe • Carl Allen Peralta • Robert David Peraza • Jon A. Perconti • Alejo Perez • Angel Perez • Angela Susan Perez • Anthony Perez • Ivan Perez • Nancy E. Perez • Joseph John Perroncino • Edward J. Perrotta • Emelda Perry • Glenn C. Perry • John William Perry • Franklin Allan Pershep • Daniel Pesce • Michael J. Pescherine • Davin Peterson • William Russel Peterson • Mark Petrocelli • Philip S. Petti • Glen Kerrin Pettit • Dominick Pezzulo • Kaleen E. Pezzuti • Kevin Pfeifer • Tu-Anh Pham • Kenneth John Phelan • Eugenia Piantieri • Ludwig John Picarro • Matthew Picerno • Joseph O. Pick • Christopher Pickford • Dennis J. Pierce • Bernard T. Pietronico • Nicholas P. Pietrunti • Theodoros Pigis • Susan Elizabeth Ancona Pinto • Joseph Piskadlo • Christopher Todd Pitman • Josh Piver • Joseph Plumitallo • John M. Pocher • William Howard Pohlmann • Laurence M. Polatsch • Thomas H. Polhemus • Steve Pollicino • Susan M. Pollio • Joshua Poptean • Giovanna Porras • Anthony Portillo • James Edward Potorti • Daphne Pouletsos • Richard Poulos • Stephen E. Poulos • Brandon Jerome Powell • Shawn Edward Powell • Tony Pratt • Gregory M. Preziose • Wanda Ivelisse Prince • Vincent Princiotta • Kevin Prior • Everett Martin (Marty) Proctor • Carrie B. Progen • David Lee Pruim • Richard Prunty • John F. Puckett • Robert D. Pugliese • Edward F. Pullis • Patricia Ann Puma • Hemanth Kumar Puttur • Edward R. Pykon • Christopher Quackenbush • Lars Peter Qualben • Lincoln Quappe • Beth Ann Quigley • Michael Quilty • James Francis Quinn • Ricardo Quinn • Carol Rabalais • Christopher Peter A. Racaniello • Leonard Ragaglia • Eugene J. Raggio • Laura Marie Ragonese-Snik • Michael Ragusa • Peter F. Raimondi • Harry A. Raines • Ehtesham U. Raja • Valsa Raju • Edward Rall • Lukas (Luke) Rambousek • Julio Fernandez Ramirez • Maria Isabel Ramirez • Harry Ramos • Vishnoo Ramsaroop • Lorenzo Ramzey • A. Todd Rancke • Adam David Rand • Jonathan C. Randall • Srinivasa Shreyas Ranganath • Anne Rose T. Ransom • Faina Rapoport • Robert Arthur Rasmussen • Amenia Rasool • Roger Mark Rasweiler • David Alan James Rathkey • William Ralph Raub • Gerard Rauzi • Alexey Razuvaev • Gregory Reda • Sarah (Prothero) Redheffer • Michele Reed • Judith A. Reese • Donald J. Regan • Robert M. Regan • Thomas M. Regan • Christian Michael Otto Regenhard • Howard Reich • Gregg Reidy •

THE TOWERS

James B. Reilly • Kevin O. Reilly • Timothy E. Reilly • Joseph Reina • Thomas Barnes Reinig • Frank B. Reisman • Joshua Scott Reiss • Karen Renda • John Armand Reo • Richard Rescorla • John Thomas Resta • Eduvigis (Eddie) Reyes • Bruce A. Reynolds • John Frederick Rhodes • Francis S. Riccardelli • Rudolph N. Riccio • AnnMarie (Davi) Riccoboni • David Rice • Eileen Mary Rice • Kenneth F. Rice • Vernon Allan Richard • Claude D. Richards • Gregory Richards • Michael Richards • Venesha O. Richards • James C. Riches • Alan Jay Richman • John M. Rigo • James Riley • Theresa (Ginger) Risco • Rose Mary Riso • Moises N. Rivas • Joseph Rivelli • Carmen A. Rivera • Isaias Rivera • Juan William Rivera • Linda Rivera • David E. Rivers • Joseph R. Riverso • Paul Rizza • John Frank Rizzo • Stephen Louis Roach • Joseph Roberto • Leo A. Roberts • Michael Edward Roberts • Michael Roberts • Donald Walter Robertson • Catherina Robinson • Jeffrey Robinson • Michell Lee Robotham • Donald Robson • Antonio Augusto Tome Rocha • Raymond J. Rocha • Laura Rockefeller • John M. Rodak • Antonio Jose Carrusca Rodrigues • David B. Rodriguez-Vargas • Anthony Rodriguez • Carlos Cortez Rodriguez • Carmen Milagros Rodriguez • Gregory E. Rodriguez • Marsha A. Rodriguez • Richard Rodriguez • Matthew Rogan • Karlie Barbara Rogers • Scott Rohner • Keith Roma • Joseph M. Romagnolo • Efrain Franco Romero • Elvin Santiago Romero • James A. Romito • Sean Rooney • Eric Thomas Ropiteau • Wendy Alice Rosario Wakeford • Aida Rosario • Angela Rosario • Mark H. Rosen • Brooke David Rosenbaum • Linda Rosenbaum • Sheryl Lynn Rosenbaum • Lloyd D. Rosenberg • Mark Louis Rosenberg • Andrew I. Rosenblum • Joshua M. Rosenblum • Joshua A. Rosenthal • Richard David Rosenthal • Daniel Rossetti • Norman Rossinow • Nicholas P. Rossomando • Michael Craig Rothberg • Donna Marie Rothenberg • Nick Rowe • Timothy A. Roy • Paul G. Ruback • Ronald J. Ruben • Joanne Rubino • David Michael Ruddle • Bart Joseph Ruggiere • Susan Ann Ruggiero • Adam K. Ruhalter • Gilbert Ruiz • Stephen P. Russell • Steven Harris Russin • Michael Thomas Russo • Wayne Alan Russo • Edward Ryan • John J. Ryan • Jonathan Stephan Ryan • Matthew Lancelot Ryan • Tatiana Ryjova • Christina Sunga Ryook • Thierry Saada • Jason E. Sabbag • Thomas E. Sabella • Scott Saber • Joseph Sacerdote • Francis J. Sadocha • Jude Elias Safi • Brock Joel Safronoff • Edward Saiya • John Patrick Salamone • Hernando R. Salas • Juan Salas • Esmerlin Salcedo • John Salvatore Salerno • Richard L. Salinardi • Wayne John Saloman • Nolbert Salomon • Catherine Patricia Salter • Frank Salvaterra • Paul R. Salvio • Samuel R. Salvo • Rena Sam-Dinnoo • Carlos Samaniego • James Kenneth Samuel • Michael V. San Phillip • Sylvia San Pio Resta • Hugo Sanay-Perafiel • Alva Jeffries Sanchez • Erick Sanchez • Jacquelyn P. Sanchez • Eric Sand • Stacey Leigh Sanders • Herman Sandler • James Sands • Ayleen J. Santiago • Kirsten Santiago • Maria Theresa Santillan • Susan G. Santo • Christopher Santora • John Santore • Mario L. Santoro • Jorge Octavio Santos Anaya • Rafael Humberto Santos • Rufino Conrado F. (Roy) Santos • Kalyan K. Sarkar • Chapelle Sarker • Paul F. Sarle • Deepika Kumar Sattaluri • Gregory Thomas Saucedo • Susan Sauer • Anthony Savas • Vladimir Savinkin • Jackie Sayegh Duggan • John Sbarbaro • Robert L. Scandole • Michelle Scarpitta • Dennis Scauso • John A. Schardt • John G. Scharf • Fred Claude Scheffold • Angela Susan Scheinberg • Scott M. Schertzer • Sean Schielke • Steven Francis Schlag • Jon S. Schlissel • Karen Helene Schmidt • Ian Schneider • Thomas G. Schoales • Marisa Di Nardo Schorpp • Frank G. Schott • Gerard P. Schrang • Jeffrey Schreier • John T. Schroeder • Susan Lee Kennedy Schuler • Edward W. Schunk • Mark E. Schurmeier • Clarin Shellie Schwartz • John Schwartz • Mark Schwartz • Adriane Victoria Scibetta • Raphael Scorca • Randolph Scott • Christopher J. Scudder • Arthur Warren Scullin • Michael Seaman • Margaret Seeliger • Anthony Segarra • Carlos Segarra • Jason Sekzer • Matthew Carmen Sellitto • Howard Selwyn • Larry John Senko • Arturo Angelo Sereno • Frankie Serrano • Alena Sesinova • Adele Sessa • Sita Nermalla Sewnarine • Karen Lynn Seymour-Dietrich • Davis (Deeg) Sezna • Thomas Joseph Sgroi • Jayesh Shah • Khalid M. Shahid • Mohammed Shajahan • Gary Shamay • Earl Richard Shanahan • Shiv Shankar • Neil G. Shastri • Kathryn Anne Shatzoff • Barbara A. Shaw • Jeffrey J. Shaw • Robert J. Shay • Daniel James Shea • Joseph Patrick Shea • Linda Sheehan • Hagay Shefi • John Anthony Sherry • Atsushi Shiratori • Thomas Shubert • Mark Shulman • See-Wong Shum • Allan Shwartzstein • Johanna Sigmund • Dianne T. Signer • Gregory Sikorsky • Stephen Gerard Siller • David Silver • Craig A. Silverstein • Nasima H. Simjee • Bruce Edward Simmons • Arthur Simon • Kenneth Alan Simon • Michael John Simon • Paul Joseph Simon • Marianne Simone • Barry Simowitz • Jeff Simpson • George V. Sims • Khamladai K. (Khami) Singh • Roshan R. (Sean) Singh • Thomas Sinton • Peter A. Siracuse • Muriel F. Siskopoulos • Joseph M. Sisolak • John P. Skala • Francis J. Skidmore • Toyena Corliss Skinner • Paul A. Skrzypek • Christopher Paul Slattery • Vincent R. Slavin • Robert Sliwak • Paul K. Sloan • Stanley S. Smagala • Wendy L. Small • Catherine T. Smith • Daniel Laurence Smith • George Eric Smith • James G. Smith • Jeffrey Randall Smith • Joyce Smith • Karl Trumbull Smith • Kevin Smith • Leon Smith • Moira Smith • Rosemary A. Smith • Sandra Fajardo Smith • Bonnie S. Smithwick • Rochelle Monique Snell • Leonard J. Snyder • Astrid Elizabeth Sohan • Sushil Solanki • Ruben Solares • Naomi Leah Solomon • Daniel W. Song • Michael C. Sorresse • Fabian Soto • Timothy P. Soulas • Gregory T. Spagnoletti • Donald F. Spampinato • Thomas Sparacio • John Anthony Spataro • Robert W. Spear • Maynard S. Spence • George E. Spencer • Robert Andrew Spencer • Mary Rubina Sperando • Frank J. Spinelli • William E. Spitz • Joseph P. Spor • Klaus Johannes Sprockamp • Saranya Srinuan • Fitzroy St. Rose • Michael F. Stabile • Lawrence T. Stack • Timothy Stackpole • Richard James Stadelberger • Eric A. Stahlman • Gregory M. Stajk • Alexandru Liviu Stan • Corina Stan • Mary D. Stanley • Joyce Stanton • Patricia Stanton • Anthony M. Starita • Jeffrey Stark • Derek James Statkevicus • Craig William Staub • William V. Steckman • Eric Thomas Steen • William R. Steiner • Alexander Robbins Steinman • Andrew Stergiopoulos • Andrew Stern • Martha Jane Stevens • Michael James Stewart • Richard H. Stewart • Sanford M. Stoller • Lonny J. Stone • Jimmy Nevill Storey • Timothy Stout • Thomas S. Strada • James J. Straine • Edward W. Straub • George Strauch • Edward T. Strauss • Steven R. Strauss • Steven F. Strobert • Walwyn W. Stuart • Benjamin Suarez • David S. Suarez • Ramon Suarez • Yoichi Sugiyama • William Christopher Sugra • Daniel Suhr • David Marc Sullins • Christopher P. Sullivan • Patrick Sullivan • Thomas Sullivan • Hilario Soriano (Larry) Sumaya • James Joseph Suozzo • Colleen Supinski • Robert Sutcliffe • Selina Sutter • Claudia Suzette Sutton • John F. Swaine • Kristine M. Swearson • Brian Edward Sweeney • Kenneth J. Swensen

THE TOWERS

• Thomas F. Swift • Derek O. Sword • Kevin T. Szocik • Gina Sztejnberg • Norbert P. Szurkowski • Harry Taback • Joann Tabeek • Norma C. Taddei • Michael Taddonio • Keiichiro Takahashi • Keiji Takahashi • Phyllis Gail Talbot • Robert R. Talhami • Sean Patrick Tallon • Paul Talty • Maurita Tam • Rachel Tamares • Hector Tamayo • Michael Andrew Tamuccio • Kenichiro Tanaka • Rhondelle Cherie Tankard • Michael Anthony Tanner • Dennis Gerard Taormina • Kenneth Joseph Tarantino • Allan Tarasiewicz • Ronald Tartaro • Darryl Taylor • Donnie Brooks Taylor • Lorisa Ceylon Taylor • Michael M. Taylor • Paul A. Tegtmeier • Yeshavant Moreshwar Tembe • Anthony Tempesta • Dorothy Temple • Stanley L. Temple • David Tengelin • Brian J. Terrenzi • Lisa Marie Terry • Goumatie T. Thackurdeen • Harshad Sham Thatte • Thomas F. Theurkauf • Lesley Thomas-O'Keefe • Brian T. Thompson • Clive Thompson • Glenn Thompson • Nigel Bruce Thompson • Perry Anthony Thompson • Vanavah Alexi Thompson • William Harry Thompson • Eric Raymond Thorpe • Nichola A. Thorpe • Sal Tieri • John Patrick Tierney • Mary Ellen Tiesi • William R. Tieste • Kenneth F. Tietjen • Stephen Edward Tighe • Scott C. Timmes • Michael E. Tinley • Jennifer M. Tino • Robert Frank Tipaldi • John J. Tipping • David Tirado • Hector Luis Tirado • Michelle Titolo • John J. Tobin • Richard J. Todisco • Vladimir Tomasevic • Stephen K. Tompsett • Thomas Tong • Doris Torres • Luis Eduardo Torres • Amy E. Toyen • Christopher M. Traina • Daniel Patrick Trant • Abdoul Karim Traore • Glenn J. Travers • Walter (Wally) P. Travers • Felicia Traylor-Bass • Lisa L. Trerotola • Karamo Trerra • Michael Trinidad • Francis Joseph Trombino • Gregory J. Trost • William Tselepis • Zhanetta Tsoy • Michael Patrick Tucker • Lance Richard Tumulty • Ching Ping Tung • Simon James Turner • Donald Joseph Tuzio • Robert T. Twomey • Jennifer Tzemis • John G. Ueltzhoeffer • Tyler V. Ugolyn • Michael A. Uliano • Jonathan J. Uman • Anil Shivhari Umarkar • Allen V. Upton • Diane Maria Urban • John Damien Vaccacio • Bradley H. Vadas • William Valcarcel • Mayra Valdes-Rodriguez • Felix Antonio Vale • Ivan Vale • Benito Valentin • Santos Valentin • Carlton Francis Valvo • Erica Van Acker • Kenneth W. Van Auken • Richard Bruce Van Hine • Daniel M. Van Laere • Edward Raymond Vanacore • Jon C. Vandevander • Frederick T. Varacchi • Gopalakrishnan Varadhan • David Vargas • Scott C. Vasel • Azael Ismael Vasquez • Santos Vasquez • Arcangel Vazquez • Peter Anthony Vega • Sankara S. Velamuri • Jorge Velazquez • Lawrence Veling • Anthony M. Ventura • David Vera • Loretta A Vero • Christopher Vialonga • Matthew Gilbert Vianna • Robert A. Vicario • Celeste Torres Victoria • Joanna Vidal • John T. Vigiano • Joseph Vincent Vigiano • Frank J. Vignola • Joseph B. Vilardo • Sergio Villanueva • Chantal Vincelli • Melissa Vincent • Francine A. Virgilio • Lawrence Virgilio • Joseph G. Visciano • Joshua S. Vitale • Maria Percoco Vola • Lynette D. Vosges • Garo H. Voskerijian • Alfred Vukosa • Gregory Wachtler • Gabriela Waisman • Courtney Wainsworth Walcott • Victor Wald • Benjamin Walker • Glen J. Wall • Mitchel Scott Wallace • Peter G. Wallace • Robert F. Wallace • Roy Michael Wallace • Jean Marie Wallendorf • Matthew Blake Wallens • John Wallice • Barbara P. Walsh • James Walsh • Jeffrey Patrick Walz • Ching H. Wang • Weibin Wang • Michael Warchola • Stephen Gordon Ward • James A. Waring • Brian G. Warner • Derrick Washington • Charles Waters • James Thomas (Muddy) Waters • Patrick J. Waters • Kenneth Watson • Michael H. Waye • Todd C. Weaver • Walter E. Weaver • Nathaniel Webb • Dinah Webster • Joanne Flora Weil • Michael Weinberg • Steven Weinberg • Scott Jeffrey Weingard • Steven Weinstein • Simon Weiser • David M. Weiss • David T. Weiss • Vincent Michael Wells • Timothy Matthew Welty • Christian Hans Rudolf Wemmers • Ssu-Hui (Vanessa) Wen • Oleh D. Wengerchuk • Peter M. West • Whitfield West • Meredith Lynn Whalen • Eugene Whelan • Adam S. White • Edward James White • James Patrick White • John S. White • Kenneth W. White • Leonard Anthony White • Malissa White • Wayne White • Leanne Marie Whiteside • Mark Whitford • Michael T. Wholey • Mary Lenz Wieman • Jeffrey David Wiener • William J. Wik • Allison M. Wildman • Glenn Wilkinson • John C. Willett • Brian Patrick Williams • Crossley Williams • David Williams • Deborah Lynn Williams • Kevin Michael Williams • Louie Anthony Williams • Louis Calvin Williams • John Williamson • Cynthia Wilson • Donna Wilson • William E. Wilson • David H. Winton • Glenn J. Winuk • Thomas Francis Wise • Alan L. Wisniewski • Frank T. Wisniewski • David Wiswall • Sigrid Charlotte Wiswe • Michael R. Wittenstein • Christopher W. Wodenshek • Martin P. Wohlforth • Katherine S. Wolf • Jennifer Y. Wong • Jenny Seu Kueng Low Wong • Siu Cheung Wong • Yin Ping (Steven) Wong • Yuk Ping Wong • Brent James Woodall • James J. Woods • Patrick Woods • Richard Herron Woodwell • David Terence Wooley • John Bentley Works • Martin Michael Wortley • Rodney James Wotton • William Wren • John Wright • Neil R. Wright • Sandra Wright • Jupiter Yambem • Suresh Yanamadala • Matthew David Yarnell • Myrna Yaskulka • Shakila Yasmin • Olabisi L. Yee • Edward P. York • Kevin Patrick York • Raymond York • Suzanne Youmans • Barrington L. Young • Jacqueline (Jakki) Young • Elkin Yuen • Joseph Zaccoli • Adel Agayby Zakhary • Arkady Zaltsman • Edwin J. Zambrana • Robert Alan Zampieri • Mark Zangrilli • Ira Zaslow • Kenneth Albert Zelman • Abraham J. Zelmanowitz • Martin Morales Zempoaltecatl • Zhe (Zack) Zeng • Marc Scott Zeplin • Jie Yao Justin Zhao • Ivelin Ziminski • Michael Joseph Zinzi • Charles A. Zion • Julie Lynne Zipper • Salvatore J. Zisa • Prokopios Paul Zois • Joseph J. Zuccala • Andrew Steven Zucker • Igor Zukelman

AMERICAN AIRLINES FLIGHT 77

Paul W. Ambrose • Yeneneh Betru • Mary Jane (MJ) Booth • Bernard Curtis Brown • Charles F. Burlingame • Suzanne M. Calley • William E. Caswell • David M. Charlebois • Sara M Clark • Asia S. Cottom • James Daniel Debeuneure • Rodney Dickens • Eddie A. Dillard • Charles A. Droz • Barbara G. Edwards • Charles S. Falkenberg • Dana Falkenberg • Zoe Falkenberg • James Joseph Ferguson • Darlene E. Flagg • Wilson F. Flagg • Richard P. Gabriel • Ian J. Gray • Stanley R. Hall • Michele M. Heidenberger • Bryan C. Jack • Steven D. Jacoby • Ann C. Judge • Chandler R. Keller • Yvonne E. Kennedy • Norma Cruz Khan • Karen Ann Kincaid • Dong Chul Lee • Jennifer Lewis • Kenneth E. Lewis • Renee A. May • Dora Marie Menchaca • Christopher C. Newton • Barbara K. Olson • Ruben S. Ornedo • Robert Penninger • Robert R. Ploger • Zandra F. Ploger • Lisa J. Raines • Todd H. Reuben • John P. Sammartino • Diane M. Simmons • George W. Simmons • Mari-Rae Sopper • Robert Speisman • Norma Lang Steuerle • Hilda E. Taylor • Leonard E. Taylor • Sandra D. Teague • Leslie A. Whittington • John D. Yamnicky • Vicki C. Yancey • Shuyin Yang • Yuguang Zheng

THE PENTAGON

Craig Amundson • Melissa Rose Barnes • Max J. Beilke • Kris Romeo Bishundat • Carrie R. Blagburn • Canfield D. Boone • Diana Borrero de Padro • Donna Bowen • Allen P. Boyle • Christopher Lee Burford • Daniel Martin Caballero • Jose Orlando Calderon-Olmedo • Angelene C. Carter • Sharon A. Carver • John J. Chada • Rosa Maria (Rosemary) Chapa • Julian T. Cooper • Eric A. Cranford • Ada M. Davis • Gerald Francis DeConto • Jerry Don Dickerson • Johnnie Doctor • Robert Edward Dolan • William Howard Donovan • Patrick Dunn • Edward Thomas Earhart • Robert Randolph Elseth • Jamie Lynn Fallon • Amelia V. Fields • Gerald P. Fisher • Matthew Michael Flocco • Sandra N. Foster • Lawrence Daniel Getzfred • Cortez Ghee • Brenda C. Gibson • Ron F. Golinski • Diane M. Hale-McKinzy • Carolyn B. Halmon • Sheila M. S. Hein • Ronald John Hemenway • Wallace Cole Hogan • Jimmie Ira Holley • Angela M. Houtz • Brady K. Howell • Peggie M. Hurt • Stephen Neil Hyland • Robert J. Hymel • Lacey B. Ivory • Dennis M. Johnson • Judith L. Jones • Brenda Kegler • Michael Scott Lamana • David W. Laychak • Samantha L. Lightbourn-Allen • Stephen V. Long • James T. Lynch • Terence M. Lynch • Nehamon Lyons • Shelley A. Marshall • Teresa M. Martin • Ada L. Mason-Acker • Dean E. Mattson • Timothy J. Maude • Robert J. Maxwell • Molly L. McKenzie • Patricia E. (Patti) Mickley • Ronald D. Milam • Gerard (Jerry) P. Moran • Odessa V. Morris • Brian Anthony Moss • Teddington H. Moy • Patrick Jude Murphy • Khang Ngoc Nguyen • Michael Allen Noeth • Chin Sun Pak • Jonas Martin Panik • Clifford L. Patterson • Darin Howard Pontell • Scott Powell • Jack D. Punches • Joseph John Pycior • Deborah A. Ramsaur • Rhonda Sue Rasmussen • Marsha Dianah Ratchford • Martha M. Reszke • Cecelia E. Richard • Edward V. Rowenhorst • Judy Rowlett • Robert E. Russell • William R. Ruth • Charles E. Sabin • Marjorie C. Salamone • David M. Scales • Robert Allan Schlegel • Janice M. Scott • Michael L. Selves • Marian H. Serva • Dan Frederic Shanower • Antionette M. Sherman • Donald D. Simmons • Cheryle D. Sincock • Gregg Harold Smallwood • Gary F. Smith • Patricia J. Statz • Edna L. Stephens • Larry L. Strickland • Kip P. Taylor • Sandra C. Taylor • Karl W. Teepe • Tamara C. Thurman • Otis Vincent Tolbert • Willie Q. Troy • Ronald James Vauk • Karen J. Wagner • Meta L. Waller • Maudlyn A. White • Sandra L. White • Ernest M. Willcher • David Lucian Williams • Dwayne Williams • Marvin R. Woods • Kevin Wayne Yokum • Donald McArthur Young • Edmond G. Young • Lisa L. Young

UNITED FLIGHT 93

Christian Adams • Lorraine G. Bay • Todd Beamer • Alan Beaven • Mark K. Bingham • Deora Frances Bodley • Sandra W. Bradshaw • Marion Britton • Thomas E. Burnett • William Joseph Cashman • Georgine Rose Corrigan • Patricia Cushing • Jason Dahl • Joseph Deluca • Patrick Joseph Driscoll • Edward P. Felt • Jane C. Folger • Colleen Laura Fraser • Andrew Garcia • Jeremy Glick • Lauren Grandcolas • Wanda Anita Green • Donald F. Greene • Linda Gronlund • Richard Jerry Guadagno • LeRoy Wilton Homer • Toshiya Kuge • CeeCee Lyles • Hilda Marcin • Waleska Martinez Rivera • Nicole Miller • Louis J. Nacke • Donald Arthur Peterson • Jean Hoadley Peterson • Mark Rothenberg • Christine Anne Snyder • John Talignani • Honor Elizabeth Wainio • Deborah Welsh • Olga Kristin Gould White